Structural and Chemical
Analysis of Materials

Structural and Chemical Analysis of Materials

X-ray, electron and neutron diffraction
X-ray, electron and ion spectrometry
Electron microscopy

J. P. EBERHART
Louis Pasteur University, Strasbourg, France

Translated by
J. P. EBERHART

JOHN WILEY & SONS
Chichester · New York · Brisbane · Toronto · Singapore

First published in 1989 ©Bordas, Paris, under the title *Analyse structurale et chimique des matériaux. Diffraction des rayons X, neutrons et électrons, spectrométrie des rayons X, électrons et ions, microscopie électronique* by J. P. Eberhart

Copyright ©1991 by John Wiley & Sons Ltd.
            Baffins Lane, Chichester
            West Sussex PO19 1UD, England

All rights reserved.

No part of this book may be reproduced by any means, or transmitted, or translated into a machine language without the written permission of the publisher.

Other Wiley Editorial Offices

John Wiley & Sons, Inc., 605 Third Avenue,
New York, NY 10158-0012, USA

Jacaranda Wiley Ltd, G.P.O. Box 859, Brisbane,
Queensland 4001, Australia

John Wiley & Sons (Canada) Ltd, 22 Worcester Road,
Rexdale, Ontario M9W 1L1, Canada

John Wiley & Sons (SEA) Pte Ltd, 37 Jalan Pemimpin 05-04,
Block B, Union Industrial Building, Singapore 2057

Library of Congress Cataloging-in-Publication Data:

Eberhart, J. P. (Jean Pierre)
 [Analyse structurale et chimique des matériaux. English]
  Structural and chemical analysis of materials : X-ray, electron
and neutron diffraction, X-ray, electron and ion spectrometry,
electron microscopy / J. P. Eberhart ; translated by J. P. Eberhart.
       p.   cm.
  Includes bibliographical references and index.
  ISBN 0 471 92977 8
  1. Materials—Microscopy. 2. Microstructure. I. Title.
TA417.23.E2413   1991
620.1′1299—dc20   90-23730
                                                  CIP

British Library Cataloguing in Publication Data:

Eberhart, J. P.
 Structural and chemical analysis of materials.
 1. Materials. Structure & properties
 I. Title II. [Analyse structurale et chimique des
matériaux. *English*]
 620.112

 ISBN 0 471 92977 8

Set in Times 10/12 pt by
Dobbie Typesetting Limited, Tavistock, Devon
Printed and bound in Great Britain by
Courier International Ltd

Contents

Introduction xxix

PART ONE
INTERACTION OF X-RAYS AND PARTICLE BEAMS WITH MATERIALS

Chapter 1
Waves, Particle Beams and Matter 3
1.1 Radiations 3
    1.1.1 Electromagnetic Waves 3
    1.1.2 Particle Beams 3
1.2 Units and Physical Quantities 4
1.3 Relation between Energy and Wavelength 4
    1.3.1 Electromagnetic Waves 4
    1.3.2 Electrons 5
    1.3.3 Neutrons 5
1.4 Range of Radiations used in Material Science 6
    1.4.1 Radiations dealt with in this Book 6
1.5 Basic Atomic Theory. Energy Levels in Atoms 7
    1.5.1 Quantum Numbers 8
    1.5.2 Electron Energy Levels 8
Bibliography 9

Chapter 2
Basics on Radiation–Matter Interactions 10
2.1 General Considerations 10
    2.1.1 Observed Effects 10
        2.1.1.1 Effects on Radiation 10
        2.1.1.2 Effects on Matter 11
    2.1.2 Radiation–Specimen Energy Transfer 11
        2.1.2.1 Elastic Interactions 11
        2.1.2.2 Inelastic Interactions 11
    2.1.3 Interaction Cross-section 12
        2.1.3.1 Scattering Cross-section 12
    2.1.4 Mean Free Path 13

Contents

- 2.2 X-ray Interaction 13
  - 2.2.1 Elastic Interaction 13
  - 2.2.2 Inelastic Interactions 13
    - 2.2.2.1 Excitation of Atomic Core Levels 13
    - 2.2.2.2 Compton Effect 14
- 2.3 Electron Interaction 15
  - 2.3.1 Elastic Interaction 15
    - 2.3.1.1 Basic Theory 15
    - 2.3.1.2 Interaction Effects 16
      - Thermal Effect 16
      - Chemical Effect 17
      - Displacement Effect 17
    - 2.3.1.3 Characteristics of Elastic Electron Scattering 18
  - 2.3.2 Inelastic Interaction 18
    - 2.3.2.1 Basic Theory 18
    - 2.3.2.2 Interaction Effects 19
      - Excitation of Valence and Conduction Levels 19
      - Excitation of Atomic Core Levels 19
      - Conversion to Electromagnetic Energy 19
    - 2.3.2.3 Characteristics of Inelastic Electron Scattering 20
- 2.4 Neutron Interaction 20
- 2.5 Ion Interaction 20
  - Ion Sputtering 21
  - Ion Implantation 21
- 2.6 Energetic Balance of Interaction 21
- Bibliography 22

Chapter 3
Basic Theory of Elastic Scattering 24
- 3.1 Definition of the Scattering Object 24
  - 3.1.1 Scattering Power 24
    - 3.1.1.1 Scattering Power as a Continuous Function 24
    - 3.1.1.2 Scattering Power as a Discontinuous Distribution 24
  - 3.1.2 Structure Function 25
  - 3.1.3 Distribution Function 25
  - 3.1.4 Size Function 27
- 3.2 Amplitude of Scattered Radiations 27
  - 3.2.1 General Expression of Scattered Amplitude 27
    - 3.2.1.1 Scattering Vector 28
    - 3.2.1.2 Scattering Amplitude 28
    - 3.2.1.3 Fourier Transform of Scattering Amplitude 29
      - Conventional Writing 29
      - Fourier Inversion 29
    - 3.2.1.4 Size Factor 31
    - 3.2.1.5 Form Factor 31
  - 3.2.2 Physical Significance of Scattering Power 31
    - 3.2.2.1 Scattering Power for X-rays 32

		3.2.2.2 Scattering Power for Electrons	33
		3.2.2.3 Scattering Power for Thermal Neutrons	34
		To Summarise	34
3.3	Intensity of Scattered Radiations		34
	3.3.1	General Expression of Scattered Intensity	34
		3.3.1.1 Distribution of Discrete Scattering Elements	35
		3.3.1.2 Continuous Distribution of Scattering Matter	36
		3.3.1.3 Size Effects	36
		3.3.1.4 Interference Function	38
		3.3.1.5 Forward Scattered Intensity	38
		Distribution of Discrete Scattering Elements	38
		Continuous Distribution of Scattering Power	38
		3.3.1.6 Intensity Weighted Reciprocal Space	39
		3.3.1.7 Angular Variation of Scattered Intensity	39
		Non-crystalline Object	39
		Crystalline Object	40
		Small Objects	40
		To Summarise	41
	3.3.2	Fourier Transform of Scattered Intensity	40
Reference Chapters			41
Bibliography			41

Chapter 4
Elastic Scattering by Individual Atoms. Atomic Scattering Amplitude 42
4.1 Atomic Scattering Factor for X-rays 42
4.2 Atomic Scattering Amplitude for Electrons 44
4.3 Atomic Scattering Amplitude for Thermal Neutrons 46
4.4 Temperature Effect. Debye–Waller Factor 47
Reference Chapters 47
Bibliography 47

Chapter 5
Diffraction by a Crystal 48
5.1 Definitions: Lattice, Crystal, Diffraction............................ 48
5.2 Geometrical Diffraction Theory 50
    5.2.1 Diffraction Conditions Expressed in the Direct Lattice 50
        5.2.1.1 Laue Condition 50
        5.2.1.2 Bragg Condition 51
    5.2.2 Diffraction Conditions Expressed in the Reciprocal Lattice 53
        5.2.2.1 Laue Condition in the Reciprocal Lattice 53
        5.2.2.2 Ewald Construction...................................... 54
5.3 Kinematic Diffraction Theory 55
    5.3.1 Diffracted Amplitude 55
        5.3.1.1 General Expression 55
        5.3.1.2 Structure Factor 56
            Structure Considered as a Continuous Distribution of Scattering Matter 56

			Structure Considered as a Distribution of Discrete Atoms	57
		5.3.1.3	Form Factor	59
	5.3.2	Diffracted Intensity		59
		5.3.2.1	General Expression	59
		5.3.2.2	Interference Function Expressed in Reciprocal Space	59
			One-dimensional Analysis	60
			Three-dimensional Generalisation. Intensity Weighted Reciprocal Lattice	60
		5.3.2.3	Diffraction Deviation	63
		5.3.2.4	Friedel Law	64
		5.3.2.5	Fourier Transform of Diffracted Intensity. Patterson Function	65
	5.3.3	Relationship Between Structure Factor and Space Group Symmetry		67
		5.3.3.1	Basics	67
		5.3.3.2	Translations Corresponding to Lattice Types	68
			Primitive Lattice	68
			Body-centred Lattice	68
			All-face Centred Lattice	68
			One-face Centred Lattice	70
		5.3.3.3	Translational Space Group Operations	70
			Screw Axis	70
			Glide Planes	71
		5.3.3.4	Examples of Simple Crystal Structures	74
			NaCl-type Structure	74
			Diamond-type Structure	75

5.4 Dynamic Diffraction Theory 75
Reference Chapters 76
Bibliography 76

Chapter 6
Basic Theory of Electron Diffraction 77
6.1 Specific Features of Electron Diffraction 77
6.2 Diffraction Condition Relaxation 77
6.3 Diffracted Amplitude and Intensity. Kinematic Theory 79
    6.3.1 Diffraction at Infinite Distance 79
        6.3.1.1 Amplitude 80
        6.3.1.2 Intensity 80
    6.3.2 Diffraction at Finite Distance 80
        6.3.2.1 Fresnel Zones 81
        6.3.2.2 Amplitude 81
        6.3.2.3 Intensity 82
    6.3.3 Extinction Distance 82
6.4 Diffracted Amplitude and Intensity. Dynamic Theory 83
    6.4.1 Basics 83
    6.4.2 Two-wave Approximation 84
Reference Chapters 86
Bibliography 86

Chapter 7
Secondary Emission 87
7.1 Atomic Level Excitation and Relaxation 87
    7.1.1 Atomic Energy Levels 87
    7.1.2 Core Level Excitation 88
    7.1.3 Relaxation 88
7.2 Secondary Radiations Induced by Excitation 89
    7.2.1 Photoelectrons 89
    7.2.2 Secondary Electrons 90
7.3 Bremsstrahlung 91
7.4 Secondary Radiations Induced by Relaxation 92
    7.4.1 Characteristic X-rays 93
        7.4.1.1 Permitted Transitions and Notation 93
            Notation of Transitions 93
            Selection Rules 93
            Line Notation 93
        7.4.1.2 Moseley Law 94
        7.4.1.3 Line Intensities 94
            Excitation Cross-section 95
            Probability of Radiative Relaxation 95
            Transition Probability 96
        7.4.1.4 Chemical Shift 96
        7.4.1.5 Practical Use 97
    7.4.2 Characteristic Auger Electrons 98
        7.4.2.1 Auger Electron Energy 98
        7.4.2.2 Permitted Transitions 99
        7.4.2.3 Line Intensities 100
        7.4.2.4 Effect on the X-ray Spectrum 100
        7.4.2.5 Chemical Shift 100
Reference Chapters 100
Bibliography 100

Chapter 8
Absorption of Radiations in Materials 102
8.1 X-ray Absorption 102
    8.1.1 General Absorption Law 102
    8.1.2 Absorption Edges 103
    8.1.3 Absorption Additivity 104
8.2 Electron Absorption 104
    8.2.1 Energy Loss 105
    8.2.2 Intensity Loss 107
8.3 Neutron Absorption 107
Reference Chapters 108
Bibliography 108

PART TWO
RADIATION GENERATION AND MEASUREMENT

Chapter 9
Sources of X-rays, Electrons, Thermal Neutrons and Ions 111
9.1 X-ray Sources 111
    9.1.1 X-ray Tubes 111
        9.1.1.1 Hot Cathode Tubes 111
        9.1.1.2 Cold Cathode Tubes 112
        9.1.1.3 Target Choice 113
        9.1.1.4 Collimators 114
    9.1.2 Monochromatic Radiation 114
        9.1.2.1 Absorption Edge Filtering 114
        9.1.2.2 Crystal Monochromator 115
            Plane Monochromator 116
            Focusing Monochromator 116
    9.1.3 Synchrotron Radiation 118
9.2 Electron Sources 118
    9.2.1 Thermionic Emission 119
    9.2.2 Field Emission 121
    9.2.3 Operating Conditions 121
        9.2.3.1 Vacuum 121
        9.2.3.2 Energy Dispersion 121
        9.2.3.3 Focusing of Electron Beams. Electron Optics 122
            General Characteristics 122
            Aberrations 123
            Electronoptical Focusing System 125
9.3 Neutron Sources 127
9.4 Ion Sources 128
    9.4.1 Gas Discharge Source 128
    9.4.2 Hot Cathode Source 129
    9.4.3 Metal Ion Source 130
        9.4.3.1 Surface Ionisation Source 130
        9.4.3.2 Liquid Metal Source 130
Reference Chapters 131
Bibliography 131

Chapter 10
Radiation Detectors and Spectrometers 132
10.1 Instrumental Parameters 132
    10.1.1 Characteristic Detector Parameters 132
    10.1.2 Characteristic Spectrometer Parameters 133
10.2 X-ray Detection and Measurement 134
    10.2.1 Qualitative Detection 134
    10.2.2 Intensity Measurement. Ionisation Detectors 135
        10.2.2.1 Gas Ionisation Detectors 135
            Basic Principle 135

			Working Regions. Multiplication Factor	136
			Proportional Counter	138
			Geiger–Müller Counter	138
		10.2.2.2	Solid Detectors	139
			Scintillation Counter	139
			Semiconductor Counter	140
		10.2.2.3	Position Sensitive Detectors	142
			One-dimensional Localising	142
			Two-dimensional Localising	142
	10.2.3	Energy Measurements. X-ray Spectrometers		143
		10.2.3.1	Wavelength-dispersive Spectrometers	143
			Plane Analysing Crystal Spectrometer	143
			Focusing Spectrometer	143
			Analysing Crystals	144
			Main Characteristics	145
		10.2.3.2	Energy-dispersive Spectrometer	146
			Relationship between Energy Resolution and Counting Statistics	146
			Si(Li) Spectrometer	147
			Comparison of X-ray Spectrometers	148

10.3 Electron and Ion Detection and Measurement ... 149
    10.3.1 Detection and Intensity Measurement of an Electron Beam ... 149
        10.3.1.1 Qualitative Detection. Fluorescent Screen and Photographic Film ... 149
        10.3.1.2 Direct Measurement ... 149
        10.3.1.3 Indirect Measurement. Electron Detectors ... 149
            Scintillation Counter ... 149
            Semiconductor Detector ... 150
            Channeltron ... 151
    10.3.2 Electron Spectrometers ... 151
        10.3.2.1 Electron Spectrometer with Angular Dispersion ... 151
            Electrostatic Analysers ... 151
            Magnetic Prism Analyser ... 152
            Resolution ... 153
            Energy Filtering ... 153
        10.3.2.2 Electron Spectrometer without Angular Dispersion ... 153
            Solid State Detector ... 153
            Retarding Voltage Spectrometer ... 153
    10.3.3 Detection and Intensity Measurement of an Ion Beam ... 154
    10.3.4 Ion Spectrometers. Mass Spectrometry ... 155
        10.3.4.1 Electrostatic Prism Energy Selector ... 155
        10.3.4.2 Magnetic Prism Mass Analyser ... 155
        10.3.4.3 Quadrupole Mass Analyser ... 156
        10.3.4.4 Time of Flight Mass Analyser ... 156
        10.3.4.5 Mass Resolution ... 157

10.4 Neutron Detection and Measurement ... 157
10.5 Accuracy of Intensity Measurements. Count Statistics ... 158

	10.5.1	Frequency Distribution	158
	10.5.2	Probable Statistical Error	159
	10.5.3	Probable Count Error in the Presence of Background	159
	10.5.4	Dead Time Correction	160

Reference Chapters 160
Bibliography 160

PART THREE
DIFFRACTION TECHNIQUES APPLIED TO MATERIAL ANALYSIS

Chapter 11
X-ray and Neutron Diffraction Applied to Crystalline Materials 165

11.1 Laue Method. Crystal Symmetry and Orientation 166
    11.1.1 Basic Principle 166
        11.1.1.1 Primary Radiation and Sample 166
        11.1.1.2 Instrumental Layout and Diffraction Pattern 166
        11.1.1.3 Indexing the Laue Pattern 168
                Transmission Mode 168
                Back-reflection Mode 169
    11.1.2 Applications 171
        11.1.2.1 Point Group Symmetry. Laue Groups 172
        11.1.2.2 Crystal Orientation 172
11.2 Rotation Method. Crystal Structure Analysis 173
    11.2.1 Basic Principle 173
        11.2.1.1 Primary Radiation and Sample 173
        11.2.1.2 Basic Layout 174
        11.2.1.3 Display of the Diffraction Pattern 174
    11.2.2 Interpreting the Diffraction Pattern 175
        11.2.2.1 Lattice Parameters 175
        11.2.2.2 Indexing the Pattern. Intensity Weighted Reciprocal Lattice 178
        11.2.2.3 Structure Analysis 180
    11.2.3 Improved Variants 180
        11.2.3.1 Weissenberg Camera 181
        11.2.3.2 Retigraph and Precession Camera 181
        11.2.3.3 Single Crystal Diffractometer 183
        11.2.3.4 Conclusion 183
11.3 Powder Method. Analysis of Polycrystalline Materials 184
    11.3.1 Basic Principle 184
        11.3.1.1 Primary Radiation and Sample 184
        11.3.1.2 Display of the Diffraction Pattern 184
        11.3.1.3 Diffracted Intensity 185
                Symmetry-related Multiplicity 185
                Nearly Equal Interplanar Spacings 185
    11.3.2 Instrumental Layout 185
        11.3.2.1 Debye–Scherrer Camera 185

			Set-up	185
			Specimen	186
			Measurements on the Diffraction Pattern	188
			Advantages and Inconveniences	189
		11.3.2.2	Focusing Cameras	190
		11.3.2.3	Powder Diffractometer	191
			Operating Modes	191
			Specimen	192
			Developments	192
		11.3.2.4	Primary Radiation	193
	11.3.3	Applications of the Powder Method		194
		11.3.3.1	Identifying Crystalline Species. Powder Data File	194
		11.3.3.2	Quantitative Analysis	195
		11.3.3.3	Lattice Parameters	196
			Quadratic Forms	197
			Cubic Lattice Parameters	198
			Parameters of Lattices with a Main Axis	201
			Low-symmetry Lattice Parameters	201
		11.3.3.4	Crystal Structure Analysis	202
		11.3.3.5	Grain Size	203
		11.3.3.6	Preferential Orientations	204
			Fibre Texture	204
			Lamellar Texture	206
			General Texture	207
		11.3.3.7	Order–Disorder Transformations	207
	11.3.4	Analysis of Poorly Crystallised and Amorphous Materials		209
		11.3.4.1	Poorly Crystallised Materials. Mixed-layered Structures	209
		11.3.4.2	Short-distance Ordering in Non-crystalline Materials	210
		11.3.4.3	Size Distribution in Amorphous Materials. Small-angle Scattering	210
11.4	X-ray and Neutron Diffraction Applied to Crystal Structure Analysis			211
	11.4.1	Point Group Symmetry		211
		11.4.1.1	Crystal with a Polyhedral Form	211
		11.4.1.2	Crystal without any Polyhedral Form	212
	11.4.2	Lattice Parameters		212
	11.4.3	Unit Cell Content		212
	11.4.4	Space Group Symmetry		213
	11.4.5	Atomic Positions		214
		11.4.5.1	Trial-and-error Method	214
			Structure Completely Determined by the Space Group	214
			Structure Determined by the Space Group to One Parameter	215
			Comparing the Measured and Calculated Intensities	215
		11.4.5.2	Fourier Transform Methods	217
			Patterson Series	217
			Phase Determination through the Patterson Function	217
			Direct Methods. Fourier Series	218

			Structure Refinement	218
			Practical Calculations	218
		11.4.5.3	Fourier Projections	219
			Two-dimensional Fourier Series. Fourier–Bragg Projection	219
			Analogical Methods. Photosumming	219
			One-dimensional Fourier Series..................	220
	11.4.6	Applications of Neutron Diffraction		222
		11.4.6.1	Practice	222
		11.4.6.2	Application to Hydrogen Compounds............	224
		11.4.6.3	Order–Disorder Transformations	224
		11.4.6.4	Structures of Magnetic Compounds...............	224
	11.4.7	Conclusion		224
Reference Chapters				225
Bibliography..				225

Chapter 12
Electron Diffraction on Thin Crystalline Layers 226

12.1	Parallel Beam High-energy Electron Diffraction			226
	12.1.1	Particular Features of High-energy Electron Diffraction.......		226
		12.1.1.1	Diffraction Angles.............................	227
		12.1.1.2	Reflecting Sphere..............................	227
		12.1.1.3	Diffraction Condition Relaxation	227
	12.1.2	Instrumental Layouts		229
		12.1.2.1	Macrodiffraction	230
		12.1.2.2	Microdiffraction through Aperture Selection	231
			Principle	231
			Selection Errors	231
		12.1.2.3	Microdiffraction and Nanodiffraction through Beam Selection	232
		12.1.2.4	Specimen	233
	12.1.3	Diffraction Patterns from Single Crystals....................		233
		12.1.3.1	Observed Diffraction Pattern	233
			Zone Diffraction Pattern	233
			Laue Zones	233
			Double Diffraction	237
			Kikuchi Patterns	239
			Form and Size Effects	243
			Crystal Lattice Deformations and Defects	243
			Superlattice Diffraction Pattern	245
		12.1.3.2	Applications of Single Crystal Diffraction Patterns ..	245
			Crystal Orientation	245
			Crystal Lattice	248
			Space Group...................................	249
			Crystal Structure	250
			Crystal Defects and Poorly Crystallised Materials ...	250
	12.1.4	Diffraction Patterns from Polycrystalline Materials		250

		12.1.4.1	Observed Diffraction Pattern	251
			Generation of the Diffraction Pattern	251
			Ring Intensity	252
			Ring Continuity	252
			Errors of Measurement	252
		12.1.4.2	Applications of Powder Diffraction Patterns	254
			Lattice Parameters	254
			Identification of Crystalline Species	254
			Grain Size	254
			Orientation Texture	254
12.2	Convergent Beam Electron Diffraction			258
	12.2.1	Basic Principle and Observed Diffraction Patterns		258
		12.2.1.1	Instrumental Layout	258
		12.2.1.2	General Outline of the Diffraction Pattern	259
		12.2.1.3	Diffraction Disc Intensity and Fine Structure	260
	12.2.2	Instrumental Layout		261
	12.2.3	Applications of CBD		263
		12.2.3.1	Zone-axis Parameter	263
		12.2.3.2	Crystal Thickness	263
		12.2.3.3	Crystal Symmetry	264
		12.2.3.4	Identification of Crystal Species	264
12.3	Low-energy Electron Diffraction			265
	12.3.1	Basic Principle		265
	12.3.2	Instrumental Layout		265
	12.3.3	Display of the Diffraction Pattern		265
		12.3.3.1	Surface Lattice	265
		12.3.3.2	Diffraction Condition	266
		12.3.3.3	Diffracted Intensities. Surface Structure	268
	12.3.4	Application to Crystal Surface Structure Analysis		268
		12.3.4.1	Intrinsic Crystal Surface	268
		12.3.4.2	Adsorption Layers	269
		12.3.4.3	Attachments	269
12.4	Conclusion			270
Reference Chapters				270
Bibliography				271

PART FOUR
X-RAY, ELECTRON AND SECONDARY ION SPECTROMETRY APPLIED TO MATERIAL ANALYSIS

Chapter 13
Important Parameters in Spectrometry ... 275

13.1	Qualitative Elemental Analysis		276
	13.1.1	Composition Parameters of the Specimen	276
		13.1.1.1 Number of Atoms of an Element in a Volume Unit	276
		13.1.1.2 Total Number of Atoms of the Specimen in a Volume Unit	276

		13.1.1.3	Number of Atoms of an Element in a Surface Unit	277
		13.1.1.4	Mass Fraction and Atomic Fraction	277
	13.1.2	Detection Limit of an Element		277
		13.1.2.1	Definitions	277
		13.1.2.2	Detection Limits Related to the Count Rate	277
		13.1.2.3	Minimum Detectable Mass Fraction	278
13.2	Quantitative Elemental Analysis			279
	13.2.1	Parameters Characterising an Element		280
	13.2.2	Parameters Characterising the Matrix		280
		13.2.2.1	Attenuation of the Characteristic Intensity	280
		13.2.2.2	Enhancement of the Characteristic Intensity	280
		13.2.2.3	Matrix Correction	281
	13.2.3	Instrumental Parameters		281
	13.2.4	Measured Intensity		281
		13.2.4.1	Intensity from a Volume Unit	282
		13.2.4.2	Intensity from a Surface Unit	282
		13.2.4.3	Instrumental Intensity	282
Reference Chapters				283
Bibliography				283

Chapter 14
Elemental Analysis by X-ray Fluorescence 284

14.1	Basic Principle and Instrumental Layout			284
	14.1.1	Primary Excitation Radiation		284
		14.1.1.1	X-ray Excitation	285
			Excitation by the Characteristic Lines	285
			Excitation by the Continuous Background	285
			Instrumental Conditions	285
		14.1.1.2	Excitation by Radioactivity	286
	14.1.2	Spectrometer		286
		14.1.2.1	Wavelength-dispersive Spectrometer	286
		14.1.2.2	Energy-dispersive Spectrometer	287
14.2	Qualitative Analysis			287
	14.2.1	Detection Limits		287
	14.2.2	Identifying Uncertainties		288
14.3	Quantitative Analysis			290
	14.3.1	Matrix Effects		290
		14.3.1.1	Absorption Effect	291
		14.3.1.2	Secondary Fluorescence Effect	294
		14.3.1.3	Grain Effect	295
		14.3.1.4	Total Correction	295
	14.3.2	Practical Methods of Quantitative Analysis		295
		14.3.2.1	Intensity Measurements	295
		14.3.2.2	Correction Procedures	296
			Linear Approximation	296
			Matrix Correction	297
			Internal Standard Methods	298

			Mathematical Correction Methods	300
			Trace Analysis	300
		14.3.2.3	Accuracy of Analysis	301

14.4 Specimen Preparation 301
14.5 Applications to Material Analysis 302
Reference Chapters 302
Bibliography 303

Chapter 15
Electron Probe Microanalysis 304

15.1 Basic Principles and Instrumentation 304
    15.1.1 Electron Probe 305
    15.1.2 Spectrometer 305
    15.1.3 Operating Modes 306
        15.1.3.1 Stationary Probe 306
        15.1.3.2 Scanning Probe 306
15.2 Qualitative Microanalysis 307
    15.2.1 Spatial Resolution 307
        15.2.1.1 Emission Volume 307
        15.2.1.2 Emission Surface 308
            Thick Specimen 308
            Thin Specimen 309
    15.2.2 Stationary Probe Operation 309
        15.2.2.1 Detection Limits 309
        15.2.2.2 Thick Specimen Analysis by Microprobe 310
        15.2.2.3 Thick Specimen Analysis by Scanning Electron Microscope 310
        15.2.2.4 Thin Specimen Analysis 311
    15.2.3 Scanning Mode 311
        15.2.3.1 Scanning X-ray Microanalysis 311
            Linear Scanning 311
            Plane Scanning. Distribution Maps 312
        15.2.3.2 Miscellaneous Scanning Images 312
        15.2.3.3 Kossel Pattern 314
15.3 Quantitative Analysis of Thick Specimen 314
    15.3.1 Matrix Effects 315
        15.3.1.1 Atomic Number Effect 316
        15.3.1.2 Absorption Effect 318
        15.3.1.3 Fluorescence Effect 319
    15.3.2 Correction Methods 319
        15.3.2.1 ZAF Method 320
        15.3.2.2 Limitation of ZAF 320
            Light Elements 320
            Thin Specimens 321
    15.3.3 Accuracy of Analysis 321
        15.3.3.1 Instrumental Errors 321
        15.3.3.2 Errors due to the Specimen 321

		15.3.3.3	Statistical Errors	322
		15.3.3.4	Correction Errors	322
		15.3.3.5	Currently Observed Accuracy	322
	15.3.4	Specimen Preparation		322
		15.3.4.1	Specimen to be Analysed	322
		15.3.4.2	Test Samples	323
	15.3.5	Applications to Material Analysis		323
15.4	Quantitative Analysis of Thin Specimens			323
	15.4.1	Basic Formulation. Matrix Corrections		324
		15.4.1.1	Relationship between Intensities and Concentrations	324
		15.4.1.2	Thin Specimen Matrix Corrections	324
			Atomic Number Correction	324
			Absorption Correction	325
			Fluorescence Correction	325
		15.4.1.3	Thin Specimen Approximation	326
		15.4.1.4	Concentration Ratio of Two Elements	326
	15.4.2	Practical Methods of Analysis		327
		15.4.2.1	Concentration Ratios	327
			Method without a Test Sample	327
			Methods with Thin Test Sample	327
		15.4.2.2	Absolute Multi-element Analysis	328
		15.4.2.3	Accuracy of Analysis	329
			Systematic Errors	329
			Statistical Errors	330
			Currently Observed Accuracy	330
		15.4.2.4	Applications	332
Reference Chapters				332
Bibliography				332

Chapter 16
Electron Spectrometry for Surface Analysis ... 333

16.1	Photoelectron Spectrometry		334
	16.1.1	Basic Principle	334
	16.1.2	Instrumental Layout	335
		16.1.2.1 Primary Radiation	336
		16.1.2.2 Spectrometer	337
		16.1.2.3 Specimen	337
		16.1.2.4 Spectrum Acquisition and Processing	337
	16.1.3	Applications	339
		16.1.3.1 Measurement of Energy Levels	339
		16.1.3.2 Qualitative Surface Analysis	339
		16.1.3.3 Surface Bonding	340
		16.1.3.4 Quantitative Surface Analysis	340
		16.1.3.5 Applications to Materials	341
16.2	Auger Electron Spectrometry		341
	16.2.1	Basic Principle	341
	16.2.2	Instrumental Layout	342

Contents

16.2.2.1	X-ray Excitation	342
16.2.2.2	Electron Excitation	342
	Retarding Field Spectrometer Combined with LEED	342
	Angular Dispersive Spectrometers	343
	Auger Microprobe and Scanning Microscope	344
16.2.3	Specimen Preparation	344
16.2.4	Applications	344
16.2.4.1	Qualitative Surface Analysis	345
16.2.4.2	Surface Bonding	345
16.2.4.3	Quantitative Surface Analysis	345
16.2.4.4	Applications to Materials	346
16.3	Conclusion	346
Reference Chapters		349
Bibliography		349

Chapter 17
X-ray Absorption Spectrometry and Electron Energy Loss Spectrometry 350

17.1	X-ray Absorption Spectrometry (EXAFS)		351
	17.1.1	Basic Principle	351
		17.1.1.1 Absorption Spectrum	351
		17.1.1.2 Fine Structure	352
		Observed Effects	353
		Interpretation of the Extended Fine Structure. EXAFS	353
		Fine Structure near to the Absorption Edge	356
	17.1.2	Practice of EXAFS Spectrometry	356
		17.1.2.1 Instrumental Layout	357
		X-ray Source	357
		Monochromator and Focusing	357
		17.1.2.2 Setting up the Interference Function	358
		Absorption Function	359
		Interference Function	360
		17.1.2.3 Extracting Structural Data	360
		Direct Fourier Transform	360
		Fitting of Experimental and Theoretical Spectra	360
		Fourier Filtering	361
		17.1.2.4 Possibilities and Limitations	361
	17.1.3	Applications	362
17.2	Electron Energy Loss Spectrometry (EELS)		362
	17.2.1	Basic Principle	363
		17.2.1.1 Experimental Parameters	363
		17.2.1.2 Observed Spectrum	364
		Around Zero-energy-loss Region	364
		Low-energy-loss Region	365
		High-energy-loss Region	366
	17.2.2	Instrumental Layout	367
		17.2.2.1 Spectrometer and Detector	367
		17.2.2.2 Specimen	367

	17.2.3	Qualitative Analysis. Detection Limits	367
	17.2.4	Quantitative Analysis	369
		17.2.4.1 Absolute Quantitative Analysis	369
		17.2.4.2 Concentration Ratio of Two Elements	370
		17.2.4.3 Accuracy of Analysis	370
	17.2.5	Applications	370

Reference Chapters ... 371
Bibliography ... 371

Chapter 18
Secondary Ion Mass Spectrometry for Surface Analysis ... 372
18.1 Basic Principles and Instrumental Layout ... 372
    18.1.1 Purely Analytical Facility ... 373
        18.1.1.1 Primary Ion Source ... 373
        18.1.1.2 Specimen ... 374
        18.1.1.3 Mass Spectrometer ... 374
        18.1.1.4 Data Acquisition and Processing ... 374
    18.1.2 Ion Emission Microscopy ... 375
        18.1.2.1 Stigmatic Images ... 375
        18.1.2.2 Scanning Images ... 375
18.2 Specimen Analysis ... 375
    18.2.1 Analytical Modes ... 375
        18.2.1.1 Surface Analysis ... 375
        18.2.1.2 Depth Concentration Profiling ... 376
    18.2.2 Spatial Resolution ... 376
        18.2.2.1 Surface Resolution ... 376
        18.2.2.2 Depth Profiling Resolution ... 376
            Stationary Analysing Mode ... 376
            Scanning Analysing Mode ... 376
            Limiting Parameters of Depth Resolution ... 377
    18.2.3 Qualitative Analysis ... 377
        18.2.3.1 Observed Emission Spectrum ... 377
            Emission Parameters ... 377
            Ion Species ... 378
            Experimental Spectrum ... 378
        18.2.3.2 Detection Limits ... 379
        18.2.3.3 Minimum Volume to be Sputtered ... 379
        18.2.3.4 Choice of Primary Ions ... 380
    18.2.4 Quantitative Analysis ... 381
        18.2.4.1 Analytical Parameters ... 381
            Sputtering Yield ... 382
            Ionisation Degree ... 382
        18.2.4.2 Analytical Practice ... 383
        18.2.4.3 Accuracy of Analysis ... 383
18.3 Specimen Preparation ... 383

18.4	Applications		384
Reference Chapters			396
Bibliography			396

PART FIVE
TECHNIQUES OF ELECTRON MICROSCOPY

Chapter 19
Transmission Electron Microscopy 391

19.1	Basic Components and Characteristics			393
	19.1.1	Electron Source		393
	19.1.2	Electron Lenses. Aberrations and Resolution		394
		19.1.2.1	General Characteristics	394
		19.1.2.2	Point Resolution	394
			Point Imaging by a Perfect Objective Lens	395
			Resolution Consistent with Aberrations	396
		19.1.2.3	Resolution Tests	399
		19.1.2.4	Depth of Field and Depth of Focus	399
			Depth of Field	399
			Depth of Focus	401
		19.1.2.5	Specimen Stage	402
		19.1.2.6	Observing and Recording	402
		19.1.2.7	Vacuum System	402
		19.1.2.8	Attachments and Developments	403
		19.1.2.9	Summary: Main Characteristics of a TEM	404
19.2	Image Contrast. Geometrical Interpretation			404
	19.2.1	Generation of Contrast		404
		19.2.1.1	Specimen Effect	404
		19.2.1.2	Instrumental Effect	404
	19.2.2	Scattering and Diffraction Contrast		404
		19.2.2.1	Amorphous Scattering Object	404
		19.2.2.2	Crystalline Diffracting Object. Bright Field and Dark Field Image	405
			Bright Field Image	405
			Dark Field Image	407
			Multiple Dark Field Imaging	407
			Selective Dark Field Imaging	407
		19.2.2.3	Summary	408
19.3	Crystal Image Contrast. Kinematic and Dynamic Interpretation			408
	19.3.1	Column Approximation		409
	19.3.2	Image of a Perfect Crystal		409
		19.3.2.1	Thickness Fringes	411
		19.3.2.2	Bend Fringes	411
		19.3.2.3	Bright Field Complementarity. Failure of Kinematic Theory	411
		19.3.2.4	Conclusion	415
	19.3.3	Image of a Real Crystal		415
		19.3.3.1	Effect of Lattice Distortions	415

			Displacement Vector	415
			Defect-induced Contrast	416
			Extinction of Defect Contrast	417
		19.3.3.2	Stacking Faults	417
		19.3.3.3	Dislocations	418
			Screw Dislocation	418
			Edge Dislocation	419
			General Dislocation	420
		19.3.3.4	Inclusions and Precipitates	420
		19.3.3.5	Mutual Orientation of Two Phases. Moiré Pattern	422
			Rotation Moiré Pattern	424
			Parallel Moiré Pattern	425
19.4	Structure Resolution. High-resolution Electron Microscopy			428
	19.4.1	Structure Imaging		429
		19.4.1.1	Wave Transmission through the Object. Object Function	429
		19.4.1.2	Wave Function in the Back Focal Plane of a Perfect Objective Lens	429
		19.4.1.3	Effect of Objective Lens Defects. Transfer Function	430
			Objective Lens Aperture	430
			Spherical Aberration and Defocusing	430
			Phase Contrast Transfer Function	431
			Wave Function in the Back Focal Plane of a Real Objective Lens	431
		19.4.1.4	Wave Function in the Image Plane	431
		19.4.1.5	Structure Image	432
			Absorption Contrast	432
			Phase Contrast	432
	19.4.2	Structure Image of a Thin Crystal Platelet		433
	19.4.3	Direct Interpretation of Structure Images		434
		19.4.3.1	Operating Mode	434
		19.4.3.2	Two-beam Imaging. Lattice Resolution	435
		19.4.3.3	Multibeam Imaging. Structure Resolution	439
	19.4.4	Indirect Interpretation of Structure Images		440
		19.4.4.1	Image Processing	440
		19.4.4.2	Image Modelling	440
	19.4.5	Applications		441
19.5	Specimen			441
	19.5.1	Preparation Techniques		441
		19.5.1.1	Specimen Support	443
			Collodion Film by the Solution Drop Method	443
			Formvar Film by the Dipping and Flotation Method	443
			Carbon Film	444
			Microgrids	445
			Holed Foils	445
		19.5.1.2	Specimen Preparation	445
			Grinding	445
			Cleavage	446

			Ultramicrotome Cutting	446
			Thinning of Bulk Materials	446
			Direct Preparation of Thin Films	447
		19.5.1.3	Surface Topography of Bulk Specimen. Replicas	447
			Direct Replica (Negative Replica)	447
			Two-stage Replica (Positive Replica)	447
			Extraction Replica	448
	19.5.2	Effect of Electron Bombardment		448
		19.5.2.1	Basic Considerations	449
			Atom Displacement Energy	449
			Energy Transfer below Displacement Energy	449
			Energy Transfer above Displacement Energy	450
		19.5.2.2	Irradiation Effects in Electron Microscopes	450
			Thermal Effect	450
			Chemical Effects	451
			Atomic Displacements	451
19.6	High-voltage Electron Microscopy			451
	19.6.1	Resolution		453
	19.6.2	Irradiation Effects		453
	19.6.3	Applications		453
19.7	Conclusion			456
Reference Chapters				456
Bibliography				456

Chapter 20
Scanning Electron Microscopy ... 458

20.1	Basic Principle			458
20.2	Instrumental Layout			459
	20.2.1	Electron Probe		459
	20.2.2	Scanning System		461
	20.2.3	Specimen		461
	20.2.4	Image Processing		461
20.3	Emission and Reflection Images			463
	20.3.1	Imaging by Secondary Electron Emission and Backscattered Electrons		463
		20.3.1.1	Imaging Radiations	463
			Secondary Electrons	463
			Backscattered Electrons	463
		20.3.1.2	Detection	463
			Secondary Electron Mode	463
			Backscattered Electron Mode	464
		20.3.1.3	Image Contrast	466
			Topographic Contrast	466
			Composition Contrast	468
		20.3.1.4	Resolution	468
			Secondary Electrons	469
			Backscattered Electrons	470
		20.3.1.5	Depth of Field	469

		20.3.1.6 Basic Operating Mode	471
	20.3.2	Imaging by X-ray Emission	471
	20.3.3	Imaging by Auger Electron Emission	471
	20.3.4	Imaging by Light Emission. Cathodoluminescence	471
20.4	Scanning Transmission Imaging		472
20.5	Miscellaneous Imaging Modes		473
	20.5.1	Imaging by Absorbed Current	473
	20.5.2	Imaging by Potential Contrast	473
	20.5.3	Imaging by Magnetic Field Contrast	473
	20.5.4	Angular Scanning on a Crystal	474
	20.5.5	Electron-beam Induced Current	474
	20.5.6	Electroacoustic Imaging	474
20.6	Specimen		474
	20.6.1	Electron and Vacuum Resistance	475
	20.6.2	Surface Conductivity	475
		20.6.2.1 Vacuum Evaporation	475
		20.6.2.2 Cathode Sputtering	475
		20.6.2.3 Low-voltage Operating	477
20.7	Applications		477
Reference Chapters			481
Bibliography			481

Chapter 21
Scanning Transmission Electron Microscopy. Analytical Electron Microscopy . 482

21.1	Scanning Transmission Electron Imaging		483
	21.1.1	Basics of Imaging	483
		21.1.1.1 Reciprocity	483
		21.1.1.2 Resolution	483
	21.1.2	Instrumental Layout and Characteristics	484
		21.1.2.1 Dedicated Scanning Transmission Electron Microscope	484
		Components	484
		Operating Modes	486
		21.1.2.2 Transmission Electron Microscope with Scanning Attachment	487
	21.1.3	Comparing STEM and Conventional TEM	487
21.2	Analytical Electron Microscope		488
	21.2.1	Available Spectrometric Methods	488
	21.2.2	Comparing X-ray Emission Spectrometry and Electron Energy Loss Spectrometry	488
	21.2.3	Auger Spectrometry as Performed with STEM	490
	21.2.4	Elemental Distribution Maps	490
21.3	Specimen Preparation and Applications		490
Reference Chapters			492
Bibliography			492

Chapter 22
Scanning Tunnelling Microscopy 493

22.1	Principle of the Tunnel Effect		493
	22.1.1	Work Function	493

	22.1.2	Field Effect	494

		22.1.3	Tunnel Effect	494

22.2	Applications of the Tunnel Effect	496
	22.2.1 Field Emission Electron Sources	496
	22.2.2 Field Emission Microscopy	496
	22.2.2.1 Field Electron Emission Microscopy	497
	22.2.2.2 Field Ion Microscopy	497
	22.2.2.3 Tunnel Effect Microscopy	497
22.3	Scanning Tunnelling Microscopy	497
	22.3.1 Instrumental Layout	498
	22.3.1.1 Probe Pin	499
	22.3.1.2 Pin Displacement System. Control and Data Acquisition	499
	22.3.1.3 Specimen	500
	22.3.1.4 Vibration Damping	500
	22.3.2 Possibilities and Developments	502
	22.3.2.1 Resolution	502
	22.3.2.2 Spectrometric Developments	502
	22.3.2.3 Miscellaneous Developments	502
	22.3.3 Applications to Material Research	503
	22.3.3.1 Basic Tunnelling Mode	503
	22.3.3.2 Variants of STM	504
Reference Chapters	504
Bibliography	504

Appendix A
Physical Quantities. Units. Universal Constants. Notations ... 505

A.1	Main Physical Quantities and their Units	505
	A.1.1 Wavelength	505
	A.1.2 Frequency	505
	A.1.3 Electrical Potential	505
	A.1.4 Energy	506
	A.1.5 Wave Vector	506
	A.1.6 Amplitude	506
	A.1.7 Intensity	506
	A.1.8 Pressure	507
A.2	Universal Constants	507
A.3	Lexicon of Notations and Symbols	507
	A.3.1 Crystallography	508
	Lattices	508
	Crystal	508
	A.3.2 Material Physics. General Radiation-Matter Interaction	509
	A.3.3 Diffraction and Scattering	509
	A.3.4 Spectrometry	510
	A.3.5 Electron Microscopy	512
Bibliography	512

Appendix B
Reciprocal Space. Reciprocal Lattice 513
- B.1.1 Relations Defining Reciprocal Space and Reciprocal Lattice 513
  - Reciprocal Space 513
  - Reciprocal Lattice 514
- B.1.2 Basic Relations between Direct and Reciprocal Lattice 514
  - Direct Lattice Vector and Reciprocal Lattice Vector 514
  - Orientation of Lattice Lines and Planes 515
  - Interplanar and Interpoint Distances 515
- B.1.3 Reciprocal Lattices with General Constants 516

Appendix C
Basic Properties of Fourier Transforms 517
- C.1.1 Definition 517
- C.1.2 Inverse Fourier Transform 517
- C.1.3 Additivity 517
- C.1.4 Translation 518
- C.1.5 Changing Sign. Symmetry and Antisymmetry 518
  - Transform of a Real Dissymmetrical Function 518
  - Transform of a Real Symmetrical Function 518
  - Transform of a Real Antisymmetrical Function 518
- C.1.6 Transform of a Product. Convolution 518
  - General Form 518
  - Alternative Form 519
  - Properties 520
  - Convolution of a Continuous Function by a Distribution 520
  - Applications of Convolution 521
- C.1.7 Convolution Square 521

Bibliography 522

Appendix D
Spectrometric Tables 523
- D.1 Comparative Performances of Spectrometric Analysing Techniques..... 523
- D.2 Electron Binding Energies in Atoms 524

Bibliography 525

Appendix E
Abbreviations and Acronyms 527
- E.1 Abbreviations 527
- E.2 Acronyms 527

Appendix F
Radioprotection 529
- F.1 Biological Action of Radiations Used in Material Analysis............. 529
  - F.1.1 Radiation Doses 530
    - Exposure Dose 530
    - Absorbed Dose 530

		Equivalent Dose	530
		Limit Dose	531
F.2	Protective Measures		531
	F.2.1	Passive Protection. Radiation Shielding	531
	F.2.2	Active Protection	531

References 533

Index 539

Introduction

Since the discovery of X-rays by Röntgen at the end of the last century, most of our knowledge about the structure of solids has been acquired by means of methods based upon radiation–matter interactions.

Those methods have undergone a tremendous development during the last decade. This development is due mainly to the progress in electronics and computer science, induced by the requirement to produce and to test more and more sophisticated materials.

This book deals with the use of radiation which is best suited for determining the structure and the composition of materials: X-rays, electrons, thermal neutrons and ions, with energies ranging from a few electronvolts to a million electronvolts.

Information on matter is provided by radiation in different ways, depending on the kind of interaction, on its nature (electromagnetic waves or matter waves) and on its energy.

Structural data are obtained by means of X-ray-, electron- and neutron scattering and diffraction, and also by electron microscopy.

Chemical data are obtained by X-ray-, electron- and ion spectrometry.

A minimum of theoretical knowledge is required in order to take the best advantage of the various techniques. The basic theory is dealt with in **Part One** of the book. A knowledge of elementary crystallography and optics is assumed.

Production, detection and measurement of radiation are common to the various techniques. In order to avoid tedious repetition, they are described together in **Part Two**.

Part Three deals with structural methods, based on diffraction, mainly of X-rays and electrons. Neutron diffraction is limited to its complementary aspect.

Part Four is dedicated to the main analytical techniques, based on spectrometry of X-rays, electrons and ions. Special attention is given to surface analysis which is becoming increasingly important in research and industry.

Part Five deals with the different aspects of electron microscopy. Imaging techniques, mainly with electrons, are probably the field where the most spectacular developments have been achieved in material science during recent years. Depending on the type of electron microscope, imaging is based on electron scattering and diffraction or on various secondary emissions.

Modern electron microscopes are increasingly fitted with spectrometric devices, whereas spectrometric techniques have increasing possibilities of displaying analytical data as images. It follows that **Parts Four** and **Five** are closely related through imaging processes.

Some additional information is given in several **appendices**, e.g. physical data, units, symbols and abbreviations used in the book, as well as some basic developments about reciprocal space, Fourier transforms and protection against radiation effects.

A **bibliography** at the end of each chapter gives some book references related to its topic.

In order to avoid continuous reference to related theoretical sections in the text, a list of **reference chapters** is given at the end of the technical chapters, before the Bibliography.

Part One

INTERACTION OF X-RAYS AND PARTICLE BEAMS WITH MATERIALS THEORETICAL ASPECTS

A good knowledge of the physical basis of radiation–solid interaction processes is essential to make the best out of the related techniques. The aim of Part One is to explain the physical aspects involved in an elementary but rigorous way, accessible to readers without a specialised knowledge of theoretical physics and mathematics. More detailed developments can be found in specialised books and review papers.

The notation and symbols have been unified for the different techniques and radiations. They are mostly in accordance with the conventions of the International Union of Crystallography and with what seems to be the most common and logical usage. Appendix A contains a lexicon of the main notation.

1

Waves, Particle Beams and Matter

1.1 RADIATIONS

The common concept of radiation includes any form of energy propagated in waves or moving particles: light, sound, cosmic rays, radioactivity, etc.

Two categories of radiation are dealt with in this book: electromagnetic waves and particle beams.

1.1.1 ELECTROMAGNETIC WAVES

An electromagnetic wave consists of an electromagnetic field, vibrating transversely and sinusoidally, propagating at light velocity c. It is characterised by its frequency ν and its wavelength λ, with $\lambda = c/\nu$. It carries quantised energy in form of *photons*. The energy of a photon is given by

$$E = h\nu \tag{1.1}$$

where h is the Planck constant.

Examples of electromagnetic radiations: light, X-rays, γ-rays.

1.1.2 PARTICLE BEAMS

Any particle of mass m moving with a velocity v has an associated wave of wavelength λ given by

$$\lambda = \frac{h}{mv} \tag{1.2}$$

This principle stated by Louis de Broglie (1925), winner of the Nobel prize for physics in 1929, is the basis of wave mechanics. The physical existence of the wave, generally called *de Broglie wave* or *matter wave*, has been demonstrated by Davisson and Germer in 1927, in their first experiment of electron diffraction.

Examples of matter waves: electron, ion and neutron beams.

Therefore, both for electromagnetic waves and for matter waves, a *wave–particle duality* appears which accounts for the similarities observed in the interaction experiments.

In some experiments, especially those involving energy transfers, the interaction effect can best be interpreted on the basis of the *particle aspect* (e.g. ionisation).

In other experiments, the *wave aspect* dominates (e.g. scattering, diffraction).

1.2 UNITS AND PHYSICAL QUANTITIES

SI (*Système International*) units are made use of throughout the book, except for energy and wavelength, where in the field covered two unconventional units seem more convenient.

The **wavelength** of radiations is preferentially expressed in *ångströms* (Å), rather than in nanometres (nm); the advantage of the ångström unit is that it has the same order of magnitude as interatomic distances.

The **energy** is expressed in *electronvolts* (eV); being in the order of magnitude of atomic electron binding energies, this unit is much more convenient than the joule.

More detailed information is found in Appendix A:

1. Appendix A.1: main physical quantities appearing in the book, their definitions, their units and their symbols.
2. Appendix A.2: useful universal constants.
3. Appendix A.3: lexicon of notation and symbols.

1.3 RELATION BETWEEN ENERGY AND WAVELENGTH

The *particle–wave duality* can be expressed by a mathematical relation between the energy and the wavelength of a radiation.

1.3.1 ELECTROMAGNETIC WAVES

The energy E of a photon is related to its wavelength λ, according to Eq. (1.1)

$$E = h\nu = \frac{hc}{\lambda} \rightarrow \lambda = \frac{hc}{E} \qquad (1.3)$$

where h is the Planck constant, ν the frequency and c the light velocity in vacuum.

When expressing the universal constants by their values, this equation can be approximated to the following very useful relation:

$$\lambda = \frac{12\,400}{E} \quad (\lambda \text{ in Å}, E \text{ in eV}) \qquad (1.4)$$

Waves, Particle Beams and Matter

The most commonly used electromagnetic radiations in crystallography and material science are **X-rays** in the range 0.2–2 Å, corresponding to energies between 60 and 6 keV respectively.

1.3.2 ELECTRONS

An electron of mass m, charge e and velocity v, accelerated through an electric potential V, reaches an energy $E = mv^2/2 = eV$. According to Eq. (1.2), the associated wavelength is

$$\lambda = \frac{h}{mv} = \frac{h}{\sqrt{2mE}} \tag{1.5}$$

where m is the relativistic mass, related to the rest mass m_0 of the electron by

$$m = \frac{m_0}{1 - v^2/c^2}$$

When expressing the universal constants h, e, m_0 and c by their values, Eq. (1.5) becomes

$$\lambda = \frac{12.26}{\sqrt{[E(1 + 0.979 \times 10^{-6} E)]}} \quad (\lambda \text{ in Å}, E \text{ in eV}) \tag{1.6}$$

For energies smaller than 100 keV, the relativistic correction can be neglected, leading to the simpler relation

$$\lambda = \frac{12.26}{\sqrt{E}} \quad (\lambda \text{ in Å}, E \text{ in eV}) \tag{1.7}$$

As can be seen from Table 1.1, the non-relativistic approximation is justified for $E < 100$ keV.

1.3.3 NEUTRONS

The neutrons used for scattering purposes in material science are the so-called *thermal neutrons* which are in thermal equilibrium with the moderator of a nuclear reactor. Their

Table 1.1. Wavelength associated with electrons as a function of energy, according to Eqs (1.6) and (1.7)

E(eV)	λ(Å) no corr.	λ(Å) corr.
10^2	1.226	1.226
10^4	0.1226	0.1220
10^5	0.0387	0.0370
10^6	0.0122	0.0087

6   *Structural and Chemical Analysis of Materials*

average kinetic energy is $Mv^2/2 = 3/2kT$ (where M is the neutron mass, k the Boltzmann constant and T the temperature in kelvin). Their associated wavelength is therefore given by

$$\lambda = \frac{h}{Mv} = \sqrt{\left(\frac{h^2}{3MkT}\right)} \qquad (1.8)$$

Due to the low velocity of thermal neutrons, the relativistic correction may be neglected and only the rest mass considered. If replacing the constants by their values, Eq. (1.8) becomes

$$\lambda = \frac{25.14}{\sqrt{T}} \quad (\lambda \text{ in Å}, T \text{ in kelvin}) \qquad (1.9)$$

For temperatures ranging between 0 and 100 °C, the wavelength varies from 1.55 to 1.35 Å, the energy from 3.5×10^{-2} to 4.8×10^{-2} eV.

Thermal neutrons have about the same wavelength as currently used X-rays. They are interesting for structure analysis as a complement to X-ray diffraction.

1.4 RANGE OF RADIATIONS USED IN MATERIAL SCIENCE

Radiation energy ranges continuously from very low to very high values. The whole range of radiations can be arbitrarily divided into four groups, according to their energy and to their mode of interaction with matter.

1. Very-high-energy radiations. With a lower limit of several MeV, they interact with the atomic nuclei by inducing nuclear reactions. They are in the field of nuclear physics.
2. High-energy radiations. Their interaction extends to the inner electron shells of atoms, their energy ranging from about 1 keV to a few MeV. It is the area of X-rays and of the so-called high-energy electrons.
3. Low-energy radiations. Their interaction extends to the external electron shells of atoms, including the valence and the conduction bands of solids. Their energy ranges from about 10 eV to 1 keV. It is the area of soft X-rays and of low-energy electrons.
4. Very-low-energy radiations. Their interaction with atoms is limited to the valence and conduction bands and to the chemical bonds, with energies in the order of 1 eV or less. It is typically the range of light, ultraviolet and infrared radiations. Interatomic vibrations induced by infrared are applied to structural analysis.

This classification in four categories of radiations is based on the localisation of the interaction at an atomic scale. It has no precise limits (Table 1.2).

1.4.1 RADIATIONS DEALT WITH IN THIS BOOK

The radiations involved in material studies in this book are mostly in the high-energy range and, for specific applications, in the low-energy range.

Table 1.2. Energy range of radiations commonly used in material science. The radiations dealt with in detail in the book are written in bold face

Energy (eV)	Electromagnetic waves	Matter waves	Classification
10^8			Very high energy
10^7	γ-rays		
10^6			
10^5		**High-energy electrons**	**High energy**
10^4	**X-rays**		
10^3			
10^2	UV	**Low-energy electrons**	**Low energy**
10^1			
1	Light		
10^{-1}			
10^{-2}	IR	**Thermal neutrons**	
10^{-3}			
10^{-4}			Very low energy
10^{-5}	Radio waves		
10^{-6}			
10^{-7}			

For *chemical analysis* of materials involving energy transfer, the high-energy radiations are the most useful, being able to act on all electron levels, up to the core levels. They include X-rays, high-energy electrons and primary ions. Low-energy electrons are specific probes for surface analysis.

For *structural analysis* by means of scattering and diffraction, the main criterion is a wavelength smaller than the interatomic distances. In addition to X-rays and high- and low-energy electrons, thermal neutrons are very useful for complementary information.

1.5 BASIC ATOMIC THEORY. ENERGY LEVELS IN ATOMS

A basic knowledge of atomic theory is required in order to understand the mechanisms of radiation–matter interaction. The following paragraph is merely a quick survey. More detailed developments are found in numerous books on physics.

An atom is defined by its atomic number Z in the periodic table of elements. It is made up of:

(a) a *nucleus* of positive charge $+Ze$ consisting of Z protons with a charge $+e$ each and of a given number of neutrons;
(b) *Z electrons* with a charge $-e$ each, arranged in orbitals.

In quantum mechanics a given orbital can be defined by its *wave function* ψ which is a characteristic solution (*Eigenfunction*) of the time-independent *Schrödinger equation*. The square $\psi\psi^*$ of the wave function represents the probability of finding an electron at a given point in the orbital.

1.5.1 QUANTUM NUMBERS

An orbital is characterised by a set of quantum numbers.

1. The *principal quantum number n* defines the electron shells K, L, M, N, ... corresponding respectively (with increasing energy) to $n = 1, 2, 3, 4, \ldots$.
2. The *azimuthal quantum number l* defines the sub-shells s, p, d, f, ... corresponding respectively to $l = 0, 1, 2, 3, \ldots$ (expressed in units of $h/2\pi$), with $l \leqslant n - 1$. It expresses the orbital angular momentum **L** and determines the form of the orbital.
3. The *magnetic quantum number m* corresponds to the possible values of the magnetic momentum of the electron in an external magnetic field. Its values are restricted to $l \leqslant m \leqslant l$.
4. Finally, an electron in an orbital is characterised by the *spin quantum number s* which corresponds to its intrinsic angular momentum (expressed in units of $h/2\pi$). It can have two values $s = \pm 1/2$.

The set of four quantum numbers n, l, m and s characterises an *electron state*. Any orbital comprises two states, with respectively $s = +1/2$ and $s = -1/2$. According to *Pauli's exclusion principle*, no two electrons can be identical in their four quantum numbers. It follows that there can be only one electron in a given state, i.e. a maximum of two spin-paired electrons per orbital.

Table 1.3 summarises these notations for the three inner shells.

1.5.2 ELECTRON ENERGY LEVELS

A discrete energy value is associated with any set of the three quantum numbers, n, l, s. It is convenient to introduce the so-called *inner quantum number j* which represents the total angular momentum $\mathbf{J} = \mathbf{L} + \mathbf{S}$. This number has the values $j = l + s = l \pm 1/2$, with $j > 0$. The energy $E(nlj)$ of an electron state is correspondingly characterised by means of the three quantum numbers n, l, j. It represents the *electron binding energy*, i.e. the amount of energy needed to move the electron to an infinite distance from the atom. It can be seen in Table 1.3 that at each s-subshell ($l = 0$) corresponds to one energy level, whereas any other subshells are split into two energy levels corresponding to opposite spins.

The concept of *electron energy levels* in an atom (also called *atomic energy levels*) is very important in radiation–matter interaction effects.

Table 1.3. Quantum notation corresponding to the three inner shells (core shells)

Shell	n	Sub-shell	l	m	Number of orbitals	Notation of orbitals	j	Number of states
K	1	s	0	0	1	1s	1/2	2
L	2	s	0	0	1	2s	1/2	8
		p	1	−1, 0, +1	3	2p	1/2, 3/2	
M	3	s	0	0	1	3s	1/2	
		p	1	−1, 0, +1	3	3p	1/2, 3/2	18
		d	2	−2, −1, 0, +1, +2	5	3d	3/2, 5/2	

BIBLIOGRAPHY

[1] Allard R. *Le système international de mesures*. Gauthier-Villars, Paris, 1965.
[2] Barford N. C. *Experimental measurements. Precision, error and truth*, 2nd edn. John Wiley, New York, Chichester.
[3] Christmann J. R. *Fundamentals of solid state physics*. John Wiley, New York, Chichester, 1988.
[4] Eisberg R., Resnick R. *Quantum physics of atoms, molecules, solids, nuclei and particles*, 2nd edn. John Wiley, New York, Chichester, 1985.
[5] Gerloch M. *Orbitals, terms and states*. John Wiley, New York, Chichester, 1986.
[6] Griffiths D. *Introduction to elementary particles*. John Wiley, New York, Chichester, 1987.
[7] Kittel C. *Introduction to solid state physics*, 6th edn. John Wiley, New York, Chichester, 1986.
[8] Kittel C. *Quantum theory of solids*, 2nd rev. printing. John Wiley, New York, Chichester.
[9] Krane K. S. *Modern physics*. John Wiley, New York, Chichester, 1983.
[10] Massey B. S. *Measures in science and engineering*. John Wiley, New York, Chichester, 1986.
[11] Mott N. F. *Elementary quantum mechanics*. The Wykeham Science Series, Wykeham, London, Winchester, 1972.
[12] Pauling L., Wilson E. B. *Introduction to quantum mechanics*. McGraw-Hill, New York, 1935.
[13] Sproull R. L., Phillips W. A. *Modern physics. The quantum physics of atoms, solids and nuclei*, 3rd edn. John Wiley, New York, Chichester, 1980.

2

Basics on Radiation–Matter Interactions

2.1 GENERAL CONSIDERATIONS

2.1.1 OBSERVED EFFECTS

According to the principle of energy and mass conservation, two complementary aspects of interaction effects may be considered: those affecting the radiation and those affecting the specimen.

2.1.1.1 Effects on Radiation

During the process of interaction with a specimen, a wave undergoes changes which can affect three of its physical quantities: *intensity*, *energy* and *wave vector*; they are noted respectively I_0, E_0 and $\mathbf{k}_0$ before interaction; I, E and $\mathbf{k}$ after interaction. The wave vector $\mathbf{k}$ indicates simultaneously the direction of wave propagation (vector direction) and the corresponding wavelength (vector modulus $k = 1/\lambda$).

The corresponding effects are respectively absorption, energy loss and scattering.

Absorption. The wave intensity I which is transmitted in the forward direction is smaller than the incident intensity I_0

$$I < I_0$$

Energy loss. A fraction of the outcoming particles have lost more or less of their initial energy

$$E \leqslant E_0 \Rightarrow \lambda \geqslant \lambda_0$$

An interaction without energy loss is called an *elastic interaction*

$$\Delta E = E_0 - E = 0$$

An interaction involving energy losses is called an *inelastic interaction*

$$\Delta E > 0$$

The energy distribution results in an energy spectrum $I(E)$.

Scattering. A fraction of the outcoming particles have been deviated from the incident direction $\mathbf{k} \neq \mathbf{k}_0$ in direction with

$\lambda = \lambda_0$ for *elastic scattering*;
$\lambda > \lambda_0$ for *inelastic scattering*.

2.1.1.2 Effects on Matter

The amount of energy lost by the incident radiation is transferred to the atoms of the specimen in various forms: excitation or ionisation energy, kinetic energy of ejected particles, thermal vibration energy, crystal defect energy. A given fraction of this energy will be re-emitted as *secondary radiation*.

2.1.2 RADIATION–SPECIMEN ENERGY TRANSFER

By analogy with classical mechanics, and considering the corpuscular aspect of radiation, it is convenient to distinguish between elastic and inelastic interactions.

2.1.2.1 Elastic Interactions

An elastic interaction takes place between the incident radiation and the atom taken as a whole. As a result, the internal structure and energy of the atom remain roughly unchanged. The relative energy transfer is zero or very small

$$\Delta E/E_0 = E_0 - E \cong 0 \Rightarrow \Delta\lambda/\lambda_0 \cong 0 \Rightarrow \lambda \cong \lambda_0; \; k \cong k_0$$

The incoming and the outgoing waves are *coherent*; the interaction results in coherent or elastic scattering giving rise to interference effects. In the particular case of crystalline solids, the phenomenon is enhanced by periodicity: the waves are diffracted by the crystal lattice. This is the fundamental effect leading to the determination of crystal structures.

2.1.2.2 Inelastic Interactions

An inelastic interaction takes place between the incident radiation and individual orbital electrons. It results in a change of the internal structure and energy of the atom concerned and therefore in a large relative *energy transfer*

$$\Delta E/E \neq 0 \Rightarrow \Delta\lambda/\lambda_0 \neq 0 \quad \lambda > \lambda_0; \; k < k_0$$

The incoming and the outgoing waves are no longer coherent; the interaction results in incoherent or inelastic scattering which cannot give rise to interference effects.

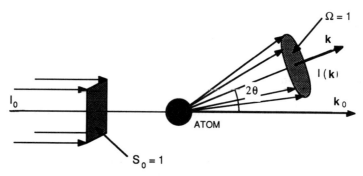

Figure 2.1. Diagram illustrating the relationship between the intensity scattered per unit solid angle and the differential scattering cross-section.

The energy transfer can be related to the nature of the atoms of the sample. This relationship is the basis of spectrometric techniques leading to chemical analysis.

2.1.3 INTERACTION CROSS-SECTION

Consider a particle of matter of radius r and apparent area $s = \pi r^2$ relative to the incident radiation. For a given interaction effect, the effective area $\sigma \leqslant s$ such that any radiation particle passing through it undergoes that interaction, is called the *interaction cross-section*. The interaction cross-section of a sample is the sum of the individual cross-sections of its atoms. Any form of interaction can be expressed by means of a corresponding cross-section, in particular the scattering effects.

2.1.3.1 Scattering Cross-section

The scattering cross-section (elastic scattering, inelastic scattering or total scattering) of a given object may be defined in three different ways.

Differential Scattering Cross-section. This is designated $\sigma(\mathbf{k})$ and defines the scattering per unit area in a direction $\mathbf{k}$ making an angle 2θ with the incident direction $\mathbf{k}_0$ (Figure 2.1).

Integrated Scattering Cross-section. This is designated $\sigma(\Omega)$ and defines the scattering in a given solid angle Ω.

Total Scattering Cross-section. This is designated $\sigma(4\pi)$ and defines the total scattering in space, over a solid angle 4π sr.

$$\sigma(4\pi) = \int_0^{4\pi} \sigma(\mathbf{k}) d\Omega$$

Relationship Between Intensity and Cross-section

Intensity and cross-section are related as follows:

$$\frac{I(\mathbf{k})}{I_0} = \frac{\sigma(\mathbf{k})}{s_0} \quad \text{and} \quad \frac{I(4\pi)}{I_0} = \frac{\sigma(4\pi)}{s_0} \qquad (2.1)$$

where I_0 is the incident intensity, $I(\mathbf{k})$ the intensity scattered per unit of solid angle in a direction $\mathbf{k}$, $I(4\pi)$ the total intensity scattered in space and s_0 the unit area.

The scattering cross-section is measured by the ratio of the scattered intensity to the incident intensity.

A cross-section has the dimensions of an area.

2.1.4 MEAN FREE PATH

The average distance covered by a given radiation between two interaction effects is called its mean free path l.

For a given interaction, the mean free path is inversely proportional to the interaction cross-section, according to the relation

$$l\sigma(4\pi)N(V) = 1 \qquad (2.2)$$

where $N(V)$ is the number of atoms or other individual particles per unit area.

2.2 X-RAY INTERACTION

The global value of the interaction cross-section of X-rays with matter is relatively small in the energy range considered.

2.2.1 ELASTIC INTERACTION

Elastic scattering of X-rays takes place, at first approximation, between a photon and the electronic cloud of an atom as a whole. The resulting coherent scattering has fundamental applications for structural analysis of materials. Its parameters will be dealt with in detail in Chapter 3.

2.2.2 INELASTIC INTERACTIONS

Inelastic interaction occurs with individual orbital electrons. Its interpretation refers to the corpuscular aspect of electromagnetic waves, the photons. There are two mechanisms of inelastic interaction: excitation or ionisation of atomic levels and the Compton effect.

2.2.2.1 Excitation of Atomic Core Levels

An incoming photon can yield its whole energy $E_0 = h\nu_0$ to an orbital electron, in one single interaction ($\Delta E = E_0$). The electron concerned can be raised to a more external vacant state (*excitation*) or expelled from the atom (*ionisation*). In the high energy range of X-rays considered, the interaction takes place mainly at *core levels* (K or L levels). The energy difference between an excitation process to a valence or conducting level and an expulsion process into the continuum is in the order of magnitude of 1 eV. This difference is negligible compared to the binding energy at core levels. Therefore, no distinction will henceforth be made between excitation and ionisation.

The depleted incident energy of the photon reappears in two ways:

1. Kinetic energy of the expelled orbital electron, called *photoelectron*;
2. Potential energy rise of the excited atom. Its de-excitation (or relaxation) leads to characteristic *secondary emissions*.

This process of energy transfer, called the *photoelectric effect*, is predominant in the energy range considered and with solids. It is of great importance for chemical analysis and will be considered in detail in Chapter 7.

2.2.2.2 Compton Effect

An incoming photon can yield a fraction of its energy in a collision process. The incident energy then reappears in two ways:

1. Kinetic energy transferred to the electron (recoil electron);
2. Electromagnetic energy of the scattered photon, with $E < E_0$, $\lambda < \lambda_0$.

This inelastic scattering process is called the *Compton effect*. The classical theory of elastic collision (energy and momentum conservation, Figure 2.2) leads to the expression of the difference $\Delta\lambda$ in wavelength before and after the interaction

$$\Delta\lambda = \lambda - \lambda_0 = \frac{h}{mc}(1 - \cos 2\theta) \qquad (2.3)$$

Substituting the universal constants by their numerical values (electron mass approximated to its rest mass m_0) leads to

$$\Delta\lambda = 0.0243 \, (1 - \cos 2\theta) \quad (\lambda \text{ in Å}) \qquad (2.4)$$

The absolute increase of the wavelength does not depend on the wavelength; it is a function of the scattering angle, with a maximum value of 0.048 Å at $2\theta = 180°$.

The Compton scattering cross-section increases with the incident energy. It is the more important the smaller the binding energy of the electrons. Its relative contribution to the total interaction process is therefore greater for gases and liquids than for solids and greater for light elements than for heavy elements (Table 2.1). It is the main ionisation effect in gas radiation detectors.

Compton scattering leads to a broadening of diffraction lines at the high angle side or to a splitting of these lines at large angles. In the techniques developed in this book,

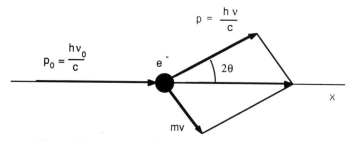

Figure 2.2. Compton effect. Momentum conservation.

Table 2.1. Share of inelastic X-ray scattering of a few elements as a function of their atomic number [1]

Element	Z	$I_{inel}/I_{el.}$
Li	3	Large
C	6	5.5
S	16	1.9
Fe	26	0.5
Cu	29	0.2
Pb	82	$\cong 0$

it has a rather negative effect. However, it can be a useful standard in X-ray fluorescence.

Compton spectrometry as a method in itself has developed recently, thanks to the new synchroton high-intensity X-ray sources. The study of Compton scattering line profiles provides information on chemical bonding in gases, liquids and solids containing light elements, e.g. metal hydrides.

2.3 ELECTRON INTERACTION

The overall cross-section of electron–matter interaction is much greater than the corresponding value for X-rays. An electron of mass m travelling at a velocity v has a kinetic energy $mv^2/2$ which can be yielded continuously in successive interaction processes. The interaction effect with an atom depends on the interaction distance r defined in Figure 2.3.

2.3.1 ELASTIC INTERACTION

2.3.1.1 Basic Theory

If the interaction distance r is much greater than the atomic radius r_0, interaction takes place between the incident electron and the atom as a whole. The electron–atom

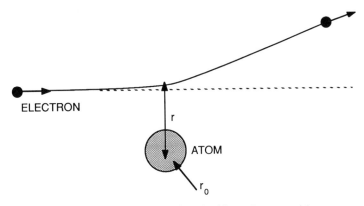

Figure 2.3. Interaction distance of an incident electron with an atom.

interaction may then be interpreted in a classical way as an elastic collision, according to the principle of energy and momentum conservation.

Consider:

(a) an incident electron of energy E_0, rest mass m_0, velocity v_0, wave vector $\mathbf{k}_0$; after the collision its velocity becomes v, its wave vector $\mathbf{k}$;
(b) an atom of mass M, initially at rest and of velocity V after the collision.

As a consequence of the small electron mass with respect to the atom mass, the relative energy transfer $\Delta E/E_0$ can be considered as insignificant and the final velocity of the electron may be assimilated to its initial velocity. It follows that the associated wavelength is conserved, $\lambda = \lambda_0$.

In a direction defined by the scattering angle 2θ or the scattering vector $\mathbf{R} = \mathbf{k} - \mathbf{k}_0$ ($k = k_0 = 1/\lambda$), the momentum variation is $2m_0 v \sin\theta$, according to Figure 2.4. The direction of the atom recoil is opposite to vector $\mathbf{R}$; the momentum conservation is written

$$2m_0 v \sin\theta = MV$$

The energy transfer ΔE to the atom is given by expressing the energy conservation. The maximum energy transfer is for $\theta = \pi/2 \Rightarrow \sin\theta = 0$. It follows that

$$\Delta E = \frac{1}{2}MV^2 = \frac{2m_0^2 v^2 \sin^2\theta}{M} \rightarrow \Delta E(\text{max}) = \frac{2m_0^2 v^2}{M}$$

$$\frac{\Delta E}{E_0} = \frac{4m_0}{M}\sin^2\theta \rightarrow \frac{\Delta E(\text{max})}{E_0} = \frac{4m_0}{M} \quad (2.5)$$

Given that $m_0/M \cong 1/1836A$ (where A is the atomic mass of the atom considered), the relative energy transfer may be approximated to

$$\frac{\Delta E(\text{max})}{E_0} = \frac{2.17 \times 10^{-3}}{A}\sin^2\theta \quad (2.6)$$

The preceding calculation is accurate for electron energies smaller than 100 keV, corresponding to the energy range generally considered in this book. For higher energies a relativistic correction has to be applied, leading to the following maximum energy transfer:

$$\Delta E(\text{max}) = \frac{2E_0(E_0 + 2mc^2)}{Mc^2} \quad (2.7)$$

2.3.1.2 Interaction Effects

Considering an increasing energy transfer, the following effects can be observed.

Thermal Effect

The energy transfer is of the order of 10^{-2}–10^{-1} eV, resulting in thermal oscillations of the atoms about their equilibrium positions. The simplest form of interaction takes

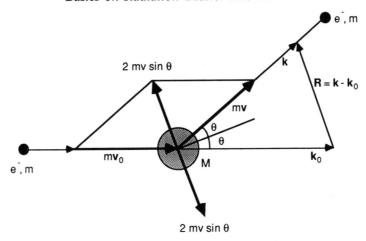

Figure 2.4. Elementary interpretation of an elastic collision.

place through the excitation of individual atomic vibrations (*Debye effect*). In crystalline solids, the thermal energy can induce the excitation of collective oscillations, with quantified energy states called *phonons*.

Thermal agitation results in a release of heat, an effect which is put to use in industry for melting refractory materials. In facilities using electron beams (X-ray tubes, electron microscopes, etc.), the thermal effect of electron bombardment generally has a negative aspect.

Chemical Effect

An energy transfer of the order of a few electronvolts can induce the breaking of chemical bonds, especially the weak bonds of organic molecules. Given the nature of electrons, this effect leads generally to reducing reactions.

Displacement Effect

The energy E_d needed to dislodge an atom from its potential well is of the order of 10–30 eV. An energy transfer greater than E_d can therefore result in permanent irradiation damage. This effect can be very important in devices based on high-intensity, high-energy electron beams, as in high-voltage electron microscopes.

Table 2.2. Energy transfer through elastic collision, for 100 keV electrons, according to Eq. (2.6)

Element	A	ΔE(eV) $\theta = 10^{-2}$ rad	ΔE(eV) $\theta = \pi/2$
Li	6.9	31.0×10^{-4}	31
O	16	13.5×10^{-4}	13.5
Mg	23.4	9.0×10^{-4}	9
Cu	63.5	3.5×10^{-4}	3.5
Au	197	1.1×10^{-4}	1.1

In the particular case of 100 keV electrons and small diffusion angles of about 10^{-2} rad, the most common values in electron microscopes, it can be seen from Table 2.2 that the energy transfer is negligible, even for light elements. For backscattered electrons ($\theta = \pi/2$), the energy transfer is maximum and becomes significant, of the order of a few electronvolts. However, the displacement energy is reached only for very light elements, such as lithium. The displacement effect becomes important for electron energies in the MeV range (Ch. 19).

2.3.1.3 Characteristics of Elastic Electron Scattering

Small-angle scattering is predominant for electrons, due to the steep decrease of the atomic elastic scattering cross-section with increasing angle θ (see Ch. 4).

The elastic scattering cross-section varies according to Z^2/E_0^2.

The typical angular half-width is some 10^{-2} rad.

Given the negligible energy loss at small angles, the wave coherence is conserved, leading to electron diffraction by crystalline solids.

2.3.2 INELASTIC INTERACTION

2.3.2.1 Basic Theory

When the interaction distance is of the same order as the atom radius, the electron–electron interaction becomes predominant. The mass of the interacting particles being the same, the relative energy transfer becomes significant, resulting in a change of the internal energy of the atoms concerned.

Consider an incident electron of energy E_0, of wave vector $\mathbf{k}_0$. After an inelastic interaction its energy and wave vector length become respectively $E = E_0 - \Delta E$ and $k = k_0 - \Delta k$, with $k = (2mE/h^2)^{1/2}$ according to Eq. (1.5).

For small relative variations ($\Delta k \ll k_0$ and $\Delta E \ll E_0$), it follows approximately that

$$\frac{\Delta k}{k_0} \cong \frac{\Delta E}{2E_0} \tag{2.8}$$

During inelastic scattering, the scattering vector $\mathbf{R} = \mathbf{k} - \mathbf{k}_0$ cannot become zero. According to Figure 2.5 it follows at the small angle approximation that

$$R^2 \cong (\Delta k)^2 + (2\theta k_0)^2$$

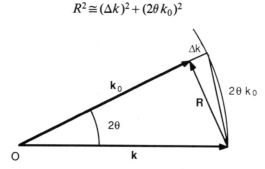

Figure 2.5. Inelastic electron scattering at the small angle approximation.

Hence, in the forward direction ($\theta = 0$), the minimum length of the scattering vector is Δk. Let $R_{min} = \Delta k = k_0 \theta_E$, where θ_E represents the so-called characteristic scattering angle for a given energy loss ΔE. According to Eq. (2.8), that angle can be written

$$\theta_E = \frac{\Delta k}{k_0} = \frac{\Delta E}{2E_0} \qquad (2.9)$$

The length R of the scattering vector is then given by

$$R^2 = k_0^2 [(2\theta)^2 + \theta_E^2] \qquad (2.10)$$

For a given energy loss ΔE, the differential scattering cross-section has a half-width of the order of θ_E.

2.3.2.2 Interaction Effects

Inelastic electron–matter interaction can cause various effects corresponding to a modification of the internal structure of the atoms in the irradiated specimen.

Excitation of Valence and Conduction Levels

The energy transfer is less than 50 eV. Conduction level excitation takes place collectively and results in quantified energy states called *plasmons*. Plasmon scattering leads to discrete energy losses, typically of the order of 10–20 eV, characteristic of the elements concerned, useful in electron spectrometry.

Excitation of Atomic Core Levels

In the same way as X-ray photons, high-energy incident electrons can cause core level excitation of atoms. The characteristic energy losses suffered by interacting electrons at core levels ($\Delta E > 50$ eV) are applied in energy loss spectrometry (Ch. 17).

The energy lost by the primary electrons reappears in two ways:

1. Kinetic energy of expelled orbital electrons, called *secondary electrons*;
2. Potential energy increase of the excited atoms. The return to the fundamental state by means of relaxation processes induces *secondary emissions* which are used for analytical purposes.

These effects are examined in more detail in Chapter 7.

Conversion to Electromagnetic Energy

The kinetic energy of electrons, and more generally of any charged particles, may be directly transformed in electromagnetic radiation energy through the braking effect they undergo when interacting with atoms. This effect, known as *bremsstrahlung*, is detailed in Chapter 7.

The importance of the inelastic electron–matter interaction in any method using electron beams has led to a great number of scientific papers (e.g. Bethe, 1930; Lenz, 1954; Colliex and Jouffrey, 1972; Jouffrey et al in [6]).

2.3.2.3 Characteristics of Inelastic Electron Scattering

As a whole, the inelastic scattering cross-section varies according to Z/E_0^2 (instead of Z^2/E_0^2 for elastic scattering). It is the greater the smaller the energy loss. According to Eq. (2.9), this entails a very small angular half-width of the order of a few milliradians, about 10 times less than for elastic scattering. As an example, the aluminium K-level excitation (1560 eV) by 100 keV electrons leads to a characteristic scattering angle $\theta_E = \Delta E/2E = 7.8$ mrad.

As a result, inelastic electron scattering is mainly concentrated about the incident direction, whereas elastic scattering is more widely spread.

2.4 NEUTRON INTERACTION

The elastic collision of neutrons with atoms can be interpreted by means of the laws of classical mechanics, in the same way as for electrons. However, the neutron mass is no longer negligible with respect to the atom mass, so that the approximation of small energy transfer which was justified for electrons is not appropriate: energy transfers are significant. Irradiation damage through atomic displacements can become very important with fast neutrons used in nuclear physics. *Thermal neutrons*, in thermal equilibrium with the moderator of a nuclear reactor, the only ones used for material analysis, have a very low energy of about 0.1 eV. They are therefore unable to cause atomic displacements and even atomic level ionisations.

Interaction with matter takes place mainly between neutrons and atom nuclei, through nuclear forces and through magnetic forces.

For large neutron–atom interaction distances, the relative energy transfers are negligible, leading to elastic and therefore coherent scattering. Elastic scattering and diffraction of thermal neutrons have important applications in material sciences as a complementary technique to X-rays.

For small interaction distances, the thermal neutrons can induce thermal vibrations of the atoms of a sample and possibly the breaking of weak chemical bonds, resulting in inelastic scattering.

2.5 ION INTERACTION

Ions are atoms with a positive or negative charge. As a result of their charge, they produce effects which are similar to those observed with electrons. However, their mass is of the same order as the mass of the atoms in the sample. It follows a very important energy transfer which can reach a maximum value equal to the kinetic energy of the incident ions when they are completely stopped (ion implantation).

Various interaction effects occur when bombarding a solid with primary ions of energy E_0:

(a) backscattering;
(b) atomic level excitation with secondary emissions (see electron interaction);
(c) bremsstrahlung emission (see electron interaction);
(d) atomic displacements leading to defects in solids (Frenkel defects);
(e) sputtering of surface layers in solids, due to the expulsion through collision of neutral atoms, of secondary positive and negative ions;
(f) implantation into the surface layers of solids.

Some of these effects are becoming more and more important in research and industry.

Ion Sputtering

This takes place when the normal component of the kinetic energy transferred to surface atoms of a solid is greater than their surface bonding energy. Ion sputtering is characterised by the so-called *sputtering rate* which represents the number of ejected atoms per incident ion. The sputtering rate depends on the operating conditions. It is maximum for an almost grazing incidence (angle of 10-20° with the surface) and for a primary ion energy of a few kiloelectronvolts. It also depends on the atomic numbers of incident ions and of sputtered atoms. The expelled particles can be neutral atoms, positive or negative ions.

Sputtering is a negative effect in many devices dealing with particle beams, such as gas-discharge tubes, ion guns, apertures of electron microscopes and of mass spectrometers, etc. However, the positive aspect of ion sputtering is becoming increasingly important, with applications such as specimen thinning for electron microscopy, metal coating by means of cathodic sputtering, and surface analysis with secondary ion emission spectrometry. Due to its growing industrial importance, ion sputtering is an important research topic and has led to numerous review papers in the last decade (e.g. in [2]).

Ion Implantation

This takes place preferentially for normal incidence and primary ion energies of more than 10 keV. The depth of implantation depends on the energy and on the nature of incident ions and on the nature of the target material. It currently reaches about 1000 Å for 50 keV ions.

Its practical importance is growing fast, e.g. for semiconductor doping and for surface treatment of materials such as steel surface hardening with nitrogen ions.

2.6 ENERGETIC BALANCE OF INTERACTION

Table 2.3 summarises the energetic balance of interaction effects dealt with in this book. For any kind of radiation, the relative importance of the various interaction processes involved depends on the incident radiation energy and on the nature of the material investigated. The most important interaction processes for material analysis in the energy range considered will be examined in more detail in the following theoretical chapters. Their applications will be developed in Parts Three, Four and Five.

Table 2.3. Summary of radiation–matter interactions and their main applications to material analysis. Principal abbreviations used in the table: atom. level excit. (atomic level excitation), charact. (characteristic), displ. (displacement), el. (electron), elast. (elastic), inel. (inelastic), scatt. (scattering), second. (secondary), spect. (spectrometry), vibr. (vibrations). The conventional abbreviations for analytical techniques are explained in Appendix E

Primary radiation	Modification of matter	Modification of radiation	Secondary radiation	Applications
X-rays	El. vibration	Elast. scatt.		X-ray diffraction Structure analysis
	Recoil el. Defects	Inel. scatt.		Compton spect.
	Atom. level excit.	Absorption		Absorption spect. (XAS, EXAFS)
			Photoelectron	photoel. spect. (XPS, ESCA)
	De-excitation		Charact. X-rays	X-ray fluorescence (XRF)
			Auger electron	Auger spect. (ESCA)
Electrons	Thermal vibr. Plasmon excit.	Elast. scatt.		El. diffraction Structure analysis
	Bond breaking Atom displ.	Inel. scatt.		
	Atom. level excit.	Absorption	Bremsstrahlung Second. el.	X-ray source imaging (SEM)
		Energy loss		Energy loss spect. (EELS)
	De-excitation		Charact. X-rays	X-ray microanal. (EPMA, EDX) X-ray source
			Auger el.	Auger spect. (AES)
Neutrons	Thermal vibr. Bond breaking Atom displ. absorption	Elast. scatt. Inel. scatt.		Neutron diffraction Structure analysis
Ions	Thermal vibr. Bond breaking	Scattering		Structure analysis
	Sputtering	Second. ions	Second. ion spect.	(SIMS)
	Ion implantation			implantation
	Atom. level excit.	Absorption	Bremsstrahlung	
	De-excitation		Charact. X-rays	

BIBLIOGRAPHY

[1] Adler I. *X-Ray Emission spectrography in geology.* Elsevier, New York, 1966.
[2] Behrish R. (Ed.) *Sputtering by particle bombardment. Topics in applied physics.* Springer-Verlag, Berlin, 1983.
[3] Billington D. S., Crawford J. H. *Radiation damage in solids.* Princeton University Press, 1961.

[4] Cauchois Y. et al *Action des rayonnements de grande énergie sur les solides.* Gauthier-Villars, Paris, 1956.
[5] Gauthe B. *Pertes d'énergie caractéristiques des électrons dans les solides.* Gauthier-Villars, Paris, 1968.
[6] Jouffrey B, Bourret A., Colliex C. (ed.) *Microscopie électronique en science des matériaux.* Editions du CNRS, Paris, 1983.

3

Basic Theory of Elastic Scattering

3.1 DEFINITION OF THE SCATTERING OBJECT

3.1.1 SCATTERING POWER

A solid can be described as a distribution of atoms at specific points in space (potential wells), each atom being made up of a nucleus and an electron cloud.

With regard to radiation scattering, any form of matter may be characterised by its *scattering power*. Depending on the specific reasoning purpose, scattering power may be regarded as a continuous function of space coordinates or as a discontinuous distribution at discrete points in space.

For a given radiation, scattering power is a characteristic function of matter.

3.1.1.1 Scattering Power as a Continuous Function

The scattering power $f(\mathbf{r})$ at a point $\mathbf{r}(xyz)$ in a volume v of matter is defined as the amplitude scattered at that point per unit of volume and per unit of solid angle, under the effect of unit incident amplitude.

Depending on the nature of radiation, the scattering agent may be the electron density or the electric potential.

Scattering power has maximum values at atomic centres; defining its repartition is therefore equivalent to defining the distribution of atoms in the scattering volume.

3.1.1.2 Scattering Power as a Discontinuous Distribution

Sometimes it is more convenient to consider a volume of matter as a distribution of discrete scattering elements.

Consider a scattering element n with its centre of gravity located at a point $\mathbf{r}_n(x_n, y_n, z_n)$ in a volume v of matter. Its scattering power f_n is defined as the amplitude it scatters into unit solid angle, under the effect of incident unit amplitude.

In a volume v containing N elements, any of them disposes of an average volume $v_0 = v/N$.

Basic Theory of Elastic Scattering

Depending on the nature of the specimen considered, a scattering element may be an atom, a group of atoms, a crystal unit cell or any solid particle.

3.1.2 STRUCTURE FUNCTION

According to the definition of matter as a continuous display or as a discontinuous distribution, the *structure function* can be defined as

(a) the space function of scattering power $f(\mathbf{r})$ at any point $\mathbf{r}$ of the volume;
(b) the distribution of scattering elements f_n at any point $\mathbf{r}_n$ of the volume.

In a given volume v of matter, the coordinates are determined by an arbitrarily chosen origin.

For a distribution of discrete scattering elements, the structure function can be represented as a sum of Dirac functions

$$f(\mathbf{r}) = \sum_n f_n \delta(\mathbf{r} - \mathbf{r}_n) \tag{3.1}$$

The sum is zero at all points, except for $\mathbf{r} = \mathbf{r}_n$ where it is infinite (with an integrated value of one in a volume v_0).

For a crystal, the scattering elements are the unit cells; their distribution on the lattice points is periodic, according to

$$\mathbf{r}_n = u\mathbf{a} + v\mathbf{b} + w\mathbf{c}$$

where $\mathbf{a}$, $\mathbf{b}$, $\mathbf{c}$ are basis vectors, u, v, w are integers and lattice point coordinates.

3.1.3 DISTRIBUTION FUNCTION

In some cases, the structure function itself cannot be determined. The scattering object can then be represented by the so-called *distribution function* which gives the averaged density distribution of scattering matter at a vectorial distance $\mathbf{r}$ of any scattering element taken arbitrarily as the origin.

In the simple instance of identical scattering elements, each with unity scattering power, the distribution function is given by

$$q(\mathbf{u}) = N\delta(\mathbf{u}) + \frac{p(\mathbf{u})}{v_0} \tag{3.2}$$

The function $q(\mathbf{u})$ represents the probability of finding a scattering element per unit volume v_0 at a vectorial distance $\mathbf{u}$ of any scattering element chosen as the origin. It has maxima for values $\mathbf{u}$ of the variable such as $\mathbf{u} = \mathbf{r}_m - \mathbf{r}_n$ is a vector joining two scattering elements m and n. It expresses a distribution function with a central peak:

(a) $\delta(\mathbf{u})$ is a Dirac function which represents the central peak $\mathbf{u} = 0$ corresponding to $m = n$;
(b) $p(\mathbf{u})$ is the distribution function in itself, without central peak, such as $m \neq n$.

26   Structural and Chemical Analysis of Materials

The distribution function expresses the order of the distribution of scattering elements in space (atoms, molecules, etc).

For a statistically disordered object, homogeneous at the scale of the wavelength, $p(\mathbf{u}) = 1$ at any point where u is greater than the radius r_0 of the scattering elements. An example is a perfect gas.

For an object which is heterogeneous with regard to the wavelength, the variations of $p(\mathbf{u})$ about unity express the mutual influence of the particles, corresponding to a certain order. This is the case for solids.

Amorphous and vitreous materials have an order at short distances only; $p(u)$ tends to one for values of u greater than a few interatomic (or interelement) distances.

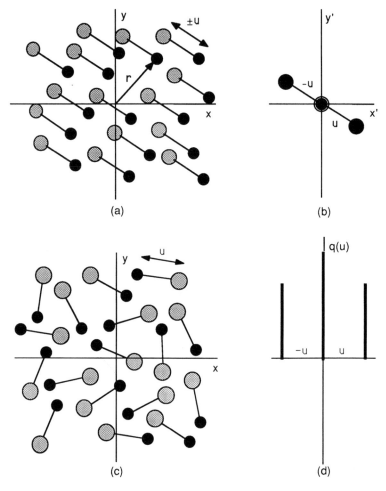

Figure 3.1. Two-dimensional illustration of the difference between the structure function $f(\mathbf{r})$ and the distribution function $q(\mathbf{r})$ in a structure made up of identical biatomic elements. The overall dimension of the structure is assumed to be infinite compared to the dimension of an element. (a) Nematic structure (same orientation of all the elements); (b) corresponding two-dimensional distribution function; (c) disordered structure; (d) corresponding radial distribution function, along an arbitrary direction.

The distribution function depends only on the scalar distance u and is called the radial distribution function (Figure 3.1c and d).

Perfect crystals have a long-distance order. The distribution function is zero everywhere, except when $\mathbf{u}$ equals a crystal lattice vector $\mathbf{r}_n$, where the function equals one. The distribution function is identical to the structure function in this case.

The distribution function $q(\mathbf{u})$ can be expressed as the convolution product of the structure function $f(\mathbf{r})$ and the inverted structure function $f(-\mathbf{r})$

$$q(\mathbf{u}) = f(\mathbf{r}) * f(-\mathbf{r})$$

This product may be called the inverted convolution square (see Figure 3.1 and Appendix C).

The relative position of any couple of scattering elements can be described by the vector $\mathbf{u}$ as well as by the vector $-\mathbf{u}$. It follows that, whatever the symmetry of the structure, the distribution function is centrosymmetrical: $q(\mathbf{u}) = q(-\mathbf{u})$.

3.1.4 SIZE FUNCTION

For a limited scattering object, the size limitation can be expressed by the *size function* $\gamma(\mathbf{r})$ such as

$\gamma(\mathbf{r}) = 1$ when $\mathbf{r}$ is included in volume v.
$\gamma(\mathbf{r}) = 0$ when $\mathbf{r}$ is exterior to volume v.

The limited structure function can be written

$$f(\mathbf{r}) = f_\infty(\mathbf{r}) \gamma(\mathbf{r}) \tag{3.3}$$

where $f_\infty(\mathbf{r})$ expresses the distribution of scattering power in an infinite volume.

The notion of limitation also applies to the distribution function $q(\mathbf{u})$ of a limited volume v: it is determined only when $\mathbf{u} = \mathbf{r}_m - \mathbf{r}_n \in v$.

In practice the size function is involved in the formulation only when the scattering object is very small (e.g. smaller than a few hundred interatomic distances). In most diffraction experiments it can be neglected.

3.2 AMPLITUDE OF SCATTERED RADIATIONS

3.2.1 GENERAL EXPRESSION OF SCATTERED AMPLITUDE

Consider a volume v of scattering matter and a point O arbitrarily chosen as the origin of coordinates and of phases.

Consider an incident plane wave of wavelength λ_0 travelling in a direction determined by the wave vector $\mathbf{k}_0$. The incident amplitude at any point of volume v is assumed to be constant. This is equivalent to neglecting the absorption and to considering the scattered amplitude to be small compared to the incident amplitude; it expresses the basis of the kinematic approximation of scattering.

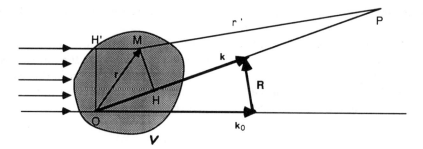

Figure 3.2. Scattering of a plane wave by a volume v of matter.

Taking into account the stationary state of coherent interferences between the scattered waves, the wave functions may be reduced to their spatial amplitudes.

Considering a continuous display of scattering matter, any point M inside the volume v, determined by its vector coordinates $\mathbf{r}(xyz)$, becomes the source of a scattered spherical wave whose amplitude is a function of the scattering power $f(\mathbf{r})$.

In most experimental settings the scattered wave will be observed at a point P at a distance r' which is great compared to the dimensions of the object ($r' \gg r$), leading to the approximation of scattering at infinite distance, the so-called Fraunhofer scattering (Figure 3.2).

3.2.1.1 Scattering Vector

Instead of defining the scattered wave through its wave vector $\mathbf{k}$ (with $k = k_0$ for elastic scattering), it is more convenient to characterise it by means of the *scattering vector* $\mathbf{R}$ such that

$$\mathbf{R} = \mathbf{k} - \mathbf{k}_0 \quad \text{with } |\mathbf{R}| = R = 2\sin\theta/\lambda \Rightarrow R \leqslant 2/\lambda$$

At a distance r' assumed to be infinite, the optical path difference δ and the phase difference ϕ of two waves scattered respectively at the origin O and at a point M of coordinates $\mathbf{r}(xyz)$ can be written

$$\delta = \text{OH} - \text{MH}' = \mathbf{r}\mathbf{k}\lambda_0 - \mathbf{r}\mathbf{k}_0\lambda_0 = \lambda_0 \mathbf{r}\mathbf{R}$$
$$\phi = 2\pi \delta/\lambda_0 = 2\pi\, \mathbf{r}\mathbf{R}$$

It follows the amplitude scattered by a differential volume dv_r surrounding M

$$dF(\mathbf{R}) = (F_0/r')\, f(\mathbf{r}) \exp(i\phi)\, dv_r$$

For a simpler formulation, the amplitude will henceforth be normalised to a unit incident amplitude F_0 and a unit distance r'.

3.2.1.2 Scattering Amplitude

The total amplitude scattered by volume v, in a direction defined by the scattering vector $\mathbf{R}$, accordingly is written

$$F(\mathbf{R}) = \int_v f(\mathbf{r}) \exp(2\pi i\, \mathbf{r}\mathbf{R})\, dv_r \tag{3.4}$$

Basic Theory of Elastic Scattering

When considering a distribution of discrete scattering elements, the integral is changed into a sum over N scattering elements contained in volume v

$$F(\mathbf{R}) = \sum_n f_n \exp(2\pi i \mathbf{r}_n \mathbf{R}) = f_1 \exp(2\pi i \mathbf{r}_1 \mathbf{R}) + \ldots + f_m \exp(2\pi i \mathbf{r}_m \mathbf{R}) + \ldots + f_n \exp(2\pi i \mathbf{r}_n \mathbf{R}) \quad (3.5)$$

3.2.1.3 Fourier Transform of Scattering Amplitude

The function expressed by Eq. (3.4) has the mathematical form of a *Fourier transform* (abbreviation FT). Its integration can be extended to infinity, according to Eq. (3.3) which expresses the zero value of the function outside the scattering volume (for more details on FT see Appendix C).

In a general set of reference axes (not necessarily Cartesian axes, e.g. crystallographic axes):

1. *Direct space* (also called object space) is determined by its basis vectors **a**, **b**, **c**; an object position vector **r** is defined by its coordinates xyz with respect to the reference vectors.
2. *Reciprocal space* (also called Fourier space) is determined by its basis vectors **a***, **b***, **c***; a scattering vector **R** is defined by its coordinates XYZ with respect to the reference vectors.

Direct space vectors and reciprocal space vectors are related through the following equations (see Appendix B):

$$\mathbf{r} = x\mathbf{a} + y\mathbf{b} + z\mathbf{c}$$
$$\mathbf{R} = X\mathbf{a}^* + Y\mathbf{b}^* + Z\mathbf{c}^*$$

with

$$\mathbf{rR} = xX + yY + zZ$$

Equation (3.4) can accordingly be written as follows:

$$F(XYZ) = v_0 \iiint_{xyx} f(xyz) \exp[2\pi i (xX + yY + zZ)] \, dx\,dy\,dz$$

with

$$v_0 = (\mathbf{abc}) = \mathbf{a}(\mathbf{b} \wedge \mathbf{c}); \quad dv_r = v_0 dx\,dy\,dz$$

Conventional Writing

Functions and physical quantities connected with scattering and diffraction will be generally written

(a) in lower-case letters in direct or object space, e.g. $f(\mathbf{r})$;
(b) in capital letters in reciprocal or Fourier space, e.g. $F(\mathbf{R})$.

Fourier Inversion

One of the most useful properties of FTs, in the field of structure analysis by means of scattering and diffraction, is the *Fourier inversion*: if $F(\mathbf{R})$ is the FT of $f(\mathbf{r})$, reciprocally $f(\mathbf{r})$ represents the FT of $F(\mathbf{R})$, according to the following relations:

$$F(\mathbf{R}) = \int_{\text{direct space}} f(\mathbf{r}) \exp(2\pi i\, \mathbf{rR})\, dv_r \to F(\mathbf{R}) = \text{FT}\{f(\mathbf{r})\} \tag{3.6}$$

$$f(\mathbf{r}) = \int_{\text{recipr. space}} F(\mathbf{R}) \exp(2\pi i\, \mathbf{rR})\, dv_R \to f(\mathbf{r}) = \text{FT}\{F(\mathbf{R})\} \tag{3.7}$$

In the case of a scattering object expressed in the form of a discontinuous distribution of scattering elements, the integral (3.6) becomes a Fourier series as described by Eq. (3.5), with the same properties.

These relations may be expressed as follows:

> *The amplitude scattered in a direction corresponding to a vector $\mathbf{R}$ in reciprocal space is proportional to the Fourier transform of the scattering power at any points $\mathbf{r}$ in direct space.*

> *The scattering power at a point $\mathbf{r}$ in direct space is proportional to the Fourier transform of the amplitude scattered in any directions corresponding to a vector $\mathbf{R}$ in reciprocal space.*

As a consequence the distribution of scattering power which represents the structure of matter can be determined if the FT (3.7) can be calculated.

For the most general form of matter, such an operation is impaired by following limitations:

1. The accessible reciprocal space is limited to a sphere of radius $2/\lambda$ ($R = |\mathbf{k} - \mathbf{k}_0| \leq 2/\lambda$); this expresses the resolution limit of any optical device; the corresponding sphere is therefore called the *resolution sphere* or limiting sphere. The accessible portion of reciprocal space is the greater, hence the resolution the better, the smaller the wavelength involved.
2. For the continuous integral (3.7) to be computed, the number of terms to be summed up is infinite; a limited sampling of the measured amplitude values leads to a further loss of information.
3. The phase component of the amplitude is out of the reach of present instrumentation; only its magnitude, in other words the scattered intensity, is available.

The first two limitations affect only the accuracy of structure analysis. The availability of powerful computers for data processing and the use of short wavelengths greatly improve the performance in this respect.

The phase uncertainty is a fundamental setback which restrains the structure analysis of matter by means of radiation scattering. The phase problem can be completely solved for crystalline materials only, for which the Fourier integral (3.7) reduces to a Fourier series with a discrete number of terms (see Ch. 5).

3.2.1.4 Size Factor

In real experiments the scattering object is always limited in size. The size effect is implicitly expressed in the preceding equations by means of the integration limited to volume v. It appears directly only in scattering experiments dealing with very small objects. The order of magnitude of the limit size may be considered to be roughly 1000 times the distance between its scattering elements, in other words 0.1 μm in solids.

In order to explain the size effect, it is convenient to write the FT of the scattering power in the form of the product $f_\infty(\mathbf{r})\gamma(\mathbf{r})$, where $\gamma(\mathbf{r})$ is the size function as defined in section 3.1.4.

Expressing the scattered amplitude of a limited object as a FT results in a convolution product

$$F(\mathbf{R}) = \text{FT}\{f_\infty(\mathbf{r})\gamma(\mathbf{r})\} = F_\infty(\mathbf{R}) * \Gamma(\mathbf{R}) \tag{3.8}$$

where $F_\infty(\mathbf{R})$ represents the amplitude scattered by the same volume of an infinite object and $\Gamma(\mathbf{R})$ the size factor, the FT of the size function.

The amplitude scattered by a limited object equals the convolution product of the amplitude scattered by the same volume of the object assumed to be infinite and the FT of the size function.

3.2.1.5 Form Factor

When considering identical scattering elements, each with scattering power f, Eq. (3.5) can be divided by f and subsequently written

$$L(\mathbf{R}) = \frac{F(\mathbf{R})}{f} = \sum_n \exp(2\pi i\, \mathbf{r}_n \mathbf{R}) \tag{3.9}$$

The function L is called the form factor. It expresses the interference effect of the waves scattered by the different elements, due to the form of their distribution in space. It may be used for an infinite or a finite object. When defined for a finite object, it includes the size factor which becomes, according to Eq. (3.8),

$$L(\mathbf{R}) = L_\infty(\mathbf{R}) * \Gamma(\mathbf{R}) \tag{3.10}$$

In the forward transmitted direction, all waves are in phase ($\mathbf{rR} = 0$); the form factor equals N, the total number of scattering elements

$$L(0) = N \tag{3.11}$$

3.2.2 PHYSICAL SIGNIFICANCE OF SCATTERING POWER

Interaction effects with matter depend on the nature of radiation. The physical meaning of the scattering power will therefore be different for the various radiations.

3.2.2.1 Scattering Power for X-rays

Elastic scattering of X-rays by matter involves only the charged particles in the atoms. The scattering cross-section of a non-polarized incident wave by a free particle of rest mass m_0 and of charge e is expressed by the following relation, established by J. J. Thomson on the basis of the classical theory of wave propagation:

$$\sigma(\mathbf{R}) = \frac{e^4}{m_0^2 c^4} \frac{1 + \cos^2 2\theta}{2} = \frac{e^4}{m_0^2 c^4} P(\mathbf{R}) \quad (3.12)$$

where 2θ is the scattering angle (see Figure 2.2); the SI unit conversion factor $1/4\pi\epsilon_0$ has been omitted.

The angle-dependent factor $P(\mathbf{R}) = P(2\theta) = (1 + \cos^2 2\theta)/2$ is called the polarization factor.

Two kinds of charged atomic particles are involved in X-ray scattering:

1. Electrons of charge $-e$ and rest mass m_0;
2. Protons of charge $+e$ and rest mass M_p.

According to Eq. (3.12), the ratio of the respective scattering cross-sections of an electron and of a proton amounts to

$$\frac{\sigma_e}{\sigma_p} = \left(\frac{M_p}{m_0}\right)^2 = 1836^2 \cong 3.4 \times 10^6 \quad (3.13)$$

Considering an atomic nucleus to act as one particle, this ratio becomes $(1836A)^2$ for an atomic mass A.

As a result, the contribution of protons, and, *a fortiori*, of atomic nuclei to elastic X-ray scattering can be neglected: scattering matter as seen by X-rays is merely a distribution of electrons.

This issue characterises X-ray scattering and determines the limits of its application to structure analysis of materials.

When replacing the physical constants in Eq. (3.12) by their values, the X-ray amplitude scattered by an electron, per unit of solid angle in a direction $\mathbf{R}$, can be written as follows:

$$T(\mathbf{R}) = 2.82 \times 10^{-15} \sqrt{[P(\mathbf{R})]} \quad (T \text{ in m}) \quad (3.14)$$

The amplitude $T(\mathbf{R})$, called the Thomson factor, can be taken as a unit for computing X-ray scattering intensities. It assumes free electrons; however, it is in fairly good agreement with experiments at low energies, but shows an increasing discrepancy at high energies, due to the increasing contribution of inelastic scattering which is not accounted for.

The scattering power for X-rays, defined as the amplitude scattered per unit of volume and of solid angle, in a direction $\mathbf{R}$ may now be written

$$f(\mathbf{r}) = T(\mathbf{R}) \rho(\mathbf{r})$$

where $\rho(\mathbf{r})$ represents the electron density at a point $\mathbf{r}$ (probability of finding an electron per unit volume at this point).

Basic Theory of Elastic Scattering

According to Eq. (3.4), the amplitude scattered by a volume v of matter becomes

$$F(\mathbf{R}) = T(\mathbf{R}) \int_v \rho(\mathbf{r}) \exp(2\pi i\, \mathbf{rR})\, dv_r \qquad (3.15)$$

> *The amplitude of X-ray scattering in a given direction, is proportional to the FT of the electron density at any points of the object.*

> *Computing the FT of the amplitude scattered in all directions provides theoretically the electron density at a given point of the object.*

In practice, a straightforward calculation of the amplitude FT is possible only in the case of crystalline materials, leading to structure analysis.

3.2.2.2 Scattering Power for Electrons

Considering the refraction of an incident electronic wave by a constant electric potential $V(\mathbf{r})$ leads to the scattering power as follows (in [5]):

$$f(\mathbf{r}) = \frac{2\pi m_0 e}{h^2} V(\mathbf{r}) \qquad (3.16)$$

(the SI unit conversion factor $1/4\pi\epsilon_0$ has been omitted).

This relation is valid for incident electron energies which are far greater than the internal electric potential which amounts to a few electronvolts. It applies therefore only to high-energy electrons.

According to Eq. (3.4), the amplitude scattered by a volume v of matter becomes

$$F(\mathbf{R}) = \frac{2\pi m_0 e}{h^2} \int_\mathbf{r} V(\mathbf{r}) \exp(2\pi i\, \mathbf{rR})\, d\mathbf{r}^3 \qquad (3.17)$$

> *The amplitude of electron scattering in a given direction is proportional to the FT of the electric potential at any points of the object.*

> *Computing the FT of the amplitude scattered in all directions provides theoretically the electric potential at a given point of the object.*

In practice, a straightforward calculation of the amplitude FT is possible only in the case of crystalline materials, but with difficulty arising from the strong electron–matter interaction (see Ch. 6).

3.2.2.3 Scattering Power for Thermal Neutrons

The scattering of thermal neutrons is due to nuclear and magnetic forces. Considering only the first process, the agents of interaction are the nuclei. Compared to the ångström scale of the neutron wavelength, a nucleus can be assimilated to a scattering point. It follows that neutron scattering takes place without phase effects and accordingly without a decrease of amplitude as a function of the direction **R**. This is in contrast with the case of X-ray and electron scattering.

With regard to thermal neutron scattering, a sample of matter can be considered as a distribution of nuclei, hence of atoms.

> *The amplitude of neutron scattering in a given direction is proportional to the FT of the distribution of atoms at any points of the object.*

> *Computing the FT of the amplitude scattered in all directions provides theoretically the distribution of atoms in the object.*

In practice, a straightforward calculation of the amplitude FT is possible only in the case of crystalline materials, for which neutron diffraction is complementary to X-ray diffraction in the field of structure analysis.

To Summarize

Computing the amplitude FT of elastically scattered X-rays, electrons and thermal neutrons leads to complementary structural information on crystal structures. In theory it provides respectively the distribution of electron density, the distribution of electric potential and the distribution of atoms in the scattering sample.

3.3 INTENSITY OF SCATTERED RADIATIONS

3.3.1 GENERAL EXPRESSION OF SCATTERED INTENSITY

An amplitude as a physical quantity is characterised both by its *magnitude* (i.e. its absolute value) and its *phase*. It can therefore be written in the form of a complex number and represented by means of a vector as shown in Figure 3.3. The amplitude resulting from scattering by a sample of volume v, in a given direction, can accordingly be represented as the geometrical sum of the contributions of all the individual scattering elements.

As a result, the scattered amplitude can be written

$$F(\mathbf{R}) = \alpha + i\beta \tag{3.18}$$

In fact, in any scattering experiment, only the magnitude of scattering is available

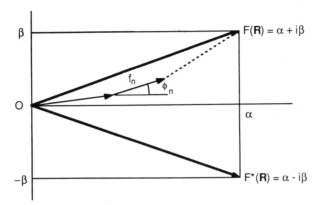

Figure 3.3. Amplitude scattered by a sample of matter, expressed as a sum of vectors in the complex plane.

for direct measurement, in the form of its square, the intensity. The information on its phase is lost. According to Eq. (3.18), the scattered intensity becomes

$$I(\mathbf{R}) = [F(\mathbf{R})]^2 = \alpha^2 + \beta^2 = (\alpha + i\beta)(\alpha - i\beta) = F(\mathbf{R})F^*(\mathbf{R}) \tag{3.19}$$

The scattered intensity is expressed by the product of the scattered amplitude and its conjugate. If the scattering power at any point $\mathbf{r}(xyz)$ is represented by a real number (which is the general case for X-ray scattering), the conjugate value of the amplitude becomes $F^*(\mathbf{R}) = F(-\mathbf{R})$ and the intensity is written

$$I(\mathbf{R}) = F(\mathbf{R})F(-\mathbf{R}) \tag{3.20}$$

3.3.1.1 Distribution of Discrete Scattering Elements

Consider a distribution of N scattering elements (atoms or any kind of structure element) located at positions $\mathbf{r}_1, \ldots \mathbf{r}_m, \ldots, \mathbf{r}_n$, with respective scattering amplitudes $f_1, \ldots f_m, \ldots f_n$ assumed to be real numbers.

The scattering amplitude in a direction $\mathbf{R}$ is given by Eq. (3.5).

The corresponding expression of the intensity is the result of carrying out the product defined by Eq. (3.20), which leads to

$$I(\mathbf{R}) = F(\mathbf{R})F(-\mathbf{R}) = \sum_m \sum_n f_m f_n \exp[2\pi i (\mathbf{r}_m - \mathbf{r}_n)\mathbf{R}] \tag{3.21}$$

An intensity is necessarily expressed by a real number, so that the phase factor may be reduced to the cosine component as follows:

$$I(\mathbf{R}) = \sum_m \sum_n f_m f_n \cos 2\pi (\mathbf{r}_m - \mathbf{r}_n)\mathbf{R} \tag{3.22}$$

Comparison of Eqs (3.5) and (3.22) leads to the following conclusion:

1. The **scattered amplitude** is determined by the absolute positions of the scattering elements;

2. The **scattered intensity** is determined by the relative positions of the scattering elements, in other words the interelement vectors (e.g. the interatomic vectors).

Let $\mathbf{r}_m - \mathbf{r}_n = \mathbf{r}_{mn}$. The different terms in Eq. (3.22) may be divided into two sets:

1. N terms such as $m = n$;
2. $N(N-1)$ terms such as $m \neq n$.

The intensity can accordingly be written

$$I(\mathbf{R}) = \sum_n f_n^2 + \sum\sum_{m \neq n} f_m f_n \cos 2\pi\, \mathbf{r}_{mn} \mathbf{R} \qquad (3.23)$$

This expression shows the scattering intensity to be the sum of two partial series:

1. The first ($m = n$) is independent of the scattering direction $\mathbf{R}$; it represents the intensity which would be observed in the absence of interference effects, the waves scattered by the various elements being completely incoherent;
2. The second ($m \neq n$) is a sum of terms depending on the scattering direction $\mathbf{R}$, the value of each oscillating periodically about zero; the total value of this sum, averaged over the whole reciprocal space, is therefore zero.

It follows that the scattered intensity, averaged over the whole reciprocal space, ($R = 2 \sin \theta / \lambda$ varying from 0 to $2/\lambda$) is equal to the intensity which would be observed if the scattering were perfectly incoherent: the result of interferences is merely to enhance the scattered intensity in specific directions without changing the total intensity.

3.3.1.2 Continuous Distribution of Scattering Matter

For various reasons it is sometimes more convenient to express the scattering power $f(\mathbf{r})$ of a volume of matter as a continuous function. The discrete sum (3.21) is then changed into an integral.

Let $\mathbf{r} - \mathbf{r}' = \mathbf{u}$ be a vector expressing the distance between any of two points determined by $\mathbf{r}$ and $\mathbf{r}'$. The integral representing the scattered intensity can then be written in the following form, similar to Eq. (3.21):

$$I(\mathbf{R}) = \int_\mathbf{r} \int_\mathbf{u} f(\mathbf{r}) f(\mathbf{r} - \mathbf{u}) \exp(2\pi i\, \mathbf{uR})\, dv_r\, dv_u \qquad (3.24)$$

In the same way as for a discrete distribution, this integral can be regarded as a sum of two terms

$$I(\mathbf{R}) = \overline{vf(\mathbf{r})^2} + \int\int_{\mathbf{r} \neq \mathbf{u}} f(\mathbf{r}) f(\mathbf{r} - \mathbf{u}) \exp(2\pi i\, \mathbf{uR})\, dv_r\, dv_u \qquad (3.25)$$

The physical significance is the same as in the discontinuous representation.

3.3.1.3 Size Effects

According to Eqs (3.8) and (3.20), the intensity scattered by a limited object can be written

Basic Theory of Elastic Scattering

$$I(\mathbf{R}) = I_\infty(\mathbf{R}) * \Gamma^2(\mathbf{R}) \quad (3.26)$$

where $\Gamma(\mathbf{R})$ is the FT of the size function. The size factor Γ depends on the respective dimensions of the scattering object.

Consider, as an example, a rectangular prism of edge length A, B, C. When limited to one dimension, e.g. the X dimension, the size function is defined by following relations:

$$\gamma(x) = 1 \quad \text{for } x \leq A$$
$$\gamma(x) = 0 \quad \text{for } x > A$$

The one-dimensional size factor is written

$$\Gamma(X) = \int_0^A \exp(2\pi i\, xX)\, dx = A\,\frac{\sin \pi AX}{\pi AX} \quad (3.27)$$

The squared three-dimensional size factor, related to intensity, then becomes

$$\Gamma^2(\mathbf{R}) = \Gamma^2(XYZ) = v^2 \left(\frac{\sin \pi AX}{\pi AX}\,\frac{\sin \pi BY}{\pi BY}\,\frac{\sin \pi CZ}{\pi CZ}\right)^2 \quad (3.28)$$

Figure 3.4 represents a cross-section of this function. It is centrosymmetric and shows a central peak of width $2/A$, $2/B$, $2/C$ (half-width $1/A$, $1/B$, $1/C$) along the respective reciprocal axes X, Y, Z. The central peak is surrounded by rapidly decreasing secondary peaks.

Each one-dimensional factor in Eq. (3.28) has the form $(\sin u/u)^2$ and therefore tends towards one when u tends towards zero, in other words when $\mathbf{R}(XYZ)$ tends towards zero, which corresponds to the forward direction.

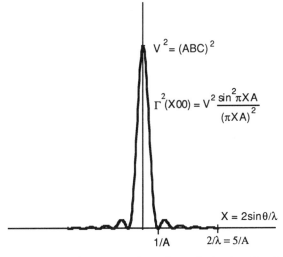

Figure 3.4. Squared size factor of a rectangular prism of edge lengths A, B, C, as a function of X in reciprocal space, for $A = 5\lambda/2$. The first minimum following the central peak occurs at $X = 1/A$.

This results in the following relation for the incident direction:

$$\Gamma^2(0) = v^2 \Rightarrow I(0) = v^2 \overline{f^2(0)}$$

The same result will be expressed hereafter by Eq. (3.32).

According to the attributes of convolution, the intensity scattered by a limited object (Eq. 3.26) can be expressed by superimposing function Γ^2 on function F^2 at any point in reciprocal space.

3.3.1.4 Interference function

For scattering elements all assumed to be identical ($f_m = f_n = f$), Eq. (3.23) can be divided throughout by f^2, leading to the squared form factor as defined in Eq. (3.9)

$$L^2(\mathbf{R}) = N + \sum_{m \neq n}\sum \cos 2\pi\, \mathbf{r}_{mn}\mathbf{R} \qquad (3.29)$$

This function expresses the interference effect and is therefore often called the *interference function*. Its value averaged over the whole reciprocal space (i.e. for any possible values of $\mathbf{R}$) equals N. Without any interference effect, it equals N whatever $\mathbf{R}$.

For a size-limited object, the interference function includes the size effect. The general form of this function can therefore be written according to Eq. (3.10)

$$L^2(\mathbf{R}) = L_\infty^2(\mathbf{R}) * \Gamma^2(\mathbf{R}) \qquad (3.30)$$

3.3.1.5 Forward Scattered Intensity

Distribution of Discrete Scattering Elements

In the incident direction ($R = 0$), the waves scattered by the various elements are in phase; the amplitudes sum up arithmetically. For elements with identical scattering powers f, the scattered intensity and the interference function become respectively

$$\begin{aligned} I(0) &= [N + N(N-1)]\, f^2 + N^2 f^2 \\ L^2(0) &= N^2 \end{aligned} \qquad (3.31)$$

It can be pointed out that in the incident direction or in any direction where all elements are scattering in phase, the resultant scattered intensity is N times the intensity scattered without any interference effect.

Continuous Distribution of Scattering Power

The forward scattered intensity can be written in a similar way as for a discontinuous distribution

$$I(0) = v^2 \overline{f(0)^2} \qquad (3.32)$$

For X-ray scattering, $I(0)$ equals the squared number of electrons in the scattering volume, the intensity scattered by one electron taken as unit.

3.3.1.6 Intensity Weighted Reciprocal Space

The distribution of scattered intensity in all directions in space may be represented by the values of the function $I(\mathbf{R})$ at any points $\mathbf{R}$ in reciprocal space. Assigning a value $I(\mathbf{R})$ at any point $\mathbf{R}$ in reciprocal space leads to defining the so-called *intensity weighted reciprocal space*.

It follows that the intensity weighted reciprocal space is *centrosymmetric*, whatever the object symmetry: $I(\mathbf{R}) = I(-\mathbf{R})$.

The intensity weighted reciprocal space is limited to the maximum value of $R = 2\sin\theta/\lambda \Rightarrow R \leq 2/\lambda$. It is therefore determined only inside a sphere of radius $2/\lambda$, called the *limiting sphere* or the *resolution sphere*. The greater the radius of this sphere, i.e. the smaller the wavelength, the more information is provided by a scattering experiment. This limitation corresponds to the general problem of resolution affecting any optical device.

1. For a **non-crystalline scattering object**, the intensity $I(\mathbf{R})$ as expressed in reciprocal space decreases continuously from its maximum value $I(0)$ at the origin where all scattering elements are in phase.
2. For a **crystalline scattering object**, the intensity $I(\mathbf{R})$ as expressed in reciprocal space is distributed in a periodic way on the reciprocal lattice points; the intensity weighted reciprocal space is therefore reduced to the *intensity weighted reciprocal lattice*.

In both cases, the total intensity scattered in a solid angle 4π rad is the same.

The photographic recording of a scattering pattern represents a section of the intensity weighted reciprocal space by a sphere of radius $1/\lambda$, called the *reflection sphere* or *Ewald sphere* (see Ch. 5).

3.3.1.7 Angular Variation of Scattered Intensity

As shown in the preceding sections, the effect of interference is to make the scattering intensity distribution in space depart from an isotropic one. The angular variation depends both on the degree of order and the size of the scattering object.

Non-crystalline Object

The scattered intensity varies continuously with $R = 2\sin\theta/\lambda$, in other words with the scattering angle 2θ for a given wavelength λ.

1. For a perfectly disordered object (e.g. monoatomic gas), each particle is scattering incoherently; the angular intensity variation is the same as for isolated atoms (Ch. 4).
2. For an object with short-distance order (e.g. molecular gases, glasses, polymers), the angular intensity distribution typically shows one or several maxima at angles corresponding to interferences between neighbouring atoms; on a photographic record this results in a diffusion halo. The angle of the first ring is given by the first maximum of Eq. (3.29). For a predominant interelement distance r_{mn}, it occurs for $r_{mn}R \cong 1.2 \Rightarrow \sin\theta \cong 0.6\lambda/r_{mn}$. The related magnitude of the scattering vector is $R = 2\sin\theta/r_{mn} \cong 1.2/r_{mn}$.

Crystalline Object

A crystalline object is characterised by long-distance ordering; as a result scattering is concentrated mainly at discrete angles which correspond to the diffraction rules (Ch. 5).

Small Objects

For very small objects, the size factor alters the angular intensity distribution.

1. With a non-crystalline object, it results in a small-angle scattering maximum. For a linear dimension d, the half-width of the central peak (Fig. 3.4) is

$$R = 1/d = 2 \sin \theta / \lambda \Rightarrow \sin \theta = \lambda/2d$$

As an illustration, for the commonly used copper Kα X-ray line $\lambda = 1.54$ Å and an object measuring 1000 Å, the half-width of the central scattering peak would be about 0.7×10^{-3} rad or 2.4 minutes of angle. It is in the range of the so-called *small-angle scattering*.

2. With a crystalline object, the size effect produces a broadening of every diffraction peak (Ch. 5).

3.3.2 FOURIER TRANSFORM OF SCATTERED INTENSITY

As mentioned above, processing the FT of the scattered amplitude $F(\mathbf{R})$ in any direction $\mathbf{R}$ results in the structure function $f(\mathbf{r})$. Due to the phase indetermination, the amplitude FT cannot be performed directly in the general case.

The FT of the scattered intensity $I(\mathbf{R})$, on the other hand, may be performed in any case within the resolution limit $R \leqslant 2/\lambda$ and within the resources of available computing facilities. Which information does it provide?

According to the properties of the FT and of the convolution square (Appendix C), and as a result of Eq. (3.7), the intensity FT can be written as follows:

$$\mathrm{FT}\{I(\mathbf{R})\} = \mathrm{FT}\{F(\mathbf{R}) F(-\mathbf{R})\} = f(\mathbf{r}) * f(-\mathbf{r})$$

> *The FT of the scattered intensity equals the convolution product of the structure function.*

This transform can be expressed according to Eq. (3.24) as follows:

$$q(\mathbf{u}) = \mathrm{FT}\{I(\mathbf{R})\} = \int_{\mathbf{r}} f(\mathbf{r}) f(\mathbf{r} - \mathbf{u}) \, dv_r \qquad (3.33)$$

Function $q(\mathbf{u})$ represents the sum of the products of the scattering powers at any couples of object points separated by a vector $\mathbf{u}$; it goes through maxima for $\mathbf{u}$ equal to vectors which are connecting points of maximum scattering power, i.e. the centres of scattering

Basic Theory of Elastic Scattering

elements. The higher the maximum values of $q(\mathbf{u})$ the greater the probability of occurrence of the vectorial interelement distance $\mathbf{u}$. If atoms are the scattering elements, the maxima of function $q(\mathbf{u})$ give access to the interatomic vectors.

Function (3.33) can be separated into two terms as follows:

$$q(\mathbf{u}) = \mathrm{FT}\{I(\mathbf{R})\} = \delta(\mathbf{u}) + \int_{\mathbf{r} \neq \mathbf{u}} f(\mathbf{r}) f(\mathbf{r}-\mathbf{u}) \, dv_r \qquad (3.34)$$

1. The first term, FT of a constant, is a Dirac function; it represents a central peak at $\mathbf{u} = 0$, corresponding to the sum of the scattering powers squared at any object points.
2. The second term corresponds to the sum of the products of the scattering powers at different object points, $\mathbf{u} \neq 0$.

Comparing Eq. (3.34) with Eq. (3.2), which determines the distribution function, shows their similarity: the FT of the scattered intensity leads to the distribution function of the scattering elements in the object.

For an amorphous material, it provides the radial distribution function. The FT of small-angle scattering leads to the size distribution function.

In the particular case of crystals, the FT of intensity, called the *Patterson function*, may generally be deconvoluted, allowing the structure function itself to be determined.

To Summarise

> *The FT of the amplitude, scattered at any point R in reciprocal space, leads to the structure function; it is not directly accessible in the general case.*

> *The FT of the intensity, scattered at any point R in reciprocal space, leads to the distribution function; it can be processed in the general case within the limits of resolution and of the computing facilities.*

REFERENCE CHAPTERS

Chapters 4, 5. Appendixes B and C.

BIBLIOGRAPHY

[1] Bacon G. E. *Neutron diffraction*. Oxford University Press, 1975.
[2] Guinier A. *Théorie et technique de la radiocristallographie*. Dunod, Paris, 1964.
[3] Guinier A., Fournet G. *Small angle scattering of X-rays*. John Wiley, New York, 1955.
[4] Hirose A., Lonngren K. E. *Introduction to wave phenomena*. John Wiley, New York, Chichester, 1985.
[5] Hirsch P. B., Howie A., Nicholson R. B., Pashley D. W., Whelan M. J. *Electron microscopy of thin crystals*. Butterworths, London, 1965.
[6] Klein M. V., Furtak T. E. *Optics*, 2nd edn. John Wiley, New York, Chichester, 1986.
[7] Krivoglaz M. A. *Théorie de la diffusion des rayons X et des neutrons thermiques par les cristaux réels*. Masson, Paris, 1969.
[8] Smith G. G., Thomson J. H. *Optics*. John Wiley, New York, Chichester, 1988.

4
Elastic Scattering by Individual Atoms. Atomic Scattering Amplitude

The amplitude $A(\mathbf{R})$ scattered by an individual atom, in a direction determined by the scattering vector $\mathbf{R} = \mathbf{k} - \mathbf{k}_0$, is proportional to the FT of the scattering power $a(\mathbf{r})$ at any point $\mathbf{r}$ in the atom. It can be written as follows, according to Eq. (3.4),

$$A(\mathbf{R}) = \int_{\text{atom}} a(\mathbf{r}) \exp(2\pi i\, \mathbf{r}\mathbf{R})\, dv \qquad (4.1)$$

4.1 ATOMIC SCATTERING FACTOR FOR X-RAYS

As far as elastic scattering is concerned, an atom is 'seen' by X-rays as a distribution of Z electrons. Let $\rho(\mathbf{r})$ be the *electron density* at a given point $\mathbf{r}$ in the atom. It represents the squared electron wave function, a solution of the Schrödinger equation defining the orbitals.

The amplitude $A_j(\mathbf{R})$ scattered by an atom j in a direction $\mathbf{R}$ is given by Eq. (3.15) which gives for an atom

$$A(\mathbf{R}) = T(\mathbf{R}) \int_{\text{atom}} \rho(\mathbf{r}) \exp(2\pi i\, \mathbf{r}\mathbf{R})\, dv \qquad (4.2)$$

Referring to the amplitude scattered by one electron, taken as a unit, results in the *atomic scattering factor* for X-rays

$$f_j(\mathbf{R}) = \frac{A_j(\mathbf{R})}{T(\mathbf{R})} = \int_{\text{atom}} \rho(\mathbf{r}) \exp(2\pi i\, \mathbf{r}\mathbf{R})\, dv \qquad (4.3)$$

This factor is a dimensionless number which expresses the interference effects between the waves scattered by the individual electrons. It can be represented as the FT of the electron distribution in the atom j.

At first approximation the atom can be considered to be spherical, with a radius r_a. According to Fig. 4.1, the integral (4.3) can then be written as follows with a differential volume $dv = 4\pi r^2 \, dr \, d(\cos \alpha)$:

$$f_j(\mathbf{R}) = \int_r 4\pi r^2 \rho(r) \, dr \int_\alpha \exp(2\pi i \, rR \cos \alpha) \, d(\cos \alpha)$$

A spherical distribution is centrosymmetric and the phase factor can accordingly be reduced to its cosine term; integrating on α leads to the following expression of the scattering factor:

$$f_j(\mathbf{R}) = \int_0^{r_a} 4\pi r^2 \rho(r) \, dr \frac{\sin 2\pi rR}{2\pi rR} = \int_0^{r_a} U(r) \, dr \frac{\sin 2\pi rR}{2\pi rR} \tag{4.4}$$

where $U(r) = 4\pi r^2 \rho(r)$.

The function $U(r)$ represents the radial electron distribution which is characteristic of the atomic structure. It has been computed by means of the Hartree–Fock self-consistent field method for light elements ($Z \leqslant 20$), and by means of the Thomas–Fermi statistical model for heavy elements ($Z > 20$).

Scattering in the Incident Direction. According to Eq. (4.4), when $R \to 0$, $\sin 2\pi rR / 2\pi rR \to 1$. It follows that in the incident direction the scattering factor becomes

$$f_j(0) = \int_{\text{atom}} U(r) \, dr = Z \tag{4.5}$$

In the incident direction, the waves scattered by all the electrons are in phase; the resulting amplitude is therefore the arithmetic sum of the individual electron scattering amplitudes.

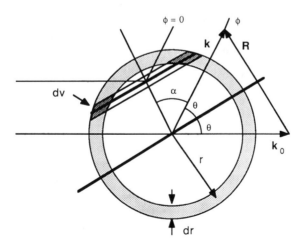

Figure 4.1. Calculation of the X-ray scattering factor of a spherical atom.

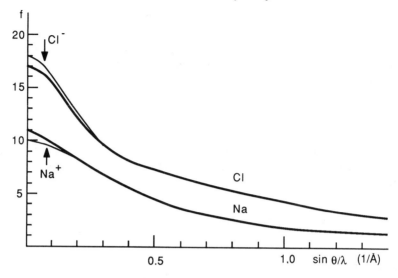

Figure 4.2. Atomic scattering factors of sodium and chlorine plotted as a function of $\sin\theta/\lambda$. For the corresponding ions, the curves differ significantly from those of the neutral atoms only at small scattering angles.

> *The scattering factor of an atom for X-rays in the incident direction equals the number of its electrons, i.e. the atomic number Z for a neutral atom.*

The values of f_j have been tabulated as a function of $R/2 = \sin\theta/\lambda$ for the whole range of elements and the corresponding ions (e.g. in [1] *Int. tables*, vol. 3, see Figure 4.2). If the actual amplitudes are needed, the tabulated factors must be multiplied by the corresponding values of the *Thomson factor T(R)*.

4.2 ATOMIC SCATTERING AMPLITUDE FOR ELECTRONS

Electron scattering by an atom involves its *electrical potential*. The amplitude scattered by an atom j, assumed to be spherical, can accordingly be written after Eq. (3.17)

$$A_j(\mathbf{R}) = \frac{2\pi m_0 e}{h^2} \int_{at.} V(\mathbf{r}) \exp(2\pi i\, \mathbf{rR})\, d\mathbf{r}^3 \qquad (4.6)$$

The atomic scattering amplitude is proportional to the FT of the electrical potential $V(\mathbf{r})$ at any points $\mathbf{r}$ in the atom. The electrical potential at a given point in the atom is the summation of the potentials produced by all the electric charges in the atom, i.e. the negative charges $-e$ of the electrons and the positive charges $+Ze$ of the nucleus. This results in a relationship between the respective atomic scattering amplitudes for electrons and for X-rays. The basis for the computation of this relationship is the Poisson equation. It leads to the following expression of the electron amplitude scattered by an atom j, in terms of the X-ray scattering factor f_j:

$$A_j(\mathbf{R}) = \frac{m_0 e^2}{2h^2} \frac{\lambda^2}{\sin^2\theta} [Z - f_j(\mathbf{R})] \qquad (4.7)$$

When substituting the universal constants by their values, the atomic scattering amplitude becomes

$$A_j(\mathbf{R}) = 2.38 \times 10^8 \frac{\lambda^2}{\sin^2\theta} [Z - f_j(\mathbf{R})] \quad \text{(in metres)} \qquad (4.8)$$

In the square brackets, the terms Z and $f_j(\mathbf{R})$ represent respectively the contribution of Rutherford scattering from the nucleus and of scattering from the electron cloud. The effect of the electron cloud is important at small angles, whereas Rutherford scattering from the nucleus is dominant at large angles. At angles greater than 5–10°, the electron cloud effect becomes negligible for 100 keV electrons. In current high-energy electron diffraction devices, the useful scattering angle extends to a few degrees; it follows that scattering is a combined electron cloud and nucleus effect.

Due to the factor $1/\sin^2\theta$, the amplitude decreases faster as a function of θ than in the case of X-rays.

For electrons of energy higher than 100 keV, relativistic corrections must be applied to the mass and the wavelength in Eqs (4.7) and (4.8).

It can be useful to express the incident electron energy eV_0 in Eq. (4.7) by means of the De Broglie relation (1.2): $\lambda^2 = h^2/2m_0 eV_0$. The amplitude can then be written as follows:

$$A_j(\mathbf{R}) = \frac{e}{4 \sin^2\theta} \frac{1}{V_0} [Z - f_j(\mathbf{R})] \qquad (4.9)$$

It can be seen that the differential elastic scattering cross-section of electrons by an atom is proportional to Z^2/V_0^2.

According to Eq. (4.9), the scattering cross-section would become infinite in the forward direction, which has no physical sense. In fact, Eq. (4.9) merely gives the general variation of the atomic scattering amplitude for not too small angles and for incident electron energies greater than 1 keV. For a *zero scattering angle*, the amplitude can be approximated to the following expression, where r_a is the mean atomic radius.

$$A_j(0) = \frac{4\pi^2 m_0 e^2}{3h^2} Z r_a^2 \qquad (4.10)$$

The values of the atomic scattering amplitudes of the elements, computed by means of the preceding methods, are tabulated in *Int. tables*, vol. 3 [1].

For atoms in solids, the effect of chemical bonds on the electron cloud must be taken into account.

This simplified theory, which regards electron scattering from an atom in the same way as X-ray scattering, assumes the transmitted electron wave in the solid to be equal to the incident wave, neglecting the absorption due to the interaction effects. However, the electron–matter interaction cross-section is about 10^8 times more intensive than the

X-ray–matter interaction, as can be seen from Table 4.1. For low-energy electrons, the ratio is even higher. The mean free path for electrons is in the ångström range, whereas for X-rays it is in the centimetre range.

The foregoing approximation, called the first Born approximation, is valid only for incident electrons with energies far greater than the mean internal potential of the scattering medium which measures commonly some 10 eV. It therefore has a limited application for high-energy electrons ($E_0 > 10$ keV).

4.3 ATOMIC SCATTERING AMPLITUDE FOR THERMAL NEUTRONS

The atomic scattering amplitude for thermal neutrons, in non-magnetic materials, is equal to the amplitude scattered by the nuclei. Neutron scattering has the following characterisic features which differentiate it from X-ray and electron scattering (see Table 4.1):

1. The amplitude is not dependent on the scattering angle, due to the point dimension of the nucleus with respect to the associated neutron wavelength.
2. The amplitude is not directly correlated to the atomic number and varies in the range of about 1–3 only. It can be very different from one element to its direct neighbour in the periodic table (e.g. S and Cl, Mg and Al). Its value for hydrogen is in the same order as for heavy elements.
3. The amplitude is a function of the isotopic state of an atom (e.g. C^{12} and C^{13}, H^1 and H^2).

These specific features determine the practical application of neutron scattering as a complementary method to X-ray and electron scattering in structure analysis.

Table 4.1. Atomic scattering amplitudes (in ångström units) for X-rays, high-energy electrons and thermal neutrons, tabulated for $\sin\theta/\lambda = 0.25$. In the case of the X-rays, the polarisation correction has been calculated for Cu-Kα. It can be seen that the respective amplitudes for X-rays, electrons and neutrons are in the order of magnitude of the ratio $1/10^4/10^{-1}$. (*Int. tables*, vol. 3 [1])

Z	Element	X-rays	Electrons	Neutrons
1	H^1	0.086×10^{-4}	0.25	3.8×10^{-5}
	H^2			6.5×10^{-5}
6	C^{12}	0.73×10^{-4}	1.15	6.6×10^{-5}
	C^{13}			6.0×10^{-5}
12	Mg	1.97×10^{-4}	1.50	5.4×10^{-5}
13	Al	2.07×10^{-4}	1.73	3.5×10^{-5}
16	S	2.44×10^{-4}	2.32	3.1×10^{-5}
17	Cl	2.59×10^{-4}	2.47	9.9×10^{-5}
25	Mn	4.20×10^{-4}	2.91	3.7×10^{-5}
26	Fe	4.39×10^{-4}	2.91	9.6×10^{-5}
27	Co	4.58×10^{-4}	2.91	2.8×10^{-5}
79	Au	15.6×10^{-4}	6.60	7.6×10^{-5}
92	U	18.5×10^{-4}	7.31	8.5×10^{-5}

4.4 TEMPERATURE EFFECT. DEBYE–WALLER FACTOR

Atoms in a solid are affected by thermal vibrations about their equilibrium positions. The amplitude of the vibrations increases with increasing temperature. The temperature effect can be expressed in two ways:

1. *Debye effect* dealing with independently vibrating atoms;
2. Collective vibration effect in a crystalline solid, considered as the propagation of a wave with quantised energy states called *phonons*.

The Debye effect will only be considered in the following development for the purposes of structure analysis.

The probability of presence of an atom j in a solid passes through a maximum at the centre of its potential well. It decreases as a function of the deviation according to a Gaussian law. The point position of an atom at $0\,K$ is replaced by a Gaussian probability cloud at temperature T. The scattered amplitude at a given temperature T is expressed as the FT of the scattering power at any point in the atom; it can be computed by multiplying the atomic scattering amplitude at rest by the FT of the probability function which is itself a Gaussian function. At a temperature T, the atomic scattering factor can accordingly be written

$$A_T(\mathbf{R}) = A(\mathbf{R}) \exp\left[-\frac{BR^2}{4}\right] = A(\mathbf{R}) \exp\left[-\frac{B\sin^2\theta}{\lambda^2}\right] \quad (4.11)$$

where $B = 8\pi^2\sigma^2$ is the *Debye–Waller factor* (sometimes the exponential factor as a whole is called the Debye-Waller factor) and σ the mean-squared deviation of the vibrating atom.

In fact this relation applies strictly for isotropic solids where vibrations are direction independent. In non-cubic crystalline solids, an anisotropic vibration factor must be used for the calculations. This factor depends necessarily on chemical bonding.

In practice, the temperature effect causes the atomic scattering amplitude to decrease more rapidly as a function of $\sin\theta/\lambda$.

The values of the temperature factors for the various elements are tabulated in *Int. tables*, vol. 3 [1].

REFERENCE CHAPTERS

Chapters 2, 3.

BIBLIOGRAPHY

See also references of Chapter 3.
[1] *International tables for X-ray crystallography*, vol. 2: *Mathematical tables* (Kasper J. S., Lonsdale K., eds); Vol. 3: *Physical and chemical tables* (MacGillvray C. H., Rieck G. D., Lonsdale K., eds); vol. 4: *Supplementary tables to vols 2, 3* (Ibers JA, Hamilton WA, eds). D. Reidel, Kluwer Acad. Publ., Dordrecht, 1985, 1983, 1982.

5

Diffraction by a Crystal

5.1 DEFINITIONS: LATTICE, CRYSTAL, DIFFRACTION

A crystal is a periodic distribution of scattering matter in space (the so-called *direct space* or *object space*).

A perfect crystalline state is determined by the crystal lattice (the so-called *direct lattice* or *object lattice*) and its structure (see Figure 5.1).

Crystal Lattice. The crystal lattice expresses the three-dimensional periodicity. A given lattice point N is located in a set of crystallographic axes of origin O by means of a vector $\mathbf{r}_n(uvw)$ such as

$$\overrightarrow{ON} = \mathbf{r}_n = u\mathbf{a} + v\mathbf{b} + w\mathbf{c} \tag{5.1}$$

where **abc** are basis lattice vectors defining the unit cell and uvw the lattice point coordinates, integers with values ranging from $-\infty$ to $+\infty$.

In the most general set of axes ($a \neq b \neq c$, lattice angles $\alpha \neq \beta \neq \gamma$), the volume v_0 of a unit cell is written

$$v_0 = \mathbf{abc} = \mathbf{a}(\mathbf{b} \wedge \mathbf{c}) \tag{5.2}$$

Crystal Structure. The crystal structure is the matter content of one unit cell; it consists of a set of one or several atoms, ions or molecules. An atom position M is determined by means of a position vector $\mathbf{r}(xyz)$ such as

$$\overrightarrow{NM} = \overrightarrow{OM} = \mathbf{r} = x\mathbf{a} + y\mathbf{b} + z\mathbf{c} \tag{5.3}$$

Real Crystal. A real crystal is limited in size and contains a certain number of faults. Hereafter it will be considered as a crystalline state without faults, but of limited size.

In order to simplify the interpretation, the crystal will be considered as limited by a parallelepiped with edges A, B, C, built up respectively of U, V, W unit cells, along

Diffraction by a Crystal

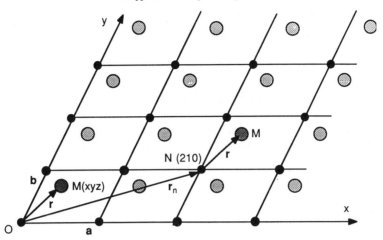

Figure 5.1. Crystal lattice and crystal structure. Locating a position **r**(*xyz*) in a unit cell and a point **r**$_n$ (*uvw*) of the lattice (lattice point 210). In order to simplify, the unit cell content is reduced to one atom.

the reference axes x, y, z. The total number of unit cells in the crystal is therefore $N = UVW$. The respective lengths of the edges along x, y, z are

$$A = Ua, \; B = Vb; \; C = Wc \\ v = \mathbf{ABC} = Nv_0 \tag{5.4}$$

where v is the crystal volume.

Diffraction. The general formulation of elastic scattering applies to scattering by a crystal, with the particular feature of a periodical distribution of identical scattering elements. This results in an intense scattering in discrete directions and is called *diffraction*.

Diffraction effects can be observed with radiations of wavelength smaller than the respective periodicities or lattice parameters a, b, c. Crystal lattice parameters have an order of magnitude of several ångströms, which is compatible with the diffraction of X-rays, electrons and neutrons of a given energy range (see Table 1.2).

The **diffraction angles** do not depend on the nature of radiation. The geometrical diffraction theory is therefore valid for X-rays as well as for electrons and neutrons.

The **diffracted intensity** depends on the mode of interaction and therefore on the nature of the interacting radiation. When dealing with weak scattering effects, which is the general case of X-ray and neutron diffraction, the *kinematic theory* is a fair approximation; for strong scattering effects, which is usually the case of electron diffraction, a more complicated interpretation based on the *dynamic theory* becomes necessary.

In the following development, barring different specifications, the incident wave is a plane wave and the observation distance is considered to be large with respect to the size of the scattering object. This results in the so-called *Fraunhofer diffraction*, which enables Eq. (3.4) to be applied to express the amplitude of diffraction. These conditions are generally met in classical diffraction layouts; they are no longer met in the process of diffraction imaging in the transmission electron microscope, where *diffraction at finite distance* prevails (the so-called *Fresnel diffraction*).

5.2 GEOMETRICAL DIFFRACTION THEORY

For an incident wave to be diffracted in a given direction, all the waves scattered by the individual unit cells or lattice points of the crystal must be in phase in that particular direction. This diffraction condition may be expressed in different, but equivalent forms, by means of the direct lattice or by means of the reciprocal lattice.

5.2.1 DIFFRACTION CONDITIONS EXPRESSED IN THE DIRECT LATTICE

5.2.1.1 Laue Condition

The direct lattice is determined by its three principal lattice lines [100], [010], [001], corresponding to the unit cell vectors **a**, **b**, **c**. It suffices to write the condition for all the points of those three lattice lines to be in phase; due to the three-dimensional periodicity, all the lattice points will then be in phase. Consider the lattice line [100]; as shown by Figure 5.2, the optical path difference δ between the waves scattered by two successive lattice points is given by

$$\delta = \mathrm{HO} + \mathrm{OK} = a(\cos\alpha - \cos\alpha_0) = \mathbf{a}(\lambda\mathbf{k} - \lambda\mathbf{k}_0) = \lambda\mathbf{a}\mathbf{R}$$

where $\mathbf{R} = \mathbf{k} - \mathbf{k}_0$ is the scattering vector.

It follows that the diffraction condition $\delta = h\lambda$ (where h is an integer) is written $\mathbf{a}\mathbf{R} = h$.

Equivalent relations result for the lattice lines [010] and [001] and the diffraction condition for the three main lines becomes

$$\mathbf{a}\mathbf{R} = h \qquad \mathbf{b}\mathbf{R} = k \qquad \mathbf{c}\mathbf{R} = l \quad (h, k, l, \text{integers}) \tag{5.5}$$

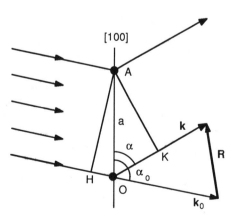

Figure 5.2. Scattering of an incident plane wave by a two-dimensional lattice.

Diffraction by a Crystal 51

If the diffraction condition is met for the three main lines, it is met for a general lattice line [uvw]. Consider $\mathbf{r}_n$ a vector of this line; according to Eqs (5.1) and (5.5) it follows that

$$\mathbf{r}_n \mathbf{R} = u\mathbf{aR} + v\mathbf{bR} + w\mathbf{cR} = uh = vk + wl = m \quad \text{(integer)} \quad (5.6)$$

The Laue condition may accordingly be written as only one relation $\mathbf{r}_n \mathbf{R} = m$, which expresses the condition for a general lattice line [uvw] to have all its lattice points in phase in the direction defined by the scattering vector $\mathbf{R}$.

5.2.1.2 Bragg Condition

Consider a set of lattice planes (hkl), of equidistance $d(hkl)$. Consider an incident plane wave of wave vector $\mathbf{k}_0$ and of incidence angle θ with respect to the planes. The incident angle θ, called the *Bragg angle*, is the complementary angle to the conventional incident angle in optics (see Figure 5.3).

Any points of one lattice plane (hkl) are scattering waves which are in phase in a direction $\mathbf{k}$ which corresponds to the classical reflection law (Descartes' law, reflection angle equals incident angle). If this condition is met, the optical path difference between the waves scattered by two adjacent planes is therefore the same whatever the position of the points in the planes, in particular between two points O and O' on a normal common to the planes; it follows easily from Figure 5.3a that $\delta = \text{HO} + \text{OK} = 2d(hkl) \sin \theta$.

The in-phase condition, called the *Bragg condition* (or Bragg law or Bragg equation) becomes accordingly

$$2d(hkl) \sin \theta = n\lambda \quad (5.7)$$

where λ is the wavelength and n an integer called diffraction (or reflection) order.

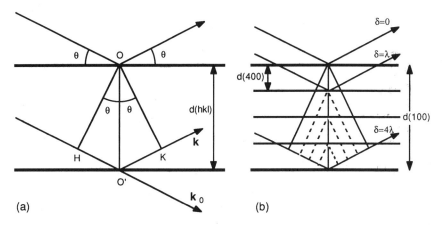

Figure 5.3. Bragg condition. (a) Setting up the condition; selective reflection on a set of lattice planes (hkl); (b) fourth-order reflection on a set of planes (100); it may be considered as a first-order reflection on a set of reflecting planes (400), of equidistance $d(hkl)/4$.

The lattice planes in diffracting position are the bisecting planes to the directions $\mathbf{k}_0$ and $\mathbf{k}$. It appears, therefore, that the diffraction effect according to Bragg's law could be considered as a reflection of the incident beam on a given set of lattice planes. However, it is not a reflection in the usual sense of light reflection. It can occur only for discrete incident directions given by Eq. (5.7); it is therefore called *selective reflection* or *Bragg reflection*.

Reflection orders. Consider a crystal rotating around an axis normal to the incident beam. With an increasing Bragg angle θ, successive selective reflections on a set of lattice planes (hkl) occur for increasing values of the reflection order n.

A reflection of order n on a set of planes (hkl) is noted as $nh\ nk\ nl$.

In the example of a set of planes (111), the reflections are as follows:

reflection order 0, $n=0$, $\delta=0$, $\sin\theta_0=0$ $\Rightarrow$ transmitted wave 000
reflection order 1, $n=1$, $\delta=\lambda$, $\sin\theta_1=\lambda/2d(hkl)$ $\Rightarrow$ reflection 111
reflection order 2, $n=2$, $\delta=2\lambda$, $\sin\theta_2=\lambda/d(hkl)$ $\Rightarrow$ reflection 222, etc.

The total number of selective reflections on a given set of planes is limited to a maximum value of 90° of the Bragg angle θ, which corresponds to $\sin\theta=1$ in Eq. (5.7)

$$n \leqslant \frac{2d(hkl)}{\lambda} \tag{5.8}$$

Reflecting Planes. A reflection of order n on a set of lattice planes (hkl) may equally be considered as a first-order reflection on a set of planes $(nh\ nk\ nl)$. Those planes are not lattice planes (Miller indices different from prime numbers). They may be called *reflecting planes* (Figure 5.6b).

Occurrence of a Reflection. Resolution Condition. For at least one reflection, i.e. the first-order reflection, on a set of planes (hkl) to occur, the condition called *resolution condition* is written according to Eq. (5.8) where $n=1$:

$$d(hkl) \geqslant \frac{\lambda}{2} \tag{5.9}$$

The difference between a conventional light reflection on a crystal face (hkl) and the Bragg reflections on the corresponding set of lattice planes (hkl) is due to the difference of magnitude of the respective wavelengths. The light wavelengths which are in the order of 1 μm are far greater than the lattice plane equidistances; it follows that the resolution condition (5.9) can never be met; lattice diffraction is impossible; optical reflection occurs according to the Descartes law. On the other hand the wavelengths of X-rays, electrons and thermal neutrons are in the ångström range and the resolution condition can be met for a number of lattice planes, depending on the value of λ.

5.2.2 DIFFRACTION CONDITIONS EXPRESSED IN THE RECIPROCAL LATTICE

5.2.2.1 Laue Condition in the Reciprocal Lattice

A given lattice point G in the reciprocal lattice can be located in a set of reciprocal axes of origin O by means of a vector $\mathbf{r}_g^* = \mathbf{r}^*(hkl)$ defined as follows:

$$\overrightarrow{OG} = \mathbf{r}_g^* = h\mathbf{a}^* + k\mathbf{b}^* + l\mathbf{c}^* \qquad (5.10)$$

where $\mathbf{a}^*, \mathbf{b}^*, \mathbf{c}^*$ are the reciprocal unit cell vectors, hkl are integers, numerical coordinates of the reciprocal lattice point G.

The product of a direct space vector $\mathbf{r}_n$ (Eq. 5.1) by a reciprocal space vector $\mathbf{r}_g^*$ is an integer m as shown by the following relation:

$$\mathbf{r}_n \mathbf{r}_g^* = uh + vk + wl = m \qquad (5.11)$$

When comparing with Eq. (5.6), the Laue condition can be expressed as follows:

$$\mathbf{R} = \mathbf{r}_g^* \Rightarrow X = h, \ Y = k, \ Z = l \qquad (5.12)$$

where X, Y, Z, are coordinates of the scattering vector $\mathbf{R}$.

It follows that the diffraction condition is met when the scattering vector is a vector of the reciprocal lattice, i.e. when its origin and its extremity are reciprocal lattice points (see Figure 5.4).

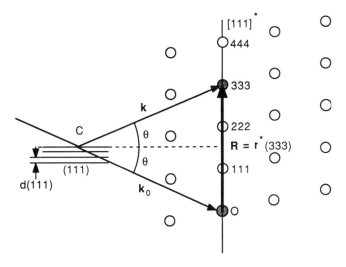

Figure 5.4. Diffraction condition expressed by means of the reciprocal lattice. Example of the third-order reflection on a set of lattice planes (111) represented by the condition $\mathbf{R} = \mathbf{r}^*(333)$.

The Bragg condition can easily be deduced from Figure 5.4: a set of planes (*hkl*) in the direct lattice is normal to the reciprocal lattice line [*hkl*] and the magnitude of the scattering vector is $1/\lambda$; it follows that:

$$\mathbf{R} = 2 \sin \theta / \lambda = \mathbf{r}_g^* = nn^*(hkl) = n/d(hkl)$$

5.2.2.2 Ewald Construction

Ewald Sphere. The diffraction condition (5.12) expressed in the reciprocal lattice may be represented by means of a convenient geometrical approach. Consider a sphere of radius $1/\lambda$ centred at the common origin C of the wave vectors $\mathbf{k}_0$ and $\mathbf{k}$. This sphere is called the *Ewald sphere* or *reflecting sphere* (abbreviation ES). The wave vectors $\mathbf{k}_0$ and $\mathbf{k}$ necessarily have their extremities on that sphere. When choosing the extremity of $\mathbf{k}_0$ as the origin O of the reciprocal lattice (the point where the incident beam intersects the ES), the diffraction condition may be given as follows:

A selective reflection of order n on a set of planes (hkl) occurs when the nth lattice point of the reciprocal line [hkl] intersects the ES; the diffracted beam passes through this point.

This form of the diffraction condition justifies quite naturally the conventional notation *nh nk nl* to express a reflection of order *n* on a set of planes (*hkl*); indeed

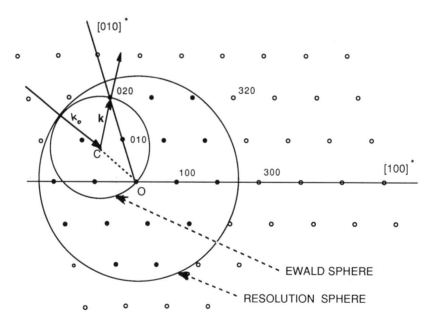

Figure 5.5. Diffraction condition and resolution condition as expressed by means of the Ewald construction. The geometrical condition for exciting a 020 reflection is displayed as an example. It can be seen that on a set of planes (100), the reflections of the first-order 100 and of the second-order 200 can occur; the third-order 300 can never be observed, the corresponding lattice point 300 being exterior to the resolution sphere. On the other hand, a set of planes (320) cannot give any reflection, the first point 320 of the lattice line [320]* already being outside the resolution sphere.

Diffraction by a Crystal

this reflection corresponds to the intersection of the reciprocal lattice point *nh nk nl* with the ES.

Resolution Condition. Resolution Sphere. The resolution condition can also be expressed by means of the Ewald construction. For a reflection to be possible the corresponding lattice point must be able to intersect the ES at a given orientation of the incident beam. The lattice point must therefore be inside a sphere of radius $2/\lambda$, centred at the origin O of the reciprocal lattice, as can be seen on Figure 5.5; this sphere is called the *resolution sphere* or *limiting sphere*. This condition can be expressed as follows:

$$\mathbf{r}^*(hkl) = nn^*(hkl) \leqslant 2/\lambda$$

For a given set of planes (*hkl*) to give at least one reflection ($n=1$), the first point of the corresponding reciprocal lattice line [*hkl*] must be situated inside the resolution sphere; this condition is expressed by

$$n^*(hkl) \leqslant 2/\lambda$$

This resolution condition is equivalent to that expressed through the direct space (Eq. 5.9), because $n^*(hkl)\, d(hkl) = 1$.

Let a crystal rotate with respect to the incident beam. Its reciprocal lattice rotates correspondingly, causing any reciprocal lattice point *hkl* within the resolution sphere eventually to intersect the Ewald sphere, producing the corresponding *hkl* reflection. Imagining a materialised fluorescent ES, one could observe light scintillations at the very points and moments of intersection. The total number of reflections equals the number of reciprocal lattice points within the resolution sphere.

For practical purposes, an ES of radius L, defined by the experimental conditions, replaces the theoretical ES of radius $1/\lambda$. For the Ewald condition to be met, the theoretical reciprocal lattice of constant 1 is then replaced by a similar reciprocal lattice of constant $K = L\lambda$.

Summary. The diffraction conditions may be expressed in several different but equivalent ways. The respective choice depends on the application. The Bragg condition is generally used for calculation purposes, whereas the Ewald construction is of great help in interpreting diffraction patterns.

5.3 KINEMATIC DIFFRACTION THEORY

5.3.1 DIFFRACTED AMPLITUDE

5.3.1.1 General Expression

Consider a unit cell at the origin of space and phase (see Figure 5.1). It will scatter an amplitude $F(\mathbf{R})$ in the direction of the scattering vector $\mathbf{R}$, according to Eq. (3.4)

$$F(\mathbf{R}) = \int_{\text{unit cell}} f(\mathbf{r}) \exp(2\pi i\, \mathbf{rR})\, dv \tag{5.13}$$

where $f(\mathbf{r})$ is scattering power at any point $\mathbf{r}(xyz)$ of a unit cell.

A crystal is a distribution of identical scattering elements for which Eq. (3.9) can be applied. In addition, the distribution is periodical.

Assuming the scattered amplitude to be small with respect to the incident amplitude, the following approximations can be made:

1. The incident amplitude is a constant for any unit cells of the crystal; absorption may be neglected;
2. The scattered wave is not rescattered; there is no secondary or multiple scattering;
3. The incident and scattered waves do not interact with each other.

These are the fundamental approximations of the *kinematic diffraction theory* which are leading to a simplified interpretation. However, their validity is limited (see Section 5.4).

With the above assumptions, the wave scattered by a unit cell at $\mathbf{r}_n$ has an absolute value of its amplitude equal to the amplitude scattered by the unit cell at the origin, but is affected by a phase shift $\phi = 2\pi\, \mathbf{r}_n \mathbf{R}$. The wave scattered by the $N = UVW$ unit cells of a whole crystal (see Eq. 5.4) can accordingly be written

$$G(\mathbf{R}) = F(\mathbf{R})\, L(\mathbf{R}) \qquad (5.14)$$

where $F(\mathbf{R})$ is the *structure factor* given by Eq. (5.13) and $L(\mathbf{R})$ the *form factor* given by the following equation:

$$L(\mathbf{R}) = \sum_{\mathbf{r}_n} \exp(2\pi i\, \mathbf{r}_n \mathbf{R}) \qquad (5.15)$$

The *structure factor* of the crystal (amplitude scattered by one unit cell) depends on the atomic content of the unit cell, i.e. the crystal structure. It accounts for the phase shifts between the atoms of the unit cell. It does not depend on the size of the crystal. It can be expressed as the *FT of the crystal structure*.

The structure factor has the dimensions of an amplitude, i.e. of a length. However, it is often measured with respect to the incident amplitude and is then expressed as a dimensionless number.

The *form factor* of the crystal (amplitude scattered by the crystal referred to the amplitude scattered by one unit cell which is taken as unity) depends on the size and the form of the crystal. It accounts for the phase shifts between the unit cells of the crystal. It does not depend on the crystal structure. It can be expressed as the *FT of the crystal lattice*.

The form factor is a dimensionless number. For a size-limited crystal, it includes the size factor, according to Eq. (3.10).

5.3.1.2 Structure Factor

The structure factor is generally expressed for a *hkl* reflection.

Structure Considered as a Continuous Distribution of Scattering Matter

For a *hkl* reflection, the scattering vector is a reciprocal lattice vector, and Eq. (5.13) is written as follows:

Diffraction by a Crystal

$$F(hkl) = \int\int\int_{\text{unit cell}} f(xyz) \exp[2\pi i(xh + yk + zl)] \, dx\,dy\,dz \qquad (5.16)$$

The structure factor is the FT of the scattering power (i.e. structure function) at any points in the unit cell.

Fourier inversion leads to the scattering power as follows:

$$f(xyz) = \frac{1}{v_0} \sum_h \sum_k \sum_l F(hkl) \exp[-2\pi i(xh + yk + zl)] \qquad (5.17)$$

Due to the limited number of discrete *hkl* reflections, the integral of Eq. (3.7) has become a *Fourier series*.

The scattering power at a given point *xyz* in the unit cell is the FT of the structure factor for any *hkl* reflections.

The fundamental limitations remain:

Phase limitation. The amplitude cannot be measured directly. The calculation is limited to the FT of the intensity. This phase problem can generally be solved in the present case, thanks to the limited number of reflections.

Resolution limitation. Only the reflections corresponding to *hkl* reciprocal lattice points inside the resolution sphere can be measured, leading to a limitation of the Fourier series (Eq. 5.17). The resolution can be improved by working with smaller wavelengths.

Equation (5.17) may be considered as the basis of crystal structure determination techniques.

Structure Considered as a Distribution of Discrete Atoms

Instead of considering a crystal unit cell as a continuous distribution of matter, of scattering power $f(xyz)$, it may be described as a distribution of discrete scattering elements, the atoms. This atomic structure layout results in an easier calculation of structure factors, e.g. for means of structure modelling.

The structure factor $F(hkl)$ can be expressed as the sum of the amplitudes scattered by the atoms in a unit cell. An atom *j* inside a unit cell is located by a position vector $\mathbf{r}_j(x_j y_j z_j)$. The individual contribution of an atom is $A_j \exp(2\pi i \mathbf{r}_j \mathbf{R})$. The resulting contribution of a unit cell for a *hkl* reflection becomes

$$F(\mathbf{R}) = F(hkl) = \sum_j A_j(hkl) \exp[2\pi i(x_j h + y_j k + z_j l)] \qquad (5.18)$$

In the general case the structure factor is expressed by a complex number. It can be represented by means of a *Fresnel construction* (Figure 5.6a).

For any reflection *hkl*, the Bragg equation (5.7) provides a value of $\sin\theta/\lambda = R/2$; the corresponding atomic scattering amplitudes A_j are tabulated. For a given structure model, Eq. (5.18) provides the theoretical amplitudes scattered per unit cell.

The atomic scattering factors are expressed differently depending on the nature of the radiation. For X-rays in particular, the structure factor is generally expressed in terms of the atomic scattering factors f_j given by the tables. Its expression then becomes

$$F(hkl) = \frac{F(hkl)}{T(hkl)} = \sum_j f_j(hkl) \exp[2\pi i(x_j h + y_j k + z_j l)] \qquad (5.19)$$

58 Structural and Chemical Analysis of Materials

In this case the structure factor is no longer an amplitude, but a dimensionless number, which is the ratio of the amplitude scattered by a unit cell to the amplitude scattered by one electron in the same direction. For absolute calculations the result has to be multiplied by the *Thomson factor*. However, in practice only the relative amplitudes *hkl* are required and the calculated values have to be corrected only by the polarisation factor $P(hkl)$ which is angle dependent and therefore varies with the reflections *hkl* (Eq. 3.12). The P-factor is given by tables together with the atomic scattering factors (e.g. [8] *Int. tab.*, vol. 2). In the particular case of a structure containing only one atomic species, of scattering amplitude $A(hkl)$, a *unitary structure factor* can be defined as follows:

$$\frac{F(hkl)}{A(hkl)} = \sum_j \exp[2\pi i (x_j h + y_j k + z_j l)] \quad (5.20)$$

This represents the FT of the atom distribution in the unit cell (also called *ponctualized structure*) and corresponds to the form factor of the unit cell.

Centrosymmetrical Structure. Consider a structure with a centre of symmetry at the origin. To any atom in $\mathbf{r}(xyz)$ there corresponds an identical atom in $-\mathbf{r}(-x-y-z)$. When developing Eq. (5.18) using $\exp(i\phi) = \cos\phi + i\sin\phi$, it can be seen that the sine terms cancel. The structure factor then becomes a *real number* (Figure 5.6b):

$$F(hkl) = \sum_j A_j(hkl) \cos 2\pi(x_j h + y_j k + z_j l) \quad (5.21)$$

Atomic positions in the unit cell are determined by the crystal space group. The existing relationship between structure factor and symmetry group is examined later in this chapter.

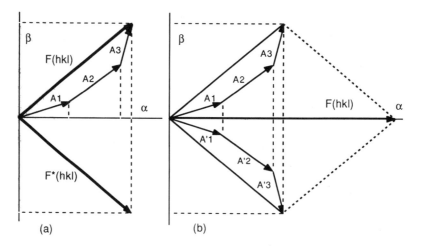

Figure 5.6. Representing the structure factor by means of an amplitude-phase diagram. (a) Crystal without centre of symmetry; three atoms per unit cell; (b) centrosymmetrical crystal; six atoms per unit cell, in pairs symmetrical about the origin.

Diffraction by a Crystal

Temperature Factor. In order to compare calculated and experimental values of the structure factor, the thermal vibrations of the atoms must be taken into account. According to Eq. (4.11), the thermal effect corrected structure factor can be written as follows by means of the *Debye-Waller temperature factor*:

$$F(\mathbf{R}) = F(hkl) = \sum_j A_j(hkl) \exp\left[-\frac{B\sin^2\theta}{\lambda^2}\right] \exp[2\pi i(x_j h + y_j k + z_j l)] \quad (5.22)$$

The effect of the temperature factor is to increase the convergence of the $F(hkl)$ series. This causes the diffracted amplitude to decrease more rapidly with the reflection order.

5.3.1.3 Form Factor

Let the scattering vector $\mathbf{R}$ be defined by its relative coordinates XYZ in reciprocal space. According to Eqs (5.1) and (5.10) it follows that

$$\mathbf{R} = X\mathbf{a}^* + Y\mathbf{b}^* + Z\mathbf{c}^* \Rightarrow \mathbf{r}_n \mathbf{R} = uX + vY = wZ$$

The form factor of Eq. (5.15) can accordingly be written

$$L(XYZ) = \sum_0^U \exp(2\pi i\, uX) \sum_0^V \exp(2\pi i\, vY) \sum_0^W \exp(2\pi i\, wZ) \quad (5.23)$$

This factor is the product of three geometrical sums which can be written as follows:

$$L(XYZ) = \frac{\sin \pi UX}{\sin \pi X} \frac{\sin \pi VY}{\sin \pi Y} \frac{\sin \pi WZ}{\sin \pi Z} \quad (5.24)$$

5.3.2 DIFFRACTED INTENSITY

Intensity is accessible to measurement on experimental diffraction patterns. It is therefore generally the basis for structure calculations.

5.3.2.1 General Expression

According to Eqs (3.19) and (5.14), the intensity diffracted by a crystal is expressed as follows:

$$I(\mathbf{R}) = |G(\mathbf{R})|^2 = |F(\mathbf{R})L(\mathbf{R})|^2 = [F(\mathbf{R})F^*(\mathbf{R})]L^2(\mathbf{R}) \quad (5.25)$$

The expressions of form and of structure will be examined separately below.

5.3.2.2 Interference Function Expressed in Reciprocal Space

In terms of intensity, the form factor of Eq. (5.24) becomes

$$L^2(XYZ) = \frac{\sin^2 \pi UX}{\sin^2 \pi X} \frac{\sin^2 \pi VY}{\sin^2 \pi Y} \frac{\sin^2 \pi WZ}{\sin^2 \pi Z} \quad (5.26)$$

This squared form factor is called the *interference function* (or diffraction function). It expresses the interference effects between waves diffracted by the unit cells of the crystal.

One-dimensional Analysis

For an easier representation of the interference function, consider its variation along a reciprocal axis, e.g. the X-axis ($Y=Z=0$). With varying X, the end point of the scattering vector $\mathbf{R}(X00)$ therefore generates the reciprocal line $[100]^*$. It follows that

$$L^2(X00) = \frac{\sin^2 \pi UX}{\sin^2 \pi X} V^2 W^2 \qquad (5.27)$$

This function has the form of the well-known interference function describing light diffraction by a one-dimensional optical grating and is analysed in books on conventional optics. Figure 5.7 shows its graphical display for an unlimited and for a limited crystal.

Unlimited Crystal. For a crystal with infinite size in the x direction ($U=\infty$), the interference function is zero for any values of X, except for integers $X=h$, where it becomes infinite. It is represented in reciprocal space by a periodical distribution of Dirac peaks located on the reciprocal lattice points of the $[100]^*$ line. Diffraction occurs only when the scattering vector $\mathbf{R}$ is a reciprocal lattice vector $\mathbf{R}(h00)$. This is equivalent to the geometrical diffraction condition established above.

Limited Crystal. For a crystal with finite size in the x direction ($U=$ finite number), the interference function is a periodic function with period a^*. It has main maxima and secondary maxima, separated by minima of zero value:

1. *Main maxima for $X=h$.* Peak height: $(UVW)^2 = N^2$ (N total number of unit cells); half-width: a^*/U (width between zero minima $2a^*/U$). With increasing values of U, the main maxima become sharper and more intense.
2. *Zero value minima for $X=h/U$.*
3. *Secondary maxima for $X=(2h+1)/2U$.*

Between successive main maxima there are $(U-2)$ secondary maxima separated by $(U-1)$ zero value minima. The height of the secondary maxima decreases rapidly when moving away from the main peaks. When the number U reaches values of the order of some 10 unit cells, the secondary peaks practically merge into the background (Figure 5.7).

Three-dimensional Generalisation. Intensity Weighted Reciprocal Lattice

For an infinite three-dimensional crystal, the interference function is a three-dimensional periodic distribution of Dirac peaks on the reciprocal lattice points hkl. It therefore represents the reciprocal lattice, the FT of the direct lattice. Henceforth the case of a limited crystal, which corresponds to the practical situation, will be considered.

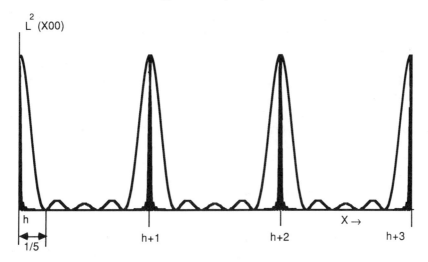

Figure 5.7. The interference function plotted against X. The graphs are computed for $U=5$ and $U=50$. The intensity scale is arbitrary.

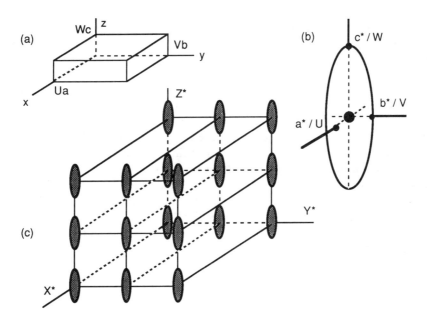

Figure 5.8. Three-dimensional form factor. (a) Crystal form in direct space, the crystal is flattened in the z direction. (b) Diffraction domains in the reciprocal space, centred around the lattice points. They are elongated along the z^* direction. (c) Periodic space distribution of the diffraction domains over the reciprocal lattice points. The diffraction domains are restricted to the main maxima, with respective dimensions $2a^*/U$, $2b^*/V$, $2c^*/W$ along x^* y^* z^* (approximate outline for $U=W=10$; $W=4$).

Diffraction Domains. Limited to the main maxima, the interference function is represented in reciprocal space by a periodical three-dimensional distribution of volumes which are centred on the reciprocal lattice points (Figure 5.8). These volumes are called *diffraction domains.* Their dimensions along the reciprocal axis are respectively $2a^*/U$, $2b^*/V$ and $2c^*/W$ when measured between zero value minima (a^*/U, b^*/V and c^*/W for the half-width).

The form of the diffraction domains depends on the crystal morphology. Consider the example of a thin crystal plate parallel to (hkl) and containing M lattice planes (hkl); its thickness is therefore $t = Md(hkl)$. The corresponding diffraction domains are rods elongated along the normal to the (hkl) planes, e.g. along the $[hkl]^*$ reciprocal lattice lines. Their length between zero value minima is $2n^*(hkl)/M = 2/t$ (see Figure 5.8).

The secondary diffraction maxima become experimentally apparent only with crystals which are very small in one direction at least, say about 10 lattice parameters.

Exact Diffraction. The diffracted intensity goes through a maximum when the end point of the scattering vector is a reciprocal lattice point, i.e. when $X = h$, $Y = k$, $Z = l$. This amounts to the geometrical diffraction condition of Eq. (5.12). In this case the N lattice points are in phase; it follows that the total scattered intensity equals the arithmetic sum of the intensities scattered by each unit cell. The diffracted intensity is therefore given by

$$I(hkl) = I(0) = (UVW)^2 F^2(hkl) = N^2 F^2(hkl) \quad (5.28)$$

This expression is equivalent to Eq. (3.31) giving the maximum scattered intensity by a general object, with the difference that in the present case it extends to all reciprocal lattice points instead of being limited to the incident direction.

Relaxation of Diffraction Conditions. For a size-limited crystal, the interference function, i.e. the intensity, does not instantly drop to zero when deviating from the exact diffraction condition of a *hkl* reflection. It decreases rapidly, but the reflection remains excited as long as the end point of the scattering vector remains inside the diffraction domain surrounding the *hkl* reflection. It follows a relaxation of the diffraction conditions which increases with decreasing crystal size.

When expressing diffraction by means of the Ewald condition, a *hkl* reflection remains excited as long as the Ewald sphere intersects the *hkl* diffraction domain.

When expressing diffraction by means of the Bragg condition, the relaxation results in an angular deviation on both sides of the exact reflection (Figure 5.9). The deviation is expressed by differentiating $R = 2 \sin \theta / \lambda$

$$d\theta = \frac{\lambda}{2 \cos \theta} dR$$

Consider as above a reflection on a set (hkl) containing M planes amounting to a thickness $t = Md(hkl)$. The whole width of the main maximum is $\Delta R = 2/t$. The corresponding angular width becomes accordingly

$$\Delta \theta = \frac{\lambda}{t \cos \theta} \quad (5.29)$$

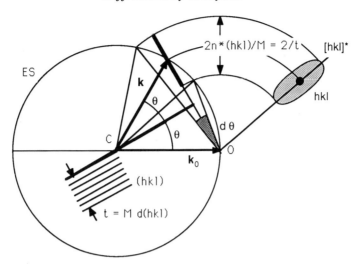

Figure 5.9. The relaxation of diffraction conditions as expressed by the Ewald construction. Example of a *hkl* reflection on a crystal of thickness *t* with *M* planes (*hkl*).

Example: $t = 1000$ Å, $\lambda = 1$ Å, $\cos\theta = 0.5 \Rightarrow \Delta\theta = 2 \times 10^{-3}$ rad.

Intensity Weighted Reciprocal Lattice. The scattering reciprocal space has been defined in section 3.3.1 in the case of a general scattering material.

Assigning to any point $\mathbf{R}(XYZ)$ in the reciprocal space of a crystal, the value of the corresponding scattering intensity $I(XYZ) = F^2(XYZ) L^2(XYZ)$ leads to the definition of the *intensity weighted reciprocal space* (or simply the *weighted reciprocal space*) of the crystal.

For not too small a crystal, the diffraction domains are practically restricted to the reciprocal lattice points. Assigning to any reciprocal lattice point *hkl* the corresponding intensity $I(hkl)$ leads to the definition of the *intensity weighted reciprocal lattice* (or simply the *weighted reciprocal lattice*) of the crystal.

The above concepts of weighted reciprocal space and reciprocal lattice will prove to be very useful for understanding and interpreting diffraction patterns which are in fact their experimental expressions.

5.3.2.3 Diffraction Deviation

For expressing diffraction outside the exact geometrical conditions, it is useful to define the *diffraction deviation* in reciprocal space. For a *hkl* reflection, this is done by means of a vector **s** such that

$$\mathbf{R} = \mathbf{r}_g^* + \mathbf{s} \\ \mathbf{s} = s_x \mathbf{a}^* + s_y \mathbf{b}^* + s_z \mathbf{c}^* \tag{5.30}$$

where **s** is often called the *deviation parameter*. Conventionally $s > 0$ if the corresponding *hkl* point is inside the ES, $s < 0$ if it is outside (Figure 5.10). The exact diffraction condition can then be stated as $\mathbf{s} = 0$. As a result of $\mathbf{r}_n \mathbf{r}_g^* = m$ being an integer, the form factor in Eq. (5.15) is written as a function of **s**

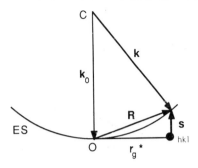

Figure 5.10. Defining the diffraction deviation.

$$L(\mathbf{s}) = \sum_{\mathbf{r}_n} \exp[2\pi i \, \mathbf{r}_n (\mathbf{r}_g^* + \mathbf{s})] = \sum_{\mathbf{r}_n} \exp(2\pi i \, \mathbf{r}_n \mathbf{s}) \qquad (5.31)$$

The interference function Eq. (5.26) around a reciprocal lattice point becomes

$$L^2(s_x s_y s_z) = \frac{\sin^2 \pi U s_x}{\sin^2 \pi s_x} \frac{\sin^2 \pi V s_y}{\sin^2 \pi s_y} \frac{\sin^2 \pi W s_z}{\sin^2 \pi s_z} \qquad (5.32)$$

In the general case of a small diffraction deviation **s** compared to the reciprocal lattice parameters, the sine functions may be approximated to the angles, leading to

$$L^2(s_x s_y s_z) = \frac{\sin^2 \pi U s_x}{(\pi s_x)^2} \frac{\sin^2 \pi V s_y}{(\pi s_y)^2} \frac{\sin^2 \pi W s_z}{(\pi s_z)^2} \qquad (5.33)$$

Comparison with Eq. (3.28) shows the interference function around any reciprocal lattice point to be equivalent to the squared size function of a general scattering particle. The interference function of a crystal amounts to the distribution of the squared size function on the reciprocal lattice points.

Note: hkl reflection. Henceforth the denomination *hkl reflection* will be used in the general case of excitation of a *hkl* diffraction domain, i.e. when the scattering vector ends inside this diffraction domain.

A *hkl* reflection is commonly abbreviated to **g**, where **g** represents the scattering vector defined as follows:

$$\mathbf{g} = \mathbf{r}_g^* + \mathbf{s} \quad \text{with } \mathbf{s} \in ((hkl)) \qquad (5.34)$$

In the following text, the notation $((hkl))$ will designate the diffraction domain surrounding the reciprocal lattice point *hkl*. This formulation will be especially useful in electron diffraction, where a deviation from exact diffraction is a common situation.

5.3.2.4 Friedel's Law

The intensity weighted reciprocal space is centrosymmetric, whatever the crystal symmetry.

Diffraction by a Crystal

This statement has already been made for a general material (section 3.3.1) and is valid in the weak interaction approximation (kinematic approximation). For crystals it corresponds to the so-called *Friedel's law* expressed by the following relations:

$$I(X\ Y\ Z) = I(\overline{X}\ \overline{Y}\ \overline{Z})$$
$$I(h\ k\ l) = I(\overline{h}\ \overline{k}\ \overline{l}) \tag{5.35}$$

It follows that the intensity diffracted on opposite crystal faces limiting a set of (*hkl*) planes has the same value.

For a centrosymmetrical crystal, the structure factor becomes a real number. Its expression in cosines is therefore insensitive to the sign of **R** (Eq. 5.21, Figure 5.6b)

$$F(-\mathbf{R}) = F(\mathbf{R}) \Rightarrow I(-\mathbf{R}) = I(\mathbf{R})$$

For a crystal without centre of symmetry, the structure factor is a complex number (Eq. 5.18, Figure 5.6a). With real atomic scattering amplitudes, intensity is expressed as follows:

$$I(\mathbf{R}) = F(\mathbf{R})F(-\mathbf{R}) \Rightarrow I(-\mathbf{R}) = I(\mathbf{R})$$

In this case changing **R** to −**R**, i.e. changing the sign of *hkl*, changes the structure factor, but does not change the intensity.

> *Consequence: the existence or absence of a centre of symmetry cannot be established by means of diffraction techniques.*

Friedel's law is experimentally confirmed with measures taking place in weak interaction conditions. It is the case with neutron diffraction. It is generally the case with X-ray diffraction when the incident photon energies differ significantly from the atomic energy levels in the material analysed. The probability of electron excitations, hence of absorption, is then small and atomic scattering takes place without further phase differences between waves scattered by different atoms. The atomic scattering amplitudes are then real numbers, thus justifying the previously recalled Eq. (3.20). If this condition is not met for some atoms of the scattering material, the atomic scattering amplitudes become complex numbers. Friedel's law no longer applies and so-called anomalous scattering occurs. For the law to apply to electron diffraction, the acceleration potential must be much higher than the internal potential of the analysed material, a condition which is met with a high-energy electron of 100 keV or more. However, recent accurate measurements have shown that Friedel's law fails to apply even with a high-energy electron. This fact enables the centre of symmetry to be characterised, in particular with convergent beam diffraction.

5.3.2.5 Fourier Transform of Diffracted Intensity. Patterson Function

Computing the FT of the amplitude scattered in any direction by a crystal leads theoretically to the crystal structure. The amplitude is not accessible to direct measurement, but the intensity can be measured in any case. According to Eqs (3.20) and (5.25), the FT of diffracted intensity is given by

$$FT\{I(\mathbf{R})\} = FT\{[F(\mathbf{R})F(-\mathbf{R})][L(\mathbf{R})]^2\}$$

The FT of the product of two functions equals the convolution product of the FT of these functions. The above equation therefore becomes

$$FT\{I(\mathbf{R})\} = FT\{F(\mathbf{R})F(-\mathbf{R})\} * FT\{[L(\mathbf{R})]^2\} \qquad (5.36)$$

The first factor represents the convolution square of the structure function. It corresponds to the distribution function of the atoms in the unit cell (see section 3.3.2). For a crystal, scattering is limited to discrete directions, the hkl reflections. The FT therefore becomes a limited Fourier series as follows:

$$FT\{F(\mathbf{R})F(-\mathbf{R})\} = q(x'y'z') = \frac{1}{v_0}\sum_h\sum_k\sum_l F(h\ k\ l)F(\bar{h}\ \bar{k}\ \bar{l})\cos 2\pi(x'h+y'k+z'l) \qquad (5.37)$$

where $x'y'z'$ are coordinates of an auxiliary variable $\mathbf{u}$ superimposed on the variable $\mathbf{r}(xyz)$ in the same set of axes.

The second factor of Eq. (5.36) represents the convolution square of the direct lattice, i.e. the lattice itself (a lattice is an invariant through the operation convolution square).

It follows that the FT of diffracted intensity represents the distribution of function (5.37) on the direct lattice points. It is called the *Patterson function*. It is centrosymmetrical and has the crystal periodicity (Figure 5.11). It corresponds to the distribution function in the case of a periodic medium.

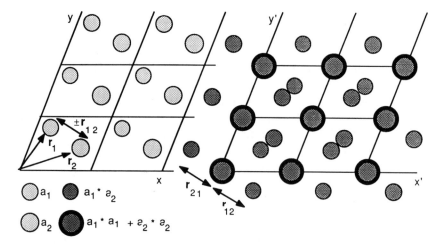

Figure 5.11. Comparing the structure function (left) and the corresponding Patterson function (right).

> **Summary:** *The maxima of the structure function represent the positions $\mathbf{r}(xyz)$ of the atoms. The maxima of the Patterson function represent the interatomic distances $\mathbf{r}_{ij} = \mathbf{r}_j - \mathbf{r}_i$ of the structure.*

The FT of the scattered intensity provides less information on a crystal structure than the FT of the amplitudes. One of the basic problems of crystal structure analysis is the deconvolution of the Patterson function.

5.3.3 RELATIONSHIP BETWEEN STRUCTURE FACTOR AND SPACE GROUP SYMMETRY

5.3.3.1 Basics

The smallest structural entity enabling the characterisation of a crystal may be called the *structure element*. It corresponds generally to the chemical formula (e.g. one Na and one Cl ion at their respective positions in NaCl). In the general case, it is a fraction of the content of the conventional unit cell **abc**. The whole structure, i.e. the content of a unit cell, is generated by submitting the structure element to the space group operations. These operations involve translations which are fractions of the unit cell translations. Those fractional translations can lead to phase oppositions of the scattered waves in particular directions. Analysing those *systematic extinctions* leads to characterisation of the space groups.

Consider a crystal with a structure element containing atoms j located by means of vectors $\mathbf{r}_j(x_j y_j z_j)$. The amplitude $F_0(hkl)$ scattered by this structure element can be written according to Eq. (5.18).

$$F_0(hkl) = \sum_j A_j(hkl) \exp[2\pi i(x_j h + y_j k + z_j l)] \qquad (5.38)$$

The complete structure (unit cell content) is generated by displacing this structure element through the translational space group operations. According to the multiplicity of operations, the structure therefore encloses one or several structure elements at positions determined by means of displacement vectors $\mathbf{r}_d(x_d y_d z_d)$. One of the positions is chosen as the origin of coordinates and of phase $\mathbf{r}_d = 0$.

The structure factor, the amplitude scattered by the atomic content of a conventional unit cell, can now be written as the geometrical sum of the amplitudes scattered by each structure element. Those amplitudes have respectively a magnitude F_0 and a phase $2\pi \mathbf{r}_d \mathbf{R}$. The structure factor becomes

$$F(hkl) = F_0(hkl) \sum_d \exp[2\pi i(x_d h + y_d k + z_d l)] \qquad (5.39)$$

Three kinds of space group operations can generate the structure from the structure element:

1. Translations related to *centred lattice types I, F, A, B, C*. Those lattices result in multiple cells chosen as conventional unit cells;

2. Rotation—translations around the *screw axis*;
3. Reflection—translations through *glide planes*.

Point group symmetry operations have no translation components and cannot therefore introduce extinction effects for atoms in general positions.

It is useful to illustrate the systematic extinction rules by means of the intensity weighted reciprocal lattice, by displaying only the lattice points corresponding to $I \neq 0$.

5.3.3.2 Translations Corresponding to Lattice Types

Primitive Lattice

In a primitive lattice (symbol P), the only translations are the basic lattice translations **a, b, c**. The structure, i.e. the conventional unit cell, contains only one structure element in a position chosen as the origin

$$\mathbf{r}_d = 0 \Rightarrow F(hkl) = F_0(hkl)$$

All *hkl* reflections occur; there is no systematic extinction.

The diffracted intensity is modulated by F_0^2. The intensity weighted reciprocal lattice includes all the reciprocal lattice points (Figure 5.12).

Body-centred Lattice

In a body-centred lattice (symbol I), the conventional unit cell contains two structure elements. With one chosen as the origin, the two positions are

$$\mathbf{r}_1(0\ 0\ 0);\ \mathbf{r}_2(1/2\ 1/2\ 1/2)$$

$$F(hkl) = F_0(hkl)\{1 + \exp[\pi i(h+k+l)]\} \tag{5.40}$$

$$h+k+l = 2n \quad \Rightarrow F(hkl) = 2F_0(hkl)$$
$$h+k+l = 2n+1 \Rightarrow F(hkl) = 0$$

Systematic extinction occurs for $(h+k+l)$ odd.

The intensity weighted reciprocal lattice is an all-face centred lattice with parameters $2a^*,\ 2b^*,\ 2c^*$ (Figure 5.12).

All-face Centred Lattice

In an all-face centred lattice (symbol F), the structure includes four structure elements. With one of them chosen as the origin, the four positions are

$$\mathbf{r}_1(0\ 0\ 0);\ \mathbf{r}_2(1/2\ 1/2\ 0);\ \mathbf{r}_3(1/2\ 0\ 1/2);\ \mathbf{r}_4(0\ 1/2\ 1/2)$$

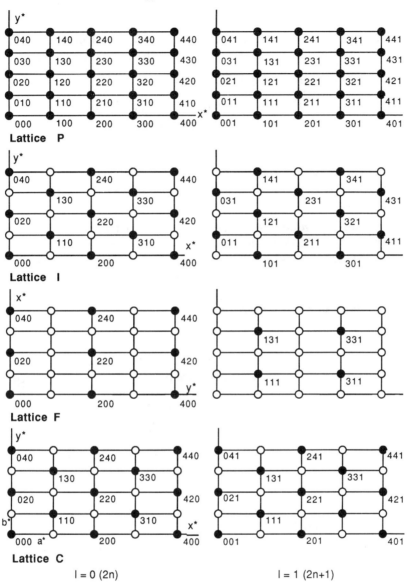

Figure 5.12. Intensity weighted reciprocal lattice for the different lattice translations. The black dots indicate the lattice points corresponding to existing reflections ($I \neq 0$).

$$F(hkl) = F_0(hkl)\{1 + \exp[i\pi(h+k)] + \exp[i\pi(k+l)] + \exp[i\pi(l+h)]\} \quad (5.41)$$

h, k, l same parity $\Rightarrow F(hkl) = 4F_0(hkl)$
h, k, l different parity $\Rightarrow F(hkl) = 0$

Systematic extinction occurs for h, k, l of different parity.

The intensity weighted reciprocal lattice is a body-centred lattice of parameters $2a^*$, $2b^*$, $2c^*$ (Figure 5.12).

One-face Centred Lattice

Consider the example of a *C*-face centred lattice; the structure includes two structure elements. With one of them chosen as the origin, the two positions are

$$\mathbf{r}_1(0\ 0\ 0);\ \mathbf{r}_2\ (1/2\ 1/2\ 0)$$

$$F(hkl) = F_0(hkl)\{1 + \exp[i\pi(h+k)]\} \tag{5.42}$$

$$\begin{aligned} h+k &= 2n &\Rightarrow F(hkl) &= 2F_0(hkl) \\ h+k &= 2n+1 &\Rightarrow F(hkl) &= 0 \end{aligned}$$

Systematic extinction occurs for $(h+k)$ odd, whatever l.

The intensity weighted reciprocal lattice is depicted in Figure 5.12.

> *Summary: Lattice translations corresponding to centred unit cells (I, F, A, B, C) lead to systematic extinctions with a three-dimensional periodicity when displayed through the intensity weighted reciprocal lattice.*

5.3.3.3 Translational Space Group Operations

Screw Axis

The screw axis operation combines rotation and translation parallel to the axis. In a crystal, a screw axis is normal to a set of lattice planes (hkl). It is characterised by its order s ($s = 2, 3, 4$ or 6) and its translation t parallel to the axis, with $t = (p/s)\,d(hkl)$ (p integer). The conventional notation is s_p (Table 5.1).

Screw axes possible in a crystal: $2_1;\ 3_1,\ 3_2;\ 4_1,\ 4_2,\ 4_3;\ 6_1,\ 6_2,\ 6_3,\ 6_4,\ 6_5$.

Example of an Axis 2_1. Space Group P 2_1. The crystal is monoclinic with a 2_1 axis normal to (010). Only the $0k0$ reflections are concerned. As shown in Figure 5.13, the structure includes two structure elements. With one of them chosen as the origin, the two positions are

$$\mathbf{r}_1(0\ 0\ 0);\ \mathbf{r}_2(-2x\ 1/2\ -2z)$$

$$F(hkl) = F_0(hkl)\left\{1 + \exp\left[2\pi i\left(-2xh + \frac{k}{2} - 2zl\right)\right]\right\} \tag{5.43}$$

Table 5.1. 001 reflection conditions introduced by a screw axis parallel to [001]. The first line corresponds to a non-translational symmetry axis

Equidistance of structure planes $\parallel$ (001)	Symmetry axis [001]	001 reflections (n integer)
$d(001)$	2 3 4 6	All
$d(001)/2$	$2_1\ 4_2\ 6_3$	$l = 2n$
$d(001)/3$	$3_1\ 3_2\ 6_2\ 6_4$	$l = 3n$
$d(001)/4$	$4_1\ 4_3$	$l = 4n$
$d(001)/6$	$6_1\ 6_5$	$l = 6n$

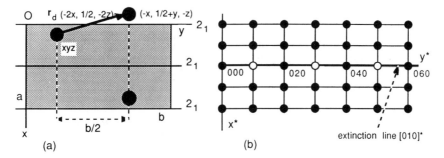

Figure 5.13. Systematic extinctions due to a screw axis. Example of space group $P\,2_1$. (a) Twofold general position of an atom in the unit cell; (b) extinction lattice line $[010]^*$ in the intensity weighted reciprocal lattice. No systematic extinctions on the other lines. The black dots indicate the lattice points corresponding to permitted reflections $(I \neq 0)$.

With atoms in general positions, systematic extinctions can take place only if the factor in square brackets can take zero values as a function of hkl, i.e. for $h = l = 0$. It follows that only the $0k0$ reflections are involved, i.e. the reflections on (010) planes. In the intensity weighted reciprocal lattice, only lattice points $0k0$ of the lattice line $[010]^*$ intervene; this line is the extinction line which is parallel to the screw axis (Figure 5.13).

For $0k0$ reflections, the structure factor becomes

$$F(0k0) = F_0(0k0)\,[\,1 + \exp(i\pi k)\,]$$
$$k = 2n \quad \Rightarrow F(0k0) = 2F_0(0k0)$$
$$k = 2n + 1 \Rightarrow F(0k0) = 0$$

Summary: A screw axis leads to systematic extinctions with a one-dimensional periodicity on the corresponding line of the intensity weighted reciprocal lattice.

Glide Planes

A glide plane in a crystal is parallel to a lattice plane. It is defined by means of a translation vector **t**. This vector is parallel to the plane and to a lattice line; its length equals the half-period along its direction.

The symbols of the different symmetry planes are as follows ([8], *Int. tabl.*, vol. A):

(a) planes without translation: *m*;
(b) axial glide planes: *a*, *b*, *c* with translation vectors **a**/2, **b**/2, **c**/2;
(c) diagonal glide planes: *n* with translation vectors (**a** + **b**)/2, (**b** + **c**)/2, (**c** + **a**)/2, (**a** + **b** + **c**)/2;
(d) diamond-type diagonal glide planes: *d* with translation vectors (**a** ± **b**)/4, (**b** ± **c**)/4, (**c** ± **a**)/4.

Example of an Axial Glide Plane c. Space Group Pc. The crystal is monoclinic with a glide plane parallel to (010) and $t = c/2$. As shown in Figure 5.14, the structure includes two structure elements. With one of them chosen as the origin, the two positions are

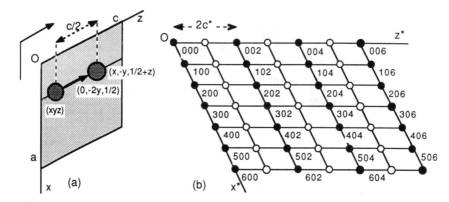

Figure 5.14. Systematic extinctions due to an axial glide plane. Example of the space group Pc. (a) Twofold general position of an atom in the unit cell; (b) extinction lattice plane (010)*, $m=0$ in the intensity weighted reciprocal lattice. No systematic extinctions on the other planes. The black dots indicate the lattice points corresponding to permitted reflections ($I \neq 0$).

$$\mathbf{r}_1(0\ 0\ 0);\ \mathbf{r}_2(0\ -2y\ 1/2)$$

$$F(hkl) = F_0(hkl) \left\{ 1 + \exp\left[2\pi i \left(-2yk + \frac{l}{2} \right) \right] \right\} \tag{5.44}$$

With atoms in general positions, systematic extinctions can take place only if the factor in braces can take zero values as a function of hkl, i.e. for $k=0$. It follows that only the $h0l$ reflections are involved, i.e. the reflections on $(h0l)$ planes which are normal to the glide plane. In the intensity weighted reciprocal lattice, only lattice points $h0l$ of the lattice plane (010)* intervene; this plane is the extinction plane which is parallel to the glide plane (Figure 5.14).

For $h0l$ reflections, the structure factor becomes

$$F(h0l) = F_0(h0l)\,[\,1 + \exp(\pi i l)\,]$$
$$l = 2n \quad \Rightarrow F(h0l) = 2F_0(h0l)$$
$$l = 2n+1 \Rightarrow F(h0l) = 0$$

Glide planes a and b result in similar rules.

Example of a Diagonal Glide Plane n. Space Group Pn. The crystal is monoclinic with a glide plane parallel to (010) and $t = (\mathbf{a} + \mathbf{c})/2$. As shown in Figure 5.15, the structure includes two structure elements. With one of them chosen as the origin, the two positions are

$$\mathbf{r}_1(0\ 0\ 0);\ \mathbf{r}_2(1/2\ -2y\ 1/2)$$

$$F(hkl) = F_0(hkl) \left\{ 1 + \exp\left[2\pi i \left(\frac{h}{2} - 2yk + \frac{l}{2} \right) \right] \right\} \tag{5.45}$$

When comparing with the above results for space group Pc, it can be seen that the

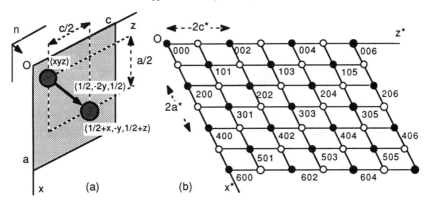

Figure 15.5. Systematic extinctions due to a diagonal glide plane. Example of Pn. (a) Twofold general position of an atom in the unit cell; (b) extinction lattice plane (010)*, $m=0$ in the intensity weighted reciprocal lattice. No systematic extinctions on the other planes. The black dots indicate the lattice points corresponding to permitted reflections ($I \neq 0$).

reflections concerned are the same, i.e. $h0l$. For those reflections, the structure factor is written

$$F(h0l) = F_0(h0l)\{1 + \exp[i\pi(h+l)]\}$$
$$h + l = 2n \quad \Rightarrow \quad F(h0l) + 2F_0(h0l)$$
$$h + l = 2n + 1 \quad \Rightarrow \quad F(h0l) = 0$$

In the case of a diamond-type glide plane with the same orientation, only reflections such as $h + l = 4n$ would be permitted.

> *Summary: A glide plane leads to systematic extinctions with a two-dimensional periodicity on the corresponding plane of the intensity weighted reciprocal lattice.*

When the space group includes more than one translational symmetry element, and possibly in addition centred lattice translations, the respective extinctions are superimposed, as can be seen in Figure 5.16.

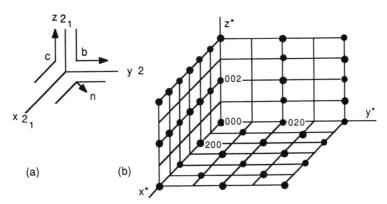

Figure 5.16. Example of the space group $P\,2_1/b\,2/c\,2_1/n$ (crystal system orthorhombic). (a) Outline of the symmetry elements; (b) intensity weighted reciprocal lattice. No extinctions outside the displayed principal planes.

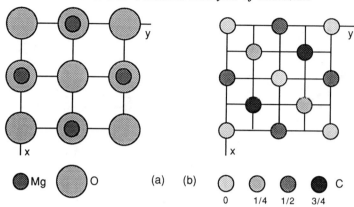

Figure 5.17. Structure projections. (a) MgO structure; (b) diamond structure.

5.3.3.4 Examples of Simple Crystal Structures

NaCl-type Structure

The point group is *m3m* (cubic, holohedral). The space group is *Fm3m*. In addition to NaCl, numerous AB compounds crystallise with the same structure, e.g. KCl, PbS, MgO.

Consider MgO (Figure 5.17a). The structure element can be chosen as including an Mg atom at the origin and an O atom at the centre of the unit cell.

The atom coordinates and the amplitude diffracted from a structure element are then respectively

$$\text{Mg}(0\ 0\ 0);\ \text{O}(1/2\ 1/2\ 1/2) \Rightarrow F_0(hkl) = A_{\text{Mg}} + A_{\text{O}}\cos(h+k+l)$$

The structure element is displaced through an *F*-lattice. According to Eq. (5.14), the structure factor is written

$$F(hkl) = [A_{\text{Mg}} + A_{\text{O}}\cos(h+k+l)][1 + \cos\pi(h+k) + \cos\pi(k+l) + \cos\pi(l+h)]$$

$$h, k, l \text{ different parity} \Rightarrow F(hkl) = 0$$
$$h, k, l \text{ same parity} \Rightarrow F(hkl) = 4F_0(hkl)\$$$
$$\text{all even } F(hkl) = 4(A_{\text{Mg}} + A_{\text{O}})$$
$$\text{all odd } F(hkl) = 4(A_{\text{Mg}} - A_{\text{O}})$$

The extinctions are those characteristic of an *F*-lattice. The permitted reflections are strong or weak, depending on the two atoms of the structure element being in phase (*hkl* all even) or out of phase (*hkl* all odd).

In the case of the two component atoms having nearly the same atomic scattering amplitudes, the all odd reflections would be very weak and could possibly not be observed. With X-ray and electron diffraction, this would occur for neighbouring elements in the periodic table. As an example KCl gives all odd reflections virtually unobservable with X-ray diffraction (the K^+ ion and the Cl^- ion have the same number of electrons). The experimental diffraction pattern simulates, therefore, a *P*-lattice with cell parameters $a/2$. In such cases it is useful to complete with neutron diffraction.

Diamond-type Structure

The diamond-type structure (Figure 5.17b) characterises elements in column four of the periodic table, with covalent tetrahedric bonds. In addition to carbon in the diamond phase, this structure occurs mainly for silicon and germanium and is of paramount importance for semiconductors.

The point group is $m3m$ (cubic holohedral). The space group is $F4_1/d\ 3\ 2/m$; it includes both 4_1 and 4_3 screw axes and d glide planes.

In the diamond structure a cubic unit cell contains eight carbon atoms. The structure element includes one carbon atom which can be placed at the origin. It is repeated:

(a) through the 4_1 screw axis operation with displacement vectors $\mathbf{r}_1(0\ 0\ 0)$ and $\mathbf{r}_2(1/4\ 1/4\ 1/4)$;
(b) through the F-lattice translations (Eq. 5.41).

The structure factor is therefore expressed as follows:

$$F(hkl) = A_C\{1 + \cos[\pi/2(h+k+l)]\}[1 + \cos\pi(h+k) + \cos\pi(k+l) + \cos\pi(l+h)]$$

h, k, l different parity $\Rightarrow F(hkl) = 0$
h, k, l same parity
$\quad h+k+l = 4n \quad\quad \Rightarrow F(hkl) = 8A_C$
$\quad h+k+l = 2n+1 \Rightarrow F(hkl) = 4A_C$
$\quad h+k+l = 4n+2 \Rightarrow F(hkl) = 0$

In this particular case, in addition to the F-lattice extinctions, there are further extinctions for $h+k+l = 4n+2$ (e.g. 200, 222), due to the translational symmetry operations.

Compiling the characteristic extinction rules for a given crystal from its diffraction pattern leads to its space group, and is an important step towards the knowledge of its structure.

Whatever the number and the positions of the atoms in a structure element, the previously described extinction rules due to the centred lattice types and to the translational symmetry operations remain valid. The effect of varying atomic positions is to modulate the diffracted intensities through phase shifts between waves scattered by the different atoms inside a unit cell.

The preceding basic extinction rules have been established considering general crystal structures, i.e. including atoms in so-called *general positions*. Particular structures, i.e. with atoms in *special positions* on point symmetry elements, may lead to extra extinctions. This is the case in the simple structure examples given previously. The general and special extinction rules for any of the 230 space groups are to be found in [8] *Int. tab.*, vol. A.

5.4 DYNAMIC DIFFRACTION THEORY

As well as in the case of elastic and inelastic scattering, the denominations kinematic and dynamic theory have been derived from classical mechanics by physicists when setting up the diffraction theory of X-rays at the beginning of the century.

The **kinematic theory**, outlined in the previous section, is so called because it is limited to the motion of waves, without taking into account their energetic interactions between themselves and with matter. It is based on the following twofold hypothesis already stated in section 5.3:

1. The diffracted intensity is small compared to the incident intensity; secondary diffraction and absorption are therefore negligible.
2. The interactions between the incident wave and the diffracted waves are negligible.

This is generally justified for X-rays and neutrons, except for large and perfect crystals where extinction effects can occur. On the other hand, it generally fails to apply for electron diffraction, because of the strong electron–matter interaction (see Table 4.1); it is, however, fairly accurate in high-energy electron diffraction on thin crystals.

The great quality of the kinematic theory is its straightforwardness and ease of application. It provides in any case a qualitative interpretation of diffraction patterns and of the image contrast in electron microscopy.

The **dynamic theory** is so called because it not only takes into account the wave motion, but considers equally their energetic interactions. The mechanical equivalent is a primary oscillator (corresponding to the incident wave) which excites secondary oscillators (the diffracted waves). The exact diffraction conditions correspond to the resonance conditions of the oscillator system. The diffraction deviation corresponds to the resonance deviation; therefore it is sometimes called resonance error.

The general dynamic theory is based on quantum mechanics. It considers the propagation of n waves in the periodic crystal field. The two-waves approximation (incident wave and one strong diffracted wave) is easier to interpret and convenient for numerous applications, and is more accurate than kinematic theory. It will be outlined for electron diffraction in Chapter 6.

REFERENCE CHAPTERS

Chapters 3, 4, 6. Appendixes B and C.

BIBLIOGRAPHY

[1] Guinier A. *Théorie et technique de la radiocristallographie*. Dunod, Paris, 1964.
[2] James R. W. *The optical principles of the diffraction of X-rays*. Bell, London, 1965.
[3] Kittel C. *Introduction to solid state physics*, 6th edn. John Wiley, New York, Chichester, 1986.
[4] Klein M. V., Furtak T. E. *Optics*, 2nd edn. John Wiley, New York, Chichester, 1986.
[5] Phillips F. C. *An introduction to crystallography*, 4th edn. John Wiley, New York, Chichester, 1971.
[6] Smith G. G., Thomson J. H. *Optics*. John Wiley, New York, Chichester.
[7] Verma A. R., Srivartava O. N., Singh G. *Crystallography for solid state physics*. John Wiley, New York, Chichester, 1982.
[8] *International tables for crystallography*; vol. A: *Space-group symmetry* (T. Hahn, ed.). D. Reidel, Kluwer Acad. Publ., Dordrecht, 1987; *Brief teaching edition of vol. A. Space-group symmetry* (available from the same publisher); *International tables for X-ray crystallography*, vol. 2: *Mathematical tables* (Kasper J. S., Lonsdale K., eds); vol. 3: *Physical and chemical tables* (MacGillvray C. H., Rieck G. D., Lonsdale K., eds); vol. 4: *Supplementary tables to vols 2, 3* (Ibers J. A., Hamilton W. A., eds). D. Reidel, Kluwer Acad. Publ., Dordrecht, 1985, 1983, 1982.

6

Basic Theory of Electron Diffraction

6.1 SPECIFIC FEATURES OF ELECTRON DIFFRACTION

The *geometrical diffraction theory* applies to electrons.

For interpreting *intensities*, it must be taken into account that the electron–matter interaction cross-section is some 10^8 times the corresponding value for X-rays and some 10^{10} times that for neutrons. It follows that the *mean free path* (Eq. 2.2) is divided by the same factor. It is currently in the range of some 100 Å for high-energy electrons (10–1000 keV) and of a few ångströms for low-energy electrons (10–1000 eV).

As a consequence of this strong interaction, the basic hypothesis of the kinematic theory fails to apply.

The kinematic theory may nevertheless be used for a qualitative or semi-quantitative interpretation in the following instances:

(a) incident energy much greater than the inner crystal potential which is of the order of some 10 eV;
(b) electron path inside the crystal much smaller than the mean free path.

These conditions are met for transmission high-energy electron diffraction from a very thin crystal.

6.2 DIFFRACTION CONDITION RELAXATION

The excitation of reflections outside the exact diffraction conditions is a current situation in electron diffraction due to the small electron penetration depth.

Transmission Mode. Diffraction in the transmission mode is possible only for high-energy electrons. Within the common range of energies (10–100 keV), the maximum object thickness is of the order of some 100 Å.

Reflection Mode. Diffraction in the reflection mode can be performed with high-energy electrons at grazing incidence or with low-energy electrons at normal incidence. In both cases the penetration depth is very low, of the order of some 10 Å. The effective useful object thickness is of the same order. These diffraction modes therefore lead to surface analysis.

Diffraction Deviation. Given the very small effective object thickness in any electron diffraction modes, diffraction condition relaxation along at least one particular direction, the normal to the object surface, is a common situation. The existence of this outstanding direction leads to particular geometrical features of the electron diffraction pattern. Those particulars are detailed in the corresponding specialised sections in Part Three. Concerning the diffracted amplitude and intensity, this one-dimensional relaxation leads to a simplified formulation of the diffraction theory.

In the following section of the present chapter, the theory is formulated exclusively in the case of high-energy electron diffraction (HEED) in the transmission mode, at normal incidence on a thin monocrystalline plate of thickness t. The incident direction is chosen as the Ox axis.

In order to simplify the formulation, consider the instance of a crystal with right angle basis vectors (e.g. cubic, tetragonal, orthorhombic) having the form of a rectangular prism with edge lengths A, B, C respectively along Ox, Oy, Oz. Its dimensions are related to the cell parameters (Figures 5.8 and 6.1a) as follows:

$$t = A = Ua; \quad B = Vb; \quad C = Wc \quad \text{with} \quad t \ll A, B$$
$$v = ABC \qquad (6.1)$$
$$S = BC = v/t$$

where S is the active surface (generally a cross-section of the incident beam) and v the volume (v_0 volume of a unit cell).

The relaxation of the diffraction conditions is limited to the incident direction $\mathbf{k}_0 // Ox$. The *diffraction domains* (restricted to the main maxima) are therefore reduced to fine rods of length $2/t$, normal to the crystal plate, e.g. parallel to $\mathbf{k}_0$ and Ox.

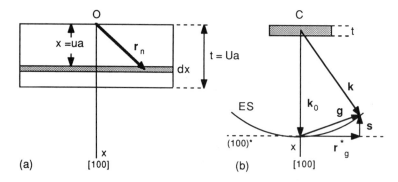

Figure 6.1. Electron diffraction on a thin crystalline plate. Transmission mode along the [001] direction. (a) Object representation; (b) Ewald construction for a reflection **g** with a diffraction deviation **s**.

It follows that the diffraction deviation vector **s** is itself parallel to the same direction (Figure 6.1b).

For a selective reflection *hkl* to be excited, the reflection sphere must intersect the corresponding diffraction domain $((hkl))$. According to Eq. (5.34), the scattering vector **g** and the diffraction deviation **s** are related as follows:

$$\mathbf{g} = \mathbf{r}_g^* + \mathbf{s} \quad \mathbf{s} \in ((hkl))$$
$$s_y = s_z = 0 \rightarrow \mathbf{s} = s_x \mathbf{a}^* \tag{6.2}$$

In order to simplify the formulation, a *hkl* reflection meeting those conditions is often written as *reflection* **g**.

6.3 DIFFRACTED AMPLITUDE AND INTENSITY. KINEMATIC THEORY

It is recalled that the kinematic theory of electron diffraction is based on the two following approximations:

1. *The diffracted intensity is assumed to be small with respect to the incident intensity; the incident intensity is therefore a constant throughout the crystal and remains unchanged when leaving the object as forward transmitted intensity.*
2. *Interactions between incident and diffracted beams are negligible.*

This theory is obviously in disagreement with the energy conservation principle: a fraction of the incoming energy being necessarily transferred to the diffracted beams, the transmitted beam cannot preserve the integrity of its initial energy. However, the kinematic theory is generally suitable when dealing with X-ray and neutron diffraction where the weak interaction with matter justifies the assumption that the diffracted intensity is infinitely small with respect to the incident intensity. For high-energy electrons in the transmission mode it provides a straightforward qualitative approach which is acceptable in many applications.

6.3.1 DIFFRACTION AT INFINITE DISTANCE

Consider a reflection **g** with a diffraction deviation **s**, observed at a great distance with respect to the object size. The scattered wave is then assimilated in a plane wave. This corresponds to the conditions of diffraction at infinity, the so-called *Fraunhofer diffraction*. In the general case the diffracted amplitude and intensity are given respectively by Eqs (5.13), (5.14), (5.31), (5.25) and (5.33).

In the particular case of electron diffraction with the conventions given in section 6.2 and illustrated in Figure 6.1, the scalar product $\mathbf{r}_n \mathbf{s}$ in Eq. (5.31) becomes simply $xs = uas$, where x is the abscissa of the unit cell located by the position vector $\mathbf{r}_n$, i.e. the projection of this vector on the Ox axis (Figure 6.1a).

6.3.1.1 Amplitude

The amplitude of a **g** reflection can be written

$$G(\mathbf{g}) = F(\mathbf{g}) VW \sum_{u=0}^{u=U} \exp(2\pi i u a s) \tag{6.3}$$

where $F(\mathbf{g})$ is the *structure factor* of the crystal for the given reflection.

An alternative formulation considers the amplitude as an integral, continuously summed over the thickness t of the crystal plate (Figure 6.1a).

$$G(\mathbf{g}) = \frac{S}{v_0} F(\mathbf{g}) \int_0^t \exp(2\pi i x s) \, dx \tag{6.4}$$

6.3.1.2 Intensity

For a small diffraction deviation, the intensity is expressed as follows, according to Eqs (5.25) and (5.33),

$$I(\mathbf{g}) = \frac{S^2}{v_0^2} F^2(\mathbf{g}) \frac{\sin^2 \pi t s}{(\pi s)^2} \tag{6.5}$$

It can be seen that for exact diffraction ($s=0$), the diffraction intensity becomes

$$I(\mathbf{g}) = [F(\mathbf{g}) v/v_0]^2 = N^2 F(\mathbf{g})^2$$

where $N = UVW$ number of unit cells in the crystal. This corresponds to the result already given by Eq. (5.28).

6.3.2 DIFFRACTION AT FINITE DISTANCE

In view of certain applications, e.g. for interpreting the diffraction contrast in transmission electron microscopy, it is more convenient to formulate diffraction at a

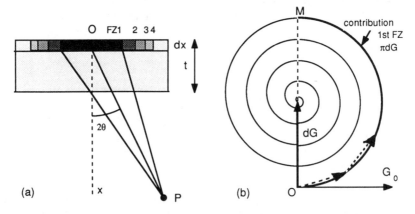

Figure 6.2. Diffraction at a small distance from a thin crystal. (a) Fresnel zones (FZ) of a thin layer dx; (b) amplitude–phase diagram of the amplitude diffracted at P by the unlimited layer dx. Arc OM represents the geometrical sum of the first Fresnel zone contribution.

point P on the bottom surface of a crystalline plate or at a small distance from this surface (Figure 6.2a). The wave scattered at P is then no longer a plane wave, but a spherical wave. This corresponds to the so-called *Fresnel diffraction*.

The foregoing amplitude and intensity formulation can easily be made consistent with finite distance diffraction by considering the so-called *Fresnel zones*, well known in conventional optics.

6.3.2.1 Fresnel Zones

Consider an elementary layer of thickness dx, parallel to the crystal plate. A plane wave incident along Ox generates scattered spherical wavelets at any points of the layer. At a point P ahead, at a distance x, the wavelet amplitudes sum up with phase shifts that increase with increasing distance of their sources from the axis.

The resulting amplitude at P can be represented by means of an amplitude–phase diagram. For this purpose the dx layer is divided into circular coaxial zones, so-called *Fresnel zones*, such that the limits of the successive zones from the axis correspond at P to a phase increment $\lambda/2$.

When considering $\lambda \ll x$, the path differences at a distance x and the corresponding radii of the Fresnel zones can be written (with n an integer)

$$\delta_1 = \frac{\lambda}{2} \Rightarrow r_1 = \sqrt{(\lambda x)}$$
$$\cdots\cdots\cdots\cdots\cdots \qquad (6.6)$$
$$\delta_n = \frac{n\lambda}{2} \Rightarrow r_n = \sqrt{(n\lambda x)}$$

This approximation is fully justified with high-energy electrons (λ order of 10^{-2} Å).

6.3.2.2 Amplitude

The diagram expressing the amplitude and phases at P, issued from zones of increasing radii, has the form of a spiral with a rapidly decreasing radius. Examining such a diagram (Figure 6.2b) leads to the following results:

1. The contribution of the successive Fresnel zones decreases rapidly and becomes generally negligible for $n > 10$.
2. The amplitude resulting at P from an unlimited plane layer equals half the amplitude due to the first Fresnel zone.
3. The amplitude resulting at P has a $\pi/2$ phase shift with respect to the incident wave.

The same result applies to waves from a *hkl* reflection, diffracted in a direction 2θ (small angle in the case of electron diffraction), due to the fact that these waves are in phase.

In order to calculate the amplitude at P, it is necessary to give the contribution of the first Fresnel zone of the dx layer. According to Eq. (6.6), its surface is $S_1 = \pi \lambda x$, its volume is therefore $v_1 = \pi \lambda x \, dx$. The arithmetic sum of its scattered amplitudes at unit distance, for a **g** reflection, is $v_1 F(\mathbf{g})/v_0$; this sum is represented on the Fresnel diagram by the length of the half-circle with a diameter equal to the amplitude scattered by the first zone. It follows the amplitude at P resulting from scattering by the dx layer

$$dG(\mathbf{g}) = \frac{i}{\pi x} v_1 \frac{F(\mathbf{g})}{v_0} = \frac{i\lambda}{v_0} F(\mathbf{g})$$

where the factor $1/x$ expresses variation versus x of spherical wave amplitude and the factor i indicates a $\pi/2$ phase rotation.

Finally the amplitude scattered at P, by a crystal plate of thickness t, for a reflection **g** corresponding to a small diffraction deviation **s**, can be written as follows (referred to incident amplitude and intensity taken as units:

$$G(\mathbf{g}) = \frac{i\lambda}{v_0} F(\mathbf{g}) \int_0^t \exp(2\pi i x s)\, dx \qquad (6.7)$$

6.3.2.3 Intensity

The corresponding intensity becomes

$$I(\mathbf{g}) = \frac{\lambda^2}{v_0^2} F^2(\mathbf{g}) \frac{\sin^2 \pi s t}{(\pi s)^2} \qquad (6.8)$$

Comparing with diffraction at infinite distance, the active surface S of the crystal in Eq. (6.5) has been replaced by $\lambda x = S_1/\pi$, where S_1 is the surface of the first Fresnel zone (taking into account the spherical wave amplitude factor $1/x$ which does not appear in the above relation).

6.3.3 EXTINCTION DISTANCE

It is common practice to introduce a parameter t_g called the extinction distance, defined as follows:

$$t_g = \frac{\pi v_0}{\lambda F(\mathbf{g})} \qquad (6.9)$$

The expressions of amplitude and intensity, scattered through a reflection **g**, can accordingly be written for **s** small

$$G(\mathbf{g}) = \frac{i\pi}{t_g} \int_0^t \exp(2\pi i\, xs)\, dx \qquad (6.10)$$

$$I(\mathbf{g}) = \frac{\pi^2}{t_g^2} \frac{\sin^2 \pi s t}{(\pi s)^2} \qquad (6.11)$$

For a given diffraction deviation s, the intensity given by this equation undergoes a sinusoidal variation of spatial period $1/s$. Through the kinematic formulation, the extinction distance represents that spatial period for a value of s such that the intensity has maxima equal to the incident intensity.

Basic Theory of Electron Diffraction

Table 6.1. Extinction distances computed for the main reflections of some elements crystallising in cubic F lattices. Metal elements show a continuous decrease of t_g with increasing atomic number (left to right) and with increasing intensity (bottom to top). Values of silicon show discrepancies for 111 and 311, due to the diamond type structure (in Hirsch et al [5])

Reflection hkl	Al	Si	Cu	Ag	Au
111	556	602	242	224	159
200	673		281	255	179
220	1057	757	416	363	248
311	1300	1349	505	433	292

This formulation fails near the exact diffraction conditions. For $s=0$ the intensity given by Eq. (6.11) becomes $I(\mathbf{g})=(\pi t/t_g)^2$; it should increase continuously with the thickness. This is denied by experience and is not physically viable, because it would mean the scattered intensity becoming greater than the incident intensity for $t>t_g/\pi$. In any case it contradicts the weak diffraction intensity hypothesis of the kinematic theory $I(\mathbf{g}) \ll I_0$. The condition $t<t_g/\pi$ can be used as a limit criterion for the kinematic theory to be valid.

Extinction distances can be calculated from Eq. (6.9). As an example, calculation for the 200 reflection of aluminium results in $t_g=673$ Å at 100 keV, which corresponds to a limit thickness $t_g/\pi=214$ Å. This illustrates the fact that the kinematic theory is suitable for very thin objects or for large diffraction deviations s (Table 6.1).

Note. Instead of integrating the scattered amplitude as a continuous function of x, it can be considered (Eq. 6.3) as a discontinuous sum over crystal layers of thickness a, with periodicity in the chosen direction. For a thickness $t=Ua$, the amplitude of Eq. (6.7) takes the following, equivalent, form:

$$G(\mathbf{g}) = \frac{i\pi a}{t_g} \sum_{u=0}^{u=U} \exp(2\pi i\, uas) \qquad (6.12)$$

6.4 DIFFRACTED AMPLITUDE AND INTENSITY. DYNAMIC THEORY

6.4.1 BASICS

It has been shown in the foregoing section that the kinematic approximation fails to apply for electron diffraction near the Bragg conditions ($s=0$), even for thin crystals commonly used in transmission electron diffraction and microscopy.

The dynamic theory takes into account the interaction between scattered and incident beams and the possibility of rediffraction of diffracted beams, leading to multiple diffraction.

The general theory considers the propagation of n waves in a periodic crystalline field and is based on quantum wave mechanics and the resolution of Schrödinger's equation.

84 Structural and Chemical Analysis of Materials

The *two-wave approximation* considers only the interaction between the transmitted beam and one strong diffracted beam. Also known as the *first-order approximation* (the kinematic theory being the zero-order approximation), it is based on conventional wave optics. It greatly simplifies the formulation and is useful in practice, e.g. for a semi-quantitative interpretation of observations in electron diffraction and electron microscopy. The following section will be limited to this approximation.

6.4.2 TWO-WAVE APPROXIMATION

Consider G_0, incident amplitude; $G(0)$, transmitted amplitude ($\mathbf{R}=0$); $G(\mathbf{g})$, diffracted amplitude ($\mathbf{R}=\mathbf{g}$). Consider also an elementary crystal layer of thickness dx.

Kinematic Approximation. The kinematic approximation (Figure 6.3a) takes the following into account:

(a) scattering of the incident wave $\quad (\mathbf{k}_0 \to \mathbf{k})$;
(b) transmission without interaction of the incident wave $\quad (\mathbf{k}_0 \to \mathbf{k}_0)$.

The variations of the diffracted amplitude $G(\mathbf{g})$ and the transmitted amplitude $G(0)$ versus x are written as follows, according to Eq. (6.10):

$$\frac{dG(\mathbf{g})}{dx} = \frac{i\pi}{t_g} G_0 \exp(2\pi i x s) \qquad (\mathbf{k}_0 \to \mathbf{k})$$

$$\frac{dG(0)}{dx} = 0 \qquad (\mathbf{k}_0 \to \mathbf{k}_0)$$

(6.13)

Dynamic Two-wave Approximation. The dynamic two-wave approximation takes the following into account:

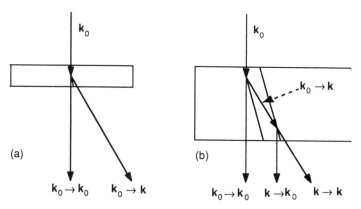

Figure 6.3. Outline of interactions between incident, transmitted and diffracted waves. (a) Kinematic approximation with a very thin object; (b) two-beam dynamic approximation with a thicker object.

Basic Theory of Electron Diffraction

(a) diffraction of the incident wave $(\mathbf{k}_0 \to \mathbf{k})$;
(b) transmission of the incident wave $(\mathbf{k}_0 \to \mathbf{k}_0)$;
(c) diffraction of the diffracted wave $(\mathbf{k} \to \mathbf{k}_0)$;
(d) transmission of the diffracted wave $(\mathbf{k} \to \mathbf{k})$.

These interactions lead to the following equations:

$$\frac{dG(\mathbf{g})}{dx} = \frac{i\pi}{t_0} G(\mathbf{g}) + \frac{i\pi}{t_g} G(0) \exp(2\pi i x s) \quad (\mathbf{k} \to \mathbf{k})(\mathbf{k}_0 \to \mathbf{k})$$

$$\frac{dG(0)}{dx} = \frac{i\pi}{t_0} G(0) + \frac{i\pi}{t_g} G(\mathbf{g}) \exp(-2\pi i x s) \quad (\mathbf{k}_0 \to \mathbf{k}_0)(\mathbf{k} \to \mathbf{k}_0)$$

(6.14)

In the kinematic theory the transmitted amplitude $G(0)$ was constant and equal to the incident amplitude G_0. Now, like the diffracted intensity $G(\mathbf{g})$, it becomes a function of the penetration path x.

The parameters t_0 and t_g have the dimensions of a length. They are inversely proportional to the amplitude scattered per unit volume respectively $f(0)$ for the incident direction and $f(\mathbf{g})$ for the diffracted direction.

This system of two first-order differential equations illustrates the mutual interactions of transmitted and diffracted waves and shows their complementarity. They have been established by Howie and Whelan (1961) as an analogy to the dynamic theory of X-ray diffraction put forward by Darwin in 1914 (in Hirsch et al [5]).

After proceeding to a change of variable and of phase origin and eliminating $G(\mathbf{g})$ and $dG(\mathbf{g})/dx$, the system (6.14) is reduced to one second-order differential equation as follows:

$$\frac{d_2 G(0)}{dx^2} - 2\pi i s \frac{dG(0)}{dx} + \frac{\pi^2}{t_g^2} G(0) = 0 \quad (6.15)$$

Introducing the limit conditions ($G(0) = G_0 = 1$ and $G(\mathbf{g}) = 0$ for $x = 0$), the solution of this equation can be expressed in terms of the diffracted intensity $I(\mathbf{g})$ and the transmitted intensity $I(0)$ when taking the incident intensity I_0 as a unit

$$I(\mathbf{g}) = \frac{\pi^2}{t_g^2} \frac{\sin^2 \pi s' x}{(\pi s')^2} \quad \text{with} \quad s' = \sqrt{(s^2 + 1/t_g^2)}$$

$$I(0) = 1 - I(\mathbf{g})$$

(6.16)

As can be seen, the diffracted intensity corresponds to that of the kinematic theory (Eq. 6.11) with the difference of the diffraction deviation s being replaced by a deviation parameter s'. This parameter tends to s when $s^2 \gg 1/t_g^2$ (kinematic approximation) and becomes equal to $1/t_g$ for $s = 0$ (exact diffraction condition).

In the exact diffraction condition, Eq. (6.16) can simply be written as a function of the depth x as

$$I(\mathbf{g}) = \sin^2 \frac{\pi x}{t_g}$$

$$I(0) = \cos^2 \frac{\pi x}{t_g}$$

(6.17)

The *extinction distance* now has its actual physical meaning, i.e. the periodicity $t_g = 1/s'$ of the sinusoidal intensity variation in the object as a function of the penetration depth, at the Bragg incidence.

Furthermore, this formulation takes into account the complementary aspect of transmitted and diffracted intensities, a fact which is experimentally observed in opposition to the kinematic theory. In this context, the observed intensity can be interpreted as resulting from complementary oscillations of the transmitted and diffracted waves. By analogy with the mutually induced oscillations of two joined pendula in mechanics, Ewald (in [1]) introduced the term *Pendellösung* (pendulum solution).

The physical significance of the above equations is apparent by comparison with the light dispersion theory. A double dispersion surface determines two slightly different wave vectors for the transmitted wave as well as for the diffracted wave. Interferences between those two waves induce beats with the same frequencies for the transmitted and for the diffracted beam, leading to the periodic intensity variations expressed by Eqs (6.16) and (6.17).

As a practical consequence of the non-validity of the kinematic theory, the diffracted intensities can no longer be directly related to the squared structure factors for the sake of crystal structure analysis. However, in the two-beam approximation, it can be shown that integrating the diffracted intensity as a function of s and averaging over the extinction distance leads to a measure which is proportional to the structure factor (instead of the squared structure factor in the kinematic theory).

A further important effect of the strong electron–matter interaction is the failure of Friedel's law in electron diffraction. As a result, the existence (or non-existence) of a centre of symmetry can be established by means of electron diffraction; the best experimental way seems to be through *convergent beam electron diffraction*.

The general dynamic theory is dealt with in specialised treatises on electron diffraction and electron microscopy, e.g. in Heidenreich [4] and Hirsch et al [5].

REFERENCE CHAPTERS

Chapters 2, 3, 5.

BIBLIOGRAPHY

[1] Ewald P. P. Die Erforschung des Aufbaues der Materie mit Roentgenstrahlen. *Handbuch der Physik*, **23** (2). Springer-Verlag, Berlin, 1931.
[2] Guinier A. *Théorie et technique de la radiocristallographie*. Dunod, Paris, 1964.
[3] Haymann P. *Théorie dynamique de la microscopie et diffraction électronique*. Presses Universitaires de France, Paris, 1974.
[4] Heidenreich R. D. *Fundamentals of electron microscopy*. John Wiley, New York, 1964.
[5] Hirsch P. B., Howie A., Nicholson R. B., Pashley D. W., Whelan M. J. *Electron microscopy of thin crystals*. Butterworths, London, 1965.

7

Secondary Emission

7.1 ATOMIC LEVEL EXCITATION AND RELAXATION

During any radiation–matter interaction process, a fraction of the total incident beam energy is transferred to atoms of the target material, thus increasing their potential energy by means of raising them from the ground state to an excited state. In the energy range considered, *excitation* concerns primarily the *core shells*; it therefore corresponds generally to an *ionisation*.

The excited state lifetime is a short one: atoms return to their ground state through a *relaxation* process. The excess potential energy is released as *secondary radiation*.

In the particular case of charged particles (electrons and ions), there is, furthermore, a direct conversion process of their kinetic energy to electromagnetic energy through a *braking effect*.

These interaction processes are interpreted by means of the corpuscular aspect of radiation, whereas the wave aspect predominates in the interpretation of scattering effects.

Excitation processes represent the main part of the inelastic radiation–matter interactions.

7.1.1 ATOMIC ENERGY LEVELS

The binding energy of electrons in an atom depends on three quantum numbers n, l and j (see section 1.5). A discrete *energy level* corresponds to any possible value of this set of numbers.

For a given sub-shell as defined by n and l, the number of energy levels is given by the possible values of the quantum number j.

Sub-shells s ($l=0$) admit one value $j=1/2$, thus leading to one energy level. Any other sub-shells ($l \neq 0$) admit two values whose difference is one, thus leading to a splitting into two energy levels. A given energy level entails $2j+1$ states, any of them accepting one electron. As a result:

1. The K-shell admits one energy level denoted K;
2. The L-shell admits three energy levels denoted L1, L2, L3;
3. The M-shell admits five energy levels denoted M1, M2, M3, M4, M5.

Table 7.1. Possible energy levels in the first three shells

Shell	n	Sub-shell	l	j	Number of states $2j+1$	Level notation	
K	1	s	0	1/2	2	K	1s
L	2	s	0	1/2	2	L1	2s
		p	1	1/2	2	L2	$2p^{1/2}$
		p	1	3/2	4	L3	$2p^{3/2}$
M	3	s	0	1/2	2	M1	3s
		p	1	1/2	2	M2	$3p^{1/2}$
		p	1	3/2	4	M3	$3p^{3/2}$
		d	2	3/2	4	M4	$3d^{3/2}$
		d	2	5/2	6	M5	$3d^{5/2}$

The corresponding energies are designated by W_K, W_{L1}, W_{L2}, etc.

An alternative quantum notation is often used. It expresses simply the three quantum numbers. As an example the K and L2 levels are respectively noted 1s and $2p^{1/2}$ (Table 7.1).

7.1.2 CORE LEVEL EXCITATION

For a core level to be excited, one of its electrons must be ejected into the continuum (*ionisation*) or be transferred to a vacant state in the valence band or in the conduction band (*excitation*), depending on the nature of the target material. The energetic balance of these processes is not significantly different (order of the eV) with respect to the energy involved (order of the keV). It is expressed experimentally by a fine structure in the corresponding spectrum.

For an X-ray level to be excited (X = K, L, M, etc), it must be provided with energy at least equal to the corresponding binding energy W_X (W_X expresses the potential energy of the excited atom, the ground state energy taken as unity). The energy is the greater the heavier the atom and the deeper the level X.

The energy W_X characterises both the level X and the atomic number of the element. Its values for any elements are tabulated (see Appendix D.2).

The excitation energy of a core level X can be supplied by an electromagnetic radiation or by a particle radiation, provided that the incident energy is greater than or equal to W_X, which is generally the case for X-rays and high-energy electrons.

Analytical data on target elements are provided by electrons ejected during the excitation process.

7.1.3 RELAXATION

The excited state is unstable, with a very short lifetime of about 10^{-16} s. The atom returns to its basic state through *electronic transitions*, noted XY: the vacancy created in the excitation process at an X level is occupied by an electron descending from a more external Y level. The resulting amount of energy $\Delta W = W_X - W_Y$ is released as an X *photon* or as an *Auger electron*.

The secondary relaxation radiations provide analytical information on the elements concerned.

7.2 SECONDARY RADIATIONS INDUCED BY EXCITATION

When an incident radiation (electromagnetic or particle radiation) with adequate energy impacts on an atom, an orbital electron of a core level may be ejected. It is imparted with a kinetic energy, the difference between the energy loss of the incident radiation and the electron binding energy.

When the incident energy is provided by an electromagnetic radiation (X-rays), the ejected electron is called a *photoelectron*.

When the incident energy is provided by a particle radiation (electrons, ions), the ejected electron is called a *secondary electron*.

The electrons issued from core levels can have sufficient energy to be able to excite themselves more external atomic levels, leading to the ejection of more secondary electrons.

Electrons with sufficient energy to leave their atom are progressively slowed down by successive interactions with other atoms of the target material. When arriving at the surface with a kinetic energy greater than the *work function* (energy required to remove an electron from a solid), they can leave the sample and be measured, thus serving for analytical purposes.

In the particular case of incident X-rays, the inelastic interaction of a photon with a weakly bound electron (*Compton effect*) results in a recoil electron and an inelastically scattered photon. This effect has a small cross-section for solids in the energy range considered, except for very light atoms. The recoil electrons have a low energy, depending on the emission angle (Eq. 2.4). Compton scattering provides information on liquids and gases, as well as data on the wave functions in solids.

7.2.1 PHOTOELECTRONS

Consider the primary excitation at an X-level of atoms through incident photons. These photons lose their whole energy $E_0 = h\nu_0$ during the interaction. As a result the initial kinetic energy of the ejected photoelectrons is given by

$$E = h\nu_0 - W_X \tag{7.1}$$

With a monochromatic incident beam, the photoelectrons acquire well-defined energies characterising their level of origin X. Before reaching the solid surface, they undergo energy losses through interaction with atoms, resulting in further excitations of more external levels and secondary electron emissions. Furthermore, to leave the sample, they must provide the extraction energy corresponding to the work function $e\Phi$ (e, electron charge, Φ, potential barrier of the order of several eV). Their energy, measured outside the sample, is therefore given by the following relation:

$$E' \leqslant h\nu_0 - W_X - e\Phi \tag{7.2}$$

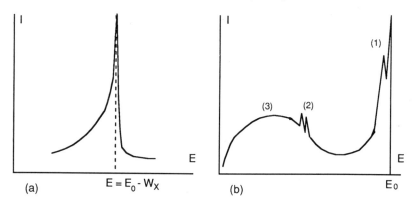

Figure 7.1. General outline of the electron emission spectrum $I(E)$. (a) Primary X-rays; photoelectron peak due to a given X-level; (b) primary electron beam: (1) elastic scattering and plasmon energy loss, (2) Auger peaks, (3) secondary electrons.

As a result, the measured energy will be dispersed within a dissymmetrical spectrum limited by a sharp peak at the high-energy end which corresponds to photoelectrons without energy loss (Figure 7.1a), i.e. with energy

$$E' = h\nu_0 - W_X - e\Phi$$

The photoelectron lines therefore correspond to electrons originating from atoms whose distances from the surface are smaller than the mean free path of the photoelectrons, i.e. coming from a very thin surface layer.

Measuring the energies of the photoelectron lines of an element leads to the direct determination of its energy levels, thanks to Eq. (7.1). Photoelectron lines are noted by the corresponding energy level, usually by means of the quantum notation (Table 7.1).

Photoelectron spectrometry is the basis of an important technique of surface analysis.

7.2.2 SECONDARY ELECTRONS

Consider primary excitation of atoms at an X-level through incident electrons. Before reaching the target atom, these electrons undergo aleatory energy losses. As a result, the emitted electrons have already from their origin a wide range of energies. Before eventually leaving the sample, they themselves undergo aleatory energy losses and generate on their way out further excitations of more external levels and secondary electron emissions. In the same way as the photoelectrons, in order to leave the sample they must provide the extraction energy. The secondary electron spectrum which finally results is therefore very broad, without any direct possibility of relating it to the chemical composition of the sample. As shown in Figure 7.1b, the overall electron spectrum consists of:

1. A relatively sharp peak due to *elastically or quasi-elastically scattered incident electrons* ($E = E_0$). Characteristic *plasmon excitation* energy losses in conducting samples lead to satellite peaks.

2. A broad maximum over a wide energy range due to the actual *secondary electrons*, with the greatest intensity at low energy, around 50 eV.
3. Small peaks barely emerging from the secondary electron background, due to *Auger electrons* in the given energy range.

Given their dominant low energy, detected secondary electrons are issued from a thin surface layer of a few ångströms thickness. The intensity of emission at a given point of a sample is a function of the surface topography, of the surface potential and of the atomic number.

Secondary electrons are not directly applied to spectrometric analysis, but they are used in the scanning electron microscope for imaging of material surfaces.

7.3 BREMSSTRAHLUNG

Any accelerated or decelerated charged particles are the sources of electromagnetic radiation. This is the case for electrons and ions. These charged particles travelling through matter are slowed down and therefore decelerate. The losses of kinetic energy occur progressively, through aleatory interactions with individual atoms. The electromagnetic radiations emitted therefore form a continuous frequency spectrum. This polychromatic or white radiation is called *braking radiation*; it is more commonly designated by the original German expression *bremsstrahlung*.

According to the law of energy conservation, the emitted frequencies have an upper limit, i.e. the wavelengths have a lower limit λ_{min}, corresponding to the incident particle losing its whole energy $E_0 = eV_0$ in one single interaction process. This is expressed by the following relations:

$$h\nu \leqslant E_0 \Rightarrow \nu \leqslant \frac{E_0}{h} \qquad \lambda \geqslant \frac{hc}{E_0} \tag{7.3}$$

Substituting universal constants h, c, e by their values leads to the following practical relation:

$$\lambda \geqslant \frac{12\,400}{E_0} \quad (\lambda \text{ in Å}; E_0 \text{ in eV}) \tag{7.4}$$

At its upper limit, this relation is identical to the equivalence relation between energy and wavelength (Eq. 1.4).

Particular Instance of Incident Electrons. For incident electrons in the energy range of about 1 keV to several MeV, the bremsstrahlung spectrum consists of X-rays (Figure 7.2). The form of the spectrum is related to electron and X-ray absorption in the target sample.

The *minimum wavelength* depends only on the incident energy. In particular, it is independent of the chemical composition of the sample (Figure 7.2a). The wavelength at maximum intensity is about $3/2\,\lambda_{min}$.

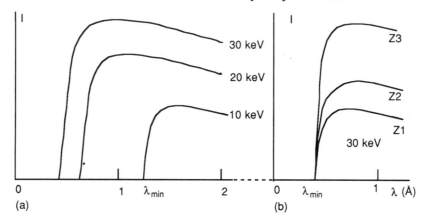

Figure 7.2. General aspect of the continuous X-ray spectrum produced by incident electrons. (a) For a given element Z; incident energies 10, 20, 30 keV; the minimum wavelengths calculated from Eq. (7.4) are respectively 1.24, 0.62 and 0.41 Å; (b) for a given incident energy of 30 keV; elements of atomic numbers $Z_1 < Z_2 < Z_3$.

The total integrated intensity is expressed by the measure of the surface limited by the graph of the $I(E)$ spectrum. It is proportional to the square of the incident energy and to the average atomic number of the target material.

The bremsstrahlung *emission yield* (fraction of the continuous emission spectrum energy over the incident electron energy) is approximated by the following empirical relation:

$$\eta = 11 \times 10^{-10} Z E_0 \quad (E_0 \text{ in eV}) \tag{7.5}$$

In the instance of a tungsten target ($Z = 74$) and 100 keV incident electrons, the continuous emission spectrum yield is 0.8%, whereas chromium ($Z = 24$) in the same conditions would give 0.26% (Figure 7.2b).

An important application of the braking emission is the production of polychromatic X-rays in X-ray tubes and in synchrotron facilities. In spectrometric techniques, bremsstrahlung is mostly a negative effect which enhances the background.

7.4 SECONDARY RADIATIONS INDUCED BY RELAXATION

The excited atoms return to their ground state through a process of cascades of electron transitions. The stored energy is released as secondary radiations which are characteristic of the sample elements.

This relaxation occurs following two different and competitive processes:

1. *Radiative mode* through emission of electromagnetic radiations (characteristic X-rays);
2. *Non-radiative mode*, through emission of electrons (characteristic Auger electrons).

7.4.1 CHARACTERISTIC X-RAYS

In the first relaxation mode, the energy $\Delta W = W_X - W_Y$ released by an XY transition is directly emitted as a photon of energy $E = \Delta W = h\nu$, of frequency $\nu = \Delta W/h$, of wavelength $\lambda = ch/\Delta W$. This mode of transition may be called a *radiative transition*. When core levels are involved, it results in emitting X-rays which are characteristic of the atom and of its energy levels.

The resulting radiation is often called *primary X-ray radiation* if the incident radiation is corpuscular (as in X-ray tubes) and secondary X-ray radiation or *fluorescence X-ray radiation* when the incident radiation is itself electromagnetic.

7.4.1.1 Permitted Transitions and Notation

Notation of Transitions

A given $Y \rightarrow X$ transition is commonly noted XY. An alternative notation is the quantum notation, but this is less frequently used for X-rays; in this notation a KL2 transition is written $2p^{1/2}$–s.

Selection Rules

All transitions corresponding to a combination of any two energy levels are not observed. Quantum mechanics leads to the following *selection rules* as a function of the quantum numbers n, l, j:

$$\Delta n \geqslant 1 \quad \Delta l = \pm 1 \quad \Delta j = 0 \text{ or } \pm 1 \tag{7.6}$$

Example of permitted transition: KL3 ($\Delta n = 1$; $\Delta l = 1$; $\Delta j = 1$); example of forbidden transition: KL1 ($\Delta l = 0$).

In fact the forbidden transitions correspond to a low probability; they may be observed under given conditions, but the related emission lines are very weak (e.g. copper KM4–5, Figure 7.3 and Table 7.3).

Line Notation

Emission lines resulting from excitation of a given level X make up the X-series. All the lines of a series are emitted if the incident radiation has an energy such as $E_0 \geqslant W_X$; at the same time, the series corresponding to less energetic levels appear *a fortiori*. In particular, for $E_0 > W_K$, any lines of the series K, L, M, etc. are emitted. In the vicinity of characteristic lines, weak satellite lines can sometimes be observed; they result from transitions in atoms undergoing a double excitation (see Auger emission). The excitation lifetime being very short, the probability of such transitions is low, hence the small line intensity.

In X-ray spectrometry the *Siegbahn notation* is most commonly used. In a given series, lines are noted by means of Greek indexed letters, according to decreasing intensities, with some exceptions (Figure 7.3).

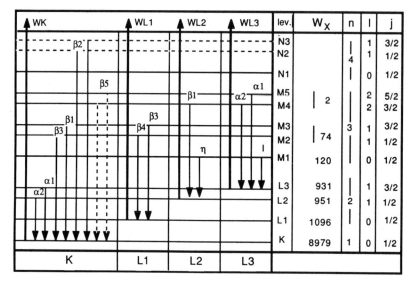

Figure 7.3. Energy levels in copper. Transitions in the K- and L-series.

Examples: Siegbahn notations in the K series:

$$KL3 \Rightarrow K\alpha_1 \qquad KM3 \Rightarrow K\beta_1$$
$$KL2 \Rightarrow K\alpha_2 \qquad KM2 \Rightarrow K\beta_3$$

Notation for a given emitting element: Cu-$K\alpha_1$ (copper); Mo-$L\beta_3$ (molybdenum), etc.

Two lines corresponding to levels which differ only by the value j (same n and l) make up a so-called *spin doublet*. The K-series for instance includes the doublets $K\alpha_1$, $K\alpha_2$ and $K\beta_1$, $K\beta_3$. For a given doublet, the small energy difference (i.e. small wavelength difference) is about constant for any emitting element; as an example, the wavelength difference between the $K\alpha_1$ and the $K\alpha_2$-lines is about 4×10^{-3} Å.

The values of characteristic wavelengths and energies for the whole of the elements are provided by the tables commonly used in X-ray spectrometry, e.g. in [7]. Some data are given in Appendix D.

7.4.1.2 Moseley Law

Given the inner level excitation, the X-ray emission lines are approximately unaffected by chemical bonds.

The frequency v of a given line in a given series is a function of the atomic number Z only, according to Moseley's empirical law.

$$\sqrt{v} = k_1(Z - k_2) \qquad (7.7)$$

where the constants k_1 and k_2 have definite values for a given line.

7.4.1.3 Line Intensities

Line intensities depend on various factors.

Excitation Cross-section

The excitation cross-section σ_x of a level X is a function of the nature of the incident radiation and of the *excitation ratio* (or *overvoltage ratio*) $U_X = E_0/W_X$.

Electron Excitation. Primary X-ray Emission. For a K-level excitation, the following empirical relation provides a fairly good agreement with experimental data

$$\sigma_K = k(E_0 - W_K)^{1.7} = k'(U_K - 1)^{1.7} \qquad (7.8)$$

In most applications it is of prime importance to achieve the highest ratio of the characteristic line intensity to the background due to bremsstrahlung. This ratio is a function of the excitation ratio and of the atomic number. As a guideline, for Kα-lines of moderately heavy elements, it has a maximum of the order of 10^2 when $U \cong 3\text{-}4$. It follows an optimal excitation energy of some 30 keV for the Cu–Kα-line emission ($W_K = 8979$ eV).

X Excitation. Secondary X-ray Emission. For a given level, the excitation cross-section is maximum when the incident energy is just above the required excitation energy W_X.

Probability of Radiative Relaxation

Given an electron vacancy at an inner level X, two relaxation modes compete. Let ω be the *probability of a radiative transition* giving rise to the emission of an X-ray photon. This probability is often called the *fluorescent yield*. The *probability of non-radiative transition* giving rise to the emission of an Auger electron is therefore $(1 - \omega)$, the total relaxation probability amounting to 1. The corresponding diagrams in Figure 7.4 show the fluorescent yield for the K-series to be small for light elements below $Z = 10$; it increases rapidly with Z and becomes nearly equal to 1 for heavy elements. The balance is situated at about $Z = 33$. For the L series, the equivalent diagram is shifted to the higher atom numbers. This emission yield is of great practical importance for the efficiency of spectrometric methods. Table 7.2 provides some data.

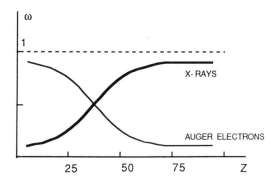

Figure 7.4. Outline of the variation of radiative and non-radiative relaxation probabilities vs atomic number for K levels.

Table 7.2. Values of fluorescent yields for some elements

Z	Element	ω_K	ω_L
4	Be	0.0004	
6	C	0.002	
13	Al	0.036	
24	Cr	0.282	0.03
29	Cu	0.445	0.056
42	Mo	0.764	0.067
92	U	0.976	0.52

Transition Probability

For a given element, the relative line intensities are proportional to the probability of the corresponding transitions. For the comparison to be relevant, it is limited to a given series, at its optimal excitation ratio. On the whole, the $K\alpha$-line intensities are of the order of 10 times the intensities of the $K\beta$ lines (Table 7.3). The K-series has the further advantage of being the simplest—whenever possible it is used in X-ray spectrometry. For analysing heavy elements the K excitation energy becomes exceedingly high, leading to the L-series which is more complicated due to three levels involved, L_1, L_2, L_3 (e.g. 115.6 keV for uranium).

7.4.1.4 Chemical Shift

Moseley's law is valid at first approximation. In fact even at inner levels the variation of the chemical bond energy in compounds, with respect to the pure element, leads to a shift δW of the energy levels which become $W_X + \delta W$ (with δW positive or negative, depending on the sample). The chemical shift is maximum for most external levels and decreases to the core levels. For a given level X the chemical effect is the more significant the smaller the number of shells which are screened from the chemical bonds,

Table 7.3. Wavelengths and relative values of intensities of the K and L emission lines of copper, with decreasing energy. The intensities of $K\alpha_1$ and $L\alpha_1$ lines are respectively equalled to 100 (in [7], ASTM)

Transition	Line	Relative intensity	Wavelength (Å)	Energy (eV)
KN2–3	$K\beta_2$	5	1.3810	8978.0
KM4–5	$K\beta_5$	1	1.3815	8975.7
KM3	$K\beta_1$	15	1.3922	8906.7
KM2	$K\beta_3$	15	1.3925	8904.8
KL3	$K\alpha_1$	100	1.5405	8049.3
KL2	$K\alpha_2$	50	1.5443	8029.5
L1M2–3	$L\beta_3$–$L\beta_4$	10	12.1154	1023.5
L2M4–5	$L\beta_1$	50	13.0794	948.0
L3M5	$L\alpha_1$	100	13.3569	928.3
L3M4	$L\alpha_2$	10	13.360	928.0
L2M1	$L\eta$	1	14.9401	829.9
L3M1	Ll	3	15.2968	810.6

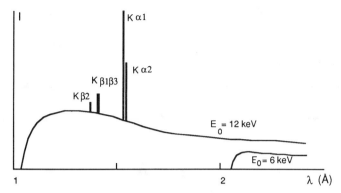

Figure 7.5. General outline of the emission spectrum of copper ($W_K = 8979$ eV) when excited by incident electrons. Main K-lines superimposed on the continuous bremsstrahlung background. For $E_0 < W_K$, the K-lines cannot appear; the L-lines are emitted, but are outside the scale of the figure (wavelength of the order of 12–15 Å). For $E_0 \geqslant W_K$, the emission conditions of the K-series are met. On this diagram the continuous spectrum is heavily exaggerated for better visibility and its K auto-absorption edge has not been featured.

i.e. the smaller the atomic number. For an XY transition the emitted energy represents the difference between the X and Y level energies; a partial compensation of the chemical line shifts can therefore occur, according to the following relations:

$$\text{Free atom:} \quad E = W_X - W_Y$$
$$\text{Bonded atom:} \quad E' = E + (\delta W_X - \delta W_Y) \quad (7.9)$$

In the X-ray range the chemical effect is virtually non-existent on the K-spectra of medium and heavy elements. It can be stated with fair accuracy that those spectra are characteristic of a given element, whatever its chemical bonding.

The chemical effect is more significant on the L- or M-spectra and for light elements (Table 7.4). The chemical effect, however small, must be taken into account in any highly accurate quantitative analysis by means of X-ray fluorescence or electron microprobe.

7.4.1.5 Practical Use

The electron-induced emission of characteristic and continuous X-rays supplies the most common *X-ray source*. The characteristic line spectrum is used for *X-ray microanalysis* (electron microprobe).

Table 7.4. Chemical shift of levels K, L2–3 and of the Kα-line of sulphur, depending on the oxidation degree (from [6], Siegbahn)

Structure	Oxidation degree	δW_K (eV)	δW_{L2-3} (eV)	$\Delta(K\alpha)$ (eV)
S_8	0	0	0	0
Na_2S	-2	-2	-1.4	-0.6
Na_2SO_3	$+4$	$+4.5$	$+3.6$	$+0.9$
Na_2SO_4	$+6$	$+5.8$	$+5.8$	0
$Na_2S_2O_3$	-2	-1.7	-1.3	-0.4
	$+6$	$+5.3$	$+4.7$	$+0.6$

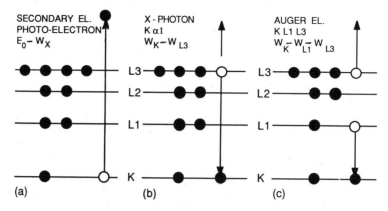

Figure 7.6. Excitation and relaxation. (a) Excitation of a level K; (b) relaxation through a radiative transition KL3 leading to the emission of the X-ray line $K\alpha_1$; (c) relaxation through a non-radiative transition KL1L3 leading to the emission of an Auger electron.

The X-ray-induced emission provides an identical characteristic spectrum, but without continuous background. This secondary emission is the basis of *X-ray fluorescence spectrometry*. Some weak continuous background is still observed in this case, due to the bremsstrahlung induced by the secondary electron emission processes (Auger electrons, photoelectrons, recoil electrons).

In both techniques, the result is primarily an elementary analysis, due to a chemical effect which is generally too weak to characterise the bonds.

7.4.2 CHARACTERISTIC AUGER ELECTRONS

In the alternative mode of atomic relaxation, the energy ΔW generated by way of an XY transition induces the ejection of a second orbital electron from a Y' level with a bonding energy smaller than ΔW. The excess energy reappears as kinetic energy of the ejected electron. This transition without photon emission, discovered by Auger in 1925 (hence the designation *Auger transition*), is often called a *non-radiative* (or *radiationless*) *transition*. In addition to the orbital electron ejected during the excitation process (photoelectron or secondary electron), a second electron, the so-called *Auger electron*, is emitted during the relaxation process.

An Auger transition involves three energy levels:

1. Excitation level X, origin of the photoelectron or the secondary electron leaving an electron vacancy;
2. Level Y, origin of the transition electron filling the vacancy;
3. Level Y', origin of the Auger electron.

This transition is noted XYY' (e.g. KL1L3, see Figure 7.6).

7.4.2.1 Auger Electron Energy

The kinetic energy of the Auger electron equals the difference between the energy released through the XY transition and the second ionisation energy $W'_{Y'}$ of level Y'. For an atom Z this energy can be written approximately as

$$W'_{Y'}(Z) = W_{Y'}(Z+1)$$

where $W_{Y'}(Z+1)$ represents the first ionisation energy Y' of the element $(Z+1)$.

For the Auger electron to be able to leave a solid sample in order to be detected, it has in addition to surmount the surface potential barrier Φ characterised by the *work function* $e\Phi$. The energy of the Auger electron (XYZ) as measured outside the sample then becomes

$$E = W_X - W_Y - W'_{Y'} - e\Phi \cong W_X - W_Y - W_{Y'}(Z+1) - e\Phi \quad (7.10)$$

The above equation is valid solely for electrons issued in the sample at a maximum depth equal to its mean free path. The electrons generated at a distance from the surface greater than this *escape depth* undergo energy losses before reaching the surface. The Auger electron energy is commonly in the low-energy range (Table 7.5) leading to a mean free path of the order of a few ångströms (see Figure 8.2). As in the case of the photoelectrons, the Auger spectrum can therefore characterise the sole atoms within a thin surface layer.

Auger spectrometry is a technique of surface analysis complementary to photoelectron spectrometry.

7.4.2.2 Permitted Transitions

The selection rules of radiative transitions do not apply to Auger transitions. The condition for an Auger transition to take place is for the energy expressed by Eq. (7.10) to be positive, which is expressed by the following relation:

$$W'_{Y'} < W_X - W_Y \quad (7.11)$$

Example. The KL1M1 transition is permitted whereas the KL1 radiative transition is forbidden ($\Delta l = 0$).

Particular transitions such as $\Delta n = 0$ are called *Koster–Kronig transitions*. In those transitions, impossible at the K level, the generated transition energy $(W_X - W_Y)$ is small (e.g. transitions L1L2M1 and L1L3M1 in Table 7.5).

Table 7.5. Non-radiative transitions in aluminium and related Auger electron energies in eV (Yasko and Whitmayer 1971)

Transition	Energy (eV)	Transition	Energy (eV)
KL1L1	1293	KL3M1	1479
KL1L2	1342	KM1M1	1551
KL1L3	1343	L1L2M1	36
KL1M1	1434	L1L3M1	37
KL2L2	1386	L1M1M1	109
KL2L3	1387	L2M1M1	65
KL2M1	1478	L3M1M1	64
KL3L3	1388		

The frequency of occurrence of Auger transitions relative to the primary excitation of s states ($l=0$) appears to be particularly high.

The number of XYY' combinations increases rapidly with increasing atomic number, leading to more and more complicated Auger spectra which are often difficult to interpret. Significant chemical shifts result in further difficulties for line identification. In addition the signal to background ratio is generally small. These drawbacks have limiting effects on the practical use of Auger spectrometry.

7.4.2.3 Line Intensities

For a given XY transition, the Auger emission intensity is complementary to the related X-ray emission intensity, due to the competition between both transition modes (Figure 7.4). The probability of Auger transition varies inversely with the electron bonding energy and accordingly increases with decreasing atomic number. The Auger emission yield is therefore important for light elements for which the fluorescent yield is small.

7.4.2.4 Effect on the X-ray Spectrum

The multiple ionisation effects due to Auger transitions affect the X-ray spectrum by changing the relative intensities of X-ray lines and by generating satellite lines caused by energy level shifts. The coexistence of Auger transitions decreases the lifetime of excited states and therefore broadens X-ray lines with respect to their theoretical width calculated for the sole radiative transitions, according to the uncertainty principle. This effect is notably significant for light elements which provide a high Auger yield; it adds to the influence of the valence band to broaden X-ray lines.

7.4.2.5 Chemical Shift

The Auger electron energy is directly related to the energy of its level of origin Y' which may in addition be relatively external. It is therefore much more sensitive than the corresponding characteristic X-ray energy to the influence of chemical bonding. In the instance of sulphur emission lines outlined in Table 7.4, the energy difference of the KL1L1 Auger lines of the two sulphur atoms in $Na_2S_2O_3$ (oxidation degrees respectively -2 and $+6$) reaches 5 eV, five times the difference of energy of the $K\alpha$ X-ray lines (in [6], Siegbahn).

REFERENCE CHAPTERS

Chapters 1, 2. Appendix D.

BIBLIOGRAPHY

[1] Adler I. *X-ray emission spectrography in geology*. Elsevier, New York, 1966.
[2] Bertin E. P. *Principles and practice of X-ray spectrometric analysis*. Plenum Press, New York, 1975.

[3] Burhop E. H. S. *The Auger effect and other radiationless transitions.* University Press, Cambridge, 1952.
[4] Compton, A. H., Allison S. K. *X-rays in theory and experiment.* Van Nostrand, New York, 1954.
[5] Chattarji D. *The theory of Auger transitions.* Academic Press, London, 1971.
[6] Siegbahn K. *E.S.C.A.—Atomic, molecular and solid state structure studied by means of electron spectroscopy.* Almquist-Wicksells, Uppsala, 1967.
[7] *X-ray emission line wavelength and two theta tables.* ASTM Data series DS 37. Philadelphia, 1965.

8
Absorption of Radiations in Materials

In the preceding chapters the physical processes of energy transfer between radiation and matter have been detailed. According to the conservation principle the total amount of energy gained by matter is exactly compensated by the total amount of energy lost by the transmitted beam during interaction. The overall result is absorption: the transmitted intensity is necessarily smaller than the incident intensity. Depending on the interaction mode (elastic or inelastic) the wave particles may in addition undergo energy losses.

The main part of the energy transferred to matter is finally released as secondary radiation. The absorption effects are therefore in some way complementary to the secondary emission effects.

8.1 X-RAY ABSORPTION

8.1.1 GENERAL ABSORPTION LAW

Ionisation is the predominant X-ray absorption process in a solid. The scattering component (elastic or Compton) is small and can be neglected.

An incoming photon loses is whole energy when exciting or ionising an atomic level. The transmitted X-ray photons which have escaped the photoelectric effect have therefore conserved their energy. This results in a simplified formulation of the absorption effect. The relative intensity variation is proportional to the mass dp of matter contained in a unit cross-section of thickness dx according to the following relations (Figure 8.1a):

$$I = I_0 \exp(-\mu p) = I_0 \exp(-\mu \rho x) \tag{8.1}$$

where μ is the mass absorption coefficient, $\mu\rho$ the linear absorption coefficient, and ρ the density. The dimension of the mass absorption coefficient μ is $L^2 M^{-2}$ (surface/mass) and is generally expressed in cm^2/g.

In order to take into account the two absorption effects, the absorption coefficient μ can be written as the sum of two terms, the *ionisation absorption coefficient* τ

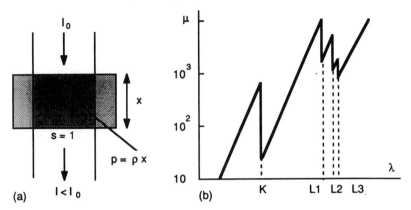

Figure 8.1. X-ray absorption. (a) Definition of involved parameters; (b) absorption as a function of wavelength (logarithmic coordinates).

Table 8.1. Approximate values in cm²/g of the respective absorption coefficients τ and σ (from [5])

Element		$\lambda = 0.41$ Å	$\lambda = 1.24$ Å
C	σ	0.2	0.4
	τ	0.05	2.2
Cu	σ	0.4	1.3
	τ	10	220

(also called true absorption coefficient) and the *scattering absorption coefficient* σ. In the X-ray range, except for light elements, the first is largely predominant as shown in Table 8.1. It will be neglected in the following parts.

8.1.2 ABSORPTION EDGES

Absorption increases rapidly with increasing atomic number and with increasing wavelength. However, this variation shows discontinuities (Figure 8.1b). For a given element these discontinuities in terms of wavelength, called *absorption edges*, correspond to the wavelengths λ_X associated with the electron energy levels W_X in the excited atoms. When the incident energy increases (the wavelength decreases), it reaches successively the critical excitation energies W_X where the excitation cross-section (and accordingly the absorption coefficient) increases abruptly. Between two discontinuities the variation of the absorption coefficient can be expressed by means of the empirical *Bragg–Pierce power law* as follows:

$$\mu = k Z^\alpha \lambda^\beta \qquad (8.2)$$

where $a \cong 4$, $\beta \cong 3$; k is a constant, changing after each absorption edge.

Measuring the edge positions leads theoretically to the energy levels of the constituent atoms. In fact the absorption spectrum does not have the simplified form depicted in Figure 8.1b; it shows a *fine structure* on and around each edge. For a given edge, the

Table 8.2. Some mass absorption coefficients in cm²/g. The example of copper illustrates the relatively small absorption of its own X-ray emission

Z	Element	μ (cm²/g) Cu–Kα 1.54 Å	μ (cm²/g) Mo–Kα 0.71 Å
4	Be	1.5	0.3
5	B	2.39	0.39
13	Al	48.6	5.16
29	Cu	52.9	50.9
82	Pb	232	120

fine structure is due to the interaction of the correspondingly emitted photoelectrons with the outer cores of the excited atoms and with the neighbouring atoms. It is applied in EXAFS spectrometry (*extended X-ray absorption fine structure*). A further application of the absorption edges is the filtering of X-rays generated in X-ray tubes.

8.1.3 ABSORPTION ADDITIVITY

Consider a compound of elements J (J = A, B, C, . . .). The mass absorption coefficient μ of the compound is approximately independent of its chemical state and is therefore additive with respect to the elementary mass absorption coefficients μ_J. This is expressed in the following equation:

$$\mu = \sum_J \mu_J c_J = \mu_A c_A + \mu_B c_B + \ldots + \mu_J c_J \tag{8.3}$$

Values of mass absorption coefficients for all elements and for different X-ray wavelengths are provided by tables. Quantitative analysis by means of X-ray spectrometry demands an increasingly accurate knowledge of the absorption coefficient (Table 8.2). It is often useful to determine the thickness of a material providing a given X-ray attenuation, e.g. X-ray protection screens, X-ray filters, X-ray tube windows. Equations (8.1) and (8.3) enable such thickness to be calculated when the composition is known (Table 8.5).

8.2 ELECTRON ABSORPTION

Electron absorption is far stronger than X-ray absorption. The main absorption effects are elastic and inelastic scattering. A further difference from X-ray absorption is that electrons undergo energy losses as well as intensity losses. During their travel through matter, electrons are subjected to multiple scattering effects accompanied by aleatory energy losses, along zigzag paths until they are finally stopped. Computer simulation based on interaction probability calculations (*Monte-Carlo method*) enables the visualisation of individual electron paths (see Figure 8.3b).

The total absorption cross-section is much greater for electrons than for X-rays. The corresponding mean free paths vary inversely (Figure 8.2).

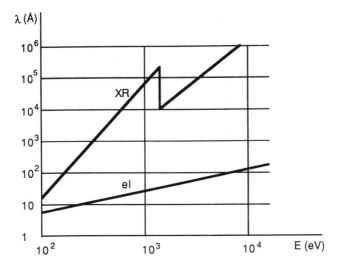

Figure 8.2. Order of magnitude of the mean free path of electrons and of X-rays in aluminium. Notice the K edge of Al at 1560 eV (from Carlson cited in [3]).

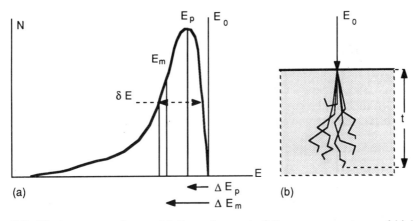

Figure 8.3. Electron energy losses. (a) General aspect of the energy spectrum of high-energy electrons transmitted through a metal foil which has a thickness of the order of the electron mean free path; (b) electron trajectories in a material; the length s of electron paths is greater than the average penetration thickness t.

8.2.1 ENERGY LOSS

Energy loss measurements in an electron beam are carried out by means of energy filters. The electron energy spectrum transmitted by a thin foil has an asymmetrical outline as illustrated in Figure 8.3a.

The *average electron energy* E_m results from integrating the function $N(E)$, the number of electrons transmitted with a given energy E; it corresponds to the *average energy loss* $\Delta E_m = E_0 - E_m$.

The *most probable electron energy* E_p corresponds to the maximum of the spectrum; it determines the *most probable energy loss* ΔE_p.

The asymmetry results in the relations $E_m < E_p \Rightarrow \Delta E_m > \Delta E_p$.

A fairly accurate theoretical treatment of electron energy loss can be performed thanks to simplified interaction models. The classical formulation due to Bethe (1933) is based on the interaction of the incident electrons with individual atomic electrons. It is written as follows, as a function of the electron velocity v and of the electron path s:

$$dE_m = \frac{4\pi e^4}{m_0 v^2} \left[\log_e \frac{1.16 m_0 v^2}{2 W_m} \right] \frac{NZ\rho}{A} ds \qquad (8.4)$$

where Z, A, ρ, W_m are respectively average atomic number, average atomic mass, density and average ionisation energy of the sample, N the Avogadro number and $NZ\rho ds/A$ the number of atoms per unit surface of thickness ds.

The Bethe relation applies to compounds through additivity relations with respect to the constituent elements, similar to Eq. (8.3), performed on $Z\rho/A$ and W_m.

The average ionisation energy W_m depends on the atomic number. For medium elements, the empirical relation W_m (eV) $= 11.5 Z$ (Springer, 1967) is generally satisfactory. More accurate variation laws, particularly for light elements, have been set up (Duncumb et al, 1969).

A non-relativistic writing of Bethe's relation is justified when $E < 100$ keV. Replacing the universal constants by their values leads then to

$$dE_m = \frac{7.84 \times 10^{10}}{E_m} \left[\log_e \frac{1.16 E_m}{W_m} \right] \frac{Z\rho}{A} ds \quad \text{(in cm, g, eV)} \qquad (8.5)$$

For a thin sample and high-energy electrons, the relative energy loss is small ($\Delta E_m \ll E_0$), the path length s can be assimilated in the average penetration thickness t and Eq. (8.5) simplifies to the following expression of the *average energy loss*:

$$\Delta E_m = \frac{7.84 \times 10^{10}}{E_0} \left[\log_e \frac{1.16 E_0}{W_m} \right] \frac{Z}{A} \rho t \quad \text{(in cm, g, eV)} \qquad (8.6)$$

The *most probable energy loss* ΔE_p, is given by the following relativistic equation established by Landau (1944):

$$dE_p = \frac{2\pi e^4}{m_0 v^2} \left[\log_e \left(\frac{4\pi e^4}{W_m^2 (1-\beta^2)} \frac{NZ\rho t}{A} \right) - \beta^2 + 10.37 \right] \frac{NZ\rho t}{A} \qquad (8.7)$$

where $\beta = v/c$.

The *half-width* δE of the energy spectrum, deduced from the foregoing relations, can be approximated as follows (Cosslett, 1969):

$$\delta E = \frac{8\pi e^4}{m_0 v^2} \frac{NZ\rho t}{A} = \frac{61.2 \times 10^4}{\beta^2} \frac{Z\rho t}{A} \quad \text{(in cm, s, g, eV)} \qquad (8.8)$$

The above laws of Bethe and Landau provide a useful order of magnitude of energy losses in solids (Table 8.3). They do not take into account plasmon excitation losses in conductors.

Table 8.3. Energy losses and half-widths in eV from the energy distribution spectrum of electrons transmitted through an aluminium foil of 100 Å thickness, as calculated by means of the Bethe and Landau equations

E_0 (eV)	β^2	ΔE_m	ΔE_p	δE
10^4	0.038	45.2	13.5	21
10^5	0.300	8.46	1.74	2.65
10^6	0.885	3.35	0.87	0.90

Computing energy losses is important for quantitative analysis by means of the electron probe as well as for interpreting images and diffraction patterns in the electron microscope.

The transmitted electron spectrum shows characteristic *energy loss edges* and a *fine structure* similar to the features observed in the X-ray absorption spectrum. Those effects are the basis of EELS (electron energy loss spectrometry).

8.2.2 INTENSITY LOSS

The relative intensity attenuation of an electron beam in the energy range 50–240 keV, transmitted in metal foils, neglecting the energy losses, can be approximated by the following empirical expression, set up as follows by Dupouy et al, 1964:

$$\frac{I}{I_0} = \exp\left[-\left(\frac{t}{CV_0^n}\right)^p\right] \tag{8.9}$$

with: $C = 3.33 \times 10^{-2}/\rho$; $n = 2.44\,(Z/A)^{1/2}$; $p = 1.51\,n/\log_{10} Z$, and where t is in μm, ρ in g/cm² and V_0 in keV. Data in Table 8.5 have been calculated by means of this relation.

8.3 NEUTRON ABSORPTION

Neutron–matter interaction is generally small in the range of thermal neutrons, and has an important proportion of scattering absorption. Absorption of neutrons is inversely

Table 8.4. Absorption cross-section and mass absorption coefficients (scattering excluded) of some elements for thermal neutrons of 1.08 Å wavelength (from [5])

Z	Element	σ_a (10^{-24} cm²)	μ (cm²/g)
4	Be	0.005	0.0003
5	B	430	24
13	Al	0.13	0.003
29	Cu	2.2	0.021
48	Cd	2650	14
64	Gd	19200	73
82	Pb	0.1	0.0003

Table 8.5. Thickness in cm of some metals providing a 99% attenuation of the incident beam. The density is indicated in g/cm³ below the element symbols

Radiation	4-Be 1.84	13-Al 2.70	29-Cu 8.93	82-Pb 11.34
Cu-Kα	1.67	0.035	0.0097	0.0017
Mo-Kα	8.3	0.33	0.010	0.0034
Thermal neutrons	8900	600	26	1430
Electrons 100 keV	39×10^{-4}	42×10^{-4}	11×10^{-4}	0.6×10^{-4}

proportional to their velocity, i.e. proportional to their associated wavelength. Mass absorption coefficients can also be defined as for X-rays. On the whole they are much smaller, barring some exceptions due to resonance effects, such as boron, cadmium and gadolinium; the last mentioned is a remarkable neutron absorber useful in the nuclear industry. There is no direct correlation with the atomic number (Table 8.4).

As a conclusion to this chapter, Table 8.5 compares for some elements, the thickness providing a 99% absorption of X-rays, electrons and neutrons.

REFERENCE CHAPTERS

Chapters 2, 4, 7, 10.

BIBLIOGRAPHY

[1] Bacon G. E. *Neutron diffraction*. Oxford University Press, 1975.
[2] Leroux J., Thinh T. P. *Revised tables of X-ray mass attenuation coefficients*. Claisse Scientific Corp., Quebec, 1977.
[3] Shirley D. A. (Ed). Electron spectroscopy. *Proc. Int. Conf. Asilomar, USA*. North-Holland, Amsterdam, 1972.
[4] Theisen R., Vollath D. *Tables for X-ray attenuation coefficients*. Stahleisen, Düsseldorf, 1967.
[5] *International tables for X-ray crystallography*. Vol. 3: *Physical and chemical tables* (MacGillavry C. H., ed.). D. Reidel, Kluwer Acad. Publ., Dordrecht, 1983.

Part Two

RADIATION GENERATION AND MEASUREMENT

Any facility designed for material analysis by means of radiation interactions requires two major components: (a) an adequate *radiation source*, (b) a radiation *measuring device*.

The methods for generating and measuring radiation are more or less similar in different techniques. In order to avoid numerous repetitions in the specialised chapters, they are reviewed together in Chapters 9 and 10.

The relevant techniques for X-rays, electrons and ions are commonly used in standard laboratories; they are therefore developed in some detail. Neutron production and measurement are localised in highly specialised facilities connected with nuclear reactors, and are therefore dealt with very briefly.

The basic principles of measurement apply in their entirety to any other radiation, e.g. radioactivity.

9

Sources of X-rays, Electrons, Thermal Neutrons and Ions

9.1 X-RAY SOURCES

9.1.1 X-RAY TUBES

The most commonly used X-ray sources are X-ray tubes in which the electromagnetic wave is generated through electron bombardment of a metal target. Depending on the kind of electron source, there is the widely used *hot cathode tube* (*Coolidge tube*), and the *cold cathode tube* (*Crookes tube*).

Synchrotron facilities provide a new X-ray source of high brilliancy.

9.1.1.1 Hot Cathode Tubes

The basic principle is illustrated in Figure 9.1a. Electrons are generated in a heated *filament* and accelerated in vacuum towards the *target anode* through a positive potential of some

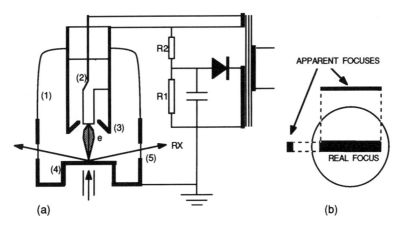

Figure 9.1. Hot cathode X-ray tube. (a) Basic diagram; (1) vacuum enclosure; (2) tungsten filament cathode; (3) Wehnelt; (4) water-cooled target anode; (5) exit windows. (b) Producing two apparent point sources and two apparent line sources.

10 kV with respect to the cathode. In order to facilitate target cooling through water, the anode is grounded and the cathode at the negative high voltage provided by a classical stabilised d.c. power supply. *Exit windows* made of thin sheets of vacuum-tight material with low absorption, mostly beryllium, transmit the X-ray beam towards the using facility.

Source Dimension. The required source dimension depends on the kind of application (Table 9.1).

In common devices, electron focusing on the target is provided by a simple electrode called a *Wehnelt*. Focusing is generally performed along a line source. Thanks to a small take-off angle (about 6°), two apparent point sources and two apparent line sources are achieved in two perpendicular directions (Figure 9.1b).

More sophisticated electron optical devices are required to produce micro sources (see electron guns).

X-ray Exit Windows. The hot cathode operating mode requires a high vacuum and therefore the presence of vacuum-tight windows. Too high an X-ray absorption in the windows prevents the use of this type of tube for generating long-wave X-rays (soft X-rays).

Special tubes with a frontal window and an annular cathode have been developed as high-intensity X-ray sources. The cathode is at high voltage, the window is grounded and therefore no longer subjected to heating through backscattered electron bombardment as in conventional tubes. In addition to a better spatial emission pattern, these tubes therefore accept very thin beryllium windows (125 μm), solely limited by their airtightness.

9.1.1.2 Cold Cathode Tubes

In the cold cathode tube, the electron source is provided by a discharge in a low-pressure gas. This Crookes tube was historically at the origin of the discovery of X-rays, but has been rapidly supplanted by the Coolidge tube which is easier to operate.

The discharge tube is nevertheless useful for specific applications. Operating at a low vacuum (order of 10^{-2} Torr), it can be directly connected without any window to a

Table 9.1. Maximum load of an X-ray tube with a copper target, as a function of the source size (from [3] Urlaub). The last two lines show the common characteristics of an electron microbeam on a sample. Notice the increase of the permissible load with decreasing source dimensions

Source size (mm)	Source surface (mm^2)	Total load (W)	Specific load (W/mm^2)	Utilization
0.05 × 7.5	0.375	500	1330	X-ray diffraction
0.4 × 8	3.2	1200	375	X-ray diffraction
0.75 × 10	7.5	750	100	X-ray diffraction
2.5 × 7.5	18.7	1400	75	X-ray diffraction
7 × 16	112	3000	27	X-ray fluorescence spectrometry
⌀ 10^{-3}	0.8×10^{-6}	4×10^{-3}	5×10^3	Electron microprobe
⌀ 4×10^{-6}	1.2×10^{-11}	4×10^{-6}	3×10^5	Scanning electron microscope

device placed in the same vacuum. It is therefore specially designed for soft X-ray facilities, up to 70 Å wavelength, by means of accelerating voltages of a few kilovolts. In addition a discharge tube is able to provide very short high-intensity X-ray pulses for analysing transient effects.

9.1.1.3 Target Choice

In order to avoid the local fusion of the target material, it must be a good heat conductor with a fairly high melting point, and in order to provide a simple spectrum, it must be a pure element. The choice is therefore limited to a few metals (Table 9.2).

X-ray Spectrum. The resulting spectrum is the addition of the continuous bremsstrahlung and of the target element's characteristic line spectrum. Depending on the application, the working parameters are set to enhance the first or the second.

Continuous spectrum. A heavy target element, excited at a voltage just below the K excitation limit, provides a strong and wide range continuous spectrum (e.g. tungsten at 69 kV).

Line spectrum. The K series is used. A medium element excited by means of an acceleration voltage of about three times the K excitation voltage provides the best intensity ratio of the line spectrum over the continuous spectrum (e.g. 27 kV for copper). The characteristic line wavelengths are a sole function of the element; in particular they do not depend on the voltage. Voltage stabilising is therefore useful only for quantitative analytical methods requiring precise intensity measures.

Power and Efficiency. The X-ray emission efficiency is only about 1%. All the remaining energy is released as heat in the target. The maximum admissible *power load* of a tube (product of the voltage and the electron beam intensity) is therefore mainly limited by the possible heat evacuation avoiding target melting. The maximum load depends on the target metal, the source size and the tube design (Table 9.1).

The maximum load is limited to some kilowatts in conventional X-ray tubes. Power values 10 times higher are possible with *rotating target tubes*, during a limited time-span.

Taking into account the 1% efficiency, the total radiation power emitted from an X-ray tube is therefore limited to about 100 W. In diffraction devices the narrowly collimated X-ray beam carries barely a small fraction of this power. An incomparably greater power is emitted from *synchrotron sources*.

Table 9.2. Characteristics of radiations and of filters for some common anode materials. Filter thicknesses are determined for a filtering ratio $I(K\alpha)/I(K\beta) = 100$

Radiation					Filter				
Element	Z	$K\alpha$ (Å)	$K\beta$ (Å)	W_K (keV)	Element	Z	W_K (keV)	λ_K (Å)	Thickness (µm)
Cr	24	2.29	2.08	5.989	V	23	5.465	2.070	11
Fe	26	1.94	1.76	7.114	Mn	25	6.539	1.896	11
Co	27	1.79	1.62	7.709	Fe	26	7.114	1.743	12
Cu	29	1.54	1.39	8.979	Ni	28	8.33	1.488	15
Mo	42	0.71	0.63	20.000	Zr	40	17.998	0.689	80
					(Nb)	41	18.986	0.653	

Absorption and Shielding. Devices using short-wave X-rays (hard X-rays) can be operated in the air, as in the instance of diffraction devices with the Cu–Kα or Mo–Kα emissions.

Operating with long-wave X-rays (soft X-rays, $\lambda > 2.5$ Å) requires the radiation path to be in vacuum.

For safety regulations to be met, the X-ray tube and the whole X-ray facility must be properly shielded by means of X-ray absorbing materials (e.g. lead, lead glass or barium glass).

9.1.1.4 Collimators

X-ray techniques often require a parallel beam. In the absence of optical devices for X-rays (except reflection focusing through crystals), the incident beam is limited by means of collimators.

A *collimator*, formed by two circular or slit-shaped diaphragms, provides a beam with a small cross-section and with an angular dispersion determined by the distance and the dimensions of the apertures. A terminal cone intercepts X-rays scattered by the exit aperture (Figure 9.2a).

Parallel absorbing tubes or sheets (*Soller slits*) provide a beam with a large cross-section; in the sheet device, an important angular spread remains in the planes parallel to the sheets (Figure 9.2b).

9.1.2 MONOCHROMATIC RADIATION

Monochromatic X-rays are needed in numerous applications. The ideal monochromatic radiation would be limited to one emission line, preferably Kα_1.

Two approaches are possible, one approximate through filtering (β-*filtering*), the second rigorous (*crystal monochromator*).

9.1.2.1 Absorption Edge Filtering

The great variation of X-ray absorption on both sides of an absorption edge is exploited in X-ray filtering. The so-called β-filter is a thin sheet of an element with the wavelength of its K absorption edge between the wavelengths of the Kα and the Kβ emission lines of the X-ray tube target element. The Kβ radiation has sufficient energy to excite the filter element K level; the Kα radiation has not. As a result, Kβ is much more absorbed than Kα.

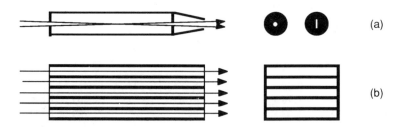

Figure 9.2. X-ray collimators. (a) Tube collimator for narrow beams; (b) sheet collimator for wide beams.

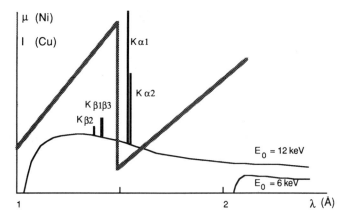

Figure 9.3. Basic principle of X-ray filtering. Nickel K absorption spectrum (tint) superimposed on the copper K emission spectrum (arbitrary scales).

For a target anode made of element Z, the accurate filter element is generally $Z-1$. For heavy elements, this rule is extended to $Z-2$. Thus the filter for the K radiation of molybdenum can be zirconium or niobium (see Table 9.2).

Example (Figure 9.3):

Copper anode: $\lambda_{K\alpha} = 1.54$ Å, $\lambda_{K\beta} = 1.39$ Å
Nickel filter: $\lambda_K = 1.49$ Å

The resulting spectrum contains mainly the $K\alpha$ doublet and the continuous spectrum attenuated around $K\beta$. The filtering efficiency $I(K\alpha)/I(K\beta)$ depends on the filter thickness; in order not to weaken $K\alpha$ excessively, the thickness is limited to the order of some 0.01 mm, leading to $I(K\alpha)/I(K\beta) \cong 100-500$, instead of 10 before filtering.

The advantage of β-filtering is its simplicity and cheapness. Its disadvantage is the persistence of the continuous background and of a weak $K\beta$ line which can be a hindrance in sensitive experiments.

9.1.2.2 Crystal Monochromator

The crystal monochromator is based on the selective reflection of the primary X-rays on a crystal plate cut parallel to a set of lattice planes (hkl), of equidistance $d(hkl)$. The crystal is set to meet the reflection condition for the $K\alpha$-line. According to Bragg's equation (Eq. 5.7), the corresponding incident angle θ is determined by

$$\sin \theta = n \lambda_{K\alpha} / 2 \, d(hkl)$$

For a strong reflection, the crystal, its plane set (hkl) and the reflection order n are chosen to result in a high structure factor $F(hkl)$.

In addition to $K\alpha$, the reflected beam in the 2θ direction includes the harmonics $\lambda_{K\alpha}/2, \lambda_{K\alpha}/3 \ldots$ contained in the continuous spectrum down to $\lambda_{\min}$. They are weak and generally not a nuisance.

116   *Structural and Chemical Analysis of Materials*

In optimum set-ups, the Kα₁ line may be selected from the Kα doublet.

Plane Monochromator

A plane crystal monochromator has the advantage of being easy to make and to adjust (Figure 9.4a). Cleavage plates such as (001) mica, LiF or graphite may be used. Its disadvantages are the low reflected intensity and the practical impossibility of resolving the Kα doublet. As an example, the angular separation of the second-order reflection of Cu–Kα on (001)-LiF ($a = 4.03$ Å) is about 3.6'; an incident beam collimated to such a small angular dispersion would have a very small intensity.

The plane monochromator is required in X-ray facilities demanding a parallel monochromatic beam with a large cross-section (e.g. X-ray fluorescence).

Focusing Monochromator

Focusing by means of curved crystals avoids the drawbacks of the plane monochromator. It provides a strong, relatively wide angle monochromatic beam (several degrees) with an excellent resolution of the Kα doublet.

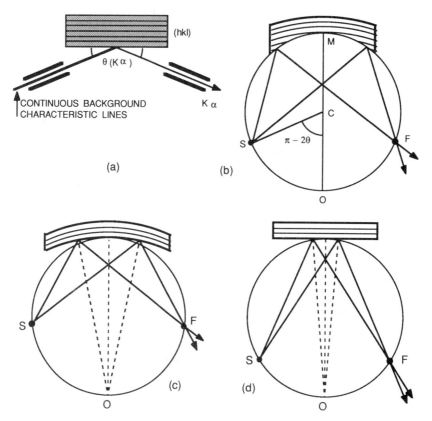

Figure 9.4. Principle of the crystal monochromator. (a) Plane monochromator without focusing; (b) Johansson symmetrical focusing monochromator; (c) Johann focusing monochromator; (d) Bragg–Brentano approximate focusing with a plane crystal.

Exact focusing. Johansson Monochromator. The *Johansson set-up* provides an exact stigmatic focusing (Figure 9.4b). A cylinder of radius $2R$ is ground into the crystal plate, tangential to the (hkl) reflecting planes. The resulting plate is bent against a cylinder of radius R. In the two-dimensional representation which is approximated in practice, the circle of radius R represents the *focusing circle* (called a *Rowland circle* in spectrometers). The focusing geometry is explained in Figure 9.4b. Any ray emitted from a point source S on the focusing circle C will be reflected on the (hkl) 'planes' and pass through the focus F. In the actual monochromator, S is the linear source of the X-ray tube and F is a focus line, the monochromatic source, both parallel to the bending axis.

With a fine-focus X-ray tube and a precision set-up, this monochromator is able to resolve the $K\alpha$ doublet and to provide a perfectly monochromatic $K\alpha_1$ radiation with an energy dispersion limited to the natural width of the spectrum line. The line width increases with decreasing atomic number, due to the increasing influence of the valence band (Table 9.3).

For an easier set-up, the *Guinier asymmetrical monochromator* uses reflecting planes which have an angular deviation of several degrees with respect to the crystal surface.

Focusing monochromators are mainly made of quartz which is easy to grind, polish and bend elastically, with the following characteristics: reflecting planes (10.0), $d = 3.34$ Å, $R = 25$ cm, aperture angle of about $3°$.

A fully focusing set-up according to Johansson provides the best available monochromatic X-ray source. It is applied in particular for highly accurate measurements of a lattice parameter. A spherical crystal provides a perfectly stigmatic monochromator.

Approximate Focusing. Johann Monochromator. A parallel-faced crystal plate is bent to a radius $2R$ and made tangential to a cylinder of radius R which materialises the focusing circle (Figure 9.4c). The effective aperture angle allowing the resolution of the $K\alpha$ doublet is about $1°$, smaller than in the Johansson device. Its lower performance is, however, balanced by a much easier method of manufacture, e.g. using elastically or plastically bent cleavage platelets. They are most commonly used as analysing crystals in X-ray spectrometers.

Approximate Focusing. Bragg–Brentano Set-up. In relation to a linear X-ray source, a plane crystal platelet provides an approximate focusing through a small aperture angle on a circle tangential to the crystal and intersecting the source (Figure 9.4d). This very simple geometry, known as the Bragg–Brentano set-up, is used in the X-ray powder diffractometer.

Table 9.3. Half-width δ and energy dispersion of some $K\alpha$ lines (from [1], Compton and Allison)

Element	Z	λ (Å)	δ (Å)	$\Delta E/E$
Ca	20	3.359	15×10^{-4}	4.5×10^{-4}
Cu	29	1.540	7.7×10^{-4}	5.0×10^{-4}
Mo	42	0.709	3.2×10^{-4}	4.5×10^{-4}
W	74	0.209	1.5×10^{-4}	7.1×10^{-4}

9.1.3 SYNCHROTRON RADIATION

Similar to the negative acceleration of electrons in an X-ray tube target the continuous centripetal acceleration of the electrons, travelling at high velocity in the storage ring of a synchrotron facility, induces the emission of polychromatic electromagnetic radiation. The resulting spectrum $I(\lambda)$ is similar to the bremsstrahlung spectrum. The wavelength at maximum intensity (*critical wavelength*) is given by

$$\lambda_c = 5.6\, R/E^3 \quad (R \text{ radius in m}, E \text{ energy in GeV})$$

Example of the ESRF (European Synchrotron Radiation Facility):

$$R = 22.36\,\text{m};\ E = 5\,\text{GeV} \Rightarrow \lambda_c = 1\,\text{Å}$$

The corresponding maximum photon energy is $E_0\,(\text{max}) \cong 50$ keV.

So-called *wigglers* are generating local high-intensity magnetic fields, resulting in trajectory bendings; their action multiplies the emission energy by a factor of 4. In the above-mentioned facility, it reaches up to 200 keV, which corresponds to hard X-rays.

Undulators submit the electrons to periodic transverse accelerations; this results in the emission of a quasi-monochromatic radiation of adjustable wavelength.

In an X-ray tube, the brutal stopping of electrons in the solid target results mainly in heat release; the radiation emission efficiency is less than 1%.

In a synchrotron facility, the continuous acceleration of electrons in vacuum leads to an emission efficiency of virtually 100%. In future ESRFs the total radiation output, in a continuous spectrum from infrared to X-rays, will be in the order of 1.5 MW. This output may be compared with the maximum effective 100 W output from a high-power rotating anode X-ray tube. This means a power increase of more than 10^4.

Moreover, the synchrotron X-ray beam is, interestingly, a well-collimated polarised beam of about 0.1 mm² cross-section at the exit, with a small angular dispersion of about 0.1 mrad (20″). Thanks to the resulting very high brightness, the improving factor with respect to a conventional X-ray tube of 10 kW is in the order of magnitude of 10^3–10^6, depending on the kind of experiment.

Numerous techniques have developed at present thanks to the availability of synchrotron sources (e.g. EXAFS, SEXAFS).

9.2 ELECTRON SOURCES

In order to obtain an electron beam of a given associated wavelength, electrons must be produced and then accelerated to the corresponding energy.

The whole electron-producing device is commonly called *electron gun*. Except for light sources, electron guns are certainly the most widely used radiation sources, for domestic applications (CRTs in TV sets, in computers, etc.) as well as for scientific applications (X-ray tubes, electron microscopes, electron microprobes, etc) and industrial applications (emission tubes, electron bombardment melting facilities, etc).

Whatever their energy, electrons are rapidly absorbed in air. Their production and their using attachments must therefore take place in secondary vacuum (max. pressure 10^{-4} Torr). Electrons cannot escape from vacuum enclosures; however, their impact on any material on their path (apertures, sample, detectors, etc.) becomes an X-ray source requiring an adequate protective shielding.

The main parts of an electron generation device are as follows:

Electron source: thermionic source or field emission source.

Electron optical focusing system: more or less sophisticated depending on the kind of facility.

Anode: At a positive voltage V_0 with respect to the cathode, it accelerates the electrons to an energy of $E_0 = eV_0$. It is stabilised in order to provide a monokinetic, i.e. a monochromatic, beam.

Main Characteristics of an Electron Gun

1. *Brilliance B*: intensity emitted per unit surface of the cathode in a unit solid angle; expressed in A/m^2 sr;
2. *Energy dispersion ΔE* of the generated electrons;
3. *Emission stability*.

9.2.1 THERMIONIC EMISSION

The hot cathode gun, based on *thermionic electron emission*, is the most commonly used electron source. When heating a metal at an adequate temperature in vacuum, some of its conduction electrons reach a kinetic energy which is high enough to cross the surface potential barrier Φ which amounts to a few electronvolts. An electron cloud then surrounds the metal surface. Applying an electric field through a voltage V_0 generates an electron beam of energy $E_0 = eV_0$.

The thermionic *emission density J* of a source is expressed by the *Richardson–Dushman* equation

$$J = A_0 T^2 \exp\left(\frac{-e\Phi}{kT}\right) \quad (9.1)$$

where $e\Phi$ is the work function, T the Kelvin temperature, k the Boltzmann constant, and $A_0 = 1.2 A/m^2 K^2$ e a universal constant.

The theoretical *brilliance B* of the source is proportional to J and to the voltage V_0, according to the *Langmuir* equation which may be written with a fair approximation

$$B = \frac{J}{\pi} \frac{eV_0}{kT} \quad (9.2)$$

According to the foregoing equations, the source is more brilliant the higher the temperature and the smaller the work function.

The *energy dispersion* of electrons at the source is due to the additional thermal energy $3/2 kT$ which is proportional to the temperature. According to the statistical electron velocity distribution following a Maxwellian law, 95% of the electrons have an energy in the interval $\Delta E = 3kT$. For a tungsten filament heated to 2800 °C, this leads to an energy dispersion of 0.8 eV, i.e. roughly 1 eV (Table 9.4).

Table 9.4. Average characteristics of electron sources at an accelerating voltage of about 50 kV

Type of cathode	Operat. temp. (°C)	Required vacuum (Torr)	Source brilliance (A/m² sr)	Energy dispersion (eV)
W-filament	2700	10^{-4}	5×10^8	1
LaB$_6$	1600	10^{-6}–10^{-7}	5×10^9–10^{10}	0.6
Field emission	Room temp.	10^{-9}–10^{-10}	10^{13}	0.2–0.4

Figure 9.5 outlines a thermionic gun configuration as it is commonly set up in an electron microscope or a similar facility.

In the most common electron guns the cathode is simply a hairpin tungsten filament heated in a secondary vacuum (10^{-4}–10^{-6} Torr), at 2500–2800 °C (limited by evaporation). The tungsten work function is about 4.5 eV. A brilliance of about 5×10^8 A/m² is reached with high-energy electrons.

Oxide-coated cathodes (ThO$_2$, SrO$_2$, BaO, CaO), or even better the *lanthanum hexaboride* cathode (LaB$_6$), provide a greater brilliance at lower temperatures, due to a smaller work function of 1.5–2 eV. However, those cathodes require a clean vacuum of at least 10^{-6}–10^{-7} Torr. Together with a higher brilliance, a reduced energy spread and a smaller source size, modern LaB$_6$ monocrystalline cathodes have a longer lifetime compared to the tungsten pin filament. Thanks to the currently improved vacuum techniques, they are increasingly used in electron microscopes (Table 9.4).

The geometrical features of the generated electron beam are dependent on the form and the respective positions of the cathode, of the focusing electrodes and of the anode. For practical reasons, a negative voltage V_0 is generally applied to the cathode and the anode is grounded. The Wehnelt (see X-ray tubes) is negatively biased with respect to the cathode; together with the anode it forms an electrostatic lens which focuses the electrons into a minimum cross-section called *cross-over* which acts as the actual electron source in the facility. The cross-over has currently a diameter of about 100 µm in a conventional electron microscope or in an electron microprobe.

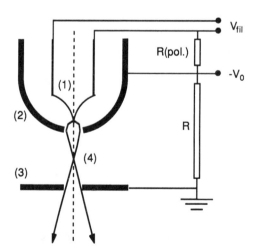

Figure 9.5. Basic diagram of a thermionic electron gun: (1) tungsten filament; (2) focusing electrode (Wehnelt) biased by the electron flow in R(pol); (3) anode; (4) cross-over.

9.2.2 FIELD EMISSION

When a metal surface is subjected to a very high electric field, of the order of 10^9–10^{10} V/m, electrons from the conduction band can cross the surface potential barrier through a tunnelling effect, even at room temperature. In order to produce the required field strength the source is a sharp-pointed tungsten cathode with a tip radius of some 100 Å, associated with an intermediate extraction anode (Wehnelt) at a positive voltage of several kilovolts.

The field emission gun requires a high vacuum of about 10^{-10} Torr in order to avoid instabilities due to residual gas ion bombardment of the tip. Even at high vacuum, contamination can lead to emission instabilities.

Compared to a conventional thermionic electron gun, the field emission provides a *brilliance* multiplied by some 10^4, an *energy dispersion* divided by five and a *source diameter* of about 100 Å (Table 9.4).

9.2.3 OPERATING CONDITIONS

9.2.3.1 Vacuum

Electron beam operation requires a vacuum ranging from 10^{-5} to 10^{-10} Torr, depending on the electron energy and on the electron source.

Vacuum is produced by a system of kinetic *vacuum pumps*, usually operating in cascade:

1. A *primary vacuum* of some 10^{-3} Torr is supplied by the fore-pump which works directly to the atmosphere. In some high-vacuum systems, the generally used *rotary pump* is sometimes replaced by a zeolite *cryosorption pump*.
2. A *secondary vacuum* down to 10^{-6}–10^{-7} Torr is commonly supplied by means of an *oil diffusion pump*. In order to prevent contamination effects, the residual hydrocarbon gas pressure is reduced through a liquid nitrogen cold trap. For the needs of an increasingly required clean vacuum, the diffusion pump is becoming supplanted by the *turbomolecular pump*; this pump works on the purely mechanical principle of moving the gas molecules to the exit pipe through a rapidly rotating cylinder.
3. A *high vacuum* down to 10^{-10}–10^{-12} Torr is supplied by an *ion pump*. In order to achieve an ultra-high vacuum below 10^{-10} Torr, great care must be taken to suppress any residual gas emitted by desorption from the various surfaces in contact with the vacuum, in particular the sample itself. Residual gas desorption may be achieved by heating.

9.2.3.2 Energy Dispersion

The electron-associated wavelength is $\lambda = 1.26 E_0^{-1/2}$ (Eq. 1.7). Any energy dispersion $\Delta E/E$ entails a wavelength dispersion

$$\Delta\lambda/\lambda = 1/2 \Delta E/E$$

Energy dispersion in an electron beam is due to instrumental and to physical causes.

Instrumental Causes. The main instrumental causes are energy dispersion in the source and high-tension supply instabilities.

By means of a stabiliser the accelerating voltage dispersion is reduced to a rate of typically 10^{-6} which makes it negligible against source dispersion.

Electrons from a conventional tungsten pin filament have a thermal energy dispersion of about 1 eV (Table 9.4). For commonly used 100 keV electrons, this leads to a relative energy dispersion $\Delta E/E = 10^{-5}$, hence to a wavelength dispersion $\Delta\lambda/\lambda = 5 \times 10^{-6}$. This highlights the importance of source temperature reduction through field emission. However, even for a tungsten hot cathode, the result is better than the spectral Kα-line dispersion of usual X-ray sources (Table 9.3). For low-energy electron beams, the result is less favourable.

Physical Causes. Aleatory electron energy losses in the sample result in an energy spread of the transmitted beam which depends on the one hand on the nature and the thickness of the sample, and on the other on the accelerating voltage. The spread can be reduced by means of *energy filtering* using electromagnetic or electrostatic prisms (e.g. energy filtering in electron microscopes).

9.2.3.3 Focusing of Electron Beams. Electron Optics

Electron beam facilities for material analysis mostly operate with small-sized electron probes on the sample. For this, a reduced image of the source cross-over is projected on the sample by means of a system of electronic lenses. Electrons are deviated through a magnetic or an electric field. Therefore either electromagnetic or electrostatic lenses can be used for focusing.

Fine focusing is limited by electron lens defects which are similar to those of conventional light lenses.

General Characteristics

The focusing systems of electron microscopes and electron microprobes consist mainly of current-stabilised electromagnetic lenses, whereas in electron and ion spectrometers they use mostly electrostatic lenses (Figures 9.6 and 9.10).

Like their conventional counterparts, electron lenses are characterised by various quantities such as focal length, aberrations, etc.

The *convergence* (reciprocal of the *focal length*) of an electromagnetic lens is given by

$$\frac{1}{f} = \frac{k}{E_0} \int_{gap} H_z^2 \, dz \qquad (9.3)$$

where E_0 is the incident energy, k the constant, and H the magnetic field component along the z-axis.

The magnetic field, and hence the convergence, vary continuously with the excitation current. Focusing of the lens is therefore very easy to operate, like zooming with a conventional lens.

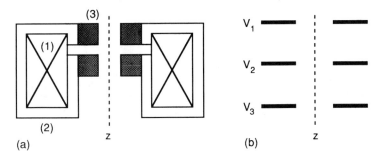

Figure 9.6. Basic outline of electronic lenses. (a) *Electromagnetic lens*. An axial magnetic field is generated by a coil (1) and a magnetic circuit (2). Electron focusing takes place in the gap of the soft iron pole pieces (3). Electrons travel through a magnetic field along a helicoidal path, thus resulting in a rotation of the electron propagation plane which increases with the lens current. (b) *Electrostatic lens*. The form and position of the coaxial electrodes vary according to the kind of source and the operating mode. A common lens consists of two or three electrodes. In the last case, outlined in the figure, the lens may be of the symmetrical type (*Einzellinse*, $V_1 = V_3$) or the asymmetrical type (*immersion lens*, $V_1 \neq V_3$). The respective electrode voltages depend on the purpose of the lens.

The theory of electron optics is found in various treatises (e.g. [5] Glaser, [6] Grivet).

Aberrations

Aberrations which do not cancel on the optical axis (*axial aberrations*) are the most important ones, due to their limitation effect on resolution. They are the *spherical aberration*, *astigmatism* (both geometrical aberrations) and the *chromatic aberration* (physical aberration).

Extra-axial aberrations are called *distortions*.

Spherical Aberration. Spherical aberration is due to a difference of focusing for axial rays and for non-axial rays. The deviation with respect to the *Gaussian plane* (ideal focusing plane) increases with increasing angle (Figure 9.7a). The theoretical stigmatic image of a point is therefore replaced by a disc of radius r'_s. When referred to the object plane, it corresponds to the so-called *spherical aberration smearing disc* with a radius $r_s = r'_s/g$ (g = magnification of the lens) which defines the aberration

$$r_s = C_s \alpha^3 \tag{9.4}$$

where α is the aperture angle and C_s the spherical aberration coefficient, the radius of the object plane smearing disc for an aperture angle of 1 rad.

In practice, spherical aberration of electron lenses cannot be corrected; it can, however, be minimised by the choice of operating parameters such as a very short focal length (a few millimetres), high magnification, and coil current near to magnetic saturation. Values of the aberration coefficient C_s are commonly of the same order of magnitude

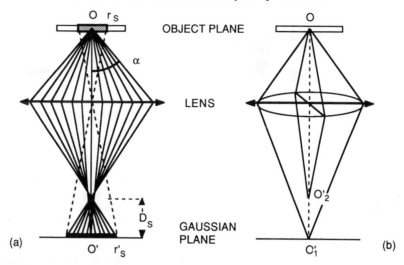

Figure 9.7. Geometrical lens aberrations. (a) Spherical aberration with Scherzer defocus D_s; (b) astigmatism.

as the focal length. As shown by Eq. (9.4), the spherical aberration increases according to the power of three of the angular aperture. This highlights the necessity of operating at small apertures.

As shown in Figure 9.7a, the minimum of smearing is actually observed at a given *defocusing* with respect to the Gaussian plane. The value of this defocus as determined by Scherzer (1949) is

$$D_s = -1.2(C_s\lambda)^{1/2} \qquad (9.5)$$

Astigmatism. An actual lens has no perfect revolution symmetry. The focal length differs in two axial planes, perpendicular to each other, thus leading to astigmatism (Figure 9.7b). This aberration has various causes: ellipticity and eccentricity of pole pieces and apertures, inhomogeneity of pole piece materials, contamination or dirt particles on the aperture which deviate the electron beam through electrostatic charges.

Astigmatism correction is straightforward by means of superimposing an elliptical electric or magnetic field, adjustable in intensity and in direction.

Chromatic Aberration. As shown by Eq. (9.3), the focal length of an electron lens is proportional to the incident electron energy. Any energy dispersion ΔE with respect to the incident energy E_0 therefore induces a *chromatic aberration smearing disc* in the Gaussian plane. Referred to the object plane, the radius r_c of this disc leads to a definition of the chromatic aberration, according to the following equation:

$$r_c = C_c \alpha \frac{\Delta E}{E_0} \qquad (9.6)$$

where α is the aperture angle and C_c the chromatic aberration coefficient.

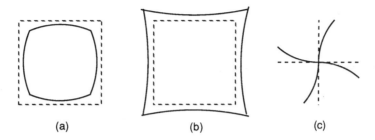

Figure 9.8. Distortions induced by an electromagnetic lens: (a) barrel distortion; (b) crescent (or pincushion) distortion; (c) tangential distortion. (Ideal image in dashed lines.)

The value of the chromatic aberration coefficient is of the order of magnitude of the focal length. The causes of energy dispersion have been outlined previously.

Distortion. Distortions are aberrations which do not cancel on the optical axis. They are characterised by a shift of extra-axial image points with respect to their theoretical positions; the shift is a function of the distance from the axis. Two kinds of distortions occur (Figure 9.8):

1. Through a radial shift leading to barrel-shaped or to crescent (or pincushion)-shaped distortion;
2. Through a tangential shift due to the electron image rotation, the angle of which increases with the distance from the axis (specific for magnetic lenses).

Distortions are minimised by operating near to the axes and by saturating the magnetic circuit of magnetic lenses.

Electronoptical Focusing System

Depending on the application, a more or less sophisticated optical system is set up. A small and easily adjustable electron probe (electron microscope, electron microprobe) is commonly produced by means of a condenser device, made of electromagnetic lenses, which project a reduced image of the cross-over on to the object plane. Depending on the minimum probe diameter to be achieved, such a condenser consists of one or several lenses.

A two-lens condenser combined with a thermionic electron source provides a micrometre-sized probe (*microprobe*) with high-energy electrons (Figure 9.9a).

Additional condenser lenses provide a further probe reduction, thus reaching a nanometre size (*nanoprobe*), especially combined with an LaB_6 or, even better, with a field emission source. This nanoprobe mode is commonly used in high-resolution scanning electron microscopes, in transmission electron microscopes operating in parallel beam or in convergent beam nanodiffraction. For this aim, the objective lens often participates in the focusing process (*condenser–objective lens*, Figure 9.9c). Combined

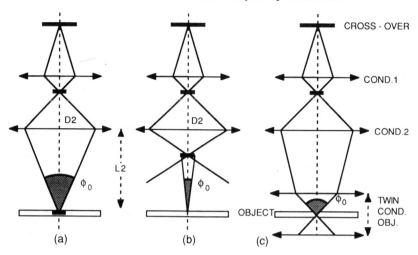

Figure 9.9. Principle of electromagnetic condenser systems. (a) Two-lens condenser system focusing on to the object plane. Aperture angle $\Phi_0 = D_2/L_2$. The first lens produces a strongly reduced image of the cross-over (reduction ratio about 1/100). This image is transferred on to the object plane through the second lens with a unit magnification. The final electron probe is the smaller, the higher the first lens excitation. (b) Two-lens condenser system operating in the over-focusing mode. Aperture angle $\Phi_0 < D_2/L_2$. (c) Two-lens condenser system combined with a condenser-objective focusing on the object plane in the nanoprobe mode. The objective system is a symmetrical twin lens. Its first part operates as an additional condenser lens with a high field near to the sample; it provides a super focusing of the beam. The second part operates as the actual objective lens to form the image. A further lens may be inserted before the twin lens in order to provide an even greater variety of settings.

with a field emission source, such a system can produce an electron probe of a diameter down to a few ångströms (Crewe, 1966; Crewe et al, 1970).

Beam Intensity. At a first approximation, the condenser system conserves the electron intensity travelling through it. It follows that the electron flow intensity I_0 transmitted by the condenser into the object plane is a sole function of the illumination aperture angle Φ_0 according to

$$\frac{I_0}{\Phi_0^2} = B \Rightarrow I_0 = B\Phi_0^2 \tag{9.7}$$

where B is the cross-over brilliance (Eq. 9.2). This results in two different condenser focusing modes.

Focusing on the Object Plane. The image of the cross-over is projected on to the sample (Figure 9.9a, c). The beam aperture angle and hence the electron flow density are maximum. The aperture angle depends on the choice of the exit aperture; it is limited to about 10^{-2} rad by lens aberrations and by the object resistance to electron bombardment.

Over-focusing. With a higher excitation of the second condenser lens (Figure 9.9b), the cross-over image is focused before the sample surface. This mode results in a small angle incident beam on the sample, down to 10^{-4}–10^{-5} rad, i.e. a nearly parallel electron beam of small flow density. It is commonly used in conventional high-energy electron diffraction.

Beam Coherence. In a *coherent* beam the individual waves are in a well-defined phase relationship to each other. The wave amplitudes add up, therefore resulting in the possibility of interference effects, i.e. electron diffraction.

In an *incoherent* beam, there is no defined phase relationship between waves. Their intensities add up. No interference takes place.

The actual beam coherence is intermediate between those extremes.

The most incoherent beam from a condenser system is provided by using a large aperture and by focusing the source image on to the sample plane. Effectively, the electron waves emitted from two distinct source points have no phase relationship, neither have those passing through the corresponding image points. This incoherent illumination mode is commonly used for imaging in a transmission electron microscope in order to minimise diffraction artefacts.

The most coherent beam is provided by over-focusing of the condenser system. This results in a very small apparent source size as viewed from the sample, thus enhancing interference effects. This coherent illumination mode is used in electron diffraction.

9.3 NEUTRON SOURCES

Thermal neutrons used for diffraction purposes are produced in specialised nuclear reactors. As an example, the high-flux nuclear reactor of the ILL (Institut Laue Langevin) in Grenoble generates a flux of 10^{15} neutrons/cm^2 from 16 working channels. This is small when compared to the flux emitted from modern X-ray and electron sources.

Like X-rays, neutrons cannot be focused by lenses. A collimator limits the exit beam in order to supply a more or less parallel beam of a given cross-section. Due to the low neutron flux at the source and the very small interaction cross-section (see Table 4.1), the beam cross-section must be a large one (some 10 cm^2) in order to provide sufficient scattering intensity for accurate measurements. Collimators and shieldings are made of neutron-absorbing materials, such as boron, cadmium, and gadolinium (see Table 8.4), e.g. in the form of cadmium-coated steel tubes embedded in boron paraffin.

The neutrons emitted from a reactor are in thermal equilibrium with the moderator atoms, with an average associated wavelength of 1.3–1.5 Å, depending on the working temperature, i.e. similar to the wavelength of commonly used X-rays. They have a wide energy spread according to a Maxwell distribution (Eq. 1.8), with a half-width of several ångströms. Their use for diffractometry therefore requires a crystal monochromator. Due to the impossibility of focusing, the efficiency is low. A lead crystal monochromator of some 10 cm in size provides an efficiency of about 1%. Furthermore the resulting wavelength selection is not very good, about 5×10^{-2} Å, corresponding to a relative chromatic dispersion of $\Delta\lambda/\lambda = 3.3 \times 10^{-2}$, some 100 times the current dispersion of monochromatic X-rays.

128 *Structural and Chemical Analysis of Materials*

Thanks to the low neutron absorption, work takes place in air. Due to the high penetration power, the protection of persons and of sensitive detectors against neutron radiation poses shielding problems.

9.4 ION SOURCES

Ion beams are required for different purposes, such as analysis through *ion sputtering* (e.g. SIMS, secondary ion mass spectrometry), *ion implanting* (e.g. for producing semiconductors), *ion abrasion* (e.g. material thinning for electron microscope samples).

Producing an ion beam of a given element and of a given energy requires an ion source associated with an attachment for extracting, accelerating and focusing the ions. The whole ion-generating system is called an *ion gun*. Its design depends on the kind of application.

Like an electron gun, the main characteristics of an ion gun are the *brilliance*, the *energy dispersion* and the *stability*.

Electron optics applies to ion beam focusing, with differences in magnetic fields due to the varying mass of the particles (Drummond, 1984).

9.4.1 GAS DISCHARGE SOURCE

The simplest and most commonly used ion gun is the cold cathode discharge gun.

Figure 9.10(a) outlines the principle of the so-called *duoplasmatron* currently set up in ion spectrometry. For analytical purposes, the ion source is an inert gas, e.g. argon or nitrogen, or a reactive gas, e.g. oxygen. The working gas is introduced into the ionisation chamber through an adjustable valve at a low pressure of about 0.3 Torr. The ionisation chamber consists of two coaxial electrodes, currently cylindrical. In high-performance sources for spectrometry, the form of the electrodes is designed for optimum intensity and focusing. Axial apertures permit the entrance of the gas and the exit of the ion beam. The discharge results from a voltage of a few kilovolts applied to the electrodes, its sign depending on the sign of the ions to be produced (e.g. Ar^+ or O^-). The discharge results in a plasma of ions and electrons multiplied by collisions. The plasma confinement, primarily due to the form of the electrodes, is increased through constriction by means of an axial magnetic field emitted from a coil. The field causes an increase of the probability of ionising collisions along helicoidal trajectories, thus resulting in a better efficiency at a lower working pressure.

A set of electrodes provides the extraction, the acceleration and the focusing of the exit beam, by means of voltages ranging from 0 to 20 kV. The ion beam reaches the working enclosure through a small aperture which permits it to operate in high vacuum through differential pumping. The same principle is used for measuring the pressure of primary vacua (*Penning gauge*).

Characteristics. The main advantages of this type of ion source are its simplicity, its high beam intensity and the possibility of generating both inert and reactive ions. Its drawbacks are a large ion energy dispersion, of the order of 10 eV, and the necessity of a relatively high gas pressure to ensure a fair stability.

The brilliance is of medium value, commonly about 10^6 A/cm^2 per sr.

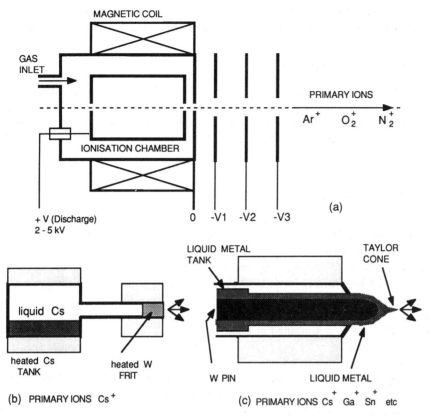

Figure 9.10. Principle of sources for positive ions. (a) Gas discharge source (cold cathode); (b) surface ionisation metal ion source; (c) liquid metal ion source, strongly magnified detail of the tungsten pin and the Taylor cone. The optical system for extracting, accelerating and focusing the ions emitted from the source is outlined solely with the gas discharge device in (a). For any ion gun, it is based on the principle of the electrostatic lens (Figure 9.6), with a display and the respective electrode voltages depending on the type of source and its usage. In current set-ups, the accelerating voltage V_2 varies from 1 to 20 kV, the voltages V_1 and V_3 vary between zero and several kilovolts, with the source enclosure, commonly grounded, taken as zero.

9.4.2 HOT CATHODE SOURCE

In a hot cathode ion gun, the gas is ionised by electrons emitted from a heated filament and accelerated through a voltage of some 100 V. Plasma constriction, ion beam extraction, acceleration and focusing are operated in the same way as in the discharge gun.

With an additional measuring electrode, this device can be used as an ionisation gauge for measuring pressures in secondary vacua (*Bayard–Alpert gauge*).

Characteristics. The main advantages are a low-energy dispersion (order of 1 eV) and a high stability at low gas pressures. The brilliance is moderate and is generally smaller than with a discharge gun. The source is limited to inert gases, the filament being destroyed by reactive gases like oxygen.

The most commonly used ion guns in spectrometry, of the duoplasmatron type, can be fitted either with a cold cathode for reactive ions or a hot cathode for inert ions.

9.4.3 METAL ION SOURCE

Various ionised elements can be generated by various techniques, such as high-temperature evaporation, electron bombardment, ion bombardment, etc. For analytical applications it is often useful to dispose of primary metal ions. Two generation modes of metal ions are presently employed: surface ionisation and liquid metal field emission.

9.4.3.1 Surface Ionisation Source

The most commonly used surface ionisation source has been designed to supply a beam of caesium ions, but it may in the same way produce different alkaline metal ions, such as sodium and potassium ions.

Caesium is the most electropositive element, with the advantage of a low melting point (28.5 °C) and a low boiling point (690 °C), together with a high vapour pressure even at moderate temperatures.

The principle of a caesium ion source is outlined in Figure 9.10b.

The liquid metal is maintained at about 260 °C in a stainless steel container, within an inert gas atmosphere. Its vapour flows through a heated molybdenum feeding pipe towards a tungsten frit heated at 1100–1200 °C. Caesium atoms diffuse through the porous tungsten towards the external surface and are ionised at a rate of 99%. The Cs^+ ions are extracted, focalised and accelerated through a voltage varying between 0 and 20 kV. A negatively biased electrode prevents negative secondary ions, emitted by primary ion sputtering on the apertures, from returning to the source which they may contaminate. A caesium charge of a few grams provides more than 1000 h of operation.

Characteristics. The beam intensity of 10–30 mA into a diameter of 1–30 μm leads to a satisfactory brilliance which depends on the emission surface (Storms et al, 1977; Williams et al, 1977) and is on average similar to that of a duoplasmatron.

The main quality is the very low energy dispersion of about 0.2 eV.

9.4.3.2 Liquid Metal Source

The principle of the liquid metal source is outlined in Figure 9.10c. Its main feature is a tungsten needle with a tip diameter of a few micrometres covered by a liquid metal layer.

The needle emerges from a metallic tube acting as the liquid metal container. The liquid metal wets the needle up to its tip through capillarity. The tip is subjected to a strong electric field generated by an extraction electrode at a negative voltage of a few kilovolts with respect to the tip. The field causes the liquid metal to form a cone tapering off into a point tip, called the *Taylor cone*, which results in a field effect metal ion source of nearly point size.

The same technique is practicable with various low-melting-point metals, like Cs, Ga, Sn, Pb, Bi, In, Au, etc.

Characteristics. The small source size of about 150 Å leads to a beam with a cross-section of about 500 Å. Due to the low intensity in such a narrow beam, this type of ion source is mainly designed for secondary ion emission imaging facilities.

The energy dispersion depends on the metal melting point.

As with field emission electron guns, the brilliance is very high, reaching values up to 10^{-10}–10^{-11} A/cm² per sr (Prewett et al, 1981, 1984; Waugh et al, 1984).

REFERENCE CHAPTERS

Chapters 5, 7, 8. Appendix F.

BIBLIOGRAPHY

X-RAYS

[1] Compton A. H., Allison S. K. *X-rays in theory and experiment.* Van Nostrand, New York, 1954.
[2] Guinier A. *Théorie et technique de la radiocristallographie.* Dunod, Paris, 1964.
[3] Urlaub J. *Roentgenanalyse.* Vol. 1: *Roentgenstrahlen und Detektoren.* Siemens Aktienges., Berlin, Munich, 1974.
[4] Winik H., Doniach S. *Synchrotron radiation research.* Plenum Press, New York, 1980.

CHARGED PARTICLES

[5] Glaser W. *Grundlagen der Elektronenoptik.* Springer-Verlag, Berlin, 1952.
[6] Grivet P. *Electron optics.* Pergamon Press, Oxford, 1965.
[7] Septier A. *Focusing of charged particles.* Academic Press, New York, 1967.
[8] Valyi L. *Atom- and ion sources.* John Wiley, London, 1977.

NEUTRONS

[9] Bacon G. E. *Neutron diffraction.* Oxford University Press, 1975.

VACUUM

[10] O'Hanlon J. F. *A user's guide to vacuum technology.* John Wiley, New York, Chichester, 1980.

10

Radiation Detectors and Spectrometers

Detection and measurement of radiation are performed through the interaction effects which they produce in a detection medium.

The various techniques for measuring X-rays, charged particle beams (electrons, ions) and neutrons are dealt with separately. However, they are similar in many respects.

Measurements may be limited to a simple qualitative detection. Quantitative determinations involve two physical quantities which characterise a given radiation: intensity and energy.

Intensity measurement consists of counting the number of photons or matter particles crossing a given surface per time unit. Their detection is based on the ionisation effects which they produce in a gas, a liquid or a solid. The ionising events are converted into electrical pulses which can easily be counted or integrated to provide an average intensity and be processed by means of electronic devices.

Energy measurement consists of evaluating the energy of the individual photons or matter particles in a beam. The result is generally expressed in the form of a *spectrum*. Due to the energy–wavelength equivalence, the spectrum can be either an intensity versus energy spectrum $I(E)$ or an intensity versus wavelength spectrum $I(\lambda)$.

10.1 INSTRUMENTAL PARAMETERS

10.1.1 CHARACTERISTIC DETECTOR PARAMETERS

The main characteristics of any radiation detector are commonly defined by means of a set of parameters.

Dead time t_m: Time interval following a detection stimulus, during which the detector is insensitive to another stimulus. The actual *recovery time* needed to deliver normal size pulses is somewhat longer.

Count rate $n = N/t_c$: Number of detection pulses per time unit (where N is total number of counts and t_c counting time).

Maximum count rate: Maximum number of pulses per second before saturating the detector. It is of the order of magnitude of $1/t_m$.

Radiation Detectors and Spectrometers

Background noise: Counting rate in the absence of any radiation to be measured. It is due to cosmic radiation, to ambient radioactivity and to the electronic noise of amplifying devices. An important quality is the signal/noise ratio.

Efficiency ε: Ratio of the intensity actually measured by the detector and the intensity reaching its window.

Mean detection threshold W_m: Average ionisation energy of atomic levels of the detection medium. Radiation particles with an energy smaller than W_m cannot be counted.

Energy resolution δE: Capability of the detector to select radiation particles with energies differing by an amount δE. This parameter is important for energy selection and energy measurement.

Lifetime: Total number of counts provided by a detector before it is to be replaced.

10.1.2 CHARACTERISTIC SPECTROMETER PARAMETERS

Any method of spectrometric analysis leads to the establishment of a radiation spectrum, i.e. of the functions $I(E)$ or $I(\lambda)$. The spectrometer facility currently includes two components:

1. The spectrometer itself, with wavelength selection or energy selection;
2. The detector for measuring the corresponding intensities.

In some cases, both functions are provided by the same component.

Resolution. The most important quality of a spectrometer is its energy resolution. Consider a perfectly monochromatic incident beam of wavelength λ_0 and of energy E_0. The actual spectrum delivered by the spectrometer consists of a peak with maximum intensity at E_0 and with a half-width δE called the *absolute energy resolution*. The ratio $\delta E/E_0$ is the *relative energy resolution* (Figure 10.1). A given spectrometer is able to separate two neighbouring spectrum lines if its absolute resolution is smaller than

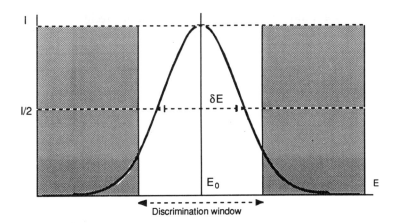

Figure 10.1. Definition of the energy resolution. Illustration of a discrimination window centred on a mean energy E_0.

the energy difference corresponding to the lines. The experimental line broadening results from various instrumental factors depending on the spectrometer.

Efficiency. Transmission Factor. For a given wavelength, the efficiency ε of the spectrometer and its associated detector, considered as a whole, can be defined in a similar way as for a detector alone. It represents the ratio between the measured intensity and the incident intensity. A given acceptance angle Ω leads to a global *transmission factor* $\tau = \varepsilon \Omega / 4\pi$.

10.2 X-RAY DETECTION AND MEASUREMENT

10.2.1 QUALITATIVE DETECTION

Fluorescent Screen. Visible X-ray fluorescence of certain crystallised materials results in the use of the fluorescent screen as a means of X-ray detection. Due to a small fluorescent yield, it is generally not possible to detect a diffracted X-ray beam in this way. The fluorescent screen is mainly used for adjusting the incident beam.

Photographic Film. In the same way as light, X-ray photons impress a photographic emulsion. This results in the oldest X-ray detection method which led Röntgen to discover X-rays in 1895 by observing the blackening of photographic plates in the vicinity of a working Crookes cathode tube.

The photographic method has several advantages over more sophisticated techniques:

1. Simultaneous recording of radiations in a large solid angle.
2. Cumulativeness, thanks to the absence of an intensity threshold (Figure 10.2). As opposed to the case of light photons, individual X-ray photons have enough energy to induce the photo-reduction process. Even very weak diffracted X-ray beams can therefore be recorded, provided that there is sufficient exposure time.
3. Cheapness and simplicity.

Because of these advantages, the photographic recording technique is still widely used for X-ray diffraction. For the highest efficiency with strongly penetrating X-rays, special

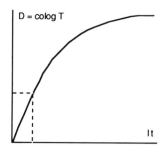

Figure 10.2. Diagram of the optical film density (colog of the light transmission factor) as a function of the cumulated incident X-ray quantity It (I = intensity, t = time of exposure). Note the absence of an intensity threshold and the linearity for limited exposure times (arbitrary scales).

films with thick emulsion layers (sometimes double layers) are used. For small incident intensities, the film density diagram remains linear (Figure 10.2). After a suitable calibration the film method can, therefore, serve even for fairly accurate intensity measurements associated with a light densitometer. Up to the 1960s, this technique was currently used for numerous crystal structure determinations.

Conventional photographic X-ray recording techniques are irreplaceable in the field of X-ray radiography in non-destructive material testing and in medicine. The film is now generally combined with an intensifying screen which converts X-ray energy into light. Therefore the film exposure is in fact due to light photons. Recent research may lead to the development of new devices for temporary storage of X-ray diagrams or images through photostimulable phosphors. The X-ray photon energy induces atom-level excitations in the phosphor material where it is stored as quasi-stable states. The latent image is retrieved through luminescence stimulated by visible or infrared radiation from a scanning laser beam (photostimulable luminescence, PSL). It is recorded on a conventional film (Sonoda et al, 1983).

10.2.2 INTENSITY MEASUREMENT. IONISATION DETECTORS

The basic principle of ionisation detectors is based on X-ray photon-induced ionisations of a gas, a liquid or a solid. The detection events are measured in the form of produced electrons or light photons. In an ideal detector, any incoming X-ray photon produces a pulse in the associated electronic measuring circuit. The pulses are counted or integrated in the form of a mean intensity representing the count rate. The actual count rate depends on the detector efficiency.

10.2.2.1 Gas Ionisation Detectors

Basic Principle

The incoming X-ray photons induce gas atom ionisations, with a predominance of the Compton effect at external levels. They lose their energy progressively through successive interactions. According to this simplified process, each photon of energy E_0 produces E_0/W_1 electron–ion pairs (where W_1 is the energy of first ionisation). However, some atoms undergo double or multiple ionisation, i.e. with a greater ionisation energy. The whole process results in an experimental mean ionisation energy W_m which represents the *mean detection threshold*.

It follows that each incoming X-ray photon produces on the average $N_m = E_0/W_m$ electron–ion pairs.

Example of argon: $Z = 18$, $W_m = 26.4$ eV. Each Cu-Kα photon, $E_0 = 8040$ eV, produces 304 electron–ion pairs.

When the incident photon energy is greater than the core level ionisation energy of the detector atoms, the photoelectric effect participates in the detection process. The accordingly emitted secondary photons may not be absorbed by the detector gas, given the fact that an element is relatively transparent to its own characteristic radiations. This leads to errors in the intensity measurements (*escape peak*, see energy measurement).

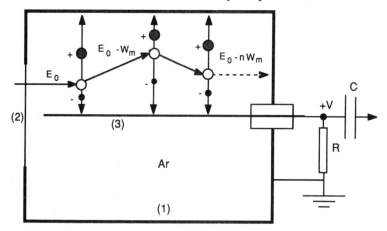

Figure 10.3. Working diagram of a gas counter. (1) Gas-filled enclosure (rare gas like argon, krypton, etc., depending on the energy range) grounded to act as the cathode; (2) radiation entrance window, (3) central anode connected to the positive detection voltage V. The electric pulse is transmitted to an electronic counter through an R–C circuit.

This effect is small for X-rays with commonly used light detector gases like neon and argon.

The resulting electrons and ions are collected by means of an electric field (Figure 10.3).

The *efficiency* of a detector is maximum for X-rays of a given energy range when the photon energy is entirely dissipated during their travel through the detector. This ideal case can be approached by using very thin entrance windows (beryllium or organic material such as mylar) and by adapting the nature of the gas to the energy range. Argon is the most commonly used detector gas for X-rays in material analysis.

Working Regions. Multiplication Factor

Depending on the detection voltage V, various working regions can be defined for gas detectors. They differ by the corresponding *multiplication factor A*, the ratio of the number of electrons collected to the number of electrons produced in the detector (Figure 10.4). The following zones are generally considered, according to an increasing value of the voltage V.

Recombination Region $(0 < A < 1)$. With a low accelerating voltage, the created electron–ion pairs remain too long in their own neighbourhood. It follows that a fraction of the electron–ion pairs do recombine before reaching the electrodes and are lost for detection.

Saturation Region. Ionisation Chamber $(A = 1)$. The detection voltage is just large enough in order to ensure the complete separation and collection of the created electron–ion pairs. In a relatively wide zone, called the saturation region, the multiplication factor remains equal to 1. This is the working region of the ionisation chamber. An incident X-ray diffraction beam induces very small electric detection currents, of the order of magnitude of $10^{-10} - 10^{-15}$ A, therefore requiring very sensitive and expensive measuring attachments (e.g. a vibrating capacitor electrometer). The ionisation chamber

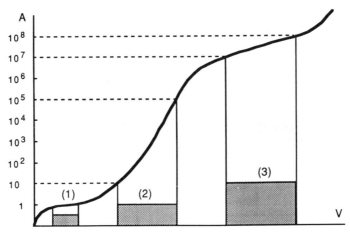

Figure 10.4. Various working conditions of gas detectors. (1) Ionisation chamber; (2) proportional counter; (3) Geiger–Müller counter.

is not applied to current material analysis, but is widely used in fundamental physics for accurate absolute radiation measurements.

Proportional Region. Proportional Counter ($A \cong 10^2$–10^5). When the voltage exceeds a given value, of the order of magnitude of some 100 V, the primary electrons induced by the incoming X-ray photons are accelerated to a kinetic energy high enough to further ionise gas atoms, producing secondary electrons which are accelerated in their turn and so on. The result is for each incoming photon to trigger off a so-called *Townsend avalanche*. Most of the accelerated electrons reach the required ionisation energy only when arriving in the vicinity of the anode, where the avalanches are therefore localised. Due to their greater mass the ions move more slowly and form a positive space charge around the anode at each photon impact; it captures electrons and results in quenching of the electron showers. The magnitudes of the avalanches (and therefore the pulse sizes in the counter) are accordingly levelled out and remain proportional to the number of ions initially produced by the incoming photon. As will be seen later, this effect permits photon *energy discrimination*. In addition, the number of pulses, leading to the measured beam intensity, is proportional to the number of incoming photons.

The self-quenching effect of the counter is improved by adding a given ratio of an organic gas, easily ionisable into heavy ions, increasing accordingly the space charge around the anode. A currently available gas is argon with 10% methane as quenching agent. A similar result may be obtained by the addition of halogen. The proportional region can thus be extended up to a multiplication factor of 10^5.

Discharge Region. Geiger–Müller Counter ($A \cong 10^7$–10^8). When the voltage exceeds a given value, depending on the design, the avalanches progressively invade the whole space. The proportional counting conditions are no longer met.

When V reaches a value just under the spontaneous discharge voltage (some 1000–1300 V), an incoming photon triggers off a discharge, due to generalised avalanches in the whole volume of the counter and enhanced by the emission of electrons and ultraviolet photons from the enclosure walls and the electrode. A rapid quenching of the discharge is provided by a high resistor in the circuit and increased by the addition of organic or halogen molecules as in the proportional counter. Any incident photon therefore produces an impulse with a magnitude some 1000 times greater than in the proportional counter. The linearity of the count rate versus the incident radiation intensity is maintained for low intensities.

Spontaneous Discharge Region. When the voltage becomes higher than the discharge threshold, a permanent discharge occurs. Detection can no longer be achieved.

The proportional counter or the Geiger–Müller counter (GM counter) are currently used for X-ray intensity measurements.

Proportional Counter

The proportional counter requires a more sensitive amplification system than the GM counter. Its main advantage is its ability to deliver pulse magnitudes proportional to the photon energies, thus resulting in the possibility of *energy discrimination*. An adjustable band-pass type electronic filter selects the pulses with magnitudes between a low level and a high level and therefore results in photons with corresponding energies inside a *discrimination window* (Figure 10.1). In diffraction facilities for instance, when centred on $E_{K\alpha}$ of the characteristic emission from an X-ray tube, the discriminator combined with a proportional counter reduces the continuous bremsstrahlung, acting in some respects like a monochromator, but with a very low resolution, incapable of separating $K\alpha$ from $K\beta$, but nevertheless greatly diminishing the background level.

Thanks to the short dead time, the linearity of the count rate versus the incident intensity, up to fairly high count rates, is ensured. It is therefore well suited for accurate intensity measurements.

The detection gas is generally continuously regenerated through an adjustable gas flow (hence the alternative name of *gas-flow counter*). The counter lifetime is therefore virtually unlimited.

Main Characteristics. Efficiency: on the average of the order of 30% for detecting medium element X-rays; with an ultra-thin window, it enables the detection of light element emissions.

Dead time: low, $t_m \cong 1\ \mu s$.
Mean detection threshold (Ar): $W_m \cong 26\ \text{eV}$.
Multiplication factor: $A \cong 10^3$–10^4.
Energy discrimination: possible.

Geiger–Müller Counter

The main advantage of the GM counter is the ability to work with a simplified amplification system. It is therefore particularly well designed for portable detecting devices. In return, it has a long dead time, resulting in a low maximum count rate and a deviation from linearity of the count rate versus the incident intensity. It is not suited for quantitative measurements. Furthermore, its output is independent of the photon energies, which excludes energy discrimination.

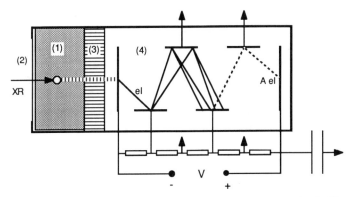

Figure 10.5. Basic outline of an X-ray scintillation counter. (1) Crystal scintillator NaI(Tl) (⌀ 2–3 cm, thickness several mm), aluminium coated on its entrance and lateral faces to improve light reflection; (2) airtight beryllium window to seal off the hygroscopic crystal; (3) light guide; (4) photomultiplier with amplification factor A.

The GM counter enclosure is generally sealed with a low-pressure gas. The lifetime of counters with organic gas quenching is limited, due to an irreversible decay. With halogens as quenching agents, the ionisation process is reversible; the gas is regenerated and the lifetime is virtually unlimited.

Main Characteristics. Efficiency: on the average of the order of 50%
  Dead time: high, $t_m \cong 200\ \mu s$
  Mean detection threshold (Ar): $W_m \cong 26$ eV
  Multiplication factor: very high, $A \cong 10^7$

10.2.2.2 Solid Detectors

The photoelectric effect is the main photon detection process in solid detectors. The generated photoelectrons in turn induce the excitation of detector atoms by means of electron transitions from the valence band to the conduction band, thus resulting in the creation of *electron–hole pairs*.

The conversion mode to a measurable electrical signal depends on the type of detector. In the scintillation counter, electron de-excitation leads to emissions of light photons which are converted into electrical pulses by a photomultiplier. In the semiconductor counter, the electron–hole pairs are collected by an applied electrical field to provide the detection pulses.

The number of electron–hole pairs generated by each incoming photon is proportional to the photon energy. Similar to the gas counter, the magnitude of the detection pulses is therefore proportional to the photon energy. This property enables the performance of energy discrimination. Energy measuring even becomes possible with semiconductor detectors.

Scintillation Counter

The active component of the detector is the scintillator. A thallium-activated NaI monocrystal is commonly used for detecting X-rays in the usual energy range. It meets the general requirements for a scintillator: good absorption of the photons to be detected, transparency towards the emitted light photons. The electrons which have been excited to the conduction band by the X-ray photons migrate towards the Tl atoms which act as point defects; there they undergo de-excitation to vacant states. The energy released is emitted as light photons which are collected by a photomultiplier (Figure 10.5). Scintillators are not necessarily mineral and crystallised. Organic scintillators are used for electron detection, but are not sufficiently absorbent for X-rays. Liquid scintillators are applied in nuclear physics.

Main Characteristics. Efficiency: very high, virtually 100% in a wavelength range between 0.3 and 4 Å, but becoming weak for light element X-ray emission ($Z \leqslant 19$-K).

Dead time: very short, ($t_m \cong 0.25\ \mu s$), resulting in a high count rate.

Mean detection threshold: high, $W_m \cong 50$ eV resulting in a poor energy resolution. However, energy discrimination is possible.

Reduced size and absence of gas supply, therefore easy to set up.

Sensitive to saturation and to moisture which result in an increase of background noise and a decrease of lifetime.

Semiconductor Counter

The most significant progress in X-ray detection has been made during recent decades with semiconductor counters. For current X-ray energies up to 30 kV (Kα of elements $Z < 55$), the most widely used material is silicon. For higher energies, germanium diodes provide a better efficiency. Photon detection is based on collecting the charges which they produce in the detection medium by means of an electrical field. A field can act only in an isolating medium. In the circumstances the detecting medium is the depletion layer of a p–n diode, operated with reverse bias under the breakdown voltage. A good X-ray detection efficiency in silicon requires a depletion width of several millimetres, much greater than that of a current p–n junction. This is provided by an intrinsic layer which is produced by means of doping the p-semiconductor with a donor element, resulting in a p–i–n diode. In the instance of silicon and the donor element lithium, the diode obtained makes up the so-called *Si(Li) detector* which is the solid equivalent of the ionisation chamber.

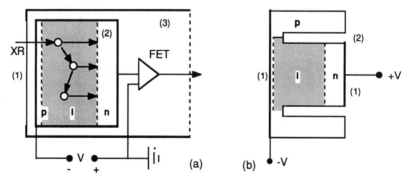

Figure 10.6. Si(Li) detector. (a) Basic diagram. (1) Beryllium window; (2) silicon p–i–n diode; (3) enclosure under vacuum and cooled with liquid nitrogen. (b) Actual shape of the diode. (1) Gold layer; (2) cylindrical delimitation slit. The active zone is grey tinted.

Basic Fabrication Process of an Si(Li) Diode. A layer of lithium is vacuum evaporated on to one face of a cylinder of p-silicon with a resistivity of a few kilo-ohms. After a first thermal diffusion (e.g. 2 min at 400 °C), an accurate compensation is obtained through voltage and temperature controlled lithium diffusion. After about 80 h of diffusion at 120 °C, under 500 V, the intrinsic layer reaches 3 mm. In order to enable a voltage to be applied, the two opposite faces of the Si cylinder are coated with gold (10–20 nm). An incident photon encounters successively (Figure 10.6):

(a) a 20 nm gold layer;
(b) an inert p zone, as thin as possible for a minimum absorption ($\cong 0.1\ \mu$m);
(c) an active i collection zone of some 3–5 mm, submitted to the field gradient;
(d) an inert n zone.

The photoelectric effect is the predominant detection process in an Si(Li) detector. The share of the Compton effect increases at higher energies. The incoming X-ray photons, together with the resulting photoelectrons, induce electrons of silicon atoms to be excited from the valence band to the conduction band, thus generating electron–hole pairs which make up the charge carriers, in many respects similar to the electron–ion pairs in the ionisation chamber. The mean energy required to create a carrier in silicon is $W_m = 3.8$ eV. Each incident photon with an energy E_0 therefore induces E_0/W_m pairs. As an example, a Cu-Kα photon generates on the average $8040/3.8 \cong 2115$ pairs, representing a flow of electricity of 3.4×10^{-16} A. In the absence of any natural multiplication effect, a high gain preamplifier is therefore required, with a field effect transistor as a main component. In order to reduce to a minimum the electronic background noise, and at the same time to prevent the lithium atoms from diffusing into the p-zone, the whole set is being cooled by means of liquid nitrogen. In order to prevent condensation, the diode and the preamplifier are in a vacuum enclosure, with a thin beryllium window for X-ray transmission.

Main Characteristics. Efficiency and detection limit. The efficiency reaches virtually 100% for photon energies between 2 and 20 keV.

The decrease below 2 kV is due to absorption in the beryllium window (absorption through the gold layer and the p-layer are not significant above 250 eV). A window of 7.5 μm thickness leaves an efficiency of some 50% for detecting Na-Kα (1.04 keV). In high-vacuum facilities, detectors may operate with retractable windows and become able to detect B-Kα. More recently, new thin window technologies enable the detection of element lines down to B-Kα without the necessity of high vacuum. In the so-called diamond technology, a 4000 Å thick crystalline carbon layer (diamond and graphite) is deposited on silicon by means of ion beam enhanced deposition. After reducing the Si-substrate to a grid, the resulting airtight window enables the detection respectively of C-Kα with a 60% efficiency and B-Kα with a 40% efficiency. A different technology, based on a boron layer deposited on a boron grid, leads to similar results.

The decrease towards high energies is due to the limited width of the active i-zone. For energies greater than 30 keV ($Z > 55$ for Kα photons), silicon may be profitably replaced by germanium.

Signal-to-noise ratio. A greater surface provides a better ratio, but with a lower resolution. A good compromise is an active entrance surface of about 12.5 mm^2 (i.e. 4 mm in diameter).

Table 10.1. Order of magnitude of the main characteristics of some X-ray detectors

Counter	Efficiency for Cu-Kα (%)	Multi-plication factor	Dead time (μs)	Count rate rate (pulses/s)	Mean detection threshold (eV)
Ionisation chamber Ar	30	1	1	10^6	26.4
Proportional counter Ar–CH$_4$	30	$10–10^5$	0.5	2×10^6	26.4
Geiger–Müller Ar–halogen	50	$>10^7$	200	5×10^3	26.4
Scintillation counter NaI(Tl)	100	10^6 (with PM)	0.2	5×10^6	50
Semiconductor counter	100	1	1	10^6	3.8

Dead time: $t_m \cong 1$ μs, equivalent to a proportional counter.

Mean detection threshold: $W_m = 3.8$ eV at 90 K. This is the lowest value of presently operational detectors. It leads to the generation of 260 electrons per keV (compared with 38 electrons per keV for an argon detector). The low detection energy threshold is the main advantage of the Si(Li) counter. Thus, such a detector may be used both for intensity measures and for energy measures, i.e. as an X-ray spectrometer.

Continuous liquid nitrogen feeding is required in order to avoid any thermal diffusion of the doping element lithium.

10.2.2.3 Position Sensitive Detectors

One-dimensional Localising

A one-dimensional localising detector, often called a linear counter, permits the measurement of the intensity of incident X-rays as a function of the position along one direction of the counter. It consists currently of a proportional counter with the anode parallel to an elongated entrance window. Localising the photons along the anode is performed by means of a delay line. The spatial resolution is mainly limited by the uncertainty on the position of the ionisation events with respect to the anode. The device may be straight or bent to form an arc. A straight counter has currently a length of some 5 cm and a spatial resolution of 70 μm. An arc-shaped counter extends currently to 120° with a resolution of some 3×10^{-4} rad.

Combining a linear detector with a multichannel analyser leads to a direct visualisation of the function $I = f(x)$ or $I = f(\theta)$.

Two-dimensional Localising

Two-dimensional localising is performed by means of a sheet of parallel conductors acting as the anode and as delay lines. A spatial resolution of about 10^{-2}–10^{-3} cm may be achieved in the present state of the technique.

A position-sensitive detector combined with a multichannel analyser has the advantages of both the photographic film and the ionisation counter. It can provide virtually without delay a display of a scattering or a diffraction pattern. It is particularly useful for research on kinetics and transient effects.

Radiation Detectors and Spectrometers

10.2.3 ENERGY MEASUREMENTS. X-RAY SPECTROMETERS

Analytical methods using X-rays require the establishment of the post-interaction spectrum, i.e. the function $I = f(E)$ or $I = f(\lambda)$. It is carried out by means of spectrometers which are of two types:

1. Wavelength-dispersive X-ray spectrometers (WDS or WDX) based on the wave aspect of the radiation;
2. Energy-dispersive X-ray spectrometer (EDS or EDX) based on the corpuscular aspect of the radiation.

10.2.3.1 Wavelength-dispersive Spectrometers

The so-called wavelength-dispersive X-ray spectrometer is in fact an angular dispersive spectrometer. The radiation to be analysed is reflected on a set of lattice planes (*hkl*) in an analysing crystal rotating about an axis parallel to its surface, under an increasing Bragg angle θ. The reflected intensity is measured by means of a counter positioned at an angle 2θ with respect to the incident beam. The selectively reflected wavelengths are given by the Bragg equation (5.7).

$$\lambda = 2\, d(hkl) \sin \theta / n$$

For a given orientation θ of the analysing crystal, the detector at 2θ collects the fundamental wavelength $\lambda = 2\, d(hkl) \sin \theta$ corresponding to the first-order reflection on the set (*hkl*), together with the successive harmonics of wavelengths λ/n corresponding to the reflections of order 2, 3, 4, ..., n. The wavelengths or energies of the incident X-ray photons are deduced from the known values of $d(hkl)$ and θ.

Plane Analysing Crystal Spectrometer

The basic principle is given by the plane monochromator (see Figure 9.4a). It has a low efficiency, even with highly reflecting lattice planes. However, it can be used in any device where the X-rays to be analysed issue from a large area (e.g. X-ray fluorescence).

Focusing Spectrometer

The principle of focusing requires the source of X-rays to be small (line or point source, see Figure 9.4b, c). The source is placed on the focusing circle, called the *Rowland circle*. In order to proceed with an increasing angle θ, both the analysing crystal and the entrance slit of the detector must move along the circle. The detector entrance slit must be at any moment on the focusing point F, in the direction 2θ with respect to the incident beam. Numerous types of more or less sophisticated spectrometers are available, depending on the focusing mode (exact or approximate) and on the design of the respective movements of the crystal and the detector. For the aim of quantitative analysis it is of prime importance for the emergence angle and the aperture angle of the beam to be measured to stay constant during the θ scanning. This condition is met in the linear focusing spectrometer outlined in Figure 10.7.

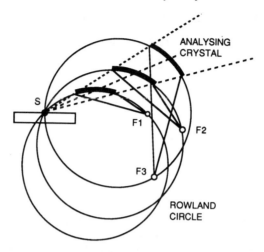

Figure 10.7. Principle of the linear focusing X-ray spectrometer; S: X-ray source. The drawing shows three positions of the analysing crystal and the related positions F1, F2, F3 of the detector slit.

Table 10.2. Characteristics and approximate range of analysis of commonly used analysing crystals

Name	Formula	Crystal system	Reflection	d (Å)	Range λ(Å)	Range Z(Kα)	Range Z(Lα)
Lithium fluoride	LiF	Cubic	220	1.42	0.25–2.7	22–68	⩾56
			200	2.01	0.35–3.8	19–58	⩾49
Quartz	α-SiO$_2$	Trigonal	10.1	3.34	0.58–6.3	15–46	⩾40
PET	C(CH$_2$OH)$_4$	Tetragonal	002	4.37	0.76–8.2	14–40	⩾36
Gypsum	SO$_4$Ca, 2H$_2$O	Monoclinic	020	7.59	1.3–14.2	11–31	29–78
Muscovite	KAl$_2$(OH)$_2$AlSi$_3$O$_{10}$	Monoclinic	002	9.9	1.7–18.6	9–27	26–69
KAP	C$_6$H$_4$(COOH)(COOK)	Hexagonal	10.0	13.31	2.3–25	8–24	23–60
PbSt	[CH$_3$(CH$_2$)$_{16}$COO]$_2$Pb	Smectic pseudocrystals	$n=1$	$\cong 50$	8.7–94	5–12	20–34

Analysing Crystals

The plane equidistance $d(hkl)$ depends on the crystal. The Bragg angle varies theoretically between 0 and 90°. For instrumental reasons it is limited to 5–70°. According to the Bragg equation (5.7), the range of wavelength to be measured by a specific analysing crystal is therefore given approximately by the relation

$$0.17\, d(hkl)/n < \lambda < 1.9\, d(hkl)/n$$

A set of three or four differently spaced analysing crystals is therefore required in order to cover the whole range of characteristic K or L lines of the elements. Most focusing spectrometers are designed according to the Johann mode, with analysing crystals bent elastically or plastically. For a crystal species to be valid for analysing purposes, it must have suitably distanced plane sets with high reflecting power and be easy to cut into thin platelets parallel to these planes and easy to bend (Table 10.2). Smectic pseudocrystals with one-dimensional periodicity (Langmuir–Blodgett crystals) are used for analysing long-wave X-rays. They are prepared by superposing several hundreds of monomolecular layers of suitable thickness. They are mostly made of organic components, such as lead stearate, PbSt. Mineral multilayers have been developed more recently. They are made by alternated evaporation of heavy element layers (e.g. tungsten) and of light elements (e.g. carbon, silicon) on a plane substrate (metal, glass, silicon, silica, etc.). Mineral analysers have a higher reflecting power and a greater stability than organic analysers at high temperature and in vacuum. They are best suited for analysing light element emissions, down to boron.

Main Characteristics

Resolution. Several effects cause a broadening of the spectrum lines as delivered by a wavelength-dispersive spectrometer.

1. Natural X-ray line width. The relative energy spread is of the order of magnitude of 10^{-4}, negligible with respect to the instrumental causes (see Table 9.3).
2. Analysing crystal imperfections (mosaic texture).
3. Focusing and collimation defects.
4. Variation of the analysing crystal plane equidistances through thermal expansion. Thermostating anneals this effect.

A relative resolution of an order of magnitude of 3×10^{-3} is achieved for Al-Kα, with good-quality analysing crystals, resulting in an absolute energy resolution of about 5 eV. Corresponding lines of two neighbour elements are largely separated on the whole wavelength range. In favourable conditions, the Kα doublet can be resolved.

Transmission Factor and Signal-to-Noise Ratio. Due to the small cross-section of X-ray diffraction and to the small available acceptance angle, the transmission factor of a crystal spectrometer is rather low, some 10^{-5}–10^{-2} depending on the device. The background noise is low. It results mainly from X-ray fluorescence and inelastic scattering effects issued from the analysing crystal and the collimators, as well as from the detector. The signal to background ratio currently reaches values up to 10^3 for the Kα lines of medium and heavy elements.

The main characteristics are summarised in Table 10.3.

Interferences Due to the Spectrometer. The spectrum issued by a crystal spectrometer is the superposition of the physical spectrum and the following instrumental effects:

1. Multiple order reflections ($n = 2, 3, \ldots$) on the analysing crystal.
2. Escape peaks due to the ionisation detector (see energy-dispersive spectrometer).
3. Continuous background issued from the analysing crystal. The Compton effect prevails for light element crystals (organic crystals). The fluorescence effect becomes significant when the spectrum to be measured contains high-energy X-rays liable to excite core levels of heavy elements of the analysing crystal (e.g. Pb in PbSt crystals). Organic crystals avoid this nuisance.

Identification Ambiguities. The superposition of neighbouring spectrum lines results in identification ambiguities. The fundamental Kα peak (first-order reflection on the analysing crystal) of a light element may coincide, to the approximation of resolution, with a high-order reflection peak or an L peak of a heavy element.

Typical Examples of Coincidences. (a) Same order: first order of As–Kα_1 (1175 Å) and first order of Pb–Lα_1 (1175 Å); (b) different orders: first order of Al–Kα_1 (8339 Å) and fourth order of Cr–Kβ (4λ = 8340 Å).

Peaks of the same order which coincide correspond to the same energies and cannot be distinguished directly. Peaks of different orders which coincide correspond to different energies; they can therefore be distinguished by means of energy discrimination (Figure 10.1) or by reducing the primary acceleration voltage to a value under the excitation limit of the heavier element. The principle of the energy-dispersive spectrometer avoids the multiple-order coincidences.

10.2.3.2 Energy-dispersive Spectrometer

The magnitude of the detection pulses delivered by certain types of detectors is proportional to the photon energy. As well as measuring the intensity, those detectors may therefore serve to measure the energy in an X-ray beam, i.e. serve as energy-dispersive spectrometers.

Relationship between Energy Resolution and Counting Statistics

The energy resolution of a detector is related to the statistics of events leading to a detection pulse. Consider a photon of energy E_0 entering a detector medium of mean detection threshold W_m. The average number of resulting detection events is therefore

$$N_m = E_0/W_m$$

The actual number N of events shows statistical fluctuations about the average number N_m. The relative fluctuations are smaller the greater the number of events, i.e. the smaller the energy threshold W_m. If the photon-induced excitations are considered as independent events, their number follows a distribution centred about N_m with a half-width $\delta N = 2.35(N_m)^{1/2}$ (Eq. 10.9 and Figure 10.13). In fact, a certain independence between excitations leads to a somewhat more favourable statistic, expressed by the *Fano factor* ($F = 0.115$–0.135). The half-width then becomes $\delta N = 2.35 (FN_m)^{1/2}$. The photon energy measurement is given by the magnitude of the pulses they induce, and is therefore proportional to N. As a result, the energy follows in the same way a distribution centred about an average energy $E_m = N_m W_m$ which may be assimilated to the actual incident energy E_0; its half-width δE_d represents the intrinsic resolution of the detector as follows:

$$\delta E_d = 2.35(FE_0 W_m)^{1/2} \qquad (10.1)$$

The instrumental detector resolution is impaired by the electronic noise which may be expressed by means of a noise half-width δE_b, leading to the actual energy resolution of the whole detector

$$\delta E = [(\delta E_d)^2 + (\delta E_b)^2]^{1/2} \qquad (10.2)$$

The absolute energy resolution is better at low energy (light element lines) than at high energy (heavy element lines). The relative energy resolution varies inversely.

The semiconductor counter is the only detector to have a detection energy threshold low enough to enable it to operate as a spectrometer with an adequate energy resolution.

In the current X-ray energy range, the best result is achieved by the Si(Li) detector ($W_m = 3.8$ eV). Noise increases with the diode capacity, therefore leading to the use of diodes with a limited surface when resolution is the main required quality. The proportional counter and the scintillation counter cannot serve as spectrometers; however, their energy-related pulse magnitude permits them to provide energy discrimination.

Si(Li) Spectrometer

Resolution. A diode surface of 12 mm² leads to a noise half-width $\delta E_b \cong 100$ eV and therefore to a spectrometer resolution of 160 eV for Mn-Kα ($E = 5894$ eV), whereas its intrinsic resolution in the absence of any noise would have been 124 eV.

Efficiency and Signal-to-Noise Ratio. The efficiency of an Si(Li) spectrometer equals virtually 100% in a wide energy range, about 2–20 keV. The signal-to-noise ratio is lowered by a factor of 2–10 over that of a wavelength-dispersive spectrometer.

Artefacts due to the Spectrometer. Escape Peak and Pile-up Peak. The escape peak effect is related to the principle of the ionisation counter. When the energy E_0 of an incident monochromatic radiation to be measured exceeds the excitation energy of the K level of detector atoms, a certain number of incident photons induce the emission of the K radiation of those atoms. A given element is relatively transparent to its proper K emission. As a result, a fraction of the generated secondary detector K photons escape without inducing any other detection event and their energy E'_K is therefore lost in the measurement. For the detector this effect simulates the existence of two incident photon energies, E_0 and $E_0 - E'_K$. The resulting spectrum therefore shows two energy distribution peaks, the normal peak centred about E_0 and a wider peak, the so-called *escape peak*, centred about $E_0 - E'_K$. This effect becomes significant when the K excitation cross-section of the detector atoms is great, i.e. for heavy detector atoms. This is not the case for the Si(Li) detector.

Examples of Strong Escape Peak Occurrences. Proportional counter with Kr($W_K = 14.32$ keV) and incident radiation Zr-Kα (15.73 keV); scintillation counter NaI($W_K = 33.15$ keV) and incident radiation La-Kα (33.24 keV).

High count rates may lead to two (or more) incident photons of respective energies E_0 and E'_0 participating in the same detection event. This *pile-up effect* results in a pulse simulating the presence of a photon of energy. $E_0 + E'_0$.

Spectrometer Layout. The associated amplifier delivers pulses of magnitudes proportional to the respectively detected photon energies. A multichannel analyser (MCA) selects the pulses by means of an analogue–digital converter which assigns to each pulse a number ranging from 1 to N proportional to its magnitude. This number is the digital expression of the related photon energy. The MCA disposes of N channels (currently 2048 or 4096) reacting to the linearly increasing numbers. All the pulses with the same assignment number N are stored in the related channel N. After a given count time, depending on the required statistical accuracy, the channel contents represent the energy spectrum in the form of a histogram which can be visualised on a display screen (Figure 10.8).

The MCA is connected to a computer which provides the processing of the data stored in the channels. Various operations can be performed: pulse processing including

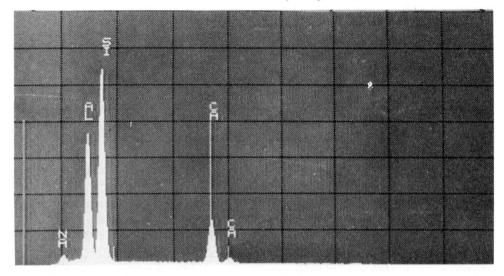

Figure 10.8. Photographic record of an I(E) spectrum from the MCA display screen of an Si(Li) spectrometer. The sample is a calcium feldspar ($CaAl_2Si_2O_8$) containing some sodium. $E_0 = 15$ keV. The pointer indicates the respective positions and intensities of the calcium K lines. The L lines are out of the scale. Notice the easy separation of Al-Kα and Si-Kα on the one hand, and the limit detection of sodium on the other (Dept. Electron Microscopy, Lab. Crystallography, Louis Pasteur University, Strasbourg).

Table 10.3. Average specifications of X-ray spectrometers. The data correspond to the K lines of medium elements in standard conditions. They may change significantly according to the instrumental layout

Spectrometer	Resolution (Kα–Mn)	Efficiency	Signal-to-noise ratio	Detection limit Z	Maximum count rate
Wavelength dispersive	20 eV	10^{-3}–10^{-2}	10^3	4-Be	10^6 (per element)
Energy-dispersive Si(Li)	160 eV	1	10^2	9-F (5-B windowless)	10^5 (whole spectrum)

pile-up rejection and dead time correction; identification of characteristic lines by comparing with stored theoretical spectra of the elements; intensity integration over a predefined energy window; deconvolution of composite peaks into the individual Gaussian components; spectrum smoothing; background subtraction and matrix correction, etc. Colour video increases the display possibilities and high-capacity storage systems such as optical discs (CD ROM) provide new possibilities for data banks.

Comparison of X-ray Spectrometers

Table 10.3 summarises the main characteristics of X-ray spectrometers. The development of new semiconductors might lead to improved specifications of the EDS. The advent of diodes working at room temperature would avoid the dependency on liquid nitrogen feeding. On the other hand, further improvement of the WDS seems improbable.

10.3 ELECTRON AND ION DETECTION AND MEASUREMENT

Electron–matter interaction and, *a fortiori*, ion–matter interaction, are much more intensive than in the case of X-rays. Owing to this fact and to their electrical charge, electrons and ions are easy to detect and to measure. The whole radiation path must be under vacuum, with a pressure sufficiently low to ensure a mean free path greater than the distances which the particles have to cover from their source to the detector.

There are some additional problems with ions, due to their sputtering effect on the detector components.

10.3.1 DETECTION AND INTENSITY MEASUREMENT OF AN ELECTRON BEAM

10.3.1.1 Qualitative Detection. Fluorescent Screen and Photographic Film

Fluorescent screens are much more sensitive to electrons than to X-rays, provided that their energy is of the order of magnitude of 1 keV or more. As a result electron diffraction patterns may be observed directly, as well as electronic images. Relative intensity measurements are sometimes performed on the screen through photometry.

Photographic films are very sensitive to electrons. Because of strong absorption, the sensitised emulsion is thin. Fine-grain emulsions are needed for imaging purposes. Organic films have too high a vapour pressure to be set up in high-vacuum enclosures. Conditions for quantitative intensity measurements are similar to those for X-rays.

10.3.1.2 Direct Measurement

An electron beam is equivalent to an electrical current. The most obvious method of measuring intensity is therefore the direct one, by means of an electrode shielded by a Faraday cylinder, centred on the beam to be measured (Figure 10.9a). Diffracted beam intensities may be smaller than 1 pA. Their measurement requires a very sensitive electrometer and an efficient shielding against stray charges (Burggraf and Goldsztaub, 1962). Except for measurements requiring high accuracy, indirect electron detection is therefore preferred for current applications.

10.3.1.3 Indirect Measurement. Electron Detectors

Electrons produce ionisation effects similar to those of X-rays in a detection medium. The same types of detectors may therefore be used, with the difference that, due to the high electron absorption (Table 8.5), even ultra-thin entrance windows are prohibited. Only solid state detectors operating in vacuum are therefore appropriate.

Scintillation Counter

Mineral or organic scintillators may be used for electron detection:

1. *Mineral scintillators* may be monocrystalline or polycrystalline. Their dead time is of the order of $3-8 \times 10^{-9}$ s, their lifetime reaches some 1000 h (e.g. cerium-doped yttrium silicate).

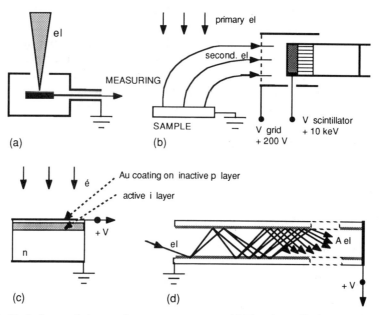

Figure 10.9. Techniques of electron beam measurement. (a) Faraday cylinder combined with an electrometer; (b) scintillation counter with a post-acceleration grid for low-energy electron detection; (c) surface junction Si semiconductor counter. A thin gold layer on the faces provides electrical contacts for biasing; (d) channeltron, working of a channel with a multiplication factor A.

2. *Organic scintillators* are made of a resin which has been doped by a scintillating compound (e.g. *p*-terphenyl, diphenyloxazole). Their advantage is a short dead time (about 10^{-9} s) and an easy manufacturing in any convenient form. Their handicap is a short lifetime of some hours only, depending on the count rate.

The scintillator is combined with a conventional photomultiplier through a light guide (Figures 10.5, 10.9b).

Low-energy electrons (< 1 keV) are absorbed by the thinnest surface layer. Their energy is generally insufficient to induce scintillations. Before detection, they have previously to be accelerated to several kiloelectronvolts (Figure 10.9b).

Semiconductor Detector

The intrinsic zone in a silicon p–n diode can act as an electron detector as well as an X-ray detector. Because of high electron–matter interaction, the active layer may be relatively thin for a good detection efficiency. In order to reduce the entrance dead zone, the junction is placed as close as possible to the surface. Depending on the energy to be measured, the detector can be a diode with a surface junction prepared through ion implantation and thermal diffusion (e.g. boron ions on n-silicon), with an active zone thickness of 100–300 μm, or a surface barrier MOS semiconductor with an active thickness of 300–1000 μm. The dead zone thickness is about 0.1 μm. Those detectors are operated at room temperature.

Channeltron

In the scintillation counter above, electron detection could have been achieved without the scintillator and the photocathode. However, the conventional diode electron multiplier would have a somewhat poor efficiency as an electron counter. A very simple and efficient electron multiplier is the so-called *channeltron* (Figure 10.9d). A straight or curved glass tube has its inner surface coated with a layer of high resistance and a high secondary electron emission yield (e.g. tin, tin oxide). Such a device plays the role of a continuous electron multiplying diode as well as of a continuous voltage dividing resistor. A voltage of about 1 kV provides the required voltage gradient along a resistance of about $10^9\,\Omega$. Electrons entering the (often cone-shaped) tube at the grounded negative side, are multiplied by successive secondary electron emissions at each impact and accelerated in zigzag paths towards the anode at the end. The *multiplication factor* depends on the length-to-diameter ratio and reaches currently 10^4–10^8. The dead time is determined by the transfer time which also depends on the length; it is of the order of 0.1 μs. Efficiency for electron detection approaches 100%.

This design of a tubular detector-multiplier can likewise be used for detecting any radiation which induces electron emission when impacting on the inner surface. This is the case for X-rays with a rather small efficiency of about 7% and for ions with an efficiency reaching 85%.

A compact parallel assembly of millimetre-long capillary tubes build up a so-called *microchannel plate*. The resulting detector can have an active surface of several square centimetres or more. The small length results in a very short dead time of about 10^{-9} s, allowing a high count rate. Such devices are particularly designed for image intensifying (electrons or X-rays).

10.3.2 ELECTRON SPECTROMETERS

Electron spectrometry requires the setting up of the $I=f(E)$ spectrum. This is provided by electron spectrometers which can operate either with or without angular dispersion.

10.3.2.1 Electron Spectrometer with Angular Dispersion

This most common spectrometer type is based on electron deviation in an electric or a magnetic field. For a good efficiency and resolution, it is mostly focusing.

Electrostatic Analysers

Cylindrical Mirror Analyser. The cylindrical mirror analyser (CMA) is a focusing electron spectrometer which has the advantage of operating in the forward direction. As illustrated in Figure 10.10a, electrons emitted from a point source S at a determined point on the cylinder axis, with a given selection energy E, determined by the negative voltage V applied to the external electrode, cross the slits and converge to a focusing point F on the axis. Electrons with energies different from E are collected by the electrodes. Scanning the voltage V and plotting the electron intensity measured at F by means of a detector (e.g. channeltron) results, after adequate calibration, in the spectrum $I=f(E)$. The source position is critical for good focusing (Sarel, 1967; Palmberg et al, 1969).

Electrostatic Prism Analyser. The electrostatic prism can be regarded as a sector of a cylindrical capacitor of angle ϕ and of mean radius r (Figure 10.10b).

A variable radial electrical field E results from the adjustable voltage supplied to the electrodes. Consider an electron or any particle of charge e and mass m travelling in a plane normal to the sector axis, entering the prism on a path tangential to the circle of radius r. Its trajectory follows that circle if its energy $E = eV = mv^2/2$ has a value such that the electrical force is balanced by the centrifugal force on the circle, i.e.

$$\frac{mv^2}{r} = eE \Rightarrow V = \frac{Er}{2} \qquad (10.3)$$

It then leaves the prism through the exit slit F, finally resulting in an energy selection $E = eV$. The system becomes focusing with a sector angle $\phi = 127.2°$. Any electron of energy eV crossing the entrance slit S, with a certain direction spread, will leave through the exit slit F.

A hemispheric capacitor (*hemispheric analyser, HSA*), operating in the same geometrical conditions, achieves a point focusing.

Scanning the voltage and plotting the intensity, measured at the exit slit F by means of an electron detector, results in setting up the spectrum $I = f(E)$ after an adequate calibration of the device.

Magnetic Prism Analyser

A magnetic prism is a sector of angle ϕ submitted to a homogeneous magnetic field H normal to its plane (Figure 10.10c). Consider an electron or any particle with a

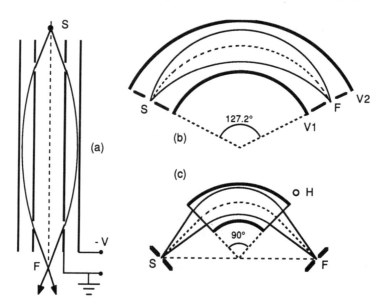

Figure 10.10. Basic outline of angular dispersive electron spectrometers. (a) Cylindrical mirror analyser (CMA); (a) focusing electrostatic prism analyser with an angle of 127°; (c) focusing magnetic prism analyser.

charge e, a mass m and a resulting energy $E = eV = mv^2/2$ entering normally to the field through the slit S. It is subjected to a force HeV directed normally to the field and to the trajectory. It therefore follows a circle of radius r such that the magnetic force is balanced by the centrifugal force. This is expressed as follows:

$$\frac{mv^2}{r} = Hev \Rightarrow V = \frac{H^2 r^2}{2} \frac{e}{m} \qquad (10.4)$$

Setting up the entrance slit S and the exit slit F in such a way as to determine a circle of radius r results in selecting electrons of energy $E = eV$. The spectrometer becomes focusing, with respect to an incident angular spread, when both slits are in a straight line with the apex of the sector. Scanning the magnetic field strength and plotting the electron intensity measured at the exit slit F by means of a detector (e.g. channeltron) results, after adequate calibration, in the spectrum $I = f(E)$.

Resolution

The energy resolution of an angle dispersive electron spectrometer depends on several instrumental factors, such as the slit aperture and the field stability. It is currently better than 10^{-2} and can reach 10^{-4} with high-resolution spectrometers, e.g. with hemispheric analysers.

Energy Filtering

When operated with a constant field at a value corresponding to a selection energy E, any of the above-mentioned electron spectrometers acts as an energy filter. Combining a magnetic prism with an electrostatic mirror leads to a device where the filtered electron beam is in line with the incident beam. Such a filter is well designed for filtering the electron beam transmitted through the sample in an electron microscope (Castaing and Henry, 1962) and in a secondary ion emission microscope.

10.3.2.2 Electron Spectrometer without Angular Dispersion

Solid State Detector

The magnitude of the pulses delivered by ionisation detectors is proportional to the incident electron energies. Those devices, e.g. semiconductor electron detectors, could therefore be used for electron spectrometry in the same way as for X-rays. In practice they serve for energy discrimination, but not for energy measurement.

Retarding Voltage Spectrometer

This type of spectrometer (Figure 10.11) is made up of concentric hemispheric electrodes with the electron source at its centre. In fact it acts as a variable high-pass filter. It has an excellent efficiency thanks to a collection angle which reaches 2π st. Increasing the retarding voltage V_r from a zero value up and recording the resulting intensity on the

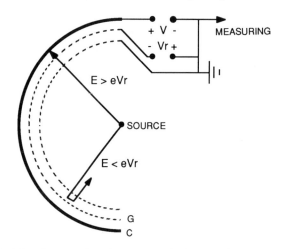

Figure 10.11. Principle of measuring the energy of electrons issued from a small source (e.g. primary electron beam) by means of a variable retarding voltage V_r. C is the electron collector (currently a fluorescent screen) at a positive voltage V, and G is the retarding grid at the negative voltage V_r.

collector electrode provides a graph of the integrated intensity of electrons with an energy $E \geqslant eV_r$

$$I(V_r) = \int_{eV_r}^{\infty} I(E) \, dE \qquad (10.5)$$

The actual energy spectrum $I(E)$ is provided by deriving the function $I(V_r)$. This is carried out by means of a sinusoidal modulation superimposed on the retarding voltage. This device is well designed in combination with a low-energy electron diffractor, which utilises the electron optics by simply adding a further grid electrode for measuring.

10.3.3 DETECTION AND INTENSITY MEASUREMENT OF AN ION BEAM

Ions are charged particles. The electron detection technology is therefore generally applicable to ions. An additional difficulty is due to the damage to electrodes caused by ion sputtering and ion implantation.

A fluorescent screen cannot be used directly. However, it can detect ions indirectly through the secondary electrons emitted by a conversion electrode preceding the screen (e.g. in the ion microscope).

Photographic recording is possible. It is widely used for the detection of fission traces in nuclear physics (ionography), as well as in mass spectrometry.

Direct intensity measurement by means of an electrometer poses no particular problems, except the previously mentioned sputtering and implantation effect.

Ion intensity measurements can be carried out by an electron multiplier, e.g. a channeltron. Impacting ions generate secondary electrons which are multiplied in the usual way. In order to avoid ion bombardment effects on the detector, indirect ion intensity measurements are often achieved by means of a conversion electrode which emits secondary electrons when subjected to ion bombardment. Those electrons are measured in a classical way, e.g. by means of a channeltron. Ion beam intensities down to 10^{-18}–10^{-20} can thus be measured.

These methods are suitable both for positive and for negative ions.

10.3.4 ION SPECTROMETERS. MASS SPECTROMETRY

Angular dispersion spectrometry performed either by an electrostatic prism or by a magnetic prism is suitable for ions as well as for electrons. Equations (10.3) and (10.4), stating the electron energy selection, are valid for ions. However, what is interesting for ion spectrometry is not the energy but the characteristic mass to charge ratio, currently called *mass ratio*. This leads to *ion mass spectrometry*.

10.3.4.1 Electrostatic Prism Energy Selector

As shown by Eq. (10.3), a given value of the electric field in an electrostatic sector of radius r selects incident ions which have been subjected to the related accelerating voltage V, whatever their mass ratio. Focusing with respect to an incident direction spread is achieved when the electrostatic prism has a sector angle of 127°.

10.3.4.2 Magnetic Prism Mass Analyser

Ions have mass ratios $M = m/e$ which vary according to their nature and which allow them to be characterised and selected in an ion beam. Mass selection can be achieved by means of a magnetic prism. Ions entering a magnetic field normal to their trajectory will be deflected along a circle, the radius of which varies according to their mass ratio, as can be seen from Eq. (10.4). For a magnetic sector of a given radius r and a given initial ion acceleration voltage V, the ions transmitted through the slits of the prism are those with a mass ratio given as follows by Eq. (10.4):

$$M = \frac{m}{e} = \frac{r^2 H^2}{2V} \qquad (10.6)$$

When scanning the magnetic field strength, an ion detector at the exit slit F provides the mass spectrum $I = f(m/e)$ of the incident beam, after adequate calibration. The peaks of the spectrum lead to characterisation of the ions in the beam.

However, Eq. (10.6) shows that the values of m/e selected in this way also depend on the acceleration voltage of the ions. With a currently observed energy spread at the ion source, the mass resolution may be insufficient to separate ions of almost the same mass ratio. In a *high-resolution mass spectrometer*, the magnetic sector is therefore preceded by an electrostatic prism *energy selector* which eliminates the ions with an accelerating voltage significantly different from the selection value V. Such a *single focusing* mass spectrometer (focusing with respect to an incident direction spread only) has a good mass resolution, but a poor transmission factor.

The *double-focusing* mass spectrometer (Figure 10.12) combines a high resolution with a high transmission factor. Its basic principle is outlined in Figure 10.12. It combines an electrostatic sector and a magnetic sector in such a way that the exit focus of the first is the entrance focus of the second, which is possible thanks to the widening of

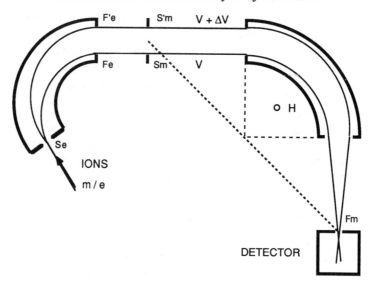

Figure 10.12. Layout of a prism mass spectrometer with double focusing. Only the focusing with respect to a spread of the ion acceleration voltage is illustrated.

the corresponding apertures. In this layout, focusing is achieved for an incident spread, both in direction and in energy, hence the denomination 'double focusing'. Most high-resolution ion spectrometers use this design (mass resolution up to 5000–10 000).

10.3.4.3 Quadrupole Mass Analyser

A quadrupole is a set-up consisting of four cylindrical electrodes parallel to the axis and displayed symmetrically around the incident ion beam in a square array. The electrodes are diagonally paired. A high-frequency voltage V_{hf}, superposed on a constant voltage V_{dc}, is applied between the two pairs. The high-frequency voltages are in phase opposition to the two pairs. Ions entering the analyser move along oscillatory trajectories. For a given value of the dc and the high-frequency voltage, ions with a certain mass ratio M are subjected to limited oscillations and can therefore cross the quadrupole along a stable trajectory for detection. The mass resolution is determined by the value of the ratio V_{dc}/V_{hf}. The mass spectrum is acquired by varying the potentials while keeping constant their ratio V_{dc}/V_{hf}. For a sufficient resolution to be achieved, an initial energy filtering is also required for this spectrometer. It has a poorer resolution (order of 2500) than the above-mentioned double focusing prism spectrometer. This drawback is balanced by a higher scanning speed.

10.3.4.4 Time of Flight Mass Analyser

Time of flight mass selection (TOF) is based on the fact that ions with the same accelerating voltage V, but with different mass ratios, travel at different speeds as shown by the following relations

$$eV = mv^2/2 \Rightarrow v = (2eV/m)^{1/2}$$

A TOF spectrometer selects incident ions by the time difference t of their flight between two electrodes separated by a distance l. For ions with the same accelerating voltage (energy prefiltering required), the flight time is proportional to the square root of the mass ratio, according to the following equation:

$$t = \frac{l}{v} = \frac{l}{(2V)^{1/2}} (m/e)^{1/2} \tag{10.7}$$

The time differences between successive mass peaks are of the order of 10^{-7}–10^{-8} s, thus requiring fast electronics.

The TOF spectrometer has a poor resolution (currently 500–600), but permits a high spectrum scanning speed.

10.3.4.5 Mass Resolution

The mass resolution determines the possibility of separating ions with an approaching mass ratio $M = m/e$. It is defined as the ratio $M/\Delta M$, where ΔM represents the smallest variation of M which can be significantly measured (e.g. the half-width of the peak from an ion of mass ratio M). Depending on the type of spectrometer, the mass resolution may vary from about 200 to values higher than 10 000 for high-resolution spectrometers. Notice that the mass resolution is expressed by a value which is greater the better the resolution. This definition is opposed to the conventional expression of the resolution (e.g. linear resolution of a microscope, energy resolution of a spectrometer) which is smaller the better the resolution.

10.4 NEUTRON DETECTION AND MEASUREMENT

Neutron detection is a matter for highly specialised laboratories. This survey is therefore limited to general principles.

Detection and Intensity Measurement. Thermal neutrons have energies which are too small to be directly detected by means of ionisation effects in the same way as electrons and X-rays. They are detected indirectly by the high-energy particles which they produce through nuclear reactions. The following reaction, based on the isotope 10 of boron, has a high cross-section:

$$n_0^1 + B_5^{10} \rightarrow Li_3^7 + He_2^4$$

The high-velocity light atoms produced are ionised when travelling through the detection medium. The resultant high-energy charged particles are detected in a conventional way. The actual detectors are either *proportional counters* filled with gaseous $^{10}BF_3$ or *liquid scintillators* containing ^{10}B. For an efficiency of 80%. a neutron gas detector has a length of the order of 1 m and a diameter of the order of 5 cm. The shielding against stray particles is currently provided by some 10 cm of boron paraffin or other neutron-absorbing materials.

158   Structural and Chemical Analysis of Materials

Energy Measurement. Neutron Spectrometer. Neutron spectrometers are mostly based on the principle of *angular wavelength dispersion* through an analysing crystal (see neutron monochromator). An alternative technique for ion spectrometry is the previously mentioned time of flight analyser; for a beam of identical particles the *time of flight* over a given distance is only a function of the particle velocity, i.e. its energy.

10.5 ACCURACY OF INTENSITY MEASUREMENTS. COUNT STATISTICS

10.5.1 FREQUENCY DISTRIBUTION

Detection pulses induced by any radiation can be considered as having a statistical distribution in time. For a sufficiently long count time, the measurements are therefore determined by the laws of statistics.

Consider the number N of pulses measured during a constant time lapse t_c. With repeated counts, this number N fluctuates about a mean value N_m. For an N_m large enough (e.g. $N_m > 100$), the measures N form a Gaussian distribution centred about N_m (Figure 10.13). The standard deviation σ of a Gaussian distribution is

$$\sigma = [\,\overline{(\Delta N)^2}\,]^{1/2} = (N_m)^{1/2} \qquad (10.8)$$

The half-width δ is slightly greater than the standard width 2σ and equals

$$\delta = 2.35\,\sigma = 2.35\,(N_m)^{1/2} \qquad (10.9)$$

The relative standard deviation ϵ is defined as follows:

$$\epsilon = \frac{\sigma}{N_m} = (N_m)^{-1/2} \Rightarrow \epsilon(\%) = 100\,(N_m)^{-1/2} \qquad (10.10)$$

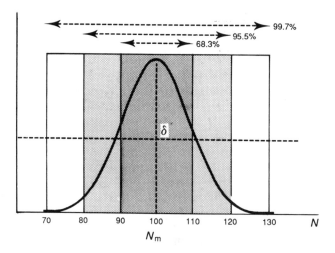

Figure 10.13. Gaussian distribution function for $N = 100$ and $\sigma = 10$. The area limited by the intervals $\pm\sigma$, $\pm 2\sigma$, $\pm 3\sigma$ is grey tinted, and the corresponding error function (in %). The ordinates express the frequency of a given count.

10.5.2 PROBABLE STATISTICAL ERROR

The probability p for a single measure N to fall inside the interval $N_m \pm \Delta N$ surrounding the mean value N_m (called *error function, ERF*) equals the area which is delimited by this interval under the Gaussian curve (Figure 10.13), with the total area of the curve taken as a unit. It can be calculated that the probability is only 68.3% ($p=0.683$) for a measure to fall inside the interval defined by the standard deviation σ. A measure given with a statistical error σ is therefore not reliable enough. A commonly adopted interval of 3σ corresponds to an error function of 99.7%.

In practice the statistical error is estimated on the basis of an experimental count N, considering its value to be great enough to be assimilated to the theoretical mean value N_m.

For an error function of 99.7%, the absolute and relative statistical count errors become

$$\Delta N = 3N^{1/2} \Rightarrow \frac{\Delta N}{N} = 3N^{-1/2} \tag{10.11}$$

Conversely, for a given relative statistical count error not to be exceeded, with the error function above, the *minimum number of counts* is

$$N = \frac{9}{(\Delta N/N)^2} \tag{10.12}$$

Note that when multiplying the total count by 100, the error is divided by 10 only.

10.5.3 PROBABLE COUNT ERROR IN THE PRESENCE OF BACKGROUND

The signal of a spectrum line, e.g. the characteristic X-ray line of an element, is always superimposed on a radiation background which has various origins. It is important for spectrometric analysis to estimate the error on the intensity measure in these conditions. Let N be the experimental total count on the line and N_0 the background count. The significant *proper count* on the line is therefore $N_c = N - N_0$. In order to evaluate the actual *background count*, the measure should be made in the same conditions on a *blank*, i.e. an object identical to the sample except for the considered element which is absent. This is difficult in practice. The usual way is to make two counts on both sides of the line, the mean value being taken as N_0. With respective standard deviations σ and σ_0 on N and N_0, the standard deviation on the proper signal $N_c = N - N_0$ becomes

$$\sigma_c = (\sigma^2 + \sigma_0^2)^{1/2} = (N + N_0)^{1/2} \tag{10.13}$$

With an error function of 99.7%, the statistical error estimated on the proper count is therefore finally

$$\Delta N_c = 3(N+N_0)^{1/2} \Rightarrow \frac{\Delta N_c}{N_c} = 3\frac{(N+N_0)^{1/2}}{N-N_0} \tag{10.14}$$

Example

Experimental count $N = 10^6$
Without background $N_0 = 0$ $\Delta N/N = 3 \times 10^{-3} = 0.3\%$
With background $N_0 = 10^5$ $\Delta N/N = 3.5 \times 10^{-3} = 0.35\%$
                         $N_0 = 950\,000$ $\Delta N/N = 28 \times 10^{-3} = 2.8\%$

When comparing with the example without background, note that in the first case (low background $N/N_0 = 10$) the accuracy is virtually unaltered, whereas in the second case (high background, limit of detection $N/N_0 \cong 1.05$), the uncertainty is multiplied by a factor of 10.

The above results are important for estimating the *lower detection limit* of an element in qualitative spectrometry and the *accuracy of analysis* in quantitative spectrometry. They also apply to the measurement of diffracted intensities in view of crystal structure analysis.

According to Eq. (10.14), it would seem that the statistical errors could be minimised *ad libitum* by simply increasing correspondingly the number of pulses to be counted, i.e. by increasing the count time. Even without a time problem, the count duration is, in practice, limited by long-term fluctuations (instrumental variations, sample deterioration through irradiation) which add *systematic errors* which eventually exceed the statistical errors.

10.5.4 DEAD TIME CORRECTION

The intensity to be measured is expressed by a count rate, the number of detection pulses per second. For high count rates near to the counter's maximum count rate, a dead time correction must be carried out. Each pulse is followed by a dead time τ. With a measured count rate $n_m = N_m/t$, there is a dead time $n_m \tau$ per second. The available detection time per second is therefore reduced to $(1 - n_m \tau)$ and the significant count rate n for intensity measurement becomes

$$n = \frac{n_m}{1 - \tau n_m} \tag{10.15}$$

instead of n_m.

Example. $\tau = 10^{-6}$ s; $n_m = 10^{-5} \Rightarrow n = 1.1 \times 10^5$. The error due to dead time is 10%.

REFERENCE CHAPTERS

Chapters 2, 5, 7, 9.

BIBLIOGRAPHY

[1] Cox C. P. *A handbook of introductory statistical methods.* John Wiley, New York, Chichester, 1986.

[2] Davis R., Frearson M. *Mass spectrometry*. John Wiley, New York, Chichester, 1987.
[3] Hawkes P. W. *Quadrupole optics*. Springer Tracts in Modern Physics. Springer-Verlag, Berlin, 1966.
[4] Mahesk K., Vij D. R. *Techniques of radiation dosimetry*. John Wiley, New York, Chichester, 1985.
[5] Richardson J. H., Peterson R. V. (eds) *Systematic materials analysis*. Vol. 4. *Mass spectrometry*. Academic Press, New York, 1978.
[6] Urlaub J. *Roentgenanalyse*. Vol. 1: *Roentgenstrahlen und Detektoren*. Siemens Aktienges., Berlin, Munich, 1974.

Part Three

DIFFRACTION TECHNIQUES APPLIED TO MATERIAL ANALYSIS

Diffraction methods are based on data provided by coherent scattering of radiation by materials, in particular those from diffraction data on crystallised solids.

1. Directions of scattering depend on the position and the size of the scattering elements.
2. Intensities of scattering depend on the nature of the scattering elements.

As a result, the information provided on materials refers to their structure.

The smallest dimensions liable to be observed by any diffraction method are limited by the wavelength of the radiation used. Diffraction experiments carried out by means of X-rays, electrons and neutrons provide structural data which cover a size span ranging from less than 1 Å to several micrometres, where the light optical techniques take over. The corresponding applications range, therefore, from the atomic structure to the micromorphology and the texture of materials.

The wave aspect of radiations predominates in diffraction techniques.

11

X-ray and Neutron Diffraction Applied to Crystalline Materials

X-ray diffraction techniques have been developing since 1912 when Max von Laue (Nobel prize-winner in 1914) and his assistants Friedrich and Knipping succeeded in achieving the first X-ray diffraction photograph of a crystal. Their experiment brilliantly confirmed the hypothesis of the lattice structure of crystals. The first crystal structures were soon analysed by Sir William Henry Bragg and his son Sir William Laurence Bragg (both Nobel prize-winners in 1915). Since then, the technique has continuously been developed, mainly thanks to improvements of intensity measurement techniques and to the progress of electronics and of computer calculation.

The experimental parameters are determined by Bragg's equation (Eq. 5.7)

$$2d(hkl)\sin\theta = n\lambda$$

The wavelength λ and the Bragg angle θ may be varied experimentally. For the diffraction conditions to be met on a set of planes (hkl), only one of these parameters may be independently fixed. The choice of the variable parameter determines two groups of X-ray diffraction methods:

θ set, λ varying: ***Laue method***
λ set, θ varying: ***Rotation method***
                      Powder method

Any of these three fundamental methods can be performed in the basic photographic set-up or in more or less sophisticated variants.

With the corresponding modification, the X-ray diffraction techniques also apply to ***neutron diffraction*** which is complementary to X-ray diffraction.

11.1 LAUE METHOD. CRYSTAL SYMMETRY AND ORIENTATION

11.1.1 BASIC PRINCIPLE

11.1.1.1 Primary Radiation and Sample

In the conventional set-up with an X-ray tube, the primary X-ray source is the continuous bremsstrahlung radiation from a heavy element target excited by means of an accelerating voltage as high as possible, but below the K excitation voltage.

A common target element is tungsten excited at 50–69 kV ($W_K = 69.5$ keV). The scale of wavelengths ranges accordingly from the cut-off wavelength ($\lambda_{min} = 12.4/E_0$, where E_0 in keV), to wavelengths of several ångströms limited by absorption in air.

The polychromatic synchrotron radiation is well designed for the Laue method and results in very short exposure times.

The sample is a stationary single crystal.

11.1.1.2 Instrumental Layout and Diffraction Pattern

The classical set-up is illustrated in Figure 11.1. It may be operated in the transmission mode or in the back-reflection mode (or both).

Any set of lattice planes (hkl) of the sample crystal selects, in the incident beam, the discrete wavelengths meeting the Bragg condition, i.e. a fundamental wavelength λ_0 and its n-order harmonics according to the relation

$$\lambda = \frac{2d(hkl)\sin\theta}{n} = \frac{\lambda_0}{n} \tag{11.1}$$

The set of planes reflects these selected wavelengths in the same way as a mirror reflects light waves. To any set (hkl) will therefore correspond a spot on the film. Each set of

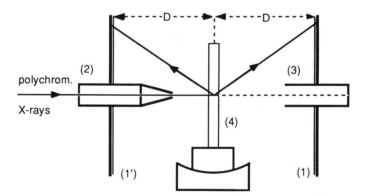

Figure 11.1. Basic diagram of the Laue method: (1) and (1′) photographic plate or flat film (respectively transmission mode and back-reflection mode); (2) beam limiting collimator; (3) exit aperture (absorption pit, sometimes reduced to a small lead disc pasted on the film pack); (4) specimen mounted on a goniometer stage with axes Φ (vertical, parallel to the drawing plane) and Ψ (normal to the drawing plane). In the transmission mode the sample is a thin crystal plate parallel to the film, with its median plane at a distance D. In the back-reflection mode, the sample may be bulk, with a plane surface parallel to the film, at a distance D.

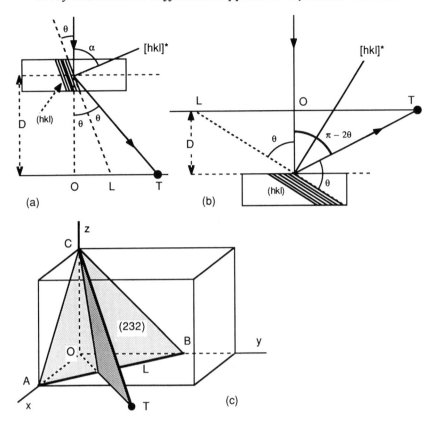

Figure 11.2. Diagram outlining the geometric relationship determining the position of a spot T on a Laue photograph. (a) Transmission mode; (b) back-reflection mode; (c) transmission mode; example of a reflecting plane (232). $OA = a$; $OB = 3/2b$; $OC = c = D$.

planes selects a different series of wavelengths; the related spots are therefore made up of different wavelengths and have accordingly different 'colours', which are not actually differentiated on the film. The geometry of forming a spot T(hkl) is outlined in Figure 11.2a.

Example. Consider the normal incidence on a (100) cleavage plate of a NaCl crystal (cubic, $a = 5.63$ Å) and the reflection by a set of lattice planes (111). The reciprocal lattice line [111]* normal to (111) makes an angle α with the incident beam (Figure 11.2a). Crystallographic calculation for a cubic crystal in the given orientation results in the following values:

$$\sin \theta = \cos \alpha = 1/\sqrt{3}; \quad d(111) = a/\sqrt{3} \Rightarrow \lambda_0 = 3.75 \text{ Å}; \quad \lambda = 3.75/n \text{ Å}$$

It can be seen that when operating at 69 keV, the set of planes (hkl) reflects the wavelengths 3.75 Å, 1.87 Å, 1.25 Å . . . with the cut-off limit of the tube $\lambda_{min} = 12.4/69 = 0.18$ Å. The limit corresponds to $n_{max} = 20$. There are therefore 20 wavelengths contributing to the spot T(111). For a set of planes to reflect at least one wavelength, the resolution condition $d(hkl) \geqslant \lambda/2$ must be met, i.e. $d \geqslant \lambda_{min}/2 = 0.09$ Å.

11.1.1.3 Indexing the Laue Pattern

Indexing of the Laue pattern consists of assigning to any spot T the indices hkl of the corresponding set of lattice planes to which it has given rise. Indexing is straightforward only for specific orientations, with two crystallographic axes parallel to the film.

Transmission Mode

The reflection angle is limited in practice to $\theta < 30°$. For a thin crystal and a well-collimated beam, the active planes (hkl) can be assimilated in a unique reflecting plane, a bisector of the angle formed by the incident and the reflected beam (Figure 11.2a). Indexing a spot $T(hkl)$ consists of finding the intersections of the reflecting plane with the crystallographic axes xyz determined by the basis vectors **abc**. The intercepts are proportional to $a/h, b/k, c/l$.

Consider the simple instance of a crystal with a right-angled lattice, the xy-axes plane parallel to the plane of the diffraction pattern, the incident beam parallel to the z-axis. The (hkl) indices do not depend on the absolute parameter values a, b, c; they are determined solely by their ratios. One may therefore consider a homothetic unit cell such as $c = CO = D$ (Figure 11.2c). The reflecting plane intersects the x, y, z-axes in A, B, C. With A and B in the plane of the diffraction pattern, the intercepts OA and OB can be directly measured on the photograph, i.e. the indices hkl determined, provided that one is able to materialise on its plane the impact point O of the incident beam, the lattice axes xyz and the intersection L of the reflecting plane.

Crystallographic Axes. In the above example of an incidence parallel to a crystallographic axis, the origin O coincides with the beam impact. The crystallographic axes are determined by considerations on the symmetry of the pattern. In the instance of an orthorhombic lattice, they are the twofold symmetry axes. High-order symmetries result in ambiguities which can be lifted *a posteriori* by choosing the orientation resulting in the smallest Miller indices. With low-order symmetries (monoclinic, triclinic) the problem is more difficult to resolve (Mauguin, 1928).

Intersection of the Reflecting Plane. Mauguin Abacus. The intersection L may be positioned by means of the so-called Mauguin abacus. The operation of this abacus is outlined in Figure 11.3 which shows that

$$OL = D \tan \theta; \quad LT = D (\tan 2\theta - \tan \theta) \tag{11.2}$$

Measuring the intercepts of the line L with the crystallographic axes leads to the Miller indices hkl as follows:

$$OA = \frac{a}{h}; \quad OB = \frac{b}{k}; \quad OC = \frac{c}{l} = D \Rightarrow \frac{h}{l} = \frac{D}{OA} \frac{a}{c}; \quad \frac{k}{l} = \frac{D}{OB} \frac{b}{c} \tag{11.3}$$

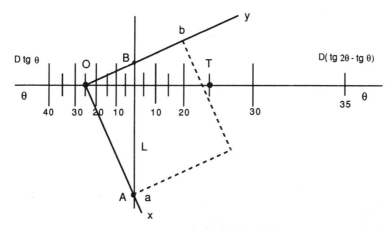

Figure 11.3. Indexing a Laue pattern by means of the Mauguin abacus. In the example shown the reflection T corresponds to an incident angle $\theta = 25°$ and to a set of planes (121). Lengths equal respectively to $D \tan \theta$ (left) and $D(\tan 2\theta - \tan \theta)$ (right) are placed on both sides of an axis perpendicular to the origin line L for increasing values of the angle θ. As outlined in Figure 11.2a, the points O (left) and T (right) correspond to equal values of θ. For indexing a given spot T on a Laue photograph on which the crystallographic axes have been drawn, the axis of the abacus is made to coincide with the line OT and translated until equal values of θ are read on both sides of the axis for O and T. The origin line L then materialises on the trace of the intersection of the reflecting plane and provides, in addition, the intercepts OA and AB with the crystallographic axes x, y, thus leading to the indices according to Eq. (11.3).

The crystal parameters a, b, c are given by other diffraction techniques, e.g. powder method or rotation method. The Laue reflections close to the beam impact are those with indices such as $h + k \geqslant l$.

Zone Axis. The locus of the rays reflected by the lattice plane sets which have a lattice line $[uvw]$ as a common zone axis, is the circular cone of half-angle ω represented in Figure 11.4. The corresponding spots are therefore located on the intersection of that cone with the film, i.e. an ellipse for $\omega < 45°$, a hyperbola for $\omega > 45°$. For a set of planes (hkl) to belong to a zone determined by two sets $(h_1 k_1 l_1)$ and $(h_2 k_2 l_2)$, the determinant of the matrix of those nine indices must be equal to zero

$$\begin{vmatrix} h & k & l \\ h_1 & k_1 & l_1 \\ h_2 & k_2 & l_2 \end{vmatrix} = 0$$

Back-reflection Mode

Due to the plane film set-up, the angular range is virtually limited to $\theta > 60°$ (Figure 11.2b). Reflections near the centre of the pattern correspond to an angle close to $\pi/2$, i.e. to indices such as $h + k \leqslant l$. The intersection L of the reflecting plane is therefore exterior to OT. The relations for the abacus become accordingly

$$\text{OL} = D \tan \theta; \qquad \text{OT} = D (\tan \pi - 2\theta) \tag{11.4}$$

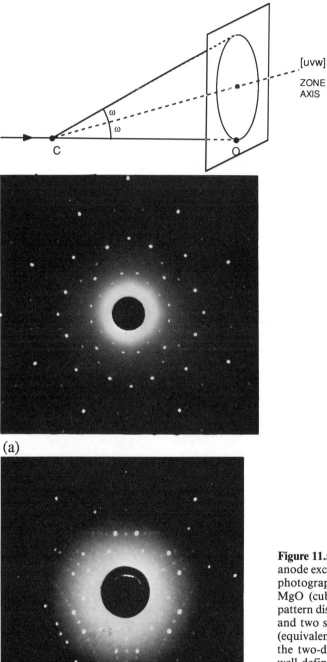

Figure 11.4. Locus of reflections on planes with zone axis $[uvw]$.

Figure 11.5. Laue photographs. Tungsten anode excited at 60 kV. (a) Transmission photograph of a cleavage plate (001) of MgO (cubic) in normal incidence. The pattern displays a fourfold symmetry axis and two sets of symmetry planes at 45° (equivalent to twofold symmetry axes on the two-dimensional pattern). Note the well-defined zone ellipses. (b) Back-reflection photograph, in normal incidence, on a crystal of tourmaline (trigonal) cut normally to the threefold axis. Note the threefold symmetry. (Lab. Crystallography, Louis Pasteur University, Strasbourg).

The abacus is operated in the same way as for transmission patterns. The loci of zone reflections are hyperbolas.

11.1.2 APPLICATIONS

The Laue method provides data on the spatial orientation of the sets of lattice planes in a crystal, with respect to the incident beam, whatever the external form of the sample. The data are applied to determining point symmetry and crystal orientation.

Table 11.1. Laue groups. International notations

Crystal system	Point group symmetry	Laue groups	No.
Triclinic	1 $\bar{1}$	$\bar{1}$	1
Monoclinic	2 m $2/m$	$2/m$	2
Orthorhombic	2 2 2 $m\,m\,2$ $m\,m\,m$	$m\,m\,m$	3
Trigonal	3 $\bar{3}$	$\bar{3}$	4
	3 2 $3\,m$ $\bar{3}\,m$	$\bar{3}\,m$	5
Tetragonal	4 $\bar{4}$ $4/m$	$4/m$	6
	4 2 2 $4\,m\,m$ $\bar{4}\,2\,m$ $4/m\,mm$	$4/m\,mm$	7
Hexagonal	6 $\bar{6}$ $6/m$	$6/m$	8
	6 2 2 $6\,m\,m$ $\bar{6}\,2\,m$ $6/m\,mm$	$6/m\,mm$	9
Cubic	2 3 $m\,3$	$m\,3$	10
	4 3 2 $\bar{4}\,3\,m$ $m\,3\,m$	$m\,3\,m$	11

11.1.2.1 Point Group Symmetry. Laue Groups

The spatial orientation of lattice planes provided by a Laue pattern is related to the point group symmetry (crystal class) of the crystal concerned. Due to the weak interaction, the Friedel law (5.3.2) generally applies to X-ray diffraction. The symmetry displayed by an X-ray diffraction pattern results, therefore, from adding a centre of symmetry to the point group symmetry, thus leading to the definition of 11 groups, the so-called *Laue groups* (Table 11.1).

11.1.2.2 Crystal Orientation

A crystal with a polyhedral form is easily oriented by means of its external symmetry, i.e. with an optical goniometer. In the more general instance of a crystal without any crystallographic form, its orientation requires a diffraction technique which analyses the orientations of its lattice planes. The Laue method is best suited for this aim.

Stereographic Projection. The direction of the set of normals to the lattice planes in a crystal is best represented by means of stereographic projection. For a known crystal species, the orientation is completely determined by the normals to two sets of lattice planes (*hkl*), i.e. by the corresponding reciprocal lattice lines [*hkl*]*. The geometrical relationship between the position T of a Laue reflection *hkl* and the stereographic projection S of the set of planes (*hkl*) is outlined in Figure 11.6 and expressed by the following equations:

Transmission

$$OS = R \tan \omega = R \tan \frac{\pi - \tan^{-1}(OT/D)}{4} \qquad (11.5)$$

Back reflection

$$OS = R \tan \omega = R \tan \frac{\tan^{-1}(OT/D)}{4} \qquad (11.6)$$

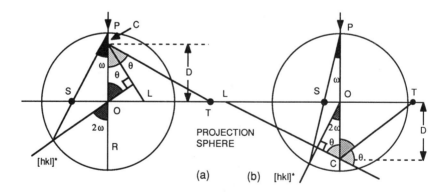

Figure 11.6. Geometric relationship between a Laue photograph (spot T) and the related stereographic projection (point S). (a) Transmission pattern; (b) back-reflection pattern. The points S, O and T are aligned. Therefore an abacus set up for a given value of the distance D and of the radius R of the projection circle (e.g. $R = 10$ cm) results in an easy construction of the stereographic projection corresponding to the spots of the diffraction pattern.

Operational Mode

1. Setting up the crystal on a goniometric stage with two perpendicular axes Φ and Ψ according to Figure 11.1.
2. Recording the Laue pattern.
3. Constructing the stereographic projections related to a few strong spots of the diffraction pattern.
4. Measuring the angles between those stereographic projections by means of a Wulff net.
5. Comparison of the resulting data with a table of precalculated angles between the main lattice plane sets of the crystal species, thus leading to identifying and therefore indexing the projections and determining the crystal orientation (Figure 11.7). For cubic crystals (e.g. most metallic crystals), the set of angles does not depend on the lattice parameters and has been tabulated (e.g. [9], *Int. tab.*, vol. 2).
6. Rotating the sample through the respective angles ϕ and ψ read on the projection in order to bring the chosen direction parallel to the incident direction.

Crystal orientation has a growing importance in various fields: cutting of gemstones and of synthetic single crystals (e.g. silicon or germanium for semiconductors) in a given crystallographic direction; determining the orientations of crystals in rocks, in metals, in ceramics.

Lattice plane deformations result in Laue spot distortions. The form of Laue spots can therefore be related to crystal deformations.

11.2 ROTATION METHOD. CRYSTAL STRUCTURE ANALYSIS

11.2.1 BASIC PRINCIPLE

11.2.1.1 Primary Radiation and Sample

The *monochromatic X-ray beam* consists of the characteristic Kα emission of a target

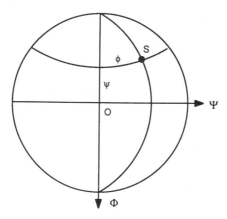

Figure 11.7. Locating the stereographic projection S of a set of planes (*hkl*) with respect to the goniometer axes Φ and Ψ. Respective rotation of angles ϕ and ψ about these axes brings the point S to the centre of the projection circle and therefore the normal to the (*hkl*) planes parallel to the incident X-ray beam.

174   *Structural and Chemical Analysis of Materials*

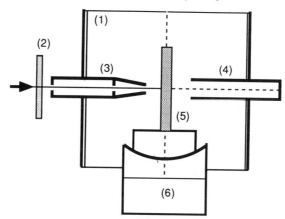

Figure 11.8. Rotation method. (1) Cylindrical camera; (2) β-filter; (3) beam limiting collimator; (4) exit pit; (5) crystal sample on a goniometric stage; (6) rotation motor.

element, filtered by means of the absorption edge or delivered by a crystal monochromator.

The *single crystal* is being rotated or oscillated. In order to record a reflection of order n on a set of lattice planes (hkl), the incidence angle θ must meet Bragg's condition, $\sin \theta = n\lambda/2d(hkl)$. Setting the crystal to successive Bragg incidence angles is achieved by rotating it about an axis.

11.2.1.2 Basic Layout

In the basic photographic set-up outlined in Figure 11.8, the sample crystal is rotated or oscillated about the axis of a cylindrical camera. More sophisticated devices, with or without film, have been designed for specific applications.

11.2.1.3 Display of the Diffraction Pattern

Interpreting the make-up of the diffraction pattern is straightforward through the Ewald construction (Figure 11.9). The crystal is rotated about a lattice line $[uvw]$. The set of reciprocal planes $(uvw)^*$ which includes all reciprocal lattice points hkl is therefore normal to the rotation axis. The indices are related through the equation of these planes as follows:

$$uh + vk + wl = m \quad (m \text{ integer, see Eq. 5.6})$$

The set of reciprocal planes $(uvw)^*$ intersects the reflecting sphere (or ES) along parallel and equidistant circles. As a result the locus of diffracted rays is a set of circular cones coaxial about the rotation axis, passing through those circles. The cones intersect the film along parallel lines. When unfolded on a plane, the impressed film shows, therefore, reflection spots along parallel layer lines. Two symmetrical spots, at a distance s from the centre of the pattern, correspond to any reciprocal point hkl (i.e. any possible reflection hkl).

X-ray and Neutron Diffraction Applied to Crystalline Materials

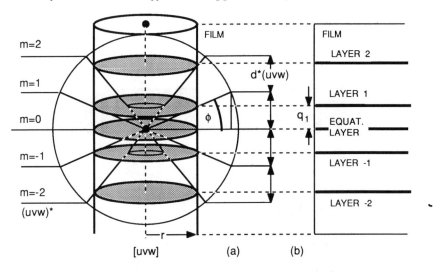

Figure 11.9. Rotation method and Ewald construction. (a) Display of the diffraction cones and layers; (b) diffraction layer lines on the plane film.

For a given crystal species, whatever the rotation axis, the total number of reflection spots equals the number of reciprocal lattice points included in the volume (torus) generated by the reflecting sphere when rotating about an axis passing through the origin of the reciprocal lattice and parallel to [uvw] (see Figure 11.12).

In normal practice, the rotation axis is chosen as a prominent zone axis [uvw] with small indices (e.g. [001]). Hence the lattice line parameter $n(uvw)$ is small and the reciprocal interplanar spacings $d^*(uvw)^* = 1/n(uvw)$ is large. The resulting pattern will thus display only a few layer lines, each including a large number of spots. The layering of the pattern is then very apparent, which is of prime importance for easy interpretation of the pattern.

11.2.2 INTERPRETING THE DIFFRACTION PATTERN

11.2.2.1 Lattice Parameters

Let q_1 be the distance between the zero-level layer line ($m = 0$) and the first-level layer line ($m = 1$). Figure 11.9b shows that

$$\tan \phi = \frac{q_1}{r}; \quad \sin \phi = \frac{d^*(uvw)}{L} = \frac{\lambda}{n(uvw)} \Rightarrow n(uvw) = \frac{\lambda}{\sin[\tan^{-1}(q_1/r)]} \quad (11.7)$$

where r is the radius of the camera and ϕ the angle of the diffraction cone with the equatorial plane.

The parameter of the rotation line [uvw] is thus determined by means of one measurement. Averaging several measurements of spot distances in symmetrical layers results in good accuracy. Rotating about successive crystallographic axes [100], [010], [001] leads to the lattice parameters a, b, c. In the most general case (triclinic,

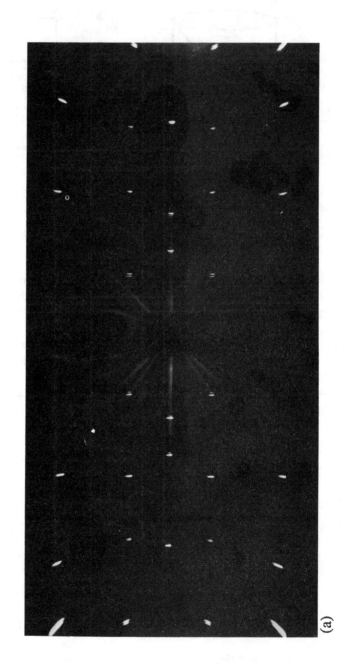

(a)

Figure 11.10. X-ray rotation diffraction photographs (Cu–Kα, Ni filter, 180 mm camera). (a) {100} Cleavage of MgO, cubic, all faces centred; rotation axis [001]; (b) rutile TiO$_2$ crystal, tetragonal, rotation axis [001]; the fog on the film is due to the Ti–K fluorescence excited by Cu–Kα. The lines radiating from the centre result from the continuous bremsstrahlung background. (Lab. Crystallography, Louis Pasteur University, Strasbourg).

monoclinic, orthorhombic lattices), three photographs are required. The number reduces to two exposures for lattices with a main axis (trigonal, tetragonal, hexagonal lattices) and to one for cubic lattices.

Adjusting the rotation axis parallel to the crystallographic axes is of prime importance. If the sample has a well-developed and indexed polyhedral form, the chosen lattice line, materialised by a crystal edge, is adjusted on the rotation axis by the vertical cross-wire of a microscope. In the general case of a single crystal without any polyhedral form (e.g. synthetic crystals), the adequate crystallographic orientation is carried out by means of the Laue method. The lattice angles are provided by the respective orientations of the crystallographic axes.

11.2.2.2 Indexing the Pattern. Intensity Weighted Reciprocal Lattice

Indexing is carried out by the Ewald construction.

Reflecting Sphere. A projection of a reflecting sphere of radius $L = 10$ cm is divided in $2\pi r$ equal parts when using a camera of radius r (normalised cameras: $2\pi r = 180$, 240 or 360 mm). The zero graduation is taken as the origin of the reciprocal lattice.

Reciprocal Lattice. Using the previously determined direct lattice parameters, a plane section diagram of the reciprocal lattice of constant $K = L\lambda$ is drawn on tracing paper, according to the rotation axis $[uvw]$ of the diffraction pattern.

Example (Figure 11.10a). MgO (cubic); rotation [001]; λ (Cu-Kα) = 1.54 Å; $2\pi r = 180$ mm.

1. Lattice parameter: $q_1 = 11.3$ mm $\rightarrow n(100) = a = 4.2$ Å.
2. Reciprocal lattice constant: $L = 10$ cm $\rightarrow K = L\lambda = 15.4 \times 10^{-8}$ cm^2.
3. Construction of the reciprocal plane (001)*, $m = 0$: indices of the reciprocal lattice points (Eq. 5.6): $hk0$; reciprocal parameters: $a^* = b^* = K/a = 3.66$ cm.

Indexing. Consider the instance of a rotation photograph about a lattice line [001] as shown in Figure 11.11.

Zero-level layer line, $m = 0$. The $hk0$ reflections on the equatorial or zero-level layer line are due to the $hk0$ lattice points of the reciprocal lattice plane $m = 0$ intersecting the equatorial circle of the reflecting sphere of radius 10 cm. When rotating the reciprocal lattice diagram about the zero graduation of the reflecting circle (materialised by a pin), its lattice points intersect the graduated circle. A given point $hk0$ determines an arc intercept S_0 whose length, expressed in terms of graduations, equals the distance s_0 in millimetres of the corresponding spot in the centre of the diffraction pattern.

m-Level layer line. Indexing an m-level layer is conducted in a similar way by setting the intersections of the reciprocal points hkm with a parallel circle of the reflecting sphere at a distance md^* (001) from the equatorial plane. A given point hkm determines the respective intercepts S_m and s_m on the circle and on the diffraction pattern.

The required reciprocal diagram is constructed by projecting the corresponding reciprocal plane hkm on the reciprocal plane $hk0$ ($m = 0$). In the particular instance of

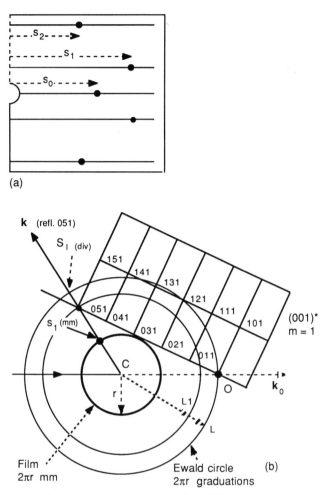

Figure 11.11. Diagram outlining the indexing of a rotation diffraction pattern about an axis [001]. Example of a right-angled lattice. (a) Half-pattern with measured spot distances s; (b) indexing a diffraction spot on the layer line $m = 1$, indices $hk1$. Example of a reflection 051.

a reciprocal lattice such as $c^* \perp a^*c^*$ (e.g. Figure 11.11), the successive projections of the reciprocal planes $hk1, hk2, \ldots hkm$ coincide with the plane $hk0$. In the general case, the successive projections are shifted.

The final indexation of the diffraction spots is carried out by comparing the constructed intercepts S_m of the reciprocal lattice points hkm with the measured distances s_m on the pattern. The theoretical diffraction pattern represented by the list of intercept values $S_m(hkl)$ corresponds to a crystal with the parameters of the sample, with a primitive lattice and without any translation symmetry. The systematic absence of specific diffraction spots with respect to that theoretical pattern leads to the setting up of the extinction rules, i.e. to gather information on the space group of the sample (see 5.3.3). An easy way to deduce the extinction rules is to draw a diagram of the intensity weighted reciprocal lattice (or diffraction reciprocal lattice) on which only the points corresponding to an existing diffraction spot are marked (e.g. see Figure 5.16).

11.2.2.3 Structure Analysis

An indexed rotation photograph contains any data required for structure determination. After adequate correction, the Fourier series of the intensities $I(hkl)$ leads to the Patterson function. Solving the phase problem leads to the crystal structure. In fact, the basic rotation method is not the ideal tool for performing accurate structure analysis of a single crystal. Since the advent of X-ray diffraction techniques, the tendency has been to develop improved variants.

11.2.3 IMPROVED VARIANTS

The basic rotation method has a number of drawbacks:

1. All diffraction spots corresponding to a *plane* of the reciprocal lattice are concentrated on a *line* of the photograph. As a result, interpreting a diffraction pattern is slow and may lead to ambiguities when spots with similar s distances superpose. This is often the case for low-symmetry crystals.
2. Rotating the crystal about a fixed axis, normal to the incident beam, results in a *blind zone* and therefore in a loss of diffraction data. The useful volume of the reciprocal lattice is reduced to a torus as shown in Figure 11.12.
3. Intensity measurements on a photograph have a poor accuracy.

Because of these deficiencies, the basic rotation method is not actually intended for structure analysis. However, thanks to its cheapness and simplicity, it remains very useful for teaching purposes.

Photographic variants have been designed to offset the first two deficiencies. All the requirements for structure analysis are met by diffraction facilities with direct intensity measurements through detectors.

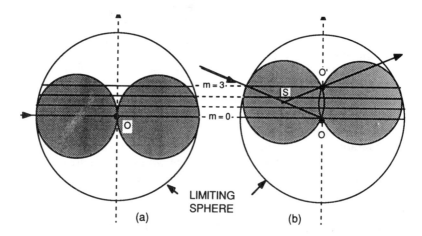

Figure 11.12. Diagram illustrating the blind zone. (a) Normal incidence resulting in a blind zone for non-equatorial layers ($m \neq 0$); (b) equi-inclination incidence for a layer of level $m = 3$, resulting in suppressing the blind zone for that layer. The section of the active volume explored by the reflecting sphere inside the limiting sphere has been displayed in grey tint.

11.2.3.1 Weissenberg Camera

The camera is cylindrical as in the basic rotation method, with two main differences:

1. A cylindrical screen selects one diffraction cone (Figure 11.9b), i.e. one given layer m.
2. The cylindrical film moves to and fro parallel to the axis, synchronised to the rotation, in such a way that a translation t corresponds to an angle of rotation ω of the crystal.

A Weissenberg photograph represents a distorted image (anamorphosis) of the distribution of lattice points in one reciprocal plane of level m. As a result, reciprocal lattice lines passing through the origin are represented on the photograph by a set of parallel spot lines, whereas any other reciprocal lattice lines are represented by a set of curves as shown in Figure 11.13. The settings of the selecting screen on the various layer levels are determined on an initial rotation photograph recorded without the translation motion. An easy indexing of the diffraction patterns is carried out by means of standard diagrams (e.g. in [9], *Int. tables*). In order to suppress the blind zone for a given recorded m-level layer, the incident beam is tilted in such a way that the incident beam and the diffracted beams make equal angles Ψ_m with respect to the equatorial plane. This equi-inclination set-up is outlined in Figure 11.12b and the tilt angle is given by the following equation:

$$\sin \psi_m = \frac{md^*(uvw)\lambda}{2} = \frac{m\lambda}{2n(uvw)} \tag{11.8}$$

11.2.3.2 Retigraph and Precession Camera

Unlike the Weissenberg camera, the retigraphs and the precession camera supply an undistorted recording of a reciprocal lattice plane.

The basic principle of such techniques can be outlined by Figure 11.12b. Recording the reflections corresponding to a zero-layer reciprocal lattice plane $(uvw)^*$, $m=0$, without a central blind zone, can be achieved by rotating the crystal about an axis $[uvw]$ parallel to OO' and selecting the related diffraction cone by means of a correspondingly positioned circular slit. Instead of a cylindrical film, consider a *plane film* positioned at O', parallel to the reciprocal lattice plane to be recorded. The crystal and the film are subjected to a joined parallel motion, respectively centred at S and O, in order for the lattice points of the reciprocal plane to intersect the reflecting sphere and to produce reflections. According to Ewald's condition, any time a reciprocal lattice point intersects the reflection sphere, a diffracted ray passes through that point and therefore impresses the film at its very position in the plane. The result is an undistorted photograph of a portion of the intensity weighted reciprocal lattice plane. The distance SO determines the scale of the photograph.

The actual techniques differ by the kind of motion applied to the crystal and the film.

In the *retigraph* (Jong and Bouman, 1938; Rimsky, 1952), the motion is simply a rotation of both the crystal and the plane film about the OO' axis. A given layer level is selected by displacing the layer screen along the crystal axis.

In the *precession camera* [4,5] a similar result is achieved with a precession motion replacing the axial rotation: in order to make the lattice points of a part of the selected

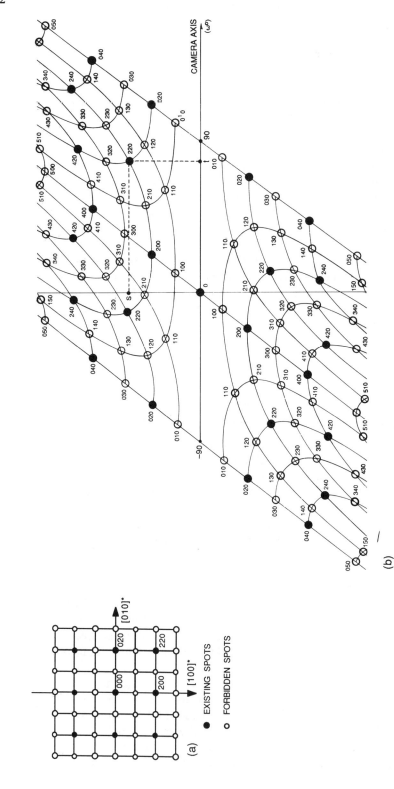

Figure 11.13. Relationship between a reciprocal lattice plane and the corresponding Weissenberg pattern. Example of MgO, cubic F. Arbitrary scales. (a) Reciprocal lattice plane (001)*, $m = 0$, containing the lattice points $hk0$; (b) theoretical Weissenberg pattern provided by a 90° rotation about [001] and 1 mm/degree translation; equatorial layer $m = 0$ containing the $hk0$ reflections. A spot is located by its coordinates s and t ($t = 0$ in a rotation pattern).

reciprocal plane intersect the reflecting sphere, the crystal with the layer selection screen on the one hand, and the film on the other, are subjected to a precession motion with the axis as the incident beam, with respective centres S and O. Translating the screen and the film results in selection of the layers.

The setting of the selecting screen on the successive m-level layers (i.e. determining the $d^*(uvw)$ distances) can be determined either from a simple rotation photograph or by recording the intersection of the diffraction cones with the film in the absence of any selecting screen and film motion. In addition to the $d^*(uvw)$ values, such a *cone–axis photograph* provides data on the crystal point symmetry, in a similar way as a Laue photograph.

The main advantage of those methods is to provide undistorted film records of reciprocal lattice planes. There still remains the problem of intensity measuring accuracy.

11.2.3.3 Single Crystal Diffractometer

Up-to-date crystal structure analysis requires accurate intensity measurements, especially when complex structures are involved. With this aim, photographic recording is replaced by detector devices associated with a goniometrical sample stage. The crystal and the counter are successively set to the respective positions meeting the diffraction conditions for all possible reflections. Any blind zone is thus suppressed. Data on the crystal lattice, previously determined by any other diffraction method (rotation or powder method), are fed into the associated computer which sets the goniometer automatically for measuring the *hkl* reflection intensities and subtracting the background intensities. The various intensity corrections (polarising, Debye–Waller and geometric corrections) are integrated in the programs.

11.2.3.4 Conclusion

Of the three basic X-ray diffraction methods, the rotation method is the most complete one for exploiting all the information provided by X-ray diffraction, especially in the form of the most advanced variants.

1. Measurement of *hkl* reflection angles leads to the determination of the lattice parameters without any ambiguity, even for low symmetry crystals.
2. Indexing of *hkl* reflections and assessing systematic absences leads to the determination of the space group.
3. Measuring the reflection intensities $I(hkl)$ leads to the establishment of the crystal structure. High-power computer programs are used to solve the phase problem and to calculate the Fourier series leading to the atomic positions.

A practical limit is set by the minimum crystal size required for adjusting it on the specimen stage and for providing a sufficient count rate for accurate intensity measurements.

More details on crystal structure analysis will be given in section 11.4.

11.3 POWDER METHOD. ANALYSIS OF POLYCRYSTALLINE MATERIALS

11.3.1 BASIC PRINCIPLE

11.3.1.1 Primary Radiation and Sample

The *monochromatic primary X-ray beam* is provided by a target anode with a medium atomic number (e.g. Cu, Mo) excited at the optimum voltage in order to achieve the highest ratio of characteristic K lines to the continuous background. In current powder facilities, an absorption edge filter (β-filter) selects Kα. In the most advanced set-ups, a crystal monochromator selects the Kα_1 line, resulting in an effectively monochromatic beam.

The sample is polycrystalline. It may be a powder or a polycrystalline solid material (e.g. metal, ceramic, rock material, etc.). Its morphology depends on the instrumental layout.

11.3.1.2 Display of the Diffraction Pattern

Consider the active volume of the sample to contain a very high number of crystallites with random orientation. There will be a certain number of them with any given

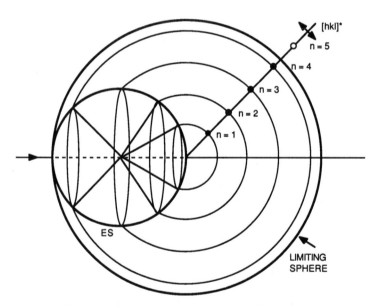

Figure 11.14. Using Ewald's construction for interpreting the diffraction cones originating from a set of lattice planes (hkl). The reciprocal space of a polycrystalline sample is the superposition of the individual reciprocal lattices with a common origin O. For a perfectly random orientation distribution, the locus of a reciprocal lattice point hkl is a sphere of centre O, of radius $r^*(hkl)$. The set of points on a reciprocal lattice line [hkl]* therefore determines a set of concentric spheres of radii $nn^*(hkl)$ which intersect the reflecting sphere along a set of parallel circles coaxial to the incident direction. As a result the locus of diffraction beams is the set of cones originating from the centre C of the reflecting sphere and passing through those circles. All spheres inside the limiting sphere ($r^* \leqslant 2/\lambda$) give rise to a diffraction cone.

orientation, such that the primary beam meets a set of planes (hkl) at the incidence angle θ required for an n-order reflection according to Bragg's equation

$$\sin \theta = \frac{n\lambda}{2d(hkl)} \qquad (11.9)$$

The resulting locus of diffracted beams is a circular cone with the incident beam as axis and with a half-angle 2θ. For the successive orders on a given set of planes, the locus is a set of coaxial cones (Figure 11.14). For all reflections meeting the limiting condition, the locus is the superposition of these sets of cones.

11.3.1.3 Diffracted Intensity

The angles of the diffraction cones are determined solely by the respective interplanar spacings. X-ray reflections on sets of lattice planes with equal interplanar spacings therefore contribute to the intensity diffracted into one diffraction cone. Multiple diffraction into the same diffraction cone results from two causes, given under the headings below.

Symmetry-related Multiplicity

From one set of lattice planes (hkl), the point group symmetry generates a form of M sets expressed by the notation $\{hkl\}$. For a given point group symmetry, the multiplicity factor M depends on the orientation of the lattice planes with respect to the symmetry elements.

Example Cubic holohedry $m3m$.

1. General form $\{hkl\}$, $M = 48$;
2. Particular forms: $\{hk0\}$, $M = 24$; $\{110\}$, $M = 12$; $\{111\}$, $M = 8$; $\{100\}$, $M = 6$.

Nearly Equal Interplanar Spacings

Low-order symmetry can result in interplanar spacings of different sets of lattice planes being nearly equal. The angular spread of the primary beam can therefore lead to an overlapping of the corresponding diffraction cones.

Those multiplicity effects must be taken into account when interpreting powder patterns in terms of intensities.

11.3.2 INSTRUMENTAL LAYOUT

11.3.2.1 Debye–Scherrer Camera

Set-up

The simplest powder diffraction device, the Debye–Scherrer camera (henceforth abbreviated DS), is outlined in Figure 11.15. When displayed on a plane, the impressed

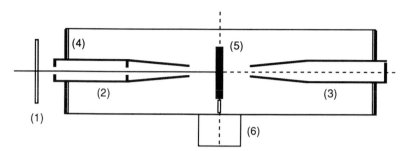

Figure 11.15. Diagram of the cross-section of a DS camera. (1) β-filter; (2) beam limiting collimator (pinhole or vertical slit apertures); (3) exit pit, (4) cylindrical film; (5) rod-shaped specimen on the camera axis; (6) rotation motor.

film shows symmetrical rings corresponding to the intersection of the diffraction cones. Due to the random crystal orientation distribution, all displayed data can be found in an equatorial plane section of the camera. As a result, a DS film may be narrower than a rotation film (Figure 11.16). However, for polycrystalline materials with preferential orientations, additional data are displayed outside the equatorial plane and a tall rotation-type camera is required.

Specimen

The specimen must meet the following more or less opposed conditions:

(a) be in the shape of a fine cylindrical rod of some tenths of millimetres in diameter;
(b) include in its irradiated volume a sufficient number of crystallites for the diffraction rings to be continuous, but of an average size large enough not to cause a broadening of the rings through the size effect.

Preparing the DS Specimen. Various techniques such as those following are commonly used to prepare such a sample:

1. Some materials are naturally in an elongated form and are directly mounted on the specimen holder (metal wires, mineral fibres).
2. Aggregating a powder, by means of a binder such as glue or resin, into a paste which can be rolled into a thin cylinder. The binder must be chemically inert with respect to the sample material.
3. Gluing a powder on to a fibre made of lithium borate glass (Lindemann glass) or of organic glass.
4. Filling a capillary made of the above-mentioned material with powder.

Grain Size. For a DS photograph to show continuous diffraction rings, a large number of crystallites must be in the active volume of the specimen, i.e. the crystals have to be small. Too large crystals lead to punctuated rings, whereas too small crystals lead to ring widening through the relaxation of the diffraction conditions (size effect).

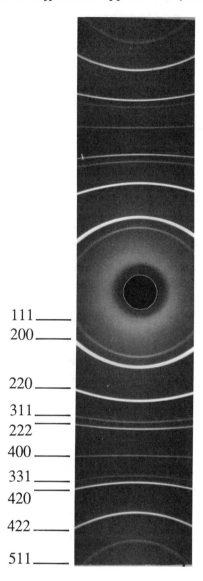

Figure 11.16. Powder photograph of MgO (cubic faces centred) recorded with Cu-Kα and a Ni filter in a 180 mm DS camera. The only diffraction rings are those for *hkl* all even or all odd. Notice that the reflections with even indices are stronger than the ones with odd indices, which is consistent with the calculated structure factor (see 5.3.3) (Lab. Crystallography, Louis Pasteur University, Strasbourg).

The optimum average grain diameter is currently of the order of 0.1–10 μm, depending on instrumental parameters (e.g. collimator aperture, sample diameter). Rotating the sample improves the statistical distribution without altering the geometry of the diffraction pattern.

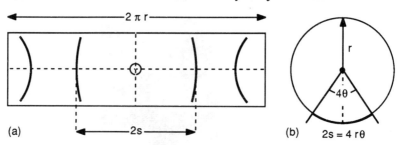

Figure 11.17. Interpreting a DS diffraction pattern. (a) General outline of a pattern; (b) cross-section of the DS camera.

Measurements on the Diffraction Pattern

Let $2s$ be the distance of a pair of symmetrical rings as measured in the equatorial plane of the diffraction pattern provided by a DS camera of radius r. Figure 11.17 shows that $2s = 4r\theta$. Measuring the distance of any pair of rings results in calculating the reflection angle θ and hence the values $d_m = d(hkl)/n$ according to Bragg's equation

$$d_m = \frac{d(hkl)}{n} = \frac{\lambda}{2 \sin \theta} \qquad (11.10)$$

Measuring the successive rings, beginning at the centre of the pattern, results in a list of decreasing values d_m which represent the interplanar spacings d in the sample material, together with their submultiples $d/2, d/3, \ldots, d/n$. However, assigning to each ring the corresponding plane indices (hkl) is not straightforward. The resolution condition (Eqs 5.8 and 5.9) results in limiting the list of values to $d_m \geq \lambda/2$.

In practice, diffraction cameras are normalised, with a circumference (film length) $2\pi r$ of 180 or 360 mm, thus enabling the easy correlation of s and θ. Corresponding tables supply directly the d_m values as a function of ring distances for currently used X-ray wavelengths.

Assigning to each d_m value the corresponding ring intensity leads to a list of pairs of values (d_m, I_m) which represent the total data which can be gathered from a powder pattern.

For certain applications, it is important that the values (d_m, I_m) be known with great accuracy. The main causes of error are surveyed below.

Errors on the Measurement of Interplanar Spacings.

Errors on distance measuring. These can be minimised by the following:

(a) Using a greater film length, but the exposure time varies according to the squared radius and the rings are widened.
(b) Replacing the simple rule by more elaborate devices, such as a travelling microscope or an optical microdensitometer. The latter device turns the set of rings along the zero line into a set of peaks; in addition to an accurate measurement of the ring distances through the half-width line distances of the peaks, it provides accurate intensities from measurements of an area of the peaks.
(c) Measuring high-order rings in order to minimise the relative error.

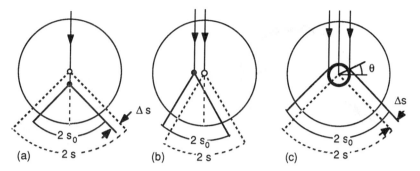

Figure 11.18. Shift of DS rings due to the specimen. (a) Eccentricity in the incident direction. A given eccentricity ϵ results in a ring shift $\Delta s = \epsilon \sin 2\theta$ which is annealed for $2\theta = 180°$. (b) eccentricity perpendicular to the incident direction; shift compensation minimises the error to the second order; (c) highly absorbing specimen, of radius ρ; diffraction is limited to a cylindrical surface layer, resulting in a maximum shift $\Delta s = \rho \cos \theta$. The actual shift is smaller and leads to dissymmetrical ring intensity profiles. Like the eccentricity error the corresponding absorption error is cancelled for $2\theta = 180°$.

Errors due to the film. The film length varies with air hygrometry and after photographic development. This cause of error can be suppressed by:

(a) recording a scale on the film;
(b) adding a convenient, well-crystallised standard to the specimen (e.g. quartz SiO_2, alumina Al_2O_3);
(c) replacing the film by a device for measuring diffraction angles and intensities (powder diffractometer).

Errors due to the specimen. A DS photograph provided by a well-crystallised specimen, with a continuous and fine diffraction ring, is subject to eccentricity and absorption errors, as shown by Figure 11.18.

Errors due to X-ray optics. The beam-limiting inlet collimator leads to a widening of the diffraction rings. For a given specimen, a compromise must be worked out between beam aperture, camera radius and exposure time.

Note that all errors are minimised at a diffraction angle of $2\theta = 180°$. Furthermore, large angle rings are often split into the $K\alpha_1$ and $K\alpha_2$ reflections, thus allowing a higher accuracy. It is therefore generally advisable to give a greater weight to the measurement of the most external rings.

Errors on Intensity Measurements. Due to the indirect measurement via the optical density, intensity measuring on DS photographs is hardly quantitative. Multiplicity effects must be taken into account.

Advantages and Inconveniences

An X-ray diffraction facility with a DS camera is cheap, easy to operate and to adjust. Specimen preparation is straightforward and requires only minute quantities of material. The diffracted beams in the equatorial plane being normal to the film, double-layered films may be used for shorter exposure times.

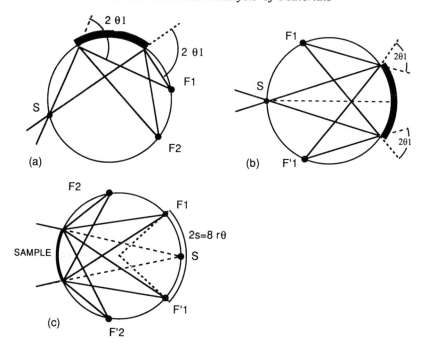

Figure 11.19. Principle of focusing powder cameras. (a) Dissymmetrical reflection set-up; (b) symmetrical reflection set-up; (c) symmetrical transmission set-up. The specimen fits an arc of the cylinder of the camera. Any rays emitted from a linear source S on the cylinder and resulting in a *hkl* reflection on the sample converge on a focusing line F such that arc SF $= \pi - 2\theta$ in the reflection mode, arc SF $= 2\theta$ in the transmission mode.

Its main drawback is the poor accuracy in measuring interplanar spacings and corresponding intensities.

11.3.2.2 Focusing Cameras

The principle of X-ray focusing by means of a crystal, previously outlined for monochromators, can be applied to the powder method, as shown in Figure 11.19. Focusing can be achieved by dissymmetrical or symmetrical, reflection or transmission set-ups. The Seemann–Bohlin camera and the Guinier camera are common focusing powder facilities.

Advantages. This kind of focusing camera is particularly well adapted to operate with a focusing monochromator. In well-adjusted set-ups, the result is a powder photograph with high intensity and fine diffraction rings, produced by the only Kα_1 line, without any continuous background. For an equal camera radius, the angular distance of two rings is twice its value in a DS camera (compare Figures 11.19 and 11.17b), resulting in an even higher accuracy.

Drawbacks. When comparing to a simple DS facility, setting up and adjusting a focusing camera is more critical. The absence of a materialised origin in the dissymmetrical reflection mode requires calibration. The specimen preparation requires a cylindrical press and a greater amount of material. One photograph does not cover the whole angular range 2θ: from 0 to $\pi/2$ in the symmetrical transmission mode; from $\pi/2$ to π in the symmetrical reflection mode.

In the double Guinier camera, the whole diffraction pattern is recorded in one operation, on two films, by juxtaposing a transmission camera (Figure 11.19c) and a reflection camera in such a way that the primary focusing point S (virtual source) of the first coincides with the source S of the second (Figure 11.19b).

Focusing cameras are mainly applied to high-accuracy lattice parameter measurement. In order to avoid errors due to the film, the use of internal standards is advisable.

11.3.2.3 Powder Diffractometer

All inconveniences due to the film are eliminated by replacing it by a detector system. In the most common layout of the resulting powder diffractometer, the ray diagram is derived from the focusing camera, as explained by Figure 11.20.

Operating Modes

Continuous Angular Scanning. The diffractometer is operated through a continuous $(\omega, 2\omega)$ rotation. The amplified and integrated detector output is fed into a chart recorder with calibrated and adjustable paper speed. In recent diffractometers, the recorder is

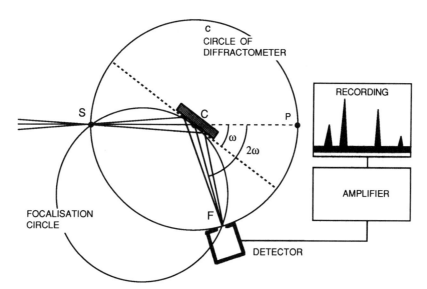

Figure 11.20. Principle of the powder diffractometer. The line source S is placed on the circle c of the diffractometer, normal to its plane. The specimen is flat, with its reference plane passing through the diffractometer axis and rotating about this axis, at a constant and adjustable angular velocity ω. The inlet slit F of a counter (currently proportional counter or scintillation counter) is made to move along the circle c with twice the angular velocity of the specimen, i.e. 2ω. At any moment the specimen plane is positioned so as to reflect the source S into the counter slit F. This results in a Bragg–Brentano focusing geometry; the focusing circle SCF has a varying radius (see Figure 9.4d).

often replaced by a video display screen which enables computer processing of the diffraction pattern. In any case, the result is a diffraction pattern I(2θ) with peaks corresponding to the *hkl* reflections. The positions of these peaks provide the Bragg reflection angles and consequently the corresponding interplanar spacings d_m (Eq. 11.10). Their areas over the background provide the corresponding intensities I_m (Figure 11.21). Disconnecting the ω and 2ω motions enables the curve $I(\theta)$ to be recorded with the detector fixed at 2θ, and is useful for analysing line profiles and preferential orientations.

Step-by-step Measurement. In this mode, the specimen and the counter are rotated through adjustable steps. Counting takes place at each step during a fixed time or up to a fixed pulse number; the latter counting mode provides equal statistical errors whatever the intensity of the peaks. Integrating the intensities over a peak, after subtracting the background over the same angular range at both sides of the peak, leads to a highly accurate value of the corresponding reflection intensity. Plotting the $I(\theta)$ graph enables accurate peak profile analysis.

Operating with Monochromator. Because of its design, the powder diffractometer can be operated with a crystal monochromator placed between the specimen and the detector, with the advantage of suppressing altogether the continuous bremsstrahlung and the possible background due to the X-ray fluorescence of specimen atoms excited by the primary beam (e.g. specimen containing Fe and Co, Cu–Kα primary radiation).

Energy Discrimination. Energy discrimination based on the energy dependence of the counter pulses (see section 10.2) leads to enhanced peak-to-background ratios when operating without a monochromator, with a simple β-filter (e.g. the diffractograms of Figure 11.21).

Specimen

The flat specimen requires more material than the rod-shaped one of the DS camera. It is, however, much easier to prepare than the cylinder-shaped specimen for the focusing camera. The main requirement is a well-defined reference plane coinciding with the diffractometer axis. Bulk samples are prepared by conventional cutting and polishing techniques (metals, ceramics, rocks). Fine-grained materials are currently screened to a given grain size, dispersed in a convenient liquid and sedimented on to a glass or a metal slide.

Developments

A linear detector (see section 10.2) provides the possibility of recording a powder diffraction pattern, in a limited angular range, without any mechanical motion. The short data acquisition time enables an easy analysis of time-dependent effects, e.g. the kinetics of phase transitions vs temperature, when combined with a heating device. A further reduction of the acquisition time of powder diffraction patterns down to the order of a second or even less is possible by replacing the conventional X-ray tube by a synchrotron X-ray source.

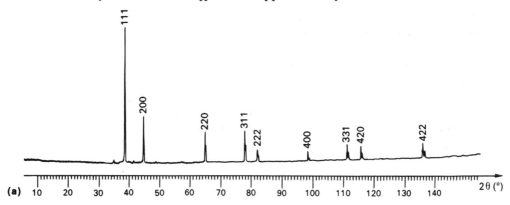

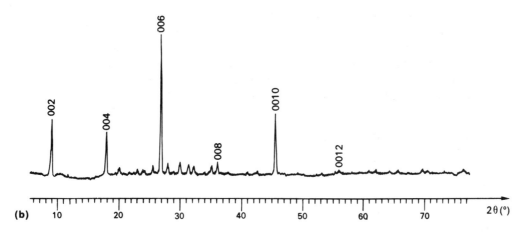

Figure 11.21. Powder diffractograms. Cu-Kα, Ni filter, energy discrimination. (a) Layer of polycrystalline gold with random orientation, ion sputtered on to a glass slide. Note the resolution of the Kα doublet at high angles, from 220 up. (b) Mica (muscovite) ground and sedimented on a glass slide. The plate morphology of the material results in a strong preferential orientation of the crystals on the reference plane, parallel to the basic layers (001). Due to the diffractometer geometry, the diffraction pattern therefore displays mainly the 00l reflections. The systematic absence of odd l reflections is due to the glide plane c of the muscovite structure.

Recent diffractometer facilities are currently associated with a microcomputer. Specialised software has been designed for monitoring the motions, for memorising and processing the diffraction patterns, e.g. determining the peak centres, their half-width, their surface, subtracting the background noise, smoothing the pattern, suppressing the $K\alpha_2$ contribution, deconvoluting the overlapping peaks, etc.

11.3.2.4 Primary Radiation

The choice of the wavelength, i.e. of the target element of the X-ray tube, depends on the specimen and on the kind of information to be gathered. The accuracy of lattice parameter measurements is favoured by large diffraction angles, i.e. by rather

194 *Structural and Chemical Analysis of Materials*

long-wave X-rays from medium elements, such as Cr–Kα (2.29 Å) or Cu–Kα (1.54 Å). A soft X-ray radiation is best suited for recording the diffraction pattern of thin surface layers.

Crystal structure analysis requires the largest possible number of reflections to be measured, i.e. a short-wave radiation, e.g. Mo–Kα (0.71 Å). High-intensity synchrotron X-ray sources provide new possibilities.

When the sample contains one or more rather heavy elements, the target element should have a smaller atomic number, in order to avoid induced fluorescence leading to a high background, unless a post-specimen monochromator is being utilised.

11.3.3 APPLICATIONS OF THE POWDER METHOD

11.3.3.1 Identifying Crystalline Species. Powder Data File

Whatever the type of powder diffraction facility, the data provided are in the form of a list of pairs of values (d_m, I_m). This list characterises a given crystal species; the d_m list is specific for the lattice, the I_m list is specific for the crystal structure. An X-ray powder pattern therefore acts as a fingerprint of the crystal species in the specimen, enabling them to be identified provided that a relevant data file exists. The first attempts to compile powder diffraction data were made in the 1930s, leading some 10 years later

d 4-0563	3.245	1.687	2.489	3.245	TiO$_2$					*
I/I$_1$ 4-0551	100	50	41	100	Titanium Dioxide			Rutile		
Rad. Cu λ 1.5405 **Filter** Ni **Dia.** Cut off **Coll.** I/I$_1$ G-M Spectrometer d corr. abs.? **Ref.** Swanson and Tatge, JC Fel. Reports, NBS 1950					d Å	I/I$_1$	hkl	d Å	I/I$_1$	hkl
					3.245	100	110	1.1485	4	400
					2.489	41	101	1.1329	1	410
					2.297	7	200	1.0933	4	222
					2.188	22	111	1.0827	4	330
					2.054	9	210	1.0424	5	411
Sys. Tetragonal **S.G.** D$_{4H}^{14}$–P4/MNM* a$_0$ 4.594 b$_0$ c$_0$ 2.958 A C 0.644 α β γ Z **Ref.** Ibid.					1.687	50	211	1.0361	4	312
					1.624	16	220	1.0273	3	420
					1.480	8	002			322
					1.453	6	310	0.9642	2	103
					1.360	16	301	0.9071	3	402
$\delta \alpha$ n $\omega \beta$ $\zeta \gamma$ Sign 2V D$_x$ 4.250 mp Color **Ref.** Ibid.					1.347	7	112	0.9007	3	510
					1.305	1	311	0.8892	5	213
					1.243	3	202	0.8773	6	431
					1.200	1	212	0.8739	5	332
*Huggins, Phys. Rev. 27, 638 (1926). Spectrographic analysis shows no impurity greater than 0.001%. Rutile obtained by heating anatase at 1000 °C for 2 hrs. Followed by slow cooling. To replace 1-1292, 2-0494 and 3-1122 at 26 °C.					1.170	4	321			422
								0.8437	5	223
								0.8290	5	303
								0.8196	8	521

Figure 11.22. ASTM card of rutile TiO$_2$ (No. 4-551).

to the well-known ASTM file (American Society for Testing Materials). An example of a conventional ASTM card is shown in Figure 11.22; in addition to the complete list of interplanar distances and the related intensities, it contains further information on the species (symmetry, optical properties, bibliographic references). Most crystalline species have been shown to be characterised by their three strongest diffraction lines, i.e. three pairs of values (d_1, I_1), (d_2, I_2), (d_3, I_3). The manual or mechanical research is based on a data book and data cards. The procedure consists of pre-identification by means of the three strongest diffraction lines, leading to one or several options. The identification is considered to be reliable when the whole set of experimental diffraction data is identical, to the admitted error of measurement, to those on the corresponding card.

Identification Difficulties. Identification of crystalline species becomes uncertain or even impossible in the following instances:

(a) new species;
(b) poorly crystallised material giving ill-defined diffraction peaks;
(c) strong preferential orientations leading to the reinforcement of some peaks and the absence of others;
(d) changes in chemical composition with respect to the compiled species (e.g. solid solutions);
(e) mixture of more than two or three species.

An accurate measurement of both the d_m and the I_m values is of great importance for solving those difficult cases. The diffractometer is best designed for this aim.

Computerised Powder Diffraction Files. Nowadays the card files are increasingly supplanted by computer databases, e.g. the JCPDS powder file [17]. In addition to monitoring the operation and to processing the diffraction pattern, the microcomputer, which is often connected to modern powder diffractometers, enables a more or less automatic identification of crystalline species by comparing their diffraction patterns with those stored in the database. The diffraction data of a specific field of research can be stored on hard disc. Optical storage discs (CD–ROM) enable a direct and rapid access to the whole powder databank, covering some 50 000 mineral and organic species. Automatic search/match programs have been developed and are continuously improved, opening up new possibilities for analysing polycrystalline materials.

Conclusion. The main advantages of the powder diffraction file method are simplicity, rapidity, small amount of required material, non-destructiveness. In addition to the chemical composition, it provides the identification of the crystalline species and its structural data (e.g. distinction between polymorphs).
  It is useless for analysing new species and difficult to apply for mixtures of species.
  It is a useful complement to spectrometric analysis.

11.3.3.2 Quantitative Analysis

Consider a mixture of identified crystalline species J ($J=A, B, \ldots$). The mass concentration c_J of a species J is related to the intensity I_J of one of its diffraction lines *hkl*. The concentration c_J is determined by comparing the intensity I_J given by the sample with the intensity $I_J(T)$ of the same reflection given by a test mixture T of known composition. In the absence of any influence of the matrix species (species other than the one to be measured), the concentration vs intensity relationship would be

$$c_J/c_J(T) = I_J/I_J(T)$$

Due to a different composition, the absorption coefficients of the unknown specimen and of the test specimen are different. As a result, the simple relation above cannot generally be applied, except for a mixture of species with similar absorption coefficients.

An example outlining the absorption effect:

Sample 1: 10% Al_2O_3 and 90% SiO_2
Sample 2: 10% Al_2O_3 and 90% Fe_2O_3

Both samples contain the same mass concentration of Al_2O_3. However, the intensity of a given Al_2O_3 reflection will be smaller in the second case, due to a higher average atomic number resulting in a higher absorption of the reflection.

Correcting the absorption effect is achieved by adding a given concentration c_S of a crystalline species S, normally absent in the specimen, serving as internal standard. The reflection intensities of the component species are now measured with respect to a reflection of the internal standard, from the specimen to be analysed as well as from the standard specimen. Owing to the same absorption effect, whatever the reflection, a linear relationship between intensities and concentrations can now be written as follows:

$$\frac{I_A}{I_A(T)} = K\frac{c_A}{c_A(T)} \ldots \frac{I_S}{I_S(T)} = K\frac{c_S}{c_S(T)} \Rightarrow K = \frac{I_S/c_S}{I_S(T)/c_S(T)} \quad (11.11)$$

An alternative correction method is based on measuring the absorption coefficient of the specimen (Norrish and Taylor, 1962; Tatlock, 1966).

The measured intensities are sensitive to any variation of the chemical composition of the species, as well as to orientation textures and to grain sizes. Those parameters must be identical in the specimen to be analysed and in the test specimen for an acceptable accuracy. When those conditions are met, the detection limit is about 1% and the accuracy 5%.

Obviously the diffraction method for quantitative analysis does not apply to amorphous and glassy materials. It can, however, be useful for determining the amount of crystalline phases in composite materials such as glass ceramics or vitreous rocks.

11.3.3.3 Lattice Parameters

A powder diffraction pattern is produced by a monochromatic radiation. It therefore contains any data which are required to measure the lattice parameters with the wavelength as a length standard. These data are in the form of a list of decreasing values of interplanar spacings d_m as given by Eq. (11.10). The general problem is to determine the direct lattice corresponding to that list of spacings, i.e. to affect a set of reflection indices *hkl* to any d_m of the list.

A similar analysis may be conducted within the reciprocal lattice: given the list of increasing values of the lengths of the reciprocal vectors $r^*_m = 1/d_m = n/d_{hkl}$, determine the corresponding reciprocal lattice. The reciprocal lattice method leads to an easier formulation and will be adopted hereafter. It requires that the so-called quadratic forms are defined for the seven crystal lattices.

Quadratic Forms

The quadratic form expresses the lengths of the vectors of a reciprocal lattice by means of their squares

$$Q(hkl) = \frac{n^2}{[d(hkl)]^2} = [r^*(hkl)]^2 = \frac{4\sin^2\theta}{\lambda^2} \tag{11.12}$$

Written as a function of the reciprocal lattice parameters, the most general quadratic form (triclinic system) becomes

$$Q(hkl) = (h\mathbf{a}^* + k\mathbf{b}^* + l\mathbf{c}^*)^2$$
$$= h^2\mathbf{a}^{*2} + k^2\mathbf{b}^{*2} + l^2\mathbf{c}^{*2} + 2hk\mathbf{a}^*\mathbf{b}^*\cos\gamma^* + 2kl\mathbf{b}^*\mathbf{c}^*\cos\alpha^* + 2lh\mathbf{c}^*\mathbf{a}^*\cos\beta^* \tag{11.13}$$

The reciprocal lattice parameters can subsequently be substituted by the direct lattice parameters with the help of the relations between the direct lattice and the reciprocal lattice.

Cubic System. $a = b = c$; $\alpha = \beta = \gamma = \pi/2$

$$Q(hkl) = a^{*2}(h^2 + k^2 + l^2) = \frac{(h^2 + k^2 + l^2)}{a^2} \tag{11.14}$$

Only one parameter to be determined: a^* (or a).

Systems with One Main Axis.

Hexagonal system: $a = b \neq c$; $\alpha = \beta = \pi/2$, $\gamma = 2\pi/3$

$$Q(hkl) = (h^2 + k^2 + hk)a^{*2} + l^2 c^{*2} = \frac{4}{3}\frac{h^2 + k^2 + hk}{a^2} + \frac{l^2}{c^2} \tag{11.15}$$

Tetragonal system: $a = b \neq c$; $\alpha = \beta = \gamma = \pi/2$

$$Q(hkl) = (h^2 + k^2)a^{*2} + l^2 c^{*2} = \frac{h^2 + k^2}{a^2} + \frac{l^2}{c^2} \tag{11.16}$$

Trigonal system: comes to the hexagonal system by means of a change of axes. For these three systems two parameters are to be determined.

Systems without Symmetry Higher than Twofold Axes.

Orthorhombic system: $a \neq b \neq c$; $\alpha = \beta = \gamma = \pi/2$

$$Q(hkl) = h^2 a^{*2} + k^2 b^{*2} + l^2 c^{*2} = \frac{h^2}{a^2} + \frac{k^2}{b^2} + \frac{l^2}{c^2} \tag{11.17}$$

Three parameters to be determined: a^*, b^*, c^* or a, b, c.
Monoclinic system: $a \neq b \neq c$; $\alpha = \gamma = \pi/2$, $\beta \neq \pi/2$

$$Q(hkl) = h^2 \mathbf{a}^{*2} + k^2 \mathbf{b}^{*2} + l^2 \mathbf{c}^{*2} + 2hl\, \mathbf{a}^* \mathbf{c}^* \cos \beta^* \qquad (11.18)$$

Four parameters to be determined: a^*, b^*, c^*, β^*.

Triclinic system: $a \neq b \neq c$; $\alpha \neq \beta \neq \gamma \neq \pi/2$. General Eq. (11.13): six parameters to be determined.

General Method. Consider an experimental list of decreasing numbers

$$Q_m = 1/d_m^2 = 4 \sin^2 \theta / \lambda^2 \quad (Q_1, Q_2, Q_3, \ldots, Q_n)$$

resulting from measurements on a powder diffraction pattern. The problem is to determine the lattice parameters such that the list of measured values Q_m be identical, to the errors of measurement, with the list of values $Q(hkl)$ as calculated from Eq. (11.13) for any sets of increasing integers hkl.

In the most general case of the triclinic system, there are six parameters to be determined. In the simplest case of the cubic system, there is only one parameter to be determined. The difficulty of solving the problem increases from the cubic to the triclinic system. Accurate values of Q_m are of prime importance for solving the low-symmetry systems.

Cubic Lattice Parameters

The parameter a^* has to be determined such that

$$Q_m \cong Q(hkl) = a^{*2} S_i \;\Rightarrow\; \frac{Q_m}{a^{*2}} \cong \frac{Q(hkl)}{a^{*2}} = S_i \;(S_i = h^2 + k^2 + l^2) \qquad (11.19)$$

The series S_i corresponding to existing hkl reflections is characteristic of the three possible lattice types (Table 11.2). For a given lattice type, the series S_i is defined by the ratios $Q(hkl)/Q_1$, where Q_1 is the first value of the list $Q(hkl)$. The experimental ratios Q_m/Q_1 therefore will have to match one of the characteristic series, to the errors of measurement:

Lattice P—first reflection 100: $Q_1 = a^{*2} \Rightarrow Q_m/Q_1 \cong S_i$
Lattice I—first reflection 110: $Q_1 = 2a^{*2} \Rightarrow 2Q_m/Q_1 \cong S_i$
Lattice F—first reflection 111: $Q_1 = 3a^{*2} \Rightarrow 3Q_m/Q_1 \cong S_i$

The following test operations result for the cubic system:

(a) compile the list Q_m ($Q_1 < Q_2 < Q_3 < \ldots < Q_n$);
(b) divide all Q_m by Q_1; if the result, to the errors of measurement, is the series of integers (except 7, 15, 23 . . .), the lattice is cubic primitive P;
(c) if not, multiply the values of Q_m/Q_1 by 2; if the result, to the errors of measurement, is the series of even integers, the lattice is cubic body centred I;

Table 11.2. Series of values of $S_i = h^2 + k^2 + l^2$ corresponding to existing reflections in the three cubic lattice types P, I, F and the diamond type ($S_i \leq 25$)

hkl	P all hkl	I $h+k+l=2n$	F hkl same parity	Diamond
100	1			
110	2	2		
111	3		3	3
200	4	4	4	
210	5			
211	6	6		
220	8	8	8	8
300, 221	9			
310	10	10		
311	11		11	11
222	12	12	12	
320	13			
321	14	14		
400	16	16	16	16
410, 322	17			
411, 330	18	18		
331	19		19	19
420	20	20	20	
421	21			
332	22	22		
422	24	24	24	24
500, 430	25			

(d) if not, multiply the values of Q_m/Q_1 by 3; if the result, to the errors of measurement, is the series of integers 3, 4, 8, 11, 12, ..., the lattice is cubic all-face centred F;

(e) if all three tests are negative, the lattice has a lower symmetry.

A positive test for one of the cubic lattices results in indexing all reflections according to Table 11.2, and therefore in calculating the lattice parameter a for each reflection from Eq. (11.14); the arithmetic mean provides an accurate value of the parameter (Table 11.3).

Note: The ratios $2Q_m/Q_1$ are identical for the lattice types P and I up to the sixth diffraction line. Whereas the lattice F is characterised from the very first lines, distinguishing between the P and I lattices therefore requires at least seven consecutive lines to be measured.

Table 11.3. Determination of the lattice type and the lattice parameter of polycrystalline MgO. The values d_m and Q_m have been compiled from a DS diffraction photograph (Figure 11.16)

d_m (Å)	Q_m (Å^{-2})	Q_m/Q_1	$2Q_m/Q_1$	$3Q_m/Q_1$		hkl	a (Å)
2.43	0.169	1	2	3		111	4.21
2.10	0.226	1.33	2.66	3.99	(4)	200	4.20
1.49	0.450	2.66	5.32	7.98	(8)	220	4.20
1.27	0.620	3.65	7.30	10.95	(11)	311	4.21
1.21	0.683	4.03	8.06	12.09	(12)	222	4.20
1.05	0.907	5.35	10.70	16.05	(16)	400	4.20
0.96	1.084	6.40	11.80	19.20	(19)	331	4.18
0.94	1.131	6.68	13.36	20.04	(20)	420	4.21
0.85	1.385	8.17	16.34	24.50	(24)	422	4.16
Lattice:				F		$a_{mean} = 4.20$ Å	

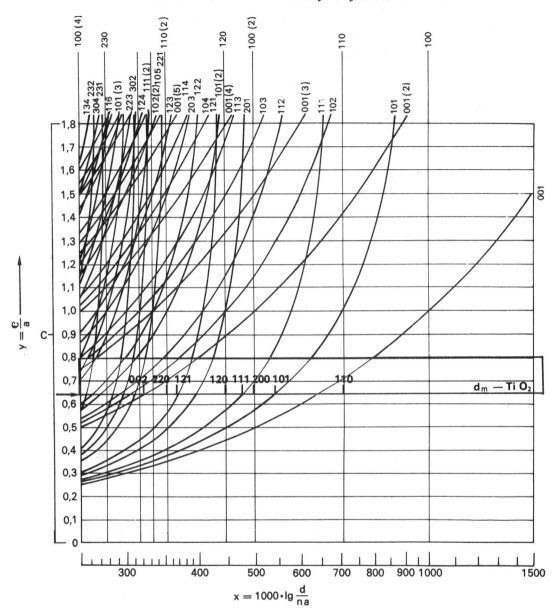

Figure 11.23. The Hull chart in practice. Example of the tetragonal system: indexing the powder diffraction pattern of TiO_2. The measured values of d_m are transferred on a strip of tracing paper by using the logarithmic x scale. Considering that $\log_{10}(d/na) = \log_{10}d - \log_{10}a - \log_{10}n$, it follows that any translation of the strip parallel to the x-axis corresponds to changing a; the scale is therefore valid whatever the value of a. In a similar way a translation parallel to the y-axis corresponds to changing c/a. The strip is displaced in both directions until its experimental marks d_m match the intersections with the hkl curves. The indices of the diffraction lines are then given by the intersections, the parameter a results from the intersection with a $h00$ curve and the ratio c/a is read on the y-axis. Possible absences due to the space group must be taken into account. For an unequivocal indexing, the d_m values must be measured with great accuracy. Even then there are often uncertainties, in particular at high orders where the curves are very dense.

Additional glide symmetry elements result in corresponding absences of reflections which must be taken into account for lattice determination. As an example, the cubic diamond type structures (F lattice, 4_1 screw axes) lead to diffraction patterns on which, in addition to the systematic absences due to the lattice, there are further absences, in particular the lines 200, 220, 420 (see 5.3.3 and Table 11.2).

Parameters of Lattices with a Main Axis

The two-parameter problem (a^* and b^* or a and b) may be solved by means of a two-dimensional abacus. The commonly used diagrams (*Hull chart* or *Bunn chart*) have been compiled on the basis of the $d(hkl)$. However, the $Q(hkl)$ values could also be used for drawing the diagrams. As shown by Eqs (11.15) and (11.16), d/na is a function of c/a. As a result a net of curves $d/na = f(c/a)$ can be plotted for successive values of the set of integers hkl. The Hull chart displays the curves $\log_{10}[d(hkl)/na]$ as a function of c/a (Figure 11.23), whereas the Bunn chart uses double logarithmic coordinates. Data for drawing them on a large scale are available in [9], *Int. tab.*, vol. 2.

Because of the development of specific computer programs, the graphical method is increasingly being replaced by trial-and-error methods derived from the general procedure, applied for indexing low-symmetry powder patterns, which is discussed below.

Low-symmetry Lattice Parameters

The graphical methods no longer work for the orthorhombic, the monoclinic and for the triclinic system which require respectively that three, four and six parameters are found. Computer programs have been designed to deal with the indexing problem which increases in difficulty with decreasing symmetry. They are based on a trial-and-error method, e.g. the Ito method [10].

A convenient method in view of computer programming is outlined below.

In the most general case, six reflections are theoretically sufficient for determining the six parameters. However, an arbitrary choice could lead to a multiple unit cell. The trial-and-error method is conducted in such a way as to lead to the most probable unit cell.

Parameters a^*, b^*, c^*. These are determined by the reflections $h00$, $0k0$ and $00l$, according to the following relations deduced from Eq. (11.13).

$$a^{*2} = \frac{Q(h00)}{h^2} \qquad b^{*2} = \frac{Q(0k0)}{k^2} \qquad c^{*2} = \frac{Q(00l)}{l^2} \qquad (11.20)$$

Corresponding reflections have to be identified in the list of Q_m. The related quadratic form is reduced to one squared element, resulting in a small value. Those reflections are therefore among the first in the list.

For determining a^*, the first-order reflection 100 is assumed to exist. The test begins with the assumption that the first reflection be $Q_1 = Q(100) = a^{*2}$. If this hypothesis is correct, there are probably higher order reflections in the list, i.e. experimental values such as $Q_m = h^2 Q_1$; multiplying Q_1 by the increasing integer squares $h^2 = 4, 9, 16, \ldots$ will then lead to matching values in the list Q_m, to the error of measurement. If the first-order 100 is absent, the same test is conducted with $Q_1 = Q(200) = 4a^{*2}$; higher orders would then lead to experimental values $Q_m = h^2 Q_1/4$, checked by multiplying $Q_1/4$ by the increasing integer squares 9, 16, 25, ... and so on. Without a result on Q_1, the same trial process is conducted with Q_2, Q_3, etc. The test leading to the best match is chosen for defining a^*, and as a result of indexing the $h00$ reflections.

Similar procedures are conducted with the as yet unindexed values for the parameters b^* and c^*. The respective assignment is arbitrary; it can be corrected *a posteriori* in accordance with existing conventions.

In the case of the orthorhombic system, the problem would be solved at this stage of trials, all reflections being indexed. A similar trial process, based on Eqs (11.15) and (11.16), would determine the two parameters of systems with a main symmetry axis.

Angular Parameters α^*, β^*, γ^*. If the test series above has left unindexed reflections, then the sample corresponds either to the monoclinic or to the triclinic system.

The lattice angle α^* is determined by means of pairs of reflections such as $0kl$ and $0k\bar{l}$ (or $0\bar{k}l$), according to the following relations deduced from Eq. (11.13):

$$Q'(0kl) = \frac{Q(0kl) + Q(0k\bar{l})}{2} = k^2 b^{*2} + l^2 c^{*2}$$

$$\cos \alpha^* = \frac{Q(0kl) - Q(0k\bar{l})}{4kl\, b^* c^*}$$
(11.21)

Using the previously determined parameters b^* and c^*, a list of values $Q'(hkl)$ is calculated for increasing values of h and k. Pairs of experimental values (Q_i, Q_j), such that $(Q_i + Q_j)/2$ coincides with values of the list $Q'(hkl)$, are sought among the as yet unindexed Q_m. Equation (11.21) then provides α^*. The lattice angles β^* and γ^* are similarly determined by means of lists $Q'(h0l)$ and $Q'(hk0)$ respectively.

The sole lattice angle β different from a right angle in the monoclinic system can generally be determined without too much difficulty. The search procedure for determining the three lattice angles in a triclinic sample is much more tricky, in particular for species with lattice angles near to a right angle (e.g. feldspars).

In any case, the accuracy of d_m measurements is of prime importance for a trial-and-error method to be successfully achieved. This requirement increases for decreasing crystal symmetry.

Checking the Unit Cell. The outcome of any search technique is a hypothetical reciprocal unit cell. In order to check its validity, the set of $Q(hkl)$ is recalculated from the hypothetical parameters. The values of the parameters are refined by means of a least-squares technique. Recent powder diffractometer facilities are directly connected to a microcomputer with adequate software to conduct the trial-and-error method on diffraction patterns which have been stored in its memory.

11.3.3.4 Crystal Structure Analysis

A powder diffraction pattern theoretically contains all data required for structure analysis, provided that all the diffraction lines can be indexed and their intensities measured with a fair accuracy and subsequently subjected to specific corrections, in particular the previously mentioned multiplicity correction. This is possible only when the lines are well defined, do not overlap and have a sufficiently high signal-to-background ratio. In practice it is therefore limited to high-symmetry crystal systems, such as the cubic system and the systems with one main symmetry axis (trigonal, tetragonal and hexagonal). Many metal and alloy structures and other high symmetry compounds have been elucidated by the powder technique which has the advantage of operating with the most common crystalline form of materials, the polycrystalline form.

11.3.3.5 Grain Size

The size of the individual crystals in a polycrystalline sample is important information related to crystal growth conditions and to mechanical properties. The powder diffraction method provides a simple way of measuring the mean grain size of a specimen that is too small to be measured through a light reflection microscope.

Microscopic Particles. With sizes of 1 μm or more, the DS photograph displays punctuated diffraction rings. The number of points on a whole ring can easily be related to the number of particles in the active volume of the sample, i.e. to their mean diameter. Grain sizes of the order of 0.1 μm may be determined by low-angle scattering.

Submicroscopic Particles. For particle sizes significantly smaller than 1 μm, the number of coherent planes which participate in a *hkl* reflection becomes sufficiently small to produce a relaxation of the diffraction conditions, i.e. a widening of the diffraction lines. For a random orientation, Eq. (5.29) provides the mean diameter t of the particles, leading to the following *Laue–Scherrer relation*

$$t = \frac{\lambda}{\epsilon \cos \theta} \quad (11.22)$$

where θ is the Bragg angle for the chosen *hkl* reflection and ϵ the angular half-width of the diffraction peak *hkl*, with $\epsilon = \Delta(2\theta)/2 = \Delta\theta$.

The peak widening in fact results from two causes: the instrumental incident beam aperture ϵ_0 and the angular widening ϵ_m due to the size factor. The value to consider in Eq. (11.22) is therefore

$$\epsilon = \epsilon_m - \epsilon_0, \quad \text{or} \quad (\epsilon_m^2 - \epsilon_0^2)^{1/2}$$

For a measure to be significant, the widening through the size effect has to be sufficiently large with respect to the instrumental widening which is determined by the inlet collimator of the diffractometer or the camera. The value of ϵ_0 is measured by means of a standard specimen made from a well-crystallised material, without size effect, such as quartz SiO_2 or alumina Al_2O_3.

Example
Clay specimen with illite particles. Interplanar spacings $d(001) = 10$ Å.
DS photograph, Cu–Kα, 360 mm camera.

003 half-width on the film: 1 mm $[d(003) = 3.3 \text{ Å}]$.
Instrumental half-width from a standard: 0.3 mm.
The significant half-width is therefore 0.7 mm, corresponding to $\epsilon = 0.7°$.
Resulting mean particle thickness: $t = (1.54.180)/(\pi.0.7.0.97) \cong 130 \text{ Å}$.
The clay particles contain on average about 13 layers of 10 Å.

The straightforward application of the Laue–Scherrer relation implies the absence of any other peak-widening effects, e.g. due to variations of the interplanar spacings or to peak overlapping. Composite peaks are separated into their Gaussian components by means of peak deconvolution programs which are now currently available with diffractometers.

11.3.3.6 Preferential Orientations

Polycrystalline materials often deviate from an ideal random orientation of their components. The preferential orientation (or orientation texture) of the crystals results either from their particular morphology (elongated or platy) or from various treatments (mechanical deformation, rolling, pressing or wire drawing of metals and of plastics and the crystallising process). It gives valuable information on the directions and the magnitudes of stress effects which affect various properties of a material.

Fibre Texture

The simplest preferential orientation in polycrystalline solids is the fibre orientation in which the individual crystals tend to have one of their important lattice lines [uvw] parallel to a common direction called the *fibre axis*. This orientation is predominant in elongated crystal morphology and is commonly related to a chain structure, e.g. amphibole asbestos, fibre axis [001] parallel to the SiO_4 chains. Due to the rod shape of the specimen, the DS technique is well designed for recording fibre diffraction patterns.

1. A stationary specimen with a totally random crystal orientation provides a powder pattern with continuous rings of constant intensity all along the specimen.
2. A stationary specimen with a perfect fibre orientation parallel to the rod axis would act, in a purely geometrical way (without considering the form factor), like a single crystal rotating continuously about the fibre axis [uvw] and would therefore provide a rotation photograph.

The actually observed diffraction pattern is generally intermediate between those extremes. It shows rings with arc-shaped reinforcements centred on the spots of the theoretical rotation pattern, i.e. at the intersections of the rings of the theoretical powder pattern with the layer lines of the theoretical rotation pattern (Figure 11.24). With well-marked reinforcements, the layers can easily be materialised on the photograph by means of lines passing through their centres. The parameter of the fibre axis $n(uvw)$ is then determined in the same way as the parameter of the rotation axis on a rotation photograph (Eq. 11.7). The fibre axis is identified by means of its indices [uvw] by comparing the observed parameter with those calculated for the high-density lattice lines of the specimen species.

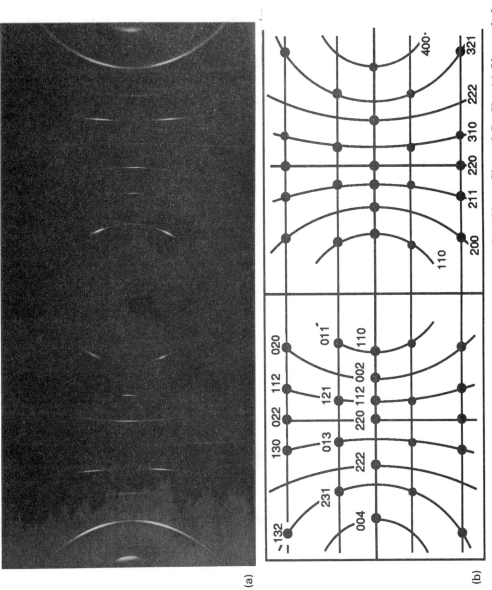

Figure 11.24. Fibre diffraction pattern from a tungsten wire. Cylindrical camera with 180 mm film and Cu–Kα. (a) Observed photograph; (b) theoretical diagram displaying the [110] rotation pattern superposed on the powder pattern.

It must be pointed out that investigating an orientation texture by means of the DS technique requires the use of a tall, rotation-type camera, in order to record the information diffracted outside the equatorial plane.

Example (Figure 11.24). Specimen: tungsten wire, cubic body-centred lattice, $a = 3.16$ Å. The lattice line parameters are related to a and uvw as follows:

$$n(uvw) = a(u^2 + v^2 + w^2)^{1/2}$$

The parameters of the three highest-density lattice lines of the tungsten lattice are therefore

$$n(100) = a = 3.16 \text{ Å}$$
$$n(110) = 4.47 \text{ Å}$$
$$n(111) = 2.74 \text{ Å}$$

The parameter of the fibre axis as measured on the DS photograph:

$$n(uvw) \cong 4.4 \text{ Å}$$

Comparison with the above values leads to the identification of the fibre axis as [110].

Lamellar Texture

The lamellar texture is related to a layered crystal structure. As a result, the individual platy crystals in a polycrystalline sample tend to have a preferential orientation as the normals to the layers become more or less parallel. Typical examples are clay minerals, talc and graphite with a common [001]* direction normal to the (001) layers. Due to its plane specimen holder, the powder diffractometer is best designed for investigating

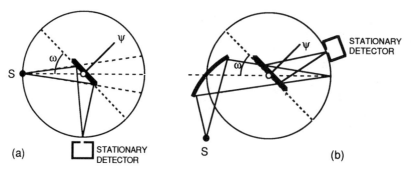

Figure 11.25. Set-ups for a texture orientation diffractometer. (a) Reflection mode with Bragg–Brentano focusing; (b) transmission mode with Johansson (or Johann) focusing. The specimen is rotated about the axis Ω (varying the incident angle ω) and about the axis Ψ normal to the specimen plane (rotation of angle ψ). Both motions are synchronised (e.g. 1° ω for 180° ψ). A lamellar texture (axial symmetry) can be investigated without ψ rotation, on a standard diffractometer, provided it is fitted with independent object and detector motions.

this kind of texture. A well-oriented lamellar specimen is easily prepared by grinding, screening and sedimentation on a plane support, e.g. a glass slide. The basal reflections are the only ones to appear (Figure 11.21). For investigating a material with lamellar orientation, the ω and 2ω motions of the diffractometer must be liable to be operated independently. With the detector fixed at an angle $2\omega = 2\theta(00l)$ for a given basal reflection $00l$, recording the curve $I(\omega)$ provides the statistical distribution of the normals to the individual lamellae around the normal to the specimen reference plane. In the reflection mode, the angle is limited to $\omega \leqslant 2\theta$ (Figure 11.25a). Operating by transmission with a thin specimen (e.g. a petrographic thin section) is possible by means of a suitable specimen holder. Focalising transmission set-ups (Figure 11.25b) have been developed for better results (Wiewiora, 1982).

General Texture

The *pole figure* is the best way to characterise a sample of material with preferential orientations in the most general case. For a given form of planes $\{hkl\}$, it is defined as stereographic projections, on the specimen plane, of the normals to these planes in the sample. In practice, it is expressed as a direction frequency plot by means of contour lines corresponding to equal densities of poles.

For this purpose the simple one-axis specimen holder of the diffractometer is replaced by a specific texture goniometer (Figures 11.21 and 11.25). The detector is set at an angle 2θ corresponding to the selected hkl reflection. The only crystallites to diffract are those with the normals to the $\{hkl\}$ planes bisecting the angle determined by the incident beam and the direction of the detector. The combined ω and ψ rotations (Figure 11.25) result in exploring theoretically the occurrence of normals in any direction, along a spiral of the stereogram. A synchronised intensity-sensitive recording device provides an automatic tracing of the pole figure (Figure 11.26). In the most common reflection mode, required with a bulk specimen, the angle is limited to $\omega \leqslant 2\theta$. A focusing transmission mode permits a wider angular range, combined with a thin specimen, as shown in Figure 11.25b.

11.3.3.7 Order–Disorder Transformations

Solid-state transformations, e.g. those of alloys or silicate minerals due to thermal treatment, may lead from a *disordered structure* (certain atoms are randomly distributed on their respective sites) to an *ordered structure* (the corresponding atoms occupy specific sites). Ordering often results in the advent of a multiple space period in one, two or the three dimensions, called a *superlattice*. A superlattice of the nth order in a given direction leads to multiplication of the corresponding period of the direct lattice by n and consequently to division of the related period of the reciprocal lattice by the same number. New diffraction domains, i.e. new reflections, will therefore appear on the diffraction pattern in the form of additional rings on the powder pattern.

Some two-component alloys show superlattice effects which are particularly easy to interpret through a simple change of their lattice type.

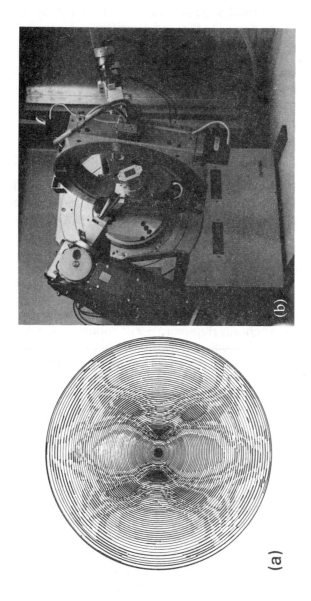

Figure 11.26. Texture analysis. (a) Pole figure 111 of a cold rolled copper sheet, recorded on a texture goniometer; (b) powder diffractometer fitted with a texture goniometer (Siemens document).

Example. Brass β-ZnCu, cubic, body-centred lattice at high temperature, primitive lattice at low temperature, with an unchanged parameter *a*.

Fast cooling. Disordered structure. The probability of occurrence of both atom species is the same on any of the structure sites.

Coordinates of positions in the unit cell: 000 and 1/2 1/2 1/2.

Statistical structure factor: $F(hkl) = (f_{Zn} + f_{Cu})/2\{1 + \cos[\pi(h+k+l)]\}$.

The diffraction pattern shows the systematic absences characteristic of an *I* lattice. Only reflections such as $h+k+l = 2n$ appear on the pattern.

Slow cooling or reheating. Coordinates of atoms in the unit cell: Zn 000; Cu 1/2 1/2 1/2.

The lattice has changed to primitive *P*.

Structure factor: $F(hkl) = f_{Zn} + f_{Cu}\cos[\pi(h+k+l)]$.

The reflections which were forbidden by the *I* lattice can appear, because $f_{Zn} \neq f_{Cu}$. Additional rings such as $h+k+l = 2n+1$ will therefore appear on the powder diffraction pattern.

When the elements concerned are close neighbours on the periodic table, as in the foregoing example, the difference of their atomic scattering factors is small, resulting in very weak superlattice reflections. Order–disorder transformations are difficult to observe in these cases by means of X-rays. Neutron diffraction is then a good alternative.

11.3.4 ANALYSIS OF POORLY CRYSTALLISED AND AMORPHOUS MATERIALS

The atomic structure of materials may cover the whole range between a perfect order (faultless single crystal) and a complete statistical disorder (amorphous material). Elastic scattering of X-rays observed by means of a powder diffractometer or any related technique can provide useful data on any such types of materials.

Performed on crystalline materials, the diffraction pattern consists of discrete diffraction peaks. The direct determination of the atomic positions is possible through the FT of diffracted amplitudes. Any deviation from the perfect crystal (crystal faults) is reflected by the diffraction pattern and can be interpreted.

Performed on non-crystalline materials, the observed scattering pattern consists of a continuous angle-dependent intensity distribution (see Eq. 3.33). Due to the spread over a wide angular range, the scattered intensity is low; accurate measurements require high-intensity X-ray sources such as synchrotron radiation.

The range of application of X-ray scattering is very wide. Its importance is highlighted by some examples below.

11.3.4.1 Poorly Crystallised Materials. Mixed-layered Structures

Mixed-layered lamellar structures provide examples of one-dimensional disorder. They consist of the stacking of two-dimensional structural units of different types in sequences varying from order to disorder. Such *mixed layering* is common in certain phyllosilicates of the *clay mineral* group. In the instance of disordered stacking of two different types of layers, the lamellar texture pattern as observed by means of a powder diffractometer (Figure 11.21b) shows widened basal 00*l* diffraction peaks leading to measured interplanar distances which have irrational ratios. Assuming the two types of layers to have equal structure factors, the FT of the unitary intensities $I(00l)/F^2(00l)$ results in a *distribution function* of the interlayer distances of the two types of layers along their normal (MacEwan et al, 1961).

11.3.4.2 Short-Distance Ordering in Non-crystalline Materials

Silicate glasses are non-crystalline solids. Their basic structure units are groups of atoms with well-defined interatomic distances. As a result, there is a short-distance order corresponding to the nearest neighbouring atoms.

The resulting scattering pattern as observed with a powder diffractometer shows a continuous intensity variation with broad maxima. The FT of the scattered intensity provides the distribution function of interatomic distances. In a silicate glass the prominent distribution peaks correspond to distances of 1.6 and 2.6 Å which equal the well-known Si–O and O–O distances in crystallised silicates. These results confirm the existence in glasses of the same basic structure unit SiO_4. Analysing the distribution function in relation to the glass composition leads to the interpretation of certain properties of glasses. The same technique applies to the investigation of the structures of high polymers and of biological materials. The same results are provided in a different way by X-ray absorption edge techniques (EXAFS).

11.3.4.3 Size Distribution in Amorphous Materials. Small-angle Scattering

The size limitation effect of an amorphous scattering object results in a central scattering peak whose width is inversely proportional to the size (Eq. 3.28 and Figure 3.4). A size of some 1000 Å leads to a width of the order of 10^{-3} rad, i.e. in the range of small-angle scattering. Measuring the width of this peak therefore provides the mean size of the scattering particles, in the same way as the width of the diffraction peaks of polycrystalline materials (Eq. 11.22). The FT of the scattered intensity provides the distribution function of the particle sizes. The main difference is that for amorphous materials the effect is limited to the vicinity of the incident direction. As a consequence, the accurate measurement of the profile of small-angle scattering peaks is made difficult by the proximity of the primary beam which has a much higher intensity. Furthermore, in current diffraction set-ups, the angular width of the primary beam is of the same order or greater than the size widening to be measured.

Instrumental Layout. The practical set-up for small-angle scattering results from the above-mentioned difficulties:

1. Very fine and well-collimated primary beam. The specific photographic or detector set-ups are combined with collimators of great length and very small apertures. A double-focusing monochromator provides a more accurate limitation of the beam. Operating in vacuum avoids additional scattering by air molecules.
2. Large specimen to film (or detector) distance, of the order of 50 cm.
3. High-intensity primary beam. The best X-ray source for small-angle scattering at present is synchrotron radiation.

Examples of Applications. Typical fields for small-angle scattering are the analysis of size distribution of micelles in colloids and gels and of macromolecules in polymers.

The principle of small-angle scattering applies similarly to electrons and neutrons, with specific instrumental layouts.

11.4 X-RAY AND NEUTRON DIFFRACTION APPLIED TO CRYSTAL STRUCTURE ANALYSIS

The amplitude (magnitude + phase) of a monochromatic radiation diffracted by a crystal is a function of the atomic positions in its unit cell.

Conversely, the measurement of the diffracted amplitudes in all directions is theoretically sufficient for determining the atomic positions in the unit cell, i.e. the crystal structure, with an accuracy depending on the wavelength. Due to the intensity being the only measurable quantity, the phase problem is predominant in structure analysis.

The first simple crystal structures (halite NaCl, diamond C, pyrite FeS_2, etc.) were determined soon after the historical diffraction experiment of Laue, by W. H. and W. L. Bragg. The fairly complicated basic structure of mica, a well-crystallised phyllosilicate, was published by Jackson and West as early as 1931. Since then the technique has been steadily improved, mainly through the development of direct and accurate intensity measuring as well as the tremendous progress in computer calculations. Thanks to automatic single-crystal diffractometers and powerful computers, complicated structures can be determined. Structures which had previously been approximated by photographic means can be refined.

The complementary character of *neutron diffraction* is often used to complete a structure already outlined by X-ray diffraction, e.g. by assessing the positions of light atoms which are inaccessible to X-rays.

Except for simple structures liable to determination by the powder method, a single crystal is required. Preparing such a crystal is therefore often the first step towards structure analysis. The minimum size for operating on a diffractometer is of the order of some tenths of a millimetre.

A similar procedure theoretically applies to *electron diffraction*, with further difficulties arising from dynamic effects.

In any structure analysis, the preliminary steps are the same: point group symmetry, lattice parameters, cell content, space group symmetry. They are outlined below.

11.4.1 POINT GROUP SYMMETRY

11.4.1.1 Crystal with a Polyhedral Form

A specimen with a recognisable crystallographic form provides symmetry data in a straightforward way through optical observation:

1. *Crystal orientation* along a crystallographic axis. Various morphological details may facilitate the operation: cleavage marks, striations, etc.

212   *Structural and Chemical Analysis of Materials*

2. *Point group symmetry* (symmetry class) limited by possible ambiguities due to the occurrence of particular forms which are not affected by a possible merihedry.

11.4.1.2 Crystal without any Polyhedral Form

The Laue method provides the orientation and the Laue group corresponding to the crystal. A rotation diffraction pattern or a powder diffraction pattern can equally provide data on point symmetry through the procedure of lattice determination. The indetermination of the symmetry centre due to Friedel's law can be clarified by means of various tests:

1. *Physical properties* incompatible with centrosymmetry (rotatory dispersion, piezoelectricity, pyroelectricity). The effect may be very weak, depending on the species; as a result the absence of observable effects does not necessarily mean centrosymmetry.
2. *Symmetry of corrosion figures* which consists of negative crystallographic forms able to identify the symmetry class.
3. *Anomalous scattering* which does not obey Friedel's law.
4. *Electron diffraction* which also does not obey Friedel's law. The convergent beam technique in particular can provide data on space group symmetry.
5. *Unit cell content*: an uneven number of atoms in the unit cell excludes a centre of symmetry.

11.4.2 LATTICE PARAMETERS

The lattice is provided by the rotation method and by the powder method. The first avoids any ambiguity, even for low symmetry, when the crystal is adequately oriented along a crystallographic direction; the latter is somewhat problematical for low symmetries, but has the advantage of operating on polycrystalline material, without preliminary sample orientation, and providing more accurate parameters with a focusing camera or a diffractometer.

11.4.3 UNIT CELL CONTENT

Given the elementary or chemical composition as determined by means of spectrometric or chemical analysis, as well as the lattice parameters, the number z of structural or chemical units per unit cell is then easily calculated.

Example. MgO:
Cubic lattice. Parameter $a = 4.21$ Å. Density $\rho = 3.55$ g/cm^3. Mass of a unit cell:

$$m = \rho a^3 = 26.5 \times 10^{-23} \text{ g}$$

The mass of one chemical unit of MgO is

$$m_0 = (A_{Mg} + A_O)/N = (24.3 + 16)/6.02 \times 10^{-23} = 6.69 \times 10^{-23} \text{ g}$$

where A is the atomic mass and N the Avogadro number. The number of chemical units per unit cell is therefore

$$z = m/m_0 = 3.98 \cong 4$$

z is necessarily an integer; as a result, a unit cell of MgO contains 4 Mg and 4 O. This reasoning implies a well-defined chemical species.

11.4.4 SPACE GROUP SYMMETRY

Systematic absences of reflections provide information on the lattice type and on the occurrence of glide symmetry elements. As a result, they lead to the space group if the point group is already known. The extinction rules are deduced from indexing any type of diffraction pattern, e.g. through the construction of the intensity weighted reciprocal lattice (see Figure 5.16).

No Systematic Absences. The intensity weighted reciprocal lattice consists of diffraction domains at any reciprocal lattice points. The direct lattice is then primitive P and the crystal structure has no glide symmetry elements. The point group determines the space group.

Systematic Absences. Three-dimensional periodicity of the absences in the reciprocal lattice, due to multiple lattice types I, F, A, B, C.

Two-dimensional periodicity of the absences in given zero-level planes of the reciprocal lattice, due to glide symmetry planes.

One-dimensional periodicity of the absences on reciprocal lattice lines passing through the origin, due to screw axes.

Diffraction Symbol. Given the complete knowledge of the point group symmetry and of the extinction rules, the space group symmetry is determined without ambiguity, with a few exceptions.

Given only the Laue class, the systematic extinctions lead to a *diffraction symbol* which consists of the successive notations of: the Laue group, the lattice type, the glide symmetry elements as identified by the systematic absences (in [9], *Int. tab.*, vol. 2). In some cases knowledge of the Laue group leads to the space group without ambiguity. The general case is an indetermination between several space groups which are compatible with the same diffraction symbol (Table 11.4). In most cases, only one of the possible groups leads to a likely structure and the indetermination can then be clarified *a posteriori*.

Simple crystal structures with all atoms in *particular positions* (on point symmetry elements) lead to additional absences. Together with the general absences, they are given for each space group in [9], *Int. tab.*, vol. A.

More recently, *convergent beam electron diffraction* has been used to determine the space group of small crystals, because of the non-validity of Friedel's law to electron diffraction.

11.4.5 ATOMIC POSITIONS

11.4.5.1 Trial-and-error Method

The possible atomic positions are limited to those compatible with the space group symmetry, with the number of atoms per unit cell and with the atomic radii.

1. Atoms do not interpenetrate. The interatomic distances must therefore be at least equal to the sum of the respective atomic radii.
2. A given atom in the unit cell is repeated through the symmetry elements of the group. If the unit cell contains j atoms of an element, their sites must correspond to those leading to j equivalent positions in the space group. In a complex structure, however, atoms of the same element may occur at different, non-equivalent sites. The equivalent positions for any of the 230 space groups are indicated in [9], *Int. tab.*, vol. A.

For some simple high-symmetry structures, the atoms occupy particular sites, e.g. in the Wyckoff positions (a) or (b), the three coordinates of all atoms may be entirely determined by the space group. In other cases their positions are determined by one or two parameters which can be worked out by means of a trial-and-error method based on the intensities. Structures with atoms in general positions (three unknown parameters) are best solved by the general FT method.

In any case a structure hypothesis is tested and adjusted by a least-squares refinement to the experimental intensities.

Structure Completely Determined by the Space Group

Consider once more the example of MgO.

1. *Point group symmetry* determined by a Laue photograph (Figure 11.5a) and the crystallographic forms: $m3m$ (cubic holohedry).
2. *Unit cell parameters* measured by a powder diffraction pattern (Figure 11.16) or by a rotation diffraction pattern (Figure 11.10a): $a = 4.21$ Å.

Table 11.4. Examples of space groups as determined by the diffraction symbol. The unequivocal examples are printed in bold face

Laue group	Diffraction symbol	Space group
$2/m$	$2/m\ P\ c$	$P\ c$
		$P\ 2/c$
	$2/m\ P\ 2_1\ c$	$P\ 2_1/c$
mmm	$mmm\ P\ n$	$P\ mn2_1$
		$P\ 2_12_12_1$
	$mmm\ C\ 2_1$	$C\ 222_1$
$4/m\ mm$	$4/m\ mm\ P\ b\ c$	$P\ 4_2bc$
		$P\ 4_2/m\ bc$
	$4/m\ mm\ P\ n\ b\ c$	$P\ 4_2/n\ bc$

3. *Number of structure units* MgO per unit cell (see above): $z = 4$.
4. *Space group symmetry.* Indexing a powder or a rotation diffraction pattern leads to the extinction rule: hkl all even or all uneven, without further extinctions. The only compatible space group is $F\,m3m$ (group No. 225).
5. *Atomic positions.* The space group provides only two sets of positions with four equivalent sites, the positions (a) and (b) in the Wyckoff notation. Placing Mg in one, O in the other (or the reverse) leads to the atomic coordinates as follows:

$$\text{Mg } (0\ 0\ 0),\ (0\ 1/2\ 1/2),\ (1/2\ 0\ 1/2),\ (1/2\ 1/2\ 0)$$

$$\text{O } (1/2\ 1/2\ 1/2),\ (1/2\ 0\ 0),\ (0\ 1/2\ 0),\ (0\ 0\ 1/2)$$

The structure hypothesis is thus imposed by the space group. It is tested by comparing the intensity as calculated from the above positions (after adequate corrections and scaling) with the experimental intensities.

Structure Determined by the Space Group to One Parameter

Consider the example of rutile TiO_2.

1. *Point group symmetry.* Determined by observing the well-developed crystallographic forms: $4/m\ mm$ (tetragonal holohedry).
2. *Unit cell parameters.* Two rotation photographs, respectively about [100] and [001], provide the parameters: $a = 4.58$ Å, $c = 2.98$ Å (Figure 11.10b).
3. *Number of structure units* TiO_2 per unit cell ($\rho = 4.2$ g/cm^3): $z = 2$. Two Ti atoms and two O atoms have to be placed in the unit cell.
4. *Space group symmetry.* Indexing a rotation diffraction pattern shows systematic absences characteristic of a 4_2 screw axis and a diagonal glide plane n and leads to the space group P $4_2/m\ nm$ (group No. 136).
5. *Atomic positions.* To place Ti, the choice is limited to the two Wyckoff positions (a) and (b). For O there are five possibilities of four equivalent sites, but three are excluded due to non-observed particular extinctions; (f) and (g) remain which are interchangeable. As a result, the atoms are placed at positions (a) and (f) with coordinates as follows:

$$\text{Ti } (0\ 0\ 0),\ (1/2\ 1/2\ 1/2)$$

$$\text{O } (x\ x\ 0),\ (-x\ -x\ 0),\ (1/2+x\ 1/2-x\ 1/2),\ (1/2-x\ 1/2+x\ 1/2)$$

The sites of Ti are fixed by the group; those of O are fixed to one parameter x. Plotting the intensities $I(hkl)$ of a few prominent reflections as a function of x results in a value of x providing the best agreement with the experimental intensities. A least-squares procedure leads to the value $x = 0.31$ and to the structure outlined by Figure 11.27.

Comparing the Measured and Calculated Intensities

To compare significant values of the measured and the calculated intensities, specific corrections must first be applied to the latter. X-ray diffraction requires the following

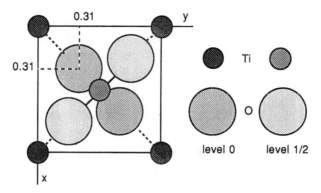

Figure 11.27. Crystal structure of rutile TiO$_2$ projected on (001).

corrections: *polarising correction*, P (Eq. 3.12); *kinematic correction*, L.

Intensities are integrated over the whole angular range corresponding to the passage of a diffraction domain through the reflecting sphere (see Figure 5.9). Depending on the instrumental layout, certain reflections are favoured with respect to others, due to a different angle-dependent intersection geometry. This is taken into account by the so-called Lorentz factor L, which is associated in the tables with the other angle-dependent factor, the polarising factor P. The calculated intensity $I_c(hkl)$ to compare with the measured intensity $I_m(hkl)$ then becomes for X-ray diffraction:

$$I_c(hkl) = LP(hkl) \left\{ \sum_j f_j(hkl) \exp\left[-\frac{B \sin^2\theta}{\lambda^2}\right] \exp\left[2\pi i(x_j h + y_j k + z_j l)\right] \right\}^2 \quad (11.23)$$

The respective values of f_j, $(B \sin^2\theta/\lambda^2)$ and LP are tabulated in [9] *Int. tab.*, vols 2 and 3.

Multiplicity Factor. When compiling the intensities from a powder diffraction pattern, the symmetry-related multiplicity factor M of the various reflecting planes must be taken into account (see 11.3.1).

Scaling. For the relative intensities to be comparable, the measured values are multiplied by a common scaling factor such as the sum of measured intensities equals the sum of calculated intensities.

Reliability Factor. Intended to assess the validity of a structure analysis, the reliability factor R expresses the relative error made on the intensity measurements as follows:

$$R = \frac{\sum_{hkl} |\sqrt{I_m} - \sqrt{I_c}|}{\sum_{hkl} \sqrt{I_m}} \quad (11.24)$$

A structure hypothesis is considered to be satisfactory when $R \leq 0.2$; in favourable conditions the factor can be as low as 0.05. The refinement operation is conducted by means of computer programs until the reliability demanded is achieved.

11.4.5.2 Fourier Transform Methods

As has been outlined previously, knowledge of the intensities and the phases of diffracted X-rays, i.e. the experimental structure factors $F_m(hkl)$, leads to the determination of the corresponding crystal structure by calculating a three-dimensional FT as follows (Eq. 5.17):

$$f(xyz) = \frac{1}{v_0} \sum_h \sum_k \sum_l F(hkl) \exp[-2\pi i(xh + yk + zl)] \qquad (11.25)$$

where $F(hkl) = F_m(hkl)/\sqrt{(LP)}$.

The result is a distribution of the electron density at any point of the unit cell. The maxima of electron density provide the atomic positions. The *limiting condition* $d(hkl) \leqslant \lambda/2$ corresponds to a limited volume of the reciprocal space inside the *limiting sphere* of radius $2/\lambda$ accessible to measurement, and therefore to a limited number of terms of the Fourier series. It determines the accuracy of the atomic positions. Using smaller wavelengths (e.g. Mo-Kα) improves the accuracy.

When the phases are unknown, the FTs of the diffracted intensities lead to the *Patterson series* (Eq. 5.37):

$$q(x'y'z') = \frac{1}{v_0} \sum_h \sum_k \sum_l I(hkl) \cos[2\pi(x'h + y'k + z'l)] \qquad (11.26)$$

where $I(hkl) = I_m(hkl)/LP$. The result is the convolution square of the structure.

A short survey of the current methods of structure analysis based on FTs is given below.

Patterson Series

The result of the Patterson series is a pseudo-structure, the *Patterson function*, expressed by a distribution of pseudo-atoms which have the interatomic vector components as coordinates (see Figure 5.11). Deconvolution of the Patterson function leading to the structure itself is facilitated by the existence of a heavy atom in the structure. In this case, the Patterson function displays a small number of large pseudo-atoms whose centres are well defined. The corresponding atomic positions can then be readily determined by deconvolution.

Phase Determination through the Patterson Function

Heavy Atom Method. As mentioned above, the positions of heavy atoms can easily be determined. They can be assumed to fix the phases of the prominent reflections. Calculating the structure factor corresponding only to the heavy atoms leads to those phases which are then attributed to the corresponding measured intensities, permitting the calculation of the corresponding Fourier series which, in turn, is used to determine the phases of further reflections. This iteration process leads to a progressive refinement of the structure. The heavy atom method can be used for structures comprising a few heavy atoms in an environment of lighter atoms, such as silicates and organometallic compounds.

Isomorphous Substitution. When no heavy atoms are naturally present, they can be introduced by substitution, provided the structure remains unchanged. This method is useful for organic crystals.

The heavy atom method is often completed by *neutron diffraction* to determine the positions of the light atoms, e.g. hydrogen in organic compounds.

Direct Methods. Fourier Series

The aim of the direct methods is to work out the phases linked to the *hkl* reflections in order to perform the Fourier series. The phase problem is easier to solve for centrosymmetrical crystals; it reduces to fixing the sign of the amplitude, the phase being either 0 or π (see Figure 5.6b). For n measured reflections, there are 2^n combinations of plus and minus signs. Attributing the signs of the prominent reflections is guided by restraints on the possible structures (necessarily positive electron density, impenetrable spherical atoms, positions in accordance to the space group), transposed into the reciprocal space in the form of relations between the signs. As in the case of the heavy atom method, a rough structure is thus outlined and then refined through iteration.

The phases related to general non-centrosymmetrical structure are worked out in a similar way.

These direct methods are now commonly used for the analysis of complicated structure.

Structure Refinement

A structure outline resulting from any method is subjected to a refinement process aimed at minimising the deviations between measured and calculated amplitudes. A currently used method is to calculate the series of differences $\delta(xyz)$ according to the following relation, written for a centrosymmetrical structure:

$$\delta(xyz) = \sum_{hkl} [F_c(hkl) - F_m(hkl)] \cos 2\pi(hx + ky + lz) \quad (11.27)$$

where $F_c(hkl)$ are the structure factors as calculated from the approximate structure and $F_m(hkl)$ the measured amplitudes, adequately corrected and scaled.

If the amplitudes are known with a perfect accuracy

$$\delta(xyz) = f_c(xyz) - f(xyz)$$

where $f_c(xyz)$ is the calculated electron density and $f(xyz)$ the real electron density.

In the ideal instance where $F_c(hkl) = F_m(hkl)$, the result would be a zero-density distribution. The actually observed density anomalies around the assumed atomic positions are used to refine the positions until the minimum of the reliability factor is reached.

Practical Calculations

For a given number of measured reflections, determining the electron density at one point *xyz* of the unit cell results in summing up the same number of terms. Consider the instance of a structure analysis with 200 reflections and an accuracy of 0.1 Å required on the positions in a cubic unit cell of 5 Å. It requires the electron density to be determined at 125 000 points, i.e. to calculate 125 000 Fourier series of 200 terms each, resulting in 2.5×10^6 operations, in addition to the correction calculations. This highlights the necessity of high-power computing techniques for resolving complex crystal structures. Dedicated computer programs have been developed for correcting the measured intensities, solving the phase problem, working out the structure outline and refining it. The final result is displayed either in the form of a list of atomic coordinates, with an accuracy reaching 0.01 Å, or in the form of a stereoscopic view of the structure.

11.4.5.3 Fourier Projections

Two-dimensional or one-dimensional projections of crystal structures are sometimes sufficient; they require much less data.

Two-dimensional Fourier Series. Fourier–Bragg Projection

The measurement is limited to reflections related to the lattice points in one reciprocal plane passing through the origin, e.g. the $hk0$ reflections in the plane (001)*. According to Eq. (5.16), it follows that

$$F(hk0) = \int \int_{xy} \left[\int_0^c f(xyz) dz \right] \exp[2\pi i (xh + yk + zl)] \, dx dy \qquad (11.28)$$

where $F(hk0)$ is the FT of the function $\int f(xyz) \, dz$ which represents the electron density in the unit cell projected on the plane (001), parallel to the direction [001]. This function is expressed by Fourier inversion

$$\int_0^c f(xyz) dz = \frac{1}{v_0} \sum_h \sum_k F(hk0) \exp[2\pi i (xh + yk)] \qquad (11.29)$$

The required number of operations is thus greatly reduced. The projection function is currently represented in the form of computer-drawn electron density contour maps (see Figure 11.29a).

Analogical Methods. Photosumming

Consider the Fourier–Bragg projection related to a centrosymmetrical structure. Equation (11.29) develops into a sum of sine-shaped density distribution curves

$$v_0 \int_0^c f(xyz) dz = F(100) \cos 2\pi x + F(200) \cos 4\pi x \ldots + F(hk0) \cos 2\pi (xh + yk) \qquad (11.30)$$

Each distribution may be displayed by means of a system of parallel and equidistant sinusoidal fringes, of amplitude $F(hk0)$, of period a/h along x and b/k along y, parallel to $(hk0)$. For a negative sign of $F(hk0)$, the fringes are shifted by a half-period (Figure 11.28).

An analogue display of this projection can therefore be performed by a photographic summing of corresponding optical fringes, with an exposure time proportional to the respective intensity.

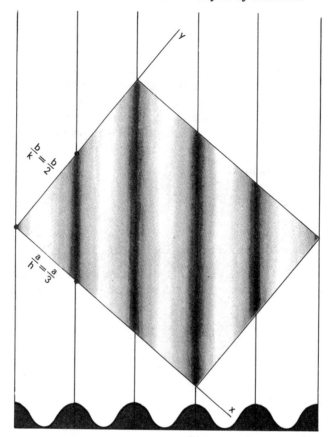

Figure 11.28. Sinusoidal electron density variation corresponding to the term $F(320)\cos 2\pi(3x+2y)$, phase 0.

Replacing the amplitudes by the intensities results in a Patterson projection. This can be used for graphical deconvolution, e.g. in connection with the heavy atom method.

Specific photosumming devices were of great use before the advent of high-power computer facilities (Figure 11.29b).

It must be pointed out that structural images provided by *high-resolution electron microscopes* are quite similar to Fourier–Bragg projections with X-rays. The electron density projection is then replaced by the electrical potential projection. The phase problem obviously does not exist: without taking into account the lens aberrations, the phases of electron waves emitted at a given point of the object are conserved along the wave path to the image plane where their amplitudes add up to form the image contrast, which therefore reproduces a potential projection of the object, to the resolution limit.

One-dimensional Fourier Series

A one-dimensional Fourier series is reduced to the reflections which correspond to the points on a reciprocal lattice line passing through the origin, i.e. to the successive orders

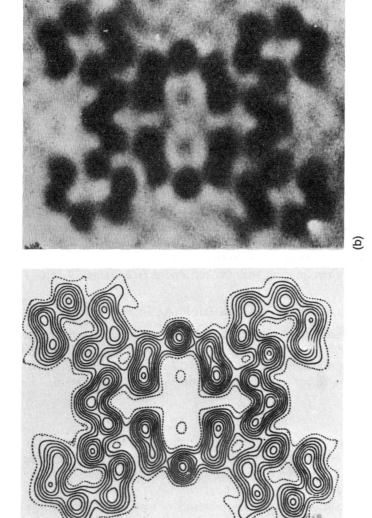

Figure 11.29. Fourier–Bragg projection of the crystal structure of copper-phthalocyanide along the [010] direction (in Von Eller, 1955). (a) Calculated electron density contour map; (b) photosummation.

of reflection on one set of lattice planes. Considering the set of planes (001), i.e. the reflections 00*l*, Eq. (5.16) becomes

$$F(00l) = \iiint_{\text{unit cell}} f(xyz)\exp(2\pi ilz)\,dx\,dy\,dz = \int_0^c \left[\iint_{xy} f(xyz)\,dx\,dy\right] \exp(2\pi ilz)\,dz \quad (11.31)$$

where $F(00l)$ is the FT of the function $\iint f(xyz)\,dx\,dy$ which represents the projection of electron density of the unit cell on the lattice line [001], parallel to the plane (001). This function is given by the Fourier inversion

$$\iint_{xy} f(xyz)\,dx\,dy = \frac{1}{v_0}\sum_l F(00l)\exp(2\pi ilz) \quad (11.32)$$

In the instance of a centrosymmetrical crystal, the following sum of sine distributions can be written:

$$\iint_{xy} f(xyz)\,dx\,dy = \frac{1}{v_0}F(00l)\cos 2\pi z + F(002)\cos 4\pi z + \ldots + F(00l)\cos 2\pi lz \quad (11.33)$$

This kind of operation can be performed with only limited means of calculation. It can be interesting for the determination of the electron density distribution along particular directions, e.g. along the normal to the layers of lamellar structures (Figure 11.30).

With the same restraints as pointed out in the previous section, the result can be compared with the corresponding one-dimensional structure projection provided by an electron microscope when imaging with 00*l* reflections (see Figure 19.24b).

11.4.6 APPLICATIONS OF NEUTRON DIFFRACTION

As has been shown earlier, X-ray diffraction fails to determine the positions of very light atoms, e.g. hydrogen, and has difficulty in discerning atoms with neighbouring numbers (e.g. Al–Mg, Fe–Co, Fe–Mn in alloys). Neutron diffraction is not directly related to the atomic number and provides complementary information.

11.4.6.1 Practice

Working out a crystal structure by means of neutron diffraction is generally limited to refining a structure outlined by X-ray diffraction, e.g. with the positions of the heavier atoms already fixed. The phases of neutron reflections can thus be determined from the partial X-ray model.

A practical problem arises from the low intensity of neutron reflections which result from two main causes:

1. A low-neutron flux from the source, when compared with the commonly used X-ray tubes and *a fortiori* with the synchrotron sources.
2. A low-scattering cross-section, about 10 times smaller than the corresponding value for X-rays which is already low (see Table 4.1).

Neutron-diffraction experiments therefore require large specimens and long counting times with highly sensitive detecting devices. High-flux nuclear reactors have been

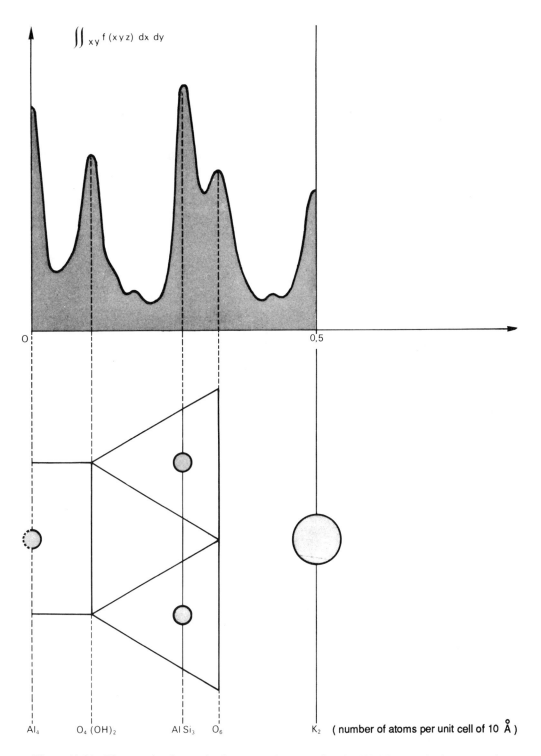

Figure 11.30. Electron density projection on to the normal to the (001) layers of mica muscovite. The display is limited to half a unit cell and is compared with the structure model (from Gatineau and Mering, 1958).

specially designed for material analysis. In Europe for instance, such a source is located at the ILL (Institut Laue–Langevin) at Grenoble with an output of 57 MW and a neutron flux per second of 10^{15} n/cm^2 into 15 lines.

11.4.6.2 Application to Hydrogen Compounds

Neutron diffraction enables the carrying out of investigations of the hydrogen bond and the determination of hydrogen sites in heavy metal hydrides, in crystalline species containing hydroxyl groups or water molecules (e.g. silicates, in particular zeolites), in organic compounds. In particular the crystal structure of solid water has been determined in this way (Peterson and Levy, 1953). Nowadays one of the most important fields of neutron diffraction is the structure analysis of biological crystallised materials.

11.4.6.3 Order–Disorder Transformations

Unlike X-ray diffraction, neutron diffraction allows easy differentiation of the sites of atoms with neighbouring numbers. This is the case in particular for most alloys and other mineral compounds. The group of spinels, for instance, is very important in magnetic materials. The properties of those minerals are sensitive to the respective positions of the two types of metal atoms, e.g. Mg–Al, Mn–Fe, Co–Fe. In addition to investigating order–disorder transformations as a function of thermal treatments, neutron diffraction allows the normal and the reverse spinel structure to be distinguished (Bacon, 1952).

11.4.6.4 Structures of Magnetic Compounds

In addition to the atomic positions, neutron diffraction is sensitive to magnetic dipole moments, thus enabling them to be located in crystals. It has led to the confirmation of the theory of Neel on antiferromagnetism and ferrimagnetism.

11.4.7 CONCLUSION

Structure analysis by means of X-ray diffraction has led to remarkable developments in solid physics and in solid chemistry. The silicates are typical examples of a structure-related chemistry, based on the permanence of the SiO_4 structure units and their linkage to form the various crystalline (or vitreous) species. The first silicate structure was established around 1930 and later refined progressively, following the progress of measuring and computing techniques. In a similar way, knowledge of the structures of crystalline biological materials is an important step towards understanding their activities.

The accuracy of measuring the atomic positions has continuously improved. In good conditions, with well-crystallised materials, the atomic coordinates can be determined to some 5×10^{-3} Å, e.g. to a relative accuracy of about 10^{-3}. Bond lengths can be measured to 10^{-2} Å.

Investigating the fine structure of X-ray absorption spectra at absorption edges (EXAFS), as well as the similar effects in the energy loss spectra of electrons (EELS), provides further important data on the structural surrounding of specific atoms in solids.

The corresponding methods have therefore become valuable complements to X-ray diffraction for structure analysis.

REFERENCE CHAPTERS

Chapters 3, 4, 5, 8, 9, 10. Appendixes B, C.

BIBLIOGRAPHY

[1] Azaroff L. V., Buerger, M. J. *The powder method in X-ray crystallography*. McGraw-Hill, New York, 1958.
[2] Bacon, G. E. *Neutron diffraction*. Oxford University Press, 1975.
[3] Bragg L. *The crystalline state*, vol. 1. Bell, London, 1949.
[4] Buerger M. J. *The photography of the reciprocal lattice*. Polycrystal Book Serv., New York, 1944.
[5] Buerger M. J. *X-ray crystallography*. John Wiley, New York, 1962.
[6] *Crystallographic computing* (IU Cr. Symposia). Oxford University Press, 1985, 1988.
[7] Gluster J. P. *Crystal structure analysis*. Oxford University Press, 1985.
[8] Guinier A. *Théorie et technique de la radiocristallographie*. Dunod, Paris, 1964.
[9] *International tables for crystallography*, vol. A: *Space-group symmetry* (Hahn T. ed.). D. Reidel, Kluwer Acad. Publ. Dordrecht, 1987; *International tables for X-ray crystallography*, vol. 2: *Mathematical tables* (Kasper J. S., Lonsdale K., eds); vol. 3: *Physical and chemical tables* (MacGillavry C. H., Rieck G. D., Lonsdale K., eds); vol. 4: *Supplementary tables to vols 2, 3* (Ibers J. A., Hamilton, W. A., eds). D. Reidel, Kluwer Acad. Publ., Dordrecht, 1985, 1983, 1982.
[10] Ito T. *X-ray studies on polymorphism*. Maruzen, Tokyo, 1950.
[11] James R. W. *The crystalline state*, vol. 2. *The optical principles of the diffraction of X-rays*. Bell, London, 1965.
[12] Klug H., Alexander L. E. *X-ray diffraction procedures for polycrystalline and amorphous materials*. John Wiley, New York, Chichester, 1987.
[13] Kovacs T. *Principles of X-ray metallurgy*. Iliffe, London, 1969.
[14] Lipson H., Steeple H. *Interpretation of X-ray powder diffraction patterns*. Macmillan, London, 1968.
[15] Lipson H. S. *The crystalline state*, vol. 3. *The determination of crystal structures*. Bell, London, 1953.
[16] Peiser H. S., Rooksby H. P., Wilson A. J. C. (ed.) *X-ray diffraction by polycrystalline materials*. The Inst. of Physics, London, 1955.
[17] *Powder diffraction file*. International Centre for Diffraction Data, Swarthmore, Pa (USA).
[18] Whiston C. *X-ray methods*. John Wiley, New York, Chichester, 1987.
[19] Woolfson M. M. *An introduction to X-ray crystallography*. Cambridge University Press, 1970.

12

Electron Diffraction on Thin Crystalline Layers

Electron diffraction is based on the principle of the particle wave as stated by Louis de Broglie in 1924 and confirmed by the first experiments of Davisson and Germer in 1927. The strong electron–matter interaction results in limiting the active thickness of involved matter to a thin layer. This in turn leads to a relaxation of the diffraction conditions in the incident direction and to specific diffraction facilities differing from those for X-rays.

The energy of electron beams commonly used in material research ranges from some electron volts (low-energy electrons) to the order of some megaelectron volts (very high-energy electrons). The most commonly used are the so-called high-energy electrons (about 10–120 keV). The low-energy electrons are widely used for crystal surface analysis.

Common abbreviations: HEED (high-energy electron diffraction), LEED (low-energy electron diffraction), TEM (transmission electron microscope), STEM (scanning transmission electron microscope).

12.1 PARALLEL BEAM HIGH-ENERGY ELECTRON DIFFRACTION

High-energy electron diffraction is currently operated in transmission electron microscopes in which diffraction and imaging are closely associated.

12.1.1 PARTICULAR FEATURES OF HIGH-ENERGY ELECTRON DIFFRACTION

The most common diffraction mode is the so-called parallel beam mode, with an angular dispersion limited to some 10^{-4}–10^{-5} rad.

For specific applications, the convergent beam mode is increasingly made use of, with an angular aperture of some 10^{-2} rad (section 12.2).

The general diffraction theory applies to electron diffraction, with two main particular features:

1. A *very small wavelength* as compared to the lattice parameters. This results in small diffraction angles and particular geometrical features.

Table 12.1. Diffraction angles and intensities for 00*l* reflections of an aluminium crystal (cubic F, $a = 4.04$ Å) and 100 keV electrons (the corresponding existing values for Cu–Kα X-rays, limited to the first two orders are in parentheses). The temperature factor is not taken into account

h00	2θ (degrees)		$I(I_{200} = 100)$	
200	1.05	(45)	100	(100)
400	2.1	(99)	16	(45)
600	3.1	...	5.3	...
800	4.2	...	2.0	...
10 00	5.2	...	0.9	...

2. A *strong interaction* with matter, which is inconsistent with the kinematic approximation. This results in difficulties for intensity interpretations. Whereas for a qualitative interpretation the kinematic theory is sufficient, a significant quantitative interpretation requires recourse to the dynamic theory.

12.1.1.1 Diffraction Angles

Due to the small wavelength (see Table 1.1) resulting in $\lambda \ll d(hkl)$, $\sin \theta$ may be approximated to θ, and the Bragg equation becomes

$$2\theta = \frac{n\lambda}{d(hkl)} \tag{12.1}$$

The diffraction angle 2θ for the first orders on current crystals is of the order of 10^{-2} rad, i.e. of the order of 1° (Table 12.1).

12.1.1.2 Reflecting Sphere

The radius L of the reflecting sphere (Ewald sphere) is very large when compared with the reciprocal lattice parameters. Taking the example of aluminium and 100 keV electrons (Table 12.1) leads to $L/a^* = d(100)/\lambda = 109$ (as compared with 2.6 for Cu–Kα X-rays). Furthermore the atomic scattering amplitude for electrons varies with $(1/\sin \theta)^2$, resulting in a much faster decrease vs λ than for X-rays (Eq. 4.7).

As a result, recording an electron diffraction pattern can be limited to a few degrees (Table 12.1), permitting the use of a plane film normal to the primary beam.

As a result of the geometrical characteristics of high-energy electron diffraction, *the reflecting sphere may be assimilated to its tangent plane*.

12.1.1.3 Diffraction Condition Relaxation

In transmission electron diffraction, the specimen is necessarily very thin. This results in a significant relaxation of the diffraction conditions along a normal to the specimen, i.e. along the incident beam in normal incidence (see section 6.2). With a thin single crystal, the resulting diffraction domains (limited to the main maxima) are in the form of fine rods normal to the specimen (Figures 6.1 and 12.1). For a *hkl* reflection to be

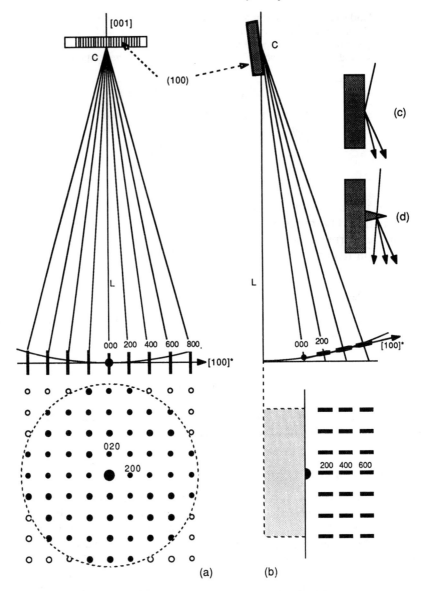

Figure 12.1. High-energy electron diffraction. Example of a [001] zone pattern with $hk0$ reflections from a cubic crystal. (a) Normal incidence transmission mode on a thin crystal. The related diffraction pattern is outlined at the foot of the diagram. Due to diffraction relaxation along the normal to the specimen, i.e. along the incident direction, the intersections of the diffraction domains with the reflecting sphere result in fine spots; (b) grazing reflection mode on a plane surface of a thick crystal. Diffraction relaxation still occurs along the normal to the specimen, i.e. in this case along the normal to the incident beam. The intersections of the diffraction domains with the reflecting sphere result in elongated spots on the diffraction pattern, normal to the shadow of the specimen which materialises the intersection of its surface with the diffraction pattern; (c) and (d) outline of the real reflection mode in (c), of the transmission–reflection mode on surface asperities in (d).

observable, the reflection sphere must intersect the corresponding diffraction domain. When the incident beam is parallel to a prominent lattice line [uvw] (i.e. an important zone axis orientation), the large radius of the reflecting sphere results in a large number of simultaneous reflections. The observed reflections are those relative to the lattice planes (hkl) which have the incident direction [uvw] as a zone axis, i.e. those which correspond to the reflecting sphere intersecting the points hkl of the reciprocal lattice plane (uvw)*, of level $m = 0$. The reflection indices hkl are therefore consistent with the following equation:

$$uh + vk + wl = 0 \qquad (12.2)$$

This is the equation of the reciprocal lattice plane (uvw)* passing through the origin ($m = 0$) and containing the lattice points hkl. The corresponding diffraction pattern is commonly called the [uvw] *zone pattern* (Figure 12.1a).

The most important particular features of high-energy electron diffraction in the parallel transmission mode can finally be summarized as follows:

> *The diffraction pattern observed on a plane screen or a plane film, normal to the incident beam, reproduces a section of the diffraction reciprocal space by a plane passing through the origin.*

Electron diffraction on a thick crystal can be performed by reflection. The active volume of the specimen is then limited to a thin surface layer. As in transmission diffraction, the diffraction condition is relaxed along the normal to the specimen surface. However, in the most common grazing reflection mode, diffraction relaxation is directed along a line perpendicular to the beam, resulting in fine rod-shaped diffraction domains parallel to the reflecting sphere (Figure 12.1b).

12.1.2 INSTRUMENTAL LAYOUTS

High-energy electron diffraction is most commonly performed in the normal incidence transmission mode on thin specimens similar to those used in transmission electron microscopy (Figure 12.1a). Thick specimens can be observed through grazing incidence diffraction (Figure 12.1b).

Because of the strong electron–matter interaction, the diffraction patterns can be directly observed on a fluorescent screen. They are recorded on plane photographic films with thin emulsion layers.

Diffraction can be performed according to two different techniques, macrodiffraction and microdiffraction. The first technique can be implemented by means of rather simple equipment, whereas the second one is generally performed by means of electron microscopes.

Energy filtering permits the separation of elastic and inelastic diffraction components and the measurement of energy losses corresponding to given details of the diffraction pattern.

Common abbreviations: HEED (high-energy electron diffraction, understood by transmission), RHEED (reflection high-energy electron diffraction).

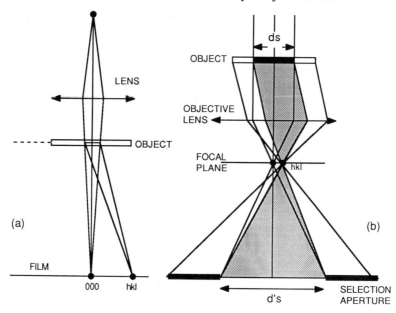

Figure 12.2. Transmission electron diffraction layouts. (a) Macrodiffraction without objective lens; the specimen–film distance materialises the radius of the reflecting sphere; (b) microdiffraction in the transmission electron microscopes by means of a selection aperture of diameter d_s' inserted into the objective image plane.

12.1.2.1 Macrodiffraction

The very simple basic layout is outlined in Figure 12.2a. An electronic lens projects the image of the electron gun cross-over on the film. The specimen is placed on a holder between the lens and the film, at a distance L from the latter. In the small angle approximation the diffracted beams are focused on the film. The active area of the specimen depends on the aperture of the incident beam and is commonly a few square millimetres. In the instance of a thin single crystal in the transmission mode, relaxation of the diffraction conditions is limited to the incident direction, therefore resulting in very fine diffraction spots. In the case of a polycrystalline sample, a large number of crystallites are in the active zone, resulting in well-defined and continuous diffraction rings. This technique leads to highly accurate measurements; it is therefore often called high-resolution electron diffraction.

The *diffraction length L* is accessible to direct measurement. It is, however, preferable to carry out an accurate determination of the *diffraction constant $K = L\lambda$* (reciprocal lattice constant) by means of a standard.

The macrodiffraction layout is well designed for reflection diffraction on a plane surface of a thick specimen. The result is a diffraction pattern, limited by the shadow cast by the specimen surface, with elongated spots normal to this limit. A specimen with tiny surface asperities can give a false reflection pattern, actually a transmission pattern through the asperities, resulting in fine diffraction spots (Figures 12.1b, c, d). The development of RHEED has been linked to the progress of high-vacuum techniques, the only means of observing contamination-free surfaces. As such, it has become a method of surface structure analysis, complementary to low-energy electron diffraction, with certain advantages, in particular a higher accuracy due to smaller diffraction spots (Cousandier, 1988).

Thanks to its spot- or ring sharpness, macrodiffraction is particularly designed for high-resolution analysis of intensity profiles, e.g. for investigating scattering outside the Bragg reflections.

12.1.2.2 Microdiffraction through Aperture Selection

Principle

The principle of *selected area electron diffraction* was proposed by Le Poole in 1947. It is commonly set up in transmission electron microscopes. With an active area of the order of 1 μm, it leads to microdiffraction.

In the parallel beam mode (Fraunhofer diffraction), the diffraction pattern is located in the *back focal plane* of the objective lens. The electron image is located in the *objective image plane* (Gaussian plane), conjugate of the object plane with respect to the objective lens. Depending on the adjustment of the post-objective lenses (intermediate lens, diffraction lens) on one of those planes, the final viewing screen or film record displays either the electron diffraction pattern or the electron image of the object.

The selection aperture inserted into the image plane limits an image selection area of diameter d_s'. The final diffraction pattern has been formed by the only rays actually passing through this aperture, i.e. those which issue from the corresponding *object selection area* of diameter $d_s = d_s'/g$, conjugate of the selection aperture with respect to the objective lens of magnification g. Choosing a given aperture results in selecting a specific area in the specimen, e.g. identifying a component of a polycrystalline material, an inclusion, a precipitate (e.g. $g = 100$, $d_s' = 50$ μm ⇒ $d_s = 0.5$ μm).

The diffraction length L is not directly expressed. The diffraction constant $K = L\lambda$ may be determined by an adequate standard specimen. It is generally adjustable through the diffraction lens.

Selection Errors

In order to avoid any wrong interpretation of a selected area diffractogram, possible selection errors, due to objective lens defocus and spherical aberration, must be taken into account.

Defocusing Error. For a perfect selection, according to the geometrical outline of Figure 12.2b, the objective lens must be exactly focused, i.e. the specimen plane must coincide with the plane Π, conjugate of the plane of the selection aperture with respect to the objective lens. A given defocus D causes the respective selection areas of the transmitted beam 000 and a diffracted beam hkl to be shifted by an amount δ_D (Figure 12.3); as a result, the two beams are emitted from different areas of the specimen. The shift direction is determined by the sign of defocusing D (*overfocusing $D > 0$; underfocusing $D < 0$*). Figure 12.3 shows that

$$\delta_D = -2D\theta \quad \text{(the direction of vector } \mathbf{r}^*(hkl) \text{ taken as positive)} \quad (12.3)$$

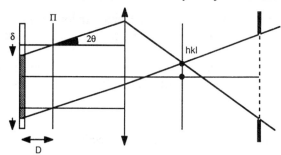

Figure 12.3. Selection error through overfocusing. The direction of the selection shift δ is opposite to the direction of the reciprocal vector $\mathbf{r}^*(hkl)$ taken as positive. The direction of the shift is reversed for underfocusing.

In order to suppress this error the plane of the image given by the objective must coincide with the plane of the selection aperture, which is obtained by carefully focusing on the viewing screen both the edge of the selection aperture and the image inside the aperture. The aperture position is an instrumental constant; the above focusing conditions are therefore met for a determined *selection magnification* G_s.

Spherical Aberration Error. The above focusing conditions being met, an exact selection requires a perfect lens. Due to the spherical aberration of the objective, the geometrical selection is valid solely for an axial beam 000. A diffracted beam *hkl* with an angle 2θ is submitted to a shift $8gC_s\theta^3$ in the selection plane (Eq. 9.4). Relative to the object plane, the spherical aberration shift is

$$\delta_s = -8C_s\theta^3 \quad (C_s \text{ spherical aberration coefficient}) \qquad (12.4)$$

The shift is proportional to the cube of the Bragg angle and its direction is opposite to the direction of the reciprocal vector $\mathbf{r}^*(hkl)$ corresponding to the reflection. Overfocusing causes both errors to be compounded; underfocusing can lead to a partial compensation.

12.1.2.3 Microdiffraction and Nanodiffraction through Beam Selection

Selecting the diffraction area on the sample simply by means of a narrowly focused and aperture-limited incident beam suppresses all selection errors. Thanks to a sophisticated condenser lens system (Figure 9.9), this technique, first proposed by Rieke (1961, 1962a, b), results in a beam cross-section on the specimen which can be as small as a few nanometres (*nanoprobe*), depending on the aperture. The related diffraction mode is therefore often called the *nanodiffraction* mode and is available in recent electron microscopes.

This selection mode enables lattice parameters to be measured at an almost unit cell scale (e.g. the variation of the lattice parameters due to substitutions, to phase separations, to local stress, etc.) as well as tiny crystals to be identified. However, such a small selection area, when limited to a few unit cells, results in an important three-dimensional diffraction relaxation effect and therefore to a poor accuracy.

The nanoprobe diffraction mode is used equally in convergent beam electron diffraction.

12.1.2.4 Specimen

For the *transmission mode*, the specimen must be very thin, of the order of some 10–100 Å, depending on the acceleration voltage. The preparation techniques and requirements are the same as for TEMs.

For the *reflection mode*, a plane surface of the specimen has to be prepared. Observing the intrinsic crystal surface requires a clean surface, produced by chemical, thermal or ion cleaning, as well as a clean and high vacuum, in order to avoid any contamination.

The preparation techniques are given in more detail in the chapters dealing respectively with electron microscopy and with surface analysis.

12.1.3 DIFFRACTION PATTERNS FROM SINGLE CRYSTALS

The main advantage of electron microdiffraction is to provide single crystal diffraction patterns of micrometre-sized particles, e.g. selected in a polycrystalline specimen. The size effect must nevertheless be taken into account.

12.1.3.1 Observed Diffraction Pattern

Zone Diffraction Pattern

The diffraction pattern is determined by the intersection of the intensity weighted reciprocal space with the reflecting sphere. Thus, for a given zone orientation $[uvw]$, the pattern approximately reproduces a reciprocal plane $(uvw)^*$ passing through the origin ($m = 0$). The actually observed diffraction spots correspond to hkl reflections, according to Eq. (12.2):

$$uh + vk + wl = 0$$

In practice, it is always advisable to operate with the incident beam down a prominent zone axis $[uvw]$, i.e. with small indices. The diffraction pattern then represents a high-density reciprocal lattice plane $(uvw)^*$, which usually has a high symmetry (e.g. Figure 12.5). A goniometric sample stage permits the exploration of the largest possible portion of the reciprocal space. Observing the successive orientation-related reciprocal lattice planes may then lead to determining the three-dimensional reciprocal lattice. Mechanical reasons limit the specimen tilt to some 60°.

Laue Zones

Assimilating the reflecting sphere to its tangent plane is a first approximation. The curvature of the sphere is actually observable on prominent zone diffraction patterns, corresponding to high-density reciprocal planes $(uvw)^*$ with sufficiently small interplanar spacings $d^*(uvw)$. The requirement is that the reflections generated by the intersection, with the reflecting sphere, of the higher level reciprocal planes $(uvw)^*$, $m \neq 0$, be inside the angular observation field. Reflections related to $m = 0$ contribute to the zero-order Laue zone (abbreviated to *ZOLZ*); those related to $m = 1$ contribute to the first-order Laue zone (*FOLZ*), etc. (Figure 12.4). Higher-order Laue zones are often abbreviated to *HOLZ*. The Laue zones are circular and concentric for an incidence parallel to the zone axis, elliptical and eccentric for an inclined incidence.

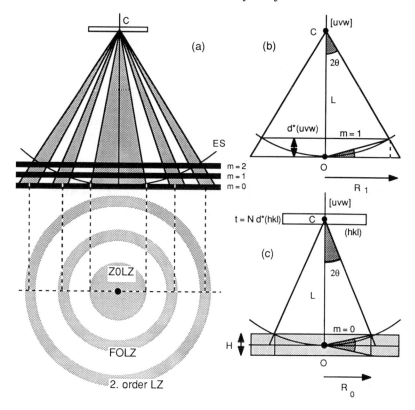

Figure 12.4. Laue zones. The sample is considered to be a thin crystal plate parallel to a set of high-density (hkl) planes, with its normal along a zone axis $[uvw]$. (a) Generation; (b) assessing the parameter $n(uvw)$ of the zone axis from the radius R_1 of the first-order Laue zone:

$$\theta = R_1/2L = d^*(uvw)/R_1 = L\lambda/R_1 n(uvw);$$

(c) assessing the thickness t of the specimen from the radius R_0 of the second-order Laue zone:

$$2\theta = R_0/L = H/R_0 = 2n^*(hkl)/NR_0 = 2L\lambda/tR_0$$

The angles are greatly exaggerated for a better readability of the figures.

Figure 12.5. *(opposite)* Electron diffraction patterns of thin cleavage platelets (001) of mica muscovite, $d(001) = 10$ Å. (a) Direct lattice orientation in normal incidence and related orientation of the reciprocal lattice intersected by the reflecting sphere; (b) diagram of the diffraction pattern with reflections $hk0$. The hexagonal symmetry of the diffraction pattern is related to the symmetry of the layers $(b = a\sqrt{3})$. (c) Normal incidence diffraction pattern with concentric zero-order ($hk0$ reflections) and first-order ($hk1$ reflections) Laue zones; (d) diffraction pattern tilted through 1°, displaying eccentric zero-order, first-order and second-order ($hk2$ reflections) Laue zones; (e) diffraction pattern tilted through 2°, showing the same, even more eccentric Laue zones and highlighting the orientation sensitivity; (f) diffraction pattern of a very thin platelet with merging Laue zones due to the important size effect (Ehret G., Dept Electron Microscopy, Lab. Crystallography, Louis Pasteur University, Strasbourg).

Electron Diffraction on Thin Crystalline Layers

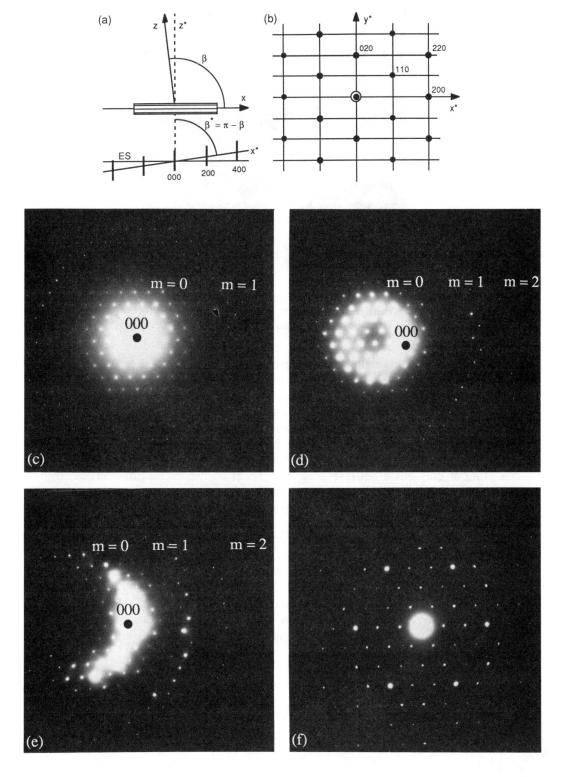

Figure 12.6. Reflection high-energy electron diffraction pattern of a clean (111) surface of silicon (cubic F). Zone [11$\bar{2}$], with zero-order and first-order Laue zones, with Kikuchi bands (Cousandier, 1968, Lab. Crystallography, Louis Pasteur University, Strasbourg).

The Laue zones can be easily observed on crystals with lamellar structures and related high-density planes with large interplanar distances (e.g. mica, Figure 12.25).

The thickness of the layers determined by the diffraction domains is inversely proportional to the number of periods along the incident beam, i.e. the thickness t of the crystal. As a result the width of the Laue zones is itself an inverse function of the crystal thickness. A relatively thick crystal gives well-marked zones; the Laue zones of a very thin crystal merge into one another; at the limit, for a two-dimensional crystal consisting of only one atom layer, the Laue zones are no longer differentiated, the diffraction domains become continuous lines hk (Figure 12.5f). The two-dimensional case is approximated in low-energy electron diffraction, due to the small electron penetration (Figure 12.25).

In a similar way as for texture patterns (12.1.4), the Laue zones $m \neq 0$ contain diffraction spots corresponding to reciprocal points outside the zero-level plane. They provide data on the third dimension of a thin crystalline platelet without any object tilt.

Systematic Absences in the Third Dimension. Systematic absences of diffraction spots in the outer Laue zones makes it possible to distinguish three-dimensional extinctions, due to multiple lattice types, from two-dimensional extinctions to a glide plane (see 5.3.3).

Period in the Third Dimension. For a given zone orientation $[uvw]$, Figure 12.4b shows that, in the small-angle approximation, the parameter $n(uvw)$ is related to the radius R_1 of the first-order Laue zone and to the diffraction constant $L\lambda$ as follows:

$$n(uvw) = \frac{2L^2\lambda}{R_1^2} \qquad (12.5)$$

This measurement is not very accurate, but can nevertheless provide an order of magnitude.

Specimen Thickness. Due to the size effect, the reciprocal planes $(uvw)^*$ become layers with a thickness H inversely proportional to the number N of planes (hkl) considered to be parallel to the crystal platelet

$$H = 2n^*(hkl)/N$$

Figure 12.4c shows that, to the small-angle approximation, the thickness t of the platelet is related to the radius R_0 of the zero-order Laue zone and to the diffraction constant $L\lambda$ as follows:

$$t = \frac{2L^2\lambda}{R_0^2} \tag{12.6}$$

Since the radius of the zero-order Laue zone is not well defined, this measurement is not very accurate either.

Observing the Laue zones is of particular importance in the convergence beam technique.

Double Diffraction

As a result of the strong electron–matter interaction and of the ensuing relaxation of the diffraction conditions, high-intensity diffracted beams may act in turn as incident beams and produce a secondary diffraction pattern, which itself can lead to third-degree diffraction, and so on. A geometrical interpretation is straightforward when limited to the secondary diffraction, resulting in double diffraction.

One-phase Specimen. The origin O of the reciprocal lattice, the end point of the wave vector $\mathbf{k}_0$, represents the primary transmitted ray, i.e. the 000 reflection. Consider a wave vector $\mathbf{k}$ of a strong primary hkl reflection. The vector $\mathbf{k}$ can act in turn as the incident wave vector. Its end point $O'(hkl)$ then becomes the secondary origin of the diffraction reciprocal space, called the *secondary diffraction origin*. A secondary diffracted beam is observed in a direction $\mathbf{k}'$. To the small-angle approximation, the secondary diffraction pattern with origin O' corresponds to a shift of the primary diffraction pattern along the vector $\mathbf{r}^*_g(hkl)$, vector of the reciprocal lattice. As a result, the secondary diffraction pattern coincides with the primary pattern, except when the crystal has glide symmetry elements which lead to systematic absences in the diffraction pattern (5.3.3); space group forbidden reflections can appear in this case (Figure 12.7). Multiple lattice forbidden reflections (lattice types $I, F, C \ldots$) cannot appear by double diffraction; the corresponding absences are simply due to the arbitrary choice of a multiple unit cell.

Due to the relaxation of diffraction conditions, more than one strong reflection can act as double diffraction origin, e.g. the whole set of reflections related to a form $\{hkl\}$ of lattice planes (hkl) generated by the point group symmetry. The corresponding translations superpose to generate the final diffraction pattern.

238   Structural and Chemical Analysis of Materials

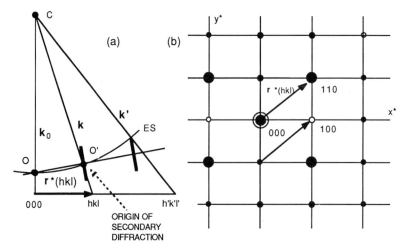

Figure 12.7. One-phase double diffraction. (a) Geometrical interpretation through the Ewald construction; (b) occurrence of diffraction spots forbidden by the space group. Instance of the zone [001] of an orthorhombic crystal, double diffraction origin 110, occurrence of a forbidden reflection 100.

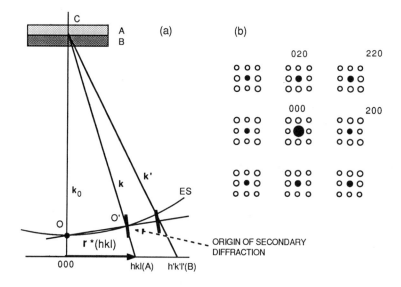

Figure 12.8. Double diffraction by two phases A and B (parameter of B larger than parameter of A). (a) Geometrical interpretation through the Ewald construction; (b) outline of the [001] zone diffraction pattern given by thin epitaxially related layers of two cubic F metals; the black spots 200, 020, 220 of phase A act as secondary diffraction origins; the white spots from phase B include the primary spots and the secondary spots generated by the translations $r^*(200)$, $r^*(020)$, $r^*(220)$, etc.

Two-phase Specimen. Consider a crystalline plate consisting of two different phases A and B, an incident beam passing successively through A and B (Figure 12.8). A strong diffracted beam from A can then act as a secondary incident beam for exciting $h'k'l'$ reflections from B. To the small-angle approximation, this corresponds to translating the primary diffraction pattern of B along a vector $\mathbf{r}^*_g(hkl)$ which connects the origin to a strong reflection *hkl* of the primary diffraction pattern given by A. The primary and secondary diffraction patterns of the A phase coincide as previously. The primary and secondary diffraction patterns of phase B do not coincide, because the translation vector does not belong to its reciprocal lattice, i.e. to its diffraction pattern. The resulting diffraction pattern shows new, double-diffraction reflections. Such double-diffraction patterns occur frequently in any case of coherent intergrowth of two or more crystal phases, e.g. epitaxy. High-symmetry phases result in a multiple secondary-diffraction origin and lead therefore to complex diffraction patterns (Figure 12.8b).

Kikuchi Patterns

When the crystal thickness reaches the order of 1000 Å or more, inelastic scattering becomes significant. As a result, the point diffraction pattern is more or less covered by a continuous background. Well-crystallised thick specimens often generate a line or band pattern contrasting with the background, the so-called *Kikuchi pattern*, first described and interpreted by Kikuchi in 1928 (Figures 12.6, 12.12). Tilting the crystal, with respect to the incident beam, causes the diffraction points to appear or disappear, virtually without moving. The Kikuchi pattern, on the contrary, is directly related to the crystal orientation.

Geometrical Interpretation. Generating a pair of Kikuchi lines K_1 and K_2 is outlined in Figure 12.9 in a plane normal to the lines. Consider a scattering point P inside the

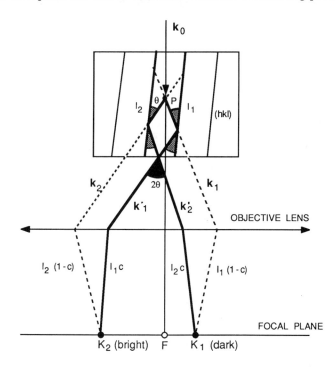

Figure 12.9. Geometrical interpretation of the Kikuchi lines.

crystal, between two lattice planes (*hkl*). It is the source of a quasi-elastically scattered spherical wave. This scattered wave generates on the diffraction pattern a continuous background whose intensity decreases rapidly with increasing angle of incidence. The scattered rays with Bragg incidence on (*hkl*) planes result in symmetrical reflections *hkl* and $\overline{hkl}$ with respect to the planes. The locus of such reflected rays is a set of circular cones with apex P and angle $(\pi - 2\theta)$. Due to aperture limitation, reflection is generally limited to the first order. To the small-angle approximation, the intercept of the double cone with the diffraction pattern is a pair of Kikuchi lines K_1 and K_2 which is symmetrical about the intercept of the reflecting plane (*hkl*). Within the angular aperture, any set of lattice planes generates a pair of parallel Kikuchi lines.

Intensity Interpretation. A complete treatment of the Kikuchi pattern involves the dynamic theory. The basic theory proposed by Kikuchi is consistent with a qualitative interpretation. It considers the line pattern as resulting from a secondary elastic scattering of electrons which have already been submitted to quasi-elastic scattering inside the crystal. When limiting to the first-order reflections, two scattered rays issued from P, with wave vectors $\mathbf{k}_1$ and $\mathbf{k}_2$ ($k_1 = k_2 \cong k_0$), with intensities I_1 and I_2 respectively, hit the (*hkl*) planes with the relevant Bragg angle θ. In the general case they make two different angles with the incident direction, resulting in $I_1 \neq I_2$; in the instance of Figure 12.9, $\mathbf{k}_2$ is more inclined than $\mathbf{k}_1$, resulting in $I_1 > I_2$. After reflection, the wave vectors become $\mathbf{k}'_1$ and $\mathbf{k}'_2$, with $\mathbf{k}'_1 \parallel \mathbf{k}_2$ and $\mathbf{k}'_2 \parallel \mathbf{k}_1$. On the diffraction pattern at infinity (or in the back focal plane of the objective lens), the rays $\mathbf{k}_1$ and $\mathbf{k}'_2$ generate the line K_1, the rays $\mathbf{k}_2$ and $\mathbf{k}'_1$ generate the line K_2. The (*hkl*) planes bisect the two directions $\mathbf{k}_1$ and $\mathbf{k}_2$. The respective line intensities are

$$I(K_1) = I_1(1-c) + I_2 c = I_1 - c(I_1 - I_2)$$
$$I(K_2) = I_2(1-c) + I_1 c = I_2 - c(I_2 - I_1)$$

where c is the reflecting factor of the planes (*hkl*), the ratio of the reflected intensity to the incident intensity.

Referring the line intensities to the background intensities I_1 and I_2 which would be measured at K_1 and K_2 without reflection on the (*hkl*) planes, the contrast of the lines with respect to the background is written

$$\begin{aligned} I(K_1) - I_1 &= -c(I_1 - I_2) < 0 \Rightarrow I(K_1) < I_1 \\ I(K_2) - I_2 &= -c(I_2 - I_1) > 0 \Rightarrow I(K_2) > I_2 \end{aligned} \quad (12.7)$$

The K_1 line is less intense than the background. It appears black on a lighter background. The K_2 line is more intense than the background. It appears light on a darker background.

As a result, each pair of Kikuchi lines consists of a negative line (electron deficit), the closest to the origin, and of a positive line (electron excess), further away from the origin.

There is an evident geometrical relationship between the Kikuchi line positions on one side and the orientation and interplanar distances of the reflecting planes on the other side.

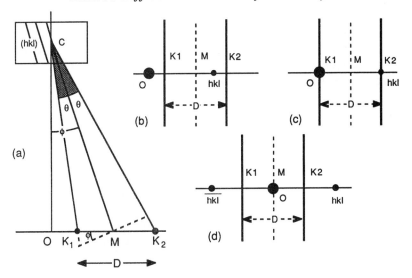

Figure 12.10. Position of a pair of Kikuchi lines from a *hkl* reflection, as a function of the crystal orientation. (a) Angular relationship; (b) general orientation of the reflecting planes; (c) Bragg orientation of the reflecting planes; (d) reflecting planes parallel to the incident beam.

Reflecting Planes with General Orientation. Consider a pair of lines K_1 and K_2 issued from first-order reflection on a set (hkl), making an angle ϕ with the incident beam. As can be seen from Figures 12.10a and b, the values of $d(hkl)$ and ϕ are given by the following relations, to the small angle approximation:

$$\tan \phi = \frac{\overline{OM}}{L} = \frac{\overline{OK_1} + \overline{OK_2}}{2L}$$

$$d(hkl) = \frac{\lambda}{2\theta} = \frac{L\lambda}{D \cos \phi}$$

(12.8)

where D is the distance between the two lines, for a diffraction constant $L\lambda$, and M the median line.

For a crystal of known species, measuring a pair of lines provides the orientation as well as the identification of the set of planes (hkl) which has generated it. The stereographic projection of the sets of planes corresponding to the observed pairs of lines can be readily drawn, knowing the angles $(\pi/2 - \phi)$ of their normals with $\mathbf{k}_0$.

Reflecting Planes Parallel to the Incident Beam. As outlined by Figure 12.10d, the above relations in this case become

$$D = \frac{L\lambda}{d(hkl)} = \mathbf{r}_g^*(hkl) \tag{12.9}$$

The lines of a given pair are now symmetrical about the origin and are the median normals to the related reciprocal vectors $\pm \mathbf{r}_g^*$ in the diffraction pattern.

242   Structural and Chemical Analysis of Materials

From the symmetry, it follows that $I_1 = I_2$; according to Eqs (12.7), there should be no contrast between the lines and the background; the lines should be invisible. They actually appear as the limits of Kikuchi bands; the area between a pair of lines appears lighter than the background. This effect is explained by the dynamic theory. A prominent zone orientation results in a Kikuchi pattern with symmetrical bands about the origin (Figure 12.6); it provides data on the Laue group and is useful for adequate orientation of a crystal along a given zone axis.

Reflecting Planes in Bragg Position. Diffraction Deviation. When the reflecting planes are in an exact Bragg orientation for a *hkl* reflection, the related Kikuchi lines pass respectively through the central spot 000 and through the *hkl* spot (Figure 12.10c). The diffraction deviation s can be easily determined from the shift x of the lines with respect to their exact Bragg position. Figure 12.11 shows that

$$s = \frac{x\,\mathbf{r}_g^*(hkl)}{L} \qquad (12.10)$$

A given deviation s, can, conversely, be produced by tilting the crystal, previously set at the exact Bragg orientation, in a convenient direction through an angle $\Delta\theta = s/\mathbf{r}_g^*$.

Observing Kikuchi patterns requires well-crystallised samples, with few lattice faults. They are generally easy to observe with non-metallic crystals (Figure 12.12), whereas plastic deformation in metals is a hindrance.

Diffraction effects similar to the Kikuchi pattern can be observed in different conditions:

1. Scattering of a convergent electron beam in a crystal results in a so-called pseudo-Kikuchi pattern (see section 12.2).

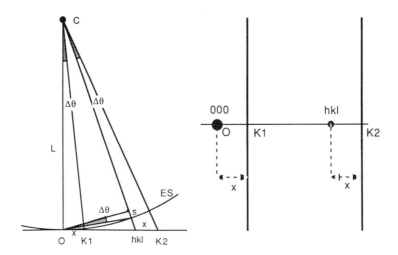

Figure 12.11. Relationship between the diffraction deviation s and the shift x of the Kikuchi lines with respect to the Bragg position.

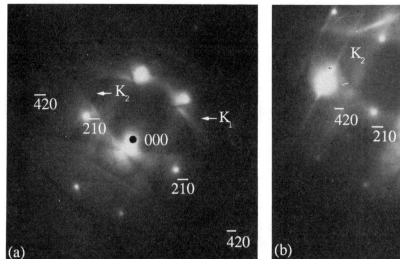

Figure 12.12. Kikuchi patterns provided by a calcite ($CaCO_3$) cleavage plate, corresponding to the [$24\bar{1}$] zone. Indexing in the hexagonal system of axes. (a) Normal incidence, incident beam parallel to the zone axis. The two lines of the pair K_1 ($\bar{4}20$) and K_2 ($4\bar{2}0$) are respectively median normals to the corresponding reciprocal vectors; as a result they pass approximately through the $\bar{2}10$ and $2\bar{1}0$ reflections; (b) exact Bragg incidence for $\bar{4}20$: the line K_1 (dark) passes through the central spot 000, the line K_2 (light) passes through the reflection $\bar{4}20$. (P. Braillon, Dept. of Material Science, Univ. Claude Bernard, Lyon).

2. X-rays generated inside a crystal by an incident electron beam are diffracted on the (*hkl*) planes and generate a so-called Kossel pattern.

Form and Size Effects

Due to the small size of coherently diffracting areas, the electron diffraction patterns are currently affected by form effects. The best way to interpret them is to model the diffraction reciprocal space and its intersection with the reflecting sphere. The configuration of the diffraction domains gives information about the object morphology through the interference function (Eq. 5.26).

Form effects are equally provided by large crystals enclosing micrometre-sized deviations from the perfect crystal (inclusions, precipitates, phase separations, etc.) Diffraction relaxation effects are then superimposed on the well-defined diffraction spots of the perfect crystal. Thus thin lamella-shaped particles parallel to (*hkl*) planes generate diffraction domains which are elongated parallel to the [*hkl*]* axis; the thinner the lamellae, the longer and finer the diffraction domains (Figure 12.13). Fine needle-shaped crystals or inclusions, parallel to a direction [*uvw*], result in diffraction domains which are flattened along reciprocal (*uvw*)* planes.

Crystal Lattice Deformations and Defects

Any deformation of the direct lattice, with respect to the perfect lattice, will result in

244   Structural and Chemical Analysis of Materials

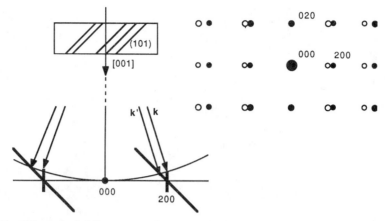

Figure 12.13. Effect of two-dimensional defects or of lamellar segregations on the [001] zone diffraction pattern of an all-faces-centred cubic crystal. The (101) defect planes are inclined at 45° to the incident beam. Occurrence of satellite spots is related to reciprocal lattice points which are not exactly on the reflecting sphere.

its reciprocal effect in the diffraction pattern, with respect to the ideal pattern. A bending deformation amounts to rotating the reciprocal lattice about the origin; any lattice point is therefore replaced by a circular arc. With a bending axis parallel to the incident beam, the point diffraction spots become circular arcs themselves; this effect is proportional to the distance from the origin. More general bending deformations result in more complicated effects.

A lattice defect, localised at a given point inside the crystal, results in a diffraction effect which is spread over the whole diffraction pattern. For the effect to be observable, the volume affected by the defects must be large enough with respect to the whole active volume of the crystal. Due to the limited size of the defects and to the locally induced stress, the resulting diffraction effects are in some ways similar to the size and form effects. Specific additional phase effects occur for lattice defects which include a *displacement vector* related to the lattice parameters, e.g. dislocations and stacking faults. In a similar way as in the image (see Ch. 19), a crystal defect with a displacement vector ρ results in the absence of the corresponding diffraction effect for reflections $\mathbf{g}(hkl)$ such as

$$\rho \, \mathbf{r}_g^* = n \quad \text{(where } n \text{ is an integer)} \tag{12.11}$$

Point defects and dislocations involve a small proportion of the crystal volume. Their occurrence results in a continuous scattering outside the diffraction domains of the perfect crystal, with an intensity proportional to the defect density which is generally small. A plane defect can be assimilated into a crystal layer with a thickness of the order of one or a few lattice parameters, inserted between two portions of a perfect crystal. It therefore involves a relatively large number of unit cells. The related diffraction effect is similar to the form effect due to a two-dimensional layer, i.e. continuous scattering streaks along reciprocal lines $[hkl]^*$ normal to the (hkl) defect plane (Figure 12.13), with the difference that the effect is absent around the reciprocal points meeting the condition (12.11).

Local Deviation of the Chemical Composition. Local atomic substitutions cause effects which are similar to those of geometric lattice defects. Randomly distributed chemical point defects (e.g. atomic substitutions, interstitial atoms) result in continuous scattering outside the diffraction domains with an intensity proportional to the density of defects. The effect of atomic substitutions along [uvw] lattice lines amounts to the form effect of one-dimensional crystals, leading to continuous scattering along (uvw)* planes which superimposes on the diffraction domains. The effect of substitutions localised along (hkl) planes is similar to the form effect of two-dimensional crystals (Figure 12.13).

Diffraction effects due to any crystal defects are generally easier to observe in the electronic image than in the diffraction pattern itself. This results from the fact that the effect of a disturbance at a point in the crystal will be localised at the related point of the image, whereas it is spread over the whole diffraction pattern.

Superlattice Diffraction Pattern

Thanks to the strong electron–matter interaction, weak superlattice reflections are easier to observe with electron diffraction than with X-ray diffraction. A superlattice can often be considered as resulting from an ordered distribution of defect planes in one, two or three dimensions, with the possibility of any intermediate state between a total disorder and a perfect order. In the diffraction pattern, the resulting new periodicity leads to additional scattering maxima which are the finer the better the ordering. A superlattice of order n in a given direction in the crystal generates a new period, a submultiple of order n of the initial period, in the corresponding direction to the reciprocal lattice. A one-dimensional example is illustrated in Figure 12.14.

12.1.3.2 Applications of Single Crystal Diffraction Patterns

Crystal Orientation

Selective electron diffraction is well designed for determining the orientation of tiny

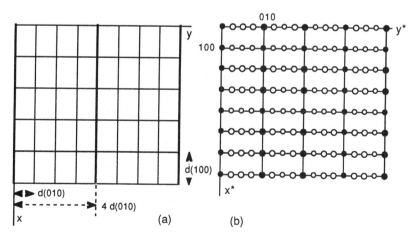

Figure 12.14. One-dimensional superlattice generated by an ordered distribution of plane defects parallel to (010), with a period $4d$(010). (a) (001) direct lattice plane. The defect planes are drawn in thick lines; (b) (001)* reciprocal lattice plane. The superlattice reflections are drawn as white dots.

crystals in a thin layer of material. Examples of applications are the analysis of structural relationships of precipitates, of crystal nuclei, of reaction products or alteration products with respect to the matrix. The mutual orientation in microtwins and in epitaxial growth can be defined.

The orientation of sufficiently thick and nearly perfect single crystals is best performed by means of the Kikuchi pattern which is very orientation sensitive.

In the general case, crystal orientation is based on indexing a point diffraction pattern. It amounts to determining the lattice line $[uvw]$ parallel to the incident direction. The observed hkl spots on the diffraction pattern are those due to reflections on the (hkl) planes of the $[uvw]$ zone. The zone axis is defined by the intersection of at least two planes $(h_1 k_1 l_1)$ and $(h_2 k_2 l_2)$ of the zone, i.e. by the indices of two spots of the diffraction pattern, except the origin. Those indices are defined by Eq. (12.2) as follows:

$$\begin{array}{ll} h_1 u + k_1 v + l_1 w = 0 & u = k_1 l_2 - l_1 k_2 \\ & \Rightarrow \quad v = l_1 h_2 - h_1 l_2 \\ h_2 u + k_2 v + l_2 w = 0 & w = h_1 k_2 - k_1 h_2 \end{array} \qquad (12.12)$$

Determining the crystal orientation amounts to indexing the observed spots on the diffraction pattern. Several methods may be used.

Indirect Method. Comparing to Standard Diffraction Pattern. The prominent zone diffraction patterns of high-symmetry lattices (cubic, tetragonal, hexagonal) have been tabulated (e.g. in [1], Andrews et al, [7] Hirsch et al). For low-symmetry crystals, corresponding sections of the reciprocal lattice are to be determined. Specific software programs have been developed for this aim.

The observed diffraction patterns are indexed by comparison with the calculated reciprocal sections. The problem may be simplified by the occurrence of preferential orientations with respect to the specimen reference plane (lamellar or fibre shape, cleavages).

Cleavage Plate. The most probable diffraction pattern corresponds to the zone normal to the cleavage plane. In the general case of low symmetry, the zone axis does not necessarily coincide with a prominent lattice line. Because of the relaxation of diffraction conditions, a diffraction pattern related to a nearby important zone axis $[uvw]$ is nevertheless observed in most cases (e.g. [001] line near to the [001]* normal to the cleavage planes (001) of clay minerals, see Figures 12.5a and b).

For a given cleavage plane (HKL), the prominent lattice line $[uvw]$ nearest to its normal, i.e. nearest to the reciprocal lattice line $[HKL]^*$, must be determined. Those lines are respectively defined by means of the vectors $\mathbf{r}_n$ and $\mathbf{R}_g^*$; their vector product is equal to zero:

$$\mathbf{r}_n \wedge \mathbf{R}_g^* = (u\mathbf{a} + v\mathbf{b} + w\mathbf{c}) \wedge (H\mathbf{a}^* + K\mathbf{b}^* + L\mathbf{c}^*) = 0 \qquad (12.13)$$

The components of the vector product lead to a set of three equations determining the three parameters u, v, w. Calculation is simplified with a system of perpendicular axes:

$$\frac{u}{v} = \frac{Hb^2}{Ka^2} \quad \frac{v}{w} = \frac{Kc^2}{Lb^2} \quad \frac{w}{u} = \frac{La^2}{Hc^2} \qquad (12.14)$$

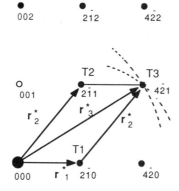

Figure 12.15. Drawing the theoretical diffraction pattern of a (210) cleavage plate of enstatite (arbitrary scale).

Example. (210) cleavage plate of enstatite MgSiO$_3$ (orthorhombic pyroxene); $a = 18.22$ Å, $b = 8.83$ Å, $c = 5.19$ Å; space group $P\,bca$. The normal incidence diffraction pattern is to be drawn.

Zone axis. This is given by relations (12.14):

$$u/v = 2b^2/a^2 = 0.48; \qquad v/w = \infty; \qquad w/u = 0$$

The nearest prominent zone axis meeting those conditions is [120].

Diffraction pattern of the [120] zone. The indices hkl of the diffraction pattern must meet the equation of zone [120]:

$$uh + vk + wl = h + 2k = 0 \Rightarrow h = -2k$$

Drawing three reciprocal vectors entirely determines the plane unit cell. Two reciprocal vectors $\mathbf{r}_1^*$ and $\mathbf{r}_2^*$, corresponding to small indices meeting the foregoing condition, are chosen arbitrarily, as well as their geometrical sum $\mathbf{r}_3^* = \mathbf{r}_1^* + \mathbf{r}_2^*$. The corresponding interplanar distances $d(hkl)$ are then calculated (orthorhombic quadratic form, Eq. 11.17). The triangle having the three vectors as its sides may now be drawn. The following vectors can be chosen:

$$r_1^*(2\bar{1}0) = \frac{L\lambda}{d(2\bar{1}0)}$$

$$r_2^*(2\bar{1}1) = \frac{L\lambda}{d(2\bar{1}1)}$$

$$r_3^*(4\bar{2}1) = \frac{L\lambda}{d(4\bar{2}1)}$$

The diffraction pattern is completed by geometric summing. The result is outlined in Figure 12.15 which represents a portion of the diffraction pattern.

Direct Method. Using the List of Interplanar Spacings. The general method consists of choosing on the diffraction pattern three neighbouring spots T_1, T_2, T_3, featuring a parallelogram of origin O and defined by the reciprocal vectors $\mathbf{r}_1^*$, $\mathbf{r}_2^*$, $\mathbf{r}_3^*$, which are measured on the diffraction pattern (Figure 12.15). The corresponding interplanar spacings are then calculated:

$$d_1 = \frac{L\lambda}{r_1^*} \qquad d_2 = \frac{L\lambda}{r_2^*} \qquad d_3 = \frac{L\lambda}{r_3^*}$$

Comparison with the list of calculated (or given by the powder data file) $d(hkl)$ leads to an indexing hypothesis. The chosen set of indices has to comply with the geometrical sum

$$\mathbf{r}_3^* = \mathbf{r}_1^* + \mathbf{r}_2^* \Rightarrow h_3 = h_1 + h_2; \qquad k_3 = k_1 + k_2; \qquad l_3 = l_1 + l_2$$

Indexing the remaining spots is carried out by summing. The zone axis $[uvw]$, i.e. the crystal orientation, is determined from two reflections by means of Eq. (12.12).

Indexing a diffraction pattern is more difficult the lower the symmetry. Computer programs have been developed (e.g. Stadelmann, 1987). The observed diffraction pattern is compared to prominent theoretical diffraction patterns of various reciprocal sections which have been calculated for relevant crystal species and stored.

Crystal Lattice

Reciprocal Lattice. An electron diffraction pattern reproduces a planar section of the intensity-weighted reciprocal space, passing through the origin (without taking into account possible high-order Laue zones). The three-dimensional reciprocal lattice of a single crystal is constructed by recording important zone patterns for various crystal orientations, by means of a goniometrical sample stage. For instrumental reasons, the tilt is generally limited to some 60°. Fine-grained material is best prepared by embedding in a resin followed by ultra-microtome cutting or ion thinning (see section 19.5). The randomly oriented particles provide various sections through the reciprocal lattice.

Direct Lattice. The direct lattice is related to the reciprocal lattice by the relevant equations (Appendix B). A possible multiple unit cell ($I, F, C \ldots$ lattice) must be taken into account. Determining the lattice parameters is easy if the diffraction patterns have been indexed. Let $R_m(hkl)$ be the measure of the distance to the origin of a spot hkl; the related interplanar distance is given, to the small-angle approximation, by

$$d(hkl) = \frac{L\lambda}{r^*(hkl)} \cong \frac{L\lambda}{R_m(hkl)} \qquad (12.15)$$

Two diffraction patterns containing reflections $h00$, $0k0$, $00l$ are theoretically sufficient to determine the lattice parameters in the most general case.

Diffraction Constant. The measurement of lattice parameters requires an accurate value of the diffraction constant $L\lambda$. It can be determined by means of the diffraction pattern of a well-crystallised standard specimen. A polycrystalline gold layer (cubic F), prepared by vacuum evaporation or by ion sputtering on to a thin organic film, provides a diffraction pattern with few fine and easily indexed rings (see Figure 12.17a); it is frequently used as a standard.

In the macrodiffraction mode, the diffraction constant is defined by the object-to-film distance and by the acceleration voltage.

In the microdiffraction mode as operated in a TEM, the diffraction constant is sensitive to instrumental parameters (e.g. the specimen position in the magnetic field of the objective, the objective excitation). The best way to account for these variations is to use an internal standard, e.g. a thin gold layer directly evaporated on to the sample.

The measurement on a diffraction pattern, i.e. in the reciprocal space, may be expressed either as the reciprocal of a length (L^{-1}) or as a length (L). The diffraction constant $L\lambda = d(hkl)R^*(hkl)$ is accordingly expressed either as a dimensionless number or as a squared number.

Accuracy of Parameter Measuring. The error of measurement depends on the operational mode. In the commonly used aperture selection microdiffraction mode (by means of an electron microscope), the main causes of error are:

1. Errors of measurement on the film. The problem is the same as for X-ray diffraction patterns.
2. Errors on the value of the diffraction constant. These can be minimized by means of an internal standard.
3. Deformation of the diffraction pattern due to aberrations of the post-specimen lenses (spherical aberration increasing with the diffraction order, astigmatism for any orders).
4. Broadening of the diffraction spots due to the size effect related to the small diffracting area.
5. Displacement of the diffraction spots through a deviation from the zone axis, due to the elongated form of the diffraction domains intersecting the reflecting sphere (Figure 12.1a). An exact zone axis orientation (e.g. by using a Kikuchi pattern) is required for good accuracy.

The currently observed relative accuracy of electron microdiffraction is of the order of 10^{-2}.

The macrodiffraction mode is better designed for accurate measurements. A relative accuracy of 10^{-4} can be attained with a good crystal.

Space Group

The principle of determining the space group is the same as for X-ray diffraction patterns, to the kinematic approximation. The possibility of double diffraction, resulting in the occurrence of normally forbidden reflections, must be taken into account and makes the problem more difficult (Figure 12.7b). New possibilities of investigating the space group of micrometre-sized crystals are provided by the convergent beam electron diffraction mode (section 12.2).

Crystal Structure

Multiple interaction effects (dynamic effects) result in difficulty of relating the intensity of observed *hkl* reflections to the squared structure factors, i.e. to the crystal structure, as in the case of X-ray diffraction. However, the greatest advantage of electron diffraction lies in the possibility of observing single crystal diffraction patterns from individual particles in microcrystalline materials. Applying the kinematic theory to compiling the intensities then enables to provide valuable data on the structure of poorly crystallised materials which give only ill-defined powder diffraction patterns with X-rays (e.g. clay minerals). The dynamic effect intervenes as soon as the crystal thickness exceeds the extinction distance (Eq. 6.9). In the two-beam approximation, it can be corrected to a certain degree by relating the measured intensities to the structure factors instead of relating them to their squares. Indeed, it can be seen from Eq. (6.16) that averaging the intensity over the extinction distance and integrating it over a whole diffraction domain leads to an expression of the intensity which is actually proportional to the structure factor. The more or less important contribution of dynamic diffraction can be expressed by the following relation between the measured intensity and the structure factor:

$$I_m = k_{kin} |F(hkl)|^2 + k_{dyn} |F(hkl)| \qquad (12.16)$$

The factors k_{kin} and k_{dyn} depend on the nature and the size of the crystals. They are adjusted to minimise the reliability factor (Eq. 11.24) after FT and refinement (e.g. in [6] Heidenreich, [15] Vainshtein, [16] Zwyagin).

Crystal Defects and Poorly Crystallised Materials

Owing to its sensitivity, electron diffraction is particularly designed for analysing individual particles of poorly crystallised materials. When operating in the nanoprobe mode, by beam selection, it permits structural and chemical analysis on an almost unit-cell scale. Crystal defects, twinning, deformations, order–disorder effects, polytypes, intergrowth, etc. may be investigated in the individual particles. It is mainly in this field that electron diffraction has the main advantage over X-ray diffraction which can only provide statistical information, averaged over a large number of particles.

12.1.4 DIFFRACTION PATTERNS FROM POLYCRYSTALLINE MATERIALS

The diffraction pattern from a polycrystalline material depends on the size of the crystals with respect to the electron probe diameter. In order to obtain a real powder-type diffraction pattern, comparable to X-ray powder patterns, the active volume of the specimen must contain a sufficiently large number of randomly oriented particles. This condition is difficult to meet in the microdiffraction mode, unless a large selection aperture is used. The best way is to operate in the macrodiffraction mode.

The applications are similar to those of X-ray powder diffraction with the advantage of displaying the diffraction pattern directly on a fluorescent screen and of short exposure times (order of a minute, instead of an hour). This makes it particularly suitable for studying transient effects, such as phase transitions.

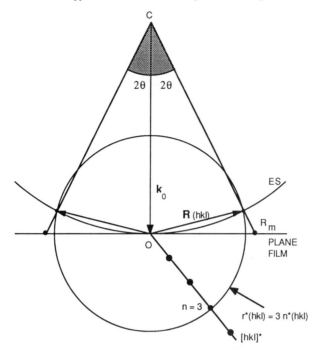

Figure 12.16. Interpreting the generation of a diffraction ring, outlined by means of the Ewald construction. Instance of an *hkl* ring of the third order.

To an even greater extent than for X-ray diffraction, reasoning by means of the reciprocal space facilitates the interpretation.

12.1.4.1 Observed Diffraction Pattern

Generation of the Diffraction Pattern

Similar to the case of the X-ray powder pattern, the active volume of the specimen is assumed to contain a sufficiently large number of crystal particles for their distribution to be statistically random. For a given form {*hkl*} of lattice planes, the locus of diffracted beams is a set of circular cones coaxial to the incident direction, with half-angles 2θ (to the small-angle approximation) as given by the Bragg equation (Eq. 12.1). Thanks to the small diffraction angles, a flat film, normal to the primary beam, permits the recording of a sufficient part of the available information. The intersection with the diffraction cones results in concentric circular diffraction rings (see Figure 12.17). Consider the reflecting sphere of radius L and the related reciprocal lattice of constant $L\lambda$. Figure 12.16 outlines the interpretation of the powder diffraction pattern from a given crystalline species by means of the Ewald representation (see 5.2.2). To the small-angle approximation, the measured radius R_m of a ring corresponding to an *hkl* reflection is related as follows to the interplanar distance (*hkl*):

$$R_m \cong r^*(hkl) = n\, n^*(hkl) = \frac{nL\lambda}{d(hkl)} \Rightarrow d_m = \frac{d(hkl)}{n} = \frac{L\lambda}{R_m} \quad (12.17)$$

where $n^*(hkl)$ is the equidistance of the points on the reciprocal lattice line $[hkl]^*$ and n the diffraction order.

Thus a value $d_m = L\lambda/R_m$ is assigned to any of the rings, resulting for the whole pattern in a list of numbers d_m which represent the existing interplanar spacings $d(hkl)$ and their submultiples $d(hkl)/2$, $d(hkl)/3$, ..., $d(hkl)/n$, in the species considered. The result is the same as for an X-ray powder diffraction pattern (see Eq. 11.10). In the case of electron diffraction, the maximum diffraction order is not given by the limiting condition $n \leqslant 2d/\lambda$ (e.g. $h00$ reflections from MgO, $a = 4.2$ Å, $\lambda = 0.037$ Å $\Rightarrow n \leqslant 227$). The number of observable rings is limited by the angular field of the flat film (adjustable by varying the diffraction constant) as well as by the rapidly decreasing ring intensity.

Ring Intensity

To the kinematic approximation, the intensity integrated along a ring is proportional:

(a) to the squared *structure factor* corrected by the temperature factor;
(b) to the *multiplicity factor* $M\{hkl\}$, the number of sets of lattice planes (hkl) contributing to a form of planes $\{hkl\}$ as generated by the point group operations (see 11.3.1);
(c) to the specimen *thickness*. When the thickness increases, at a given moment the diffracted intensity would therefore become equal to or even larger than the incident intensity which is an instrumental constant. This has no physical meaning; the kinematic approximation fails when the diffracted intensity approaches the incident intensity and must be replaced by the dynamic theory. The two-beam approximation is generally suitable for polycrystalline materials, due to the small size of the coherent scattering elements. Due to the averaging and integration of the diffracted intensity along a diffraction ring, it can be considered as proportional to the structure factor, and the correction factors of Eq. (12.26) may be used.

Ring Continuity

For a diffraction ring to be continuous, the number of involved particles must be sufficiently large. This condition is easy to meet in the macrodiffraction mode, but not in the microdiffraction mode (Figure 12.17b). In order to observe continuous rings, the incident aperture must be increased (focusing mode of the condenser lens, 9.2.3), or the particles must be very small, leading to a strong size effect. In both cases the result is a broadening of the diffraction rings and a poor measuring accuracy.

Errors of Measurement

The causes of errors on the measurements of the d_m are the same as for single-crystal diffraction patterns:

1. *Ring ellipticity*, due to astigmatism of post-specimen lenses in the microdiffraction mode, reaching up to 1%. It is compensated by averaging several diameters in different directions.

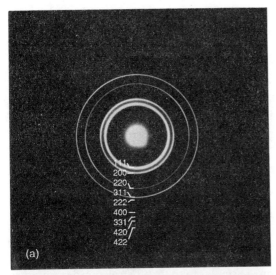

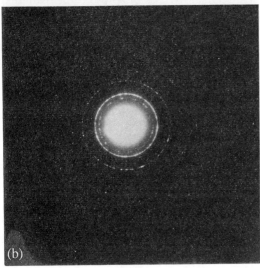

Figure 12.17. Electron microdiffraction patterns of polycrystalline materials ($V_0 = 100$ kV, selection area $d_s = 1$ µm). (a) Thin gold layer, vacuum evaporated on an organic film; continuous rings; (b) fine powdered MgO from magnesium combustion, collected on an organic film; punctuated rings. Both lattices cubic F; for the ring indexation, see Table 11.2. (Ehret G., Dept Electron Microscopy, Lab. Crystallography, Louis Pasteur University, Strasbourg).

2. *Spherical aberration* of post-specimen lenses in the microdiffraction mode. It reduces the ring diameters, the relative error increasing with the angle. It reaches the order of 10^{-3} for high-order rings.
3. *Curvature of the reflecting sphere.* Relation (12.17) is an approximation, as shown in Figure 12.16. The error is in excess, thus partially compensating for the spherical aberration error. It is small, even for high orders, some 10^{-4}. It acts in the microdiffraction mode as well as in the macrodiffraction mode without post-specimen lenses.

Example. 444-ring of aluminium at 100 keV.
Spherical aberration error: $\Delta R_m / R_m \cong -5 \times 10^{-3}$
Reflecting sphere curvature error: $\Delta R_m / R_m \cong 5 \times 10^{-4}$

4. *Diffraction constant*. The standard samples for its determination must meet the following, more or less conflicting, requirements: fine-grained, well-crystallised species with accurate parameter knowledge, giving few (high-symmetry) and well-defined (no size effect) diffraction rings, stable in the incident beam and easy to prepare in thin polycrystalline layers. The best way to minimise the above-mentioned errors is to use an internal standard and make the measurements on neighbouring rings of the standard and the sample. Vacuum-evaporated or ion-sputtered metals (Au, Al) provide a good compromise as standards.

The *accuracy to be expected* is of the same order as for a single crystal diffraction pattern, i.e. of the order of 10^{-2} in microdiffraction and of the order of 10^{-4} in macrodiffraction.

12.1.4.2 Applications of Powder Diffraction Patterns

Lattice Parameters

The basic problems of lattice determination are similar to those of X-ray powder patterns. Applying the trial-and-error method requires accurate measurements, in particular for low-symmetry species, which are only achieved with macrodiffraction.

Identification of Crystalline Species

The current powder data files are designed for X-ray diffraction uses. The interplanar spacings, however, are equally valid for electron diffraction patterns. The relative intensities are different, due to a faster decrease of the atomic scattering factor vs the diffraction angle and to dynamic effects. Specific data bases have been established for electron diffraction ([10] JCPDS–ICDD).

Grain Size

The currently used methods of grain size measurement in X-ray diffraction are easily applied to electron macrodiffraction. Application is limited in the microdiffraction mode, which is normally associated with imaging in a TEM.

Orientation Texture

A polycrystalline material with a preferential orientation gives a diffraction pattern which is intermediate between a random powder pattern and a single-crystal pattern. The most common and simplest textures are the *fibre texture* and the *lamellar texture* in which all crystallites have a preferred [uvw] direction, characteristic of the species, parallel to the fibre axis or normal to the layered crystals. In the latter case, the common normal can therefore be assimilated in a fibre axis. A perfect fibre axis would generate the same diffraction pattern as a single crystal rotating about this axis, when not taking into account the form effects.

The intensity-weighted reciprocal space of a specimen with a perfect fibre axis [uvw], without accounting for the form effect, is a set of concentric circles located in the (uvw)* planes. For fibrous crystals, the form factor results in flattened diffraction domains

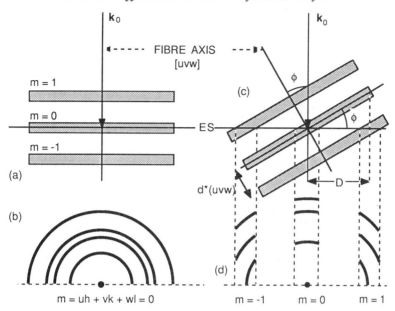

Figure 12.18. Interpreting fibre diffraction patterns. (a) Fibre axis parallel to the incident beam; (b) outline of the related diffraction pattern; circular rings such as $m = uh + vk + wl = 0$; (c) fibre axis inclined to the incident beam through an angle ϕ; (d) outline of the related diffraction pattern. Arcs of ellipses corresponding to $m = 0$ and $m = \pm 1$. The arbitrary diffraction patterns have been simplified. There is in fact no well-defined separation of the different levels m.

in a plane normal to the fibre axis; the circles are therefore replaced by flattened tori. For lamellar crystals, the form factor results in elongated diffraction domains normal to the lamellae; the circles are therefore replaced by cylindrical sections with a height equal to the width of the main diffraction maximum in this direction (Figure 12.18a and c).

In any case the diffraction pattern is interpreted by the intersection of the above-mentioned diffraction domains with the reflecting sphere, which can be assimilated into a plane. Three cases may be distinguished, according to the orientation of the fibre axis with respect to the incident beam.

Fibre Axis Parallel to the Incident Beam. Due to the small diffraction angles the reciprocal plane $(uvw)^*$, $m = 0$ coincides with the reflecting sphere. The diffraction domains intersect the reflecting sphere along concentric circles (Figures 12.18a and b). The diffraction pattern therefore consists of circular rings and seems similar to a powder diffraction pattern, with the difference that it is limited to hkl rings fulfilling Eq. (12.2) of the $[uvw]$ zone axis, i.e. such as

$$m = uh + vk + wl = 0$$

Example. Lamellar material—clay minerals (phyllosilicates) deposited by sedimentation on to the specimen support. The fibre axis is the normal to the (001) planes of the layers, i.e. the reciprocal lattice line $[001]^*$. In normal incidence the diffraction pattern is similar to that observed when rotating a single particle about this axis (diffraction pattern

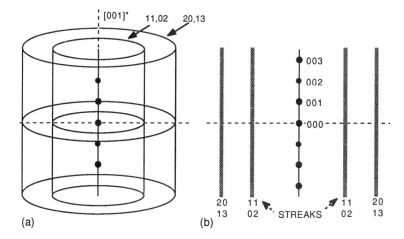

Figure 12.19. Turbostratic (smectic) texture. Diagram explaining the generation of the diffraction pattern of a smectite clay mineral with a random stacking of (001) layers. (a) Reciprocal space of one particle; (b) outline of the diffraction pattern observed with an incidence parallel to the layers. Discrete diffraction spots on the [001]* line. Continuous bands corresponding to [11] and [02], [20] and [13], etc. (experimental diffraction pattern Figure 12.20d).

Figure 12.5). It consists of fine and continuous rings related to the sole reflections $hk0$, with $h+k=2n$ (see Figure 12.20a).

Fibre Axis Inclined to the Incident Beam. The tangent plane to the reflecting sphere is inclined to the fibre axis $[uvw]$ and is intersected by a given circular diffraction domain along two symmetrical arcs of ellipses (Figures 12.18c and d). On the resulting diffraction pattern the arcs are distributed in levels $m=0$, $m=\pm 1$, etc.

Thus inclined fibre diffraction patterns provide information on the planes $m \neq 0$, i.e. outside the zero-level, notably the fibre axis parameter. Figure 12.18c outlines that

$$d^*(uvw) = D \sin \phi \Rightarrow n(uvw) = \frac{L\lambda}{D \sin \phi} \qquad (12.18)$$

where ϕ is the fibre axis inclination, and D the measured distance of level $m=1$ to the centre of the diffraction pattern. This measurement is not very accurate. However, it may be sufficient to help identify a crystallised species, e.g. a lamellar species through the parameter normal to its layers.

Inclined texture electron diffraction provides three-dimensional data, e.g. diffraction intensities on one diffraction pattern. Such diffraction patterns have therefore been applied to structure analysis (e.g. clay minerals, [16] Zwyagin).

The different m-levels are in fact ill defined. On diffraction patterns of lamellar materials with very thin particles, the important relaxation of the diffraction conditions causes the levels to merge completely (see Figure 12.20b).

Fibre Axis Normal to the Incident Beam. An incidence normal to the fibre axis results in the cylindrical diffraction rings intersecting the reflecting sphere along layer lines which, to the small-angle approximation, are parallel and equidistant. The observed diffraction

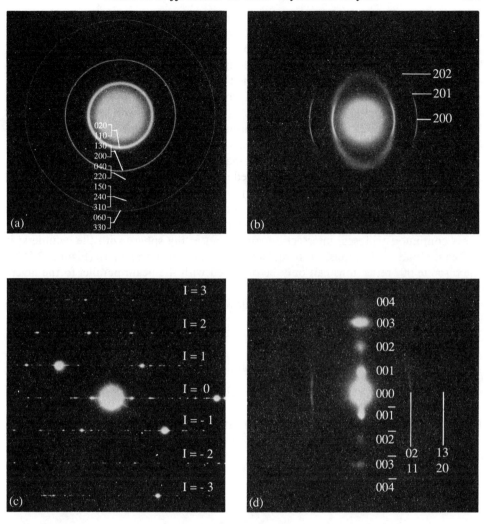

Figure 12.20. Texture electron diffraction patterns. $V_0 = 100$ kV; diameter of the selection area: 1 μm for (a, b); 0.1 μm for (c, d). (a) Montmorillonite (smectite clay mineral) deposited from a water suspension; normal incidence on the (001) layers; (b) same mineral as in (a), with incidence inclined at 50° to the layer normal; (c) crocidolite (amphibole) asbestos fibre deposited on the support film; incidence normal to the fibre axis; (d) particle of montmorillonite; same material as in (a), prepared by embedding in epoxy resin and oriented ultramicrotome cutting (Eberhart and Triki, 1972). Beam incidence parallel to the layers (Ehret G., Dept Electron Microscopy, Lab. Crystallography, Louis Pasteur University, Strasbourg).

pattern is therefore somewhat similar to an X-ray rotation pattern. This kind of diffraction pattern is usually observed with fibres deposited on the specimen support film in normal incidence (see Figure 12.20c).

Turbostratic Texture of Lamellar Minerals. A special case of lamellar texture is the so-called turbostratic texture or smectite texture, commonly encountered with certain poorly crystallised lamellar materials, e.g. clays of the smectite group or carbons. This texture occurs at the scale of the unit layer: identical two-dimensional structure units are stacked in a statistical way, with random orientations parallel to the resulting lamellar particles. The result is a one-dimensional periodicity along the normal to the layers. The reciprocal space of an individual, quasi-two-dimensional layer, consists of hk lines perpendicular to the layer planes, passing through the reciprocal lattice points $hk0$. Such a reciprocal space is somewhat similar to that of a surface lattice, as observed with low-energy electron diffraction (see Figure 12.25). For a single particle, the hk lines generate coaxial cylinders, whereas discrete $00l$ diffraction domains occur on the $[001]*$ lattice line, related to the one-dimensional periodicity along the normal to the layers (Figure 12.19). Such a diffraction pattern is, therefore, even observed in selected diffraction on a single particle.

The observed diffraction pattern is represented by the intersection of this reciprocal space with the reflecting sphere. For a normal incidence, it is similar to a corresponding fibre diffraction pattern. For inclined incidence the diffraction pattern shows continuous ellipses, intersections of the reflecting sphere with the cylinders; it is often difficult to distinguish from an inclined fibre diffraction pattern (Figure 12.20b). The discrete $00l$ reflections can only be observed with the beam parallel to the layers (Figure 12.20d).

Compiling the FT along the hk bands allows the collection of data on the stacking structure normal to the layers; thanks to the thinness of the coherent structure units the kinematic approximation undeniably applies in this case (Mering and Oberlin, 1967).

12.2 CONVERGENT BEAM ELECTRON DIFFRACTION

The basic principle of high-energy electron diffraction in a convergent incident beam was proposed by Kossel and Moellenstedt (1939). The method has developed recently, thanks to electron optical systems capable of focusing a convergent electron beam on areas of the order of a few nanometres. The resulting selection diffraction patterns therefore originate from single crystals.

12.2.1 BASIC PRINCIPLE AND OBSERVED DIFFRACTION PATTERNS

12.2.1.1 Instrumental Layout

In conventional high-energy electron diffraction (macrodiffraction and microdiffraction), the primary beam on the specimen is virtually parallel, with a convergence as small as 10^{-5}–10^{-4} rad. The diffraction pattern from a thin perfect single crystal consists of small diffraction spots related to discrete diffraction directions as determined by the diffraction conditions (Figure 12.1a).

In convergent beam electron diffraction, a given point of the crystal is subjected to an incident cone-shaped beam with a convergence (cone half-angle α) of the order of 10^{-2} rad. In the general case of a diffraction pattern localised in the back-focal plane of the objective lens in an electron microscope, a given diffraction spot hkl of the conventional diffraction pattern is therefore replaced by a *diffraction disc* of diameter $2\alpha f$ (where f is focal distance), as shown in Figure 12.21.

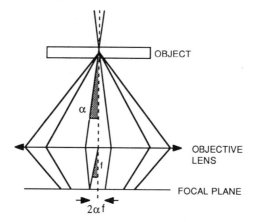

Figure 12.21. Ray diagram in convergent beam electron diffraction. Generation of a diffraction disc of radius αf.

Common abbreviations: CBD or CBED (*convergent beam electron diffraction*).

In order to observe a diffraction pattern corresponding to the intrinsic crystal structure, the active volume of the sample must have a constant thickness and no defects. This condition can easily be met only with an electron probe diameter ranging from a few nanometres to some tens of nanometres. This convergent nanoprobe mode is now currently available in TEMs with condenser–objective lenses (Figure 9.9c). Furthermore, for a straightforward interpretation, the crystal orientation must correspond to a prominent zone axis, i.e. with small indices $[uvw]$. In the following sections a $[001]$ zone will be considered; the related conventional parallel beam diffraction pattern will therefore consist of $hk0$ reflections (Eq. 12.2 for the zero Laue zone).

12.2.1.2 General Outline of the Diffraction Pattern

In the same way as in a conventional diffraction pattern, but more markedly, the reflections are distributed in Laue zones (see 12.1.3).

Zero-order Laue Zone (commonly abbreviated as ZOLZ). This forms the central part of the CBD pattern and consists of the transmitted 000 beam, surrounded by the reflections $hk0$ (see Figures 12.23a and b).

High-order Laue Zones (commonly abbreviated as HOLZ). They consist of one or more successive rings which correspond respectively to the diffracted beams $hk1$ (first-order Laue zone, commonly abbreviated as FOLZ), to the diffracted beams $hk2$ (second-order Laue zone), etc. (Figure 12.23c).

The observability of high-order Laue rings (HOLZ rings) depends on the zone axis parameter, on the diffraction constant and on the intensity diffracted at large angles. The fast intensity decrease as a function of the diffraction angle can be attenuated by cooling the crystal, thus decreasing the temperature effect (Eq. 4.11 and Figure 12.23c).

12.2.1.3 Diffraction Disc Intensity and Fine Structure

Due to the large beam aperture, the incident radiation is incoherent. The observed contrast effects can therefore be interpreted in terms of the sum of intensities. The sample thickness required for observing a significant CBD pattern is generally larger than the extinction distance (Eq. 6.9); a quantitative intensity interpretation therefore requires a dynamic treatment.

A given point in a disc is generated by a ray of the corresponding direction inside the incident cone. Each of the discs therefore represents a distribution map, either of the transmitted intensity (for the central disc 000), or of the diffracted intensities (for the diffraction discs), as a function of the polar coordinates. For the same orientation the parallel beam diffraction pattern would be reduced to the centre points of the discs ($\alpha = 0$).

With a very thin crystal, the diffraction domains are almost continuous lines with constant intensity between the reciprocal $(uvw)^*$ planes. The resulting intensity distribution in the diffraction discs is therefore virtually constant. No additional data are provided with respect to the corresponding parallel beam diffraction pattern.

With a crystal of adequate thickness to produce a Kikuchi pattern in parallel beam (section 12.1.3), two main types of contrast are observed:

1. A system of more or less diffuse *diffraction fringes* in the transmitted disc and in the other discs of the ZOLZ.
2. A system of fine lines, dark lines in the transmitted disc, related light lines in the discs of the HOLZ rings, called the *HOLZ lines* (Figure 12.23b).

Diffraction Fringes in the ZOLZ Reflections. The diffuse fringes are due to the interaction of the incident beam with the ZOLZ diffracted beams and are explained by means of the classical diffraction theory.

When the diffraction pattern corresponds to an exact Bragg incidence for a strong $hk0$ reflection (two-beam diffraction condition), the diffraction disc $hk0$ displays a set of fringes perpendicular to the related reciprocal vector $\mathbf{r}^*(hk0)$, symmetrical about a main maximum, with a periodic intensity variation as a function of the diffraction deviation, which calls to mind the interference function (Eq. 5.27, Figure 5.7). The complementary effect observed in the transmission disc is interpreted, in the two-beam approximation, by means of the interaction between diffracted $hk0$ and transmitted 000 beam (pendulum solution). This diffraction effect is equivalent to the complementary *bend contours* in the electron microscope image (Eqs 6.16, 6.17). The contrast effect can be interpreted by means of the Ewald construction, as outlined in Figure 12.22.

HOLZ lines. The fine lines observed in the discs of the HOLZ rings are similar to the Kossel lines observed with a convergent X-ray beam. They are the intersections, with the plane of the diffraction pattern, of the diffraction cones generated by the Bragg reflections of the incident rays on a set of planes (hkl). They are sometimes called *pseudo-Kikuchi lines*. The actual related Kikuchi lines (section 12.1.3) are issued from the reflection cones of quasi-elastically scattered electrons inside the crystal. The near continuity of the HOLZ lines (inside the discs) and the Kikuchi lines (on the scattered

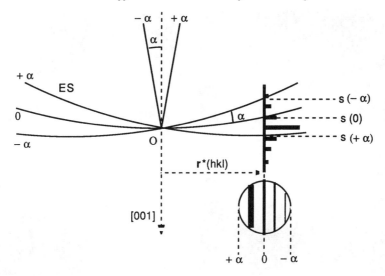

Figure 12.22. Interpreting the diffraction fringes in a diffraction disc $hk0$ by the Ewald construction. Variation of the incident angle is represented in an equivalent way by tilting the reflecting sphere about the origin. The intensity variation as a function of the angle or of the diffraction deviation s along a diffraction domain $hk0$ is outlined by bars representing the maxima of the function $I(s)$.

background) highlights their similarity; the slight discrepancy is due to the small energy loss corresponding to the latter. A pair of parallel lines (dark at small angles in the transmission disc, bright at large angles in the high-order diffraction discs) may be indexed in the same way as a pair of Kikuchi lines.

12.2.2 INSTRUMENTAL LAYOUT

Convergent beam electron diffraction is commonly implemented according to two different modes, TEM mode and STEM mode.

Stationary Convergent Beam. TEM Mode. This mode is currently practised in a TEM and is therefore often called the TEM mode. The condenser lenses are adjusted to produce an electron beam with a convergence of about 10^{-2} rad as defined by the second lens aperture, focused on to the specimen to provide a probe of a few nanometres to some ten nanometres in diameter. This setting corresponds to the nanoprobe position of some electron microscopes. The objective aperture and the selection aperture are removed.

Rocking Beam. STEM Mode. This mode is currently practised in a STEM fitted with a beam rocking device. It is therefore often called the STEM mode. The CBD pattern is generated by means of a fine focused beam with a small convergence and submitted to angular scanning in order to cover the required angular range. Thanks to this sequential mode, signal processing is easy to carry out and specimen contamination and damage are greatly reduced, in comparison to the stationary mode.

For a CBD pattern to provide significant diffraction data, the specimen must be sufficiently thick to generate a Kikuchi pattern, i.e. some 100 Å. The zone orientation

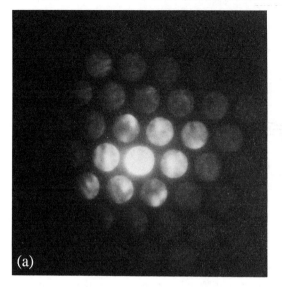
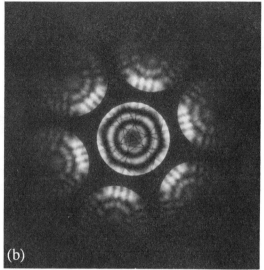
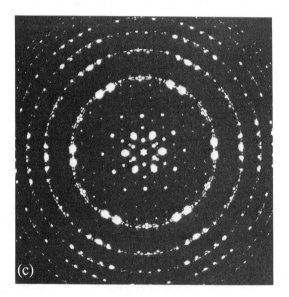

Figure 12.23. Convergent beam electron diffraction patterns. (a) ZOLZ reflections of a thin cleavage plate of mica muscovite (monoclinic) displaying a weak diffraction contrast in the discs. Orientation close to the [001] zone (compare with Figure 12.5). (Ehret G., Dept Electron Microscopy, Lab. Crystallography, Louis Pasteur University, Strasbourg); (b) ZOLZ reflections of silicon (cubic F) in a [111] zone orientation; in addition to the diffraction contrast, dark pseudo-Kikuchi lines are observable in the transmission disc; the corresponding bright HOLZ lines are not visible (JEOL document); (c) HOLZ rings of $TaSe_2$ at 80 K (orthorhombic form), oriented according to the [001] zone axis of the hexagonal room temperature polymorph (Steeds J. W., Bristol Univ., UK, in [8], Hren et al).

demanded is based on the resulting Kikuchi band pattern (Figure 12.6) and requires a perfectly centred specimen stage (eucentrical stage). The specimen must remain stable and without deformation under the strong electron bombardment, within the adjusting and exposure time. Any electrical charge of the specimen impairs its stability. If the intrinsic structure features of the crystal are to be displayed in the diffraction pattern, the selection zone must be free from any defects or deformation. This can best be checked when combining electron diffraction and electron imaging.

The *scale* of the diffraction pattern is determined by the diffraction constant. The choice depends on the kind of data to be obtained: a high diffraction constant for observing details in the ZOLZ discs, a smaller constant to include the maximum number of HOLZ rings.

The *exposure time* should be adjusted with regard to whether the transmission disc (high brilliance), the diffraction discs or the background with the Kikuchi pattern is to be recorded.

An accurate centring of the electron–optical illumination system is crucial, as is a precise adjustment of the objective–lens excitation to the diffraction setting. A criterion of adequate setting is the disappearance of any spatial information from the object (form, granularity) in the diffraction pattern. For this aim the incident beam is shifted to focus on an edge or a clearly visible object detail. If a shadow image of the detail appears in a diffraction disc, the objective excitation is adjusted in such a way that the size of the shadow image increases, then becomes inverted before decreasing. The optimum adjustment occurs at the inversion point.

12.2.3 APPLICATIONS OF CBD

A CBD pattern provides three-dimensional data on a crystal without any object tilt.

12.2.3.1 Zone-axis Parameter

Measuring the radius of the HOLZ rings leads to the parameter of the [*uvw*] zone-axis, according to Eq. (12.5). This operation can be achieved through conventional parallel-beam diffraction; however, the Laue-ring resolution is far better in CBD. An interesting application consists in determining the different polytype sequences of lamellar crystal species (e.g. phyllosilicates, graphites) in normal incidence.

12.2.3.2 Crystal Thickness

Determining the thickness of a crystal plate is a current problem in electron microscopy or in electron diffraction, e.g. in order to check the applicability of the kinematic approximation. The method which consists of assessing the diameter of the ZOLZ, already outlined in a previous section on parallel-beam diffraction (Eq. 12.6), is valid for CBD patterns. A much more accurate technique is based on measuring the fringe distance in a pair of ZOLZ discs corresponding to the two-beam incidence, as outlined in Figure 12.22. These fringes are related to the maxima of the interference function and are equivalent to the bend contours observed in the imaging mode. The interfringe distance is related to the crystal thickness t according to Eq. (6.16). The nth secondary minimum is determined by the condition $s' = n$ integer; substituting s' by its value as a function of s leads to

$$t^2\left[s^2 + \frac{1}{t_g^2}\right] = n^2 \Rightarrow \left[\frac{s}{n}\right]^2 = -\frac{1}{t_g^2}\left[\frac{1}{n}\right]^2 + \frac{1}{t^2} \qquad (12.19)$$

Plotting the values of $(1/n)^2$ against $(s/n)^2$ results in a straight line. Its slope gives $1/t_g^2$ and its intercept gives $1/t^2$. The values of s for each secondary minimum at a distance x from the main minimum are given, in the same way as for Kikuchi lines, by the relation $s = r_g^* x/L$ (Eq. 12.10).

12.2.3.3 Crystal Symmetry

Electron diffraction does not meet Friedel's law. A diffraction pattern down a [uvw] zone should therefore provide data about the symmetry along that direction.

In parallel-beam electron diffraction, the diffraction pattern symmetry can be altered by deformations and crystal defects within the rather large selection volume, as well as by a slight disorientation with respect to the zone axis.

In CBD the small active area makes it easier to select a perfect crystal portion and adjusting the zone orientation is more sensitive. Furthermore, diffraction data are integrated over a larger angular range, resulting in a three-dimensional diffraction pattern.

A zone diffraction pattern has necessarily a symmetry corresponding to one of the 10 *two-dimensional groups* (in [9], *Int. tab.*, vol. A), from which the zone-axis point group can be determined (Buxton et al, 1976; Steeds et al, 1983; Tanaka et al, 1983a). The actual crystal *point group* can then be deduced from a few zone diffraction patterns of various orientations. The space group determination is commonly based on searching for systematic absence rules (see 5.3.3), which can, however, be disturbed in electron diffraction by the occurrence of forbidden reflections. In convergent beam electron diffraction certain forbidden reflections may be identified by the occurrence of dark extinction lines (Steeds et al, 1983; Tanaka et al, 1983b).

12.2.3.4 Identification of Crystal Species

The CBD pattern of a crystal is related to its intrinsic structure. As well as a powder diffraction pattern, a CBD pattern could therefore be used to identify crystal species by means of one nanometre-sized crystal. Such diffraction patterns are, however, very sensitive to the electron energy and to the crystal thickness. Nevertheless this identifying method could be a useful complement to X-ray microanalysis and to electron energy loss spectrometry, two techniques which are liable to be operated within the same electron beam facility (TEM or STEM).

Computer software is being developed for simulating CBD patterns of perfect and defect crystal structures by means of the kinematic theory and the dynamic theory (e.g. Stadelmann and Buffat, 1989).

12.3 LOW-ENERGY ELECTRON DIFFRACTION

Davisson and Germer made use of low-energy electrons as early as 1927 in their historical electron diffraction experiment. However, for instrumental reasons, this technique was not developed until the 1950s when it progressively became a major technique for crystal surface structure analysis.

Current abbreviation: LEED (low-energy electron diffraction).

12.3.1 BASIC PRINCIPLE

The energy range of so-called low-energy electrons (or slow electrons) extends from about 10 to 1000 eV, which corresponds to wavelengths from 3.9 to 0.39 Å. Unlike high-energy electrons, their wavelength and the related diffraction angles are therefore in the same range as for commonly used X-rays.

The mean free path of low-energy electrons is of the order of a few ångströms. As a result, only a few atomic layers intervene in LEED, i.e. virtually the crystal surface. This requires operating in clean vacuum, at pressures below 10^{-9} Torr, in order to avoid any *contamination* of the crystal surface through adsorption of residual gas molecules. Surface contamination in a conventional secondary vacuum of 10^{-5} Torr causes the intrinsic surface diffraction pattern to be rapidly altered. A further instrumental requirement of the high electron absorption is the back reflection layout, whatever the thickness of the sample.

12.3.2 INSTRUMENTAL LAYOUT

In order to facilitate high-vacuum operation, the first LEED facilities were glass sealed, thus resulting in difficulties when changing the sample. Nowadays, thanks to the development of efficient vacuum techniques, they are currently manufactured in gasket-sealed stainless steel elements, with easily accessible attachments. A currently used set-up is outlined in Figure 12.24 (Goldsztaub and Lang, 1963).

12.3.3 DISPLAY OF THE DIFFRACTION PATTERN

The active portion of the single crystal specimen is limited to a few atomic layers, the thickness of which is often smaller than the corresponding periodicity. As a result the diffraction conditions are strongly relaxed, leading to a reciprocal space with *fibre-shaped diffraction domains* elongated along the normal to the specimen surface, passing through the reciprocal lattice points. The diffraction pattern is determined by the intersections of those fibres with the reflecting sphere (Figure 12.25).

12.3.3.1 Surface Lattice

The two-dimensional direct lattice of a crystal surface consists of sets of parallel and equidistant lines which are commonly defined by two Miller indices [hk] in the same way as planes in a three-dimensional lattice are defined by three indices (hkl). The indices h and k are the reciprocals of the intercepts of the [hk] lines on the Ox and Oy axes, related to the lattice parameters. They are not to be confounded with the [uvw] indices

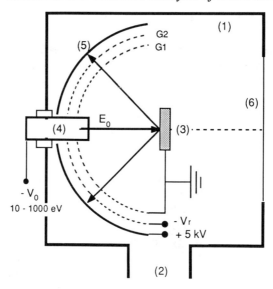

Figure 12.24. Basic diagram of a LEED facility. (1) Stainless steel enclosure with metal gasket sealings; (2) pumping system, currently with a zeolite cryosorption pump for the primary vacuum, an ion pump for the final high vacuum; the whole enclosure is generally bakeable at 200–250 °C for residual gas desorption; (3) grounded specimen holder; (4) electron gun at a negative acceleration voltage $V_0 = 10$–1000 eV, generating an electron beam of 0.1–1 mm² on the sample; (5) set of two grids and a fluorescent screen, hemispheric and centred on the electron impact on the specimen: G_1 grounded in order to provide a zero-field space for the electron trajectories; G_2 at a negative retarding voltage V_r slightly smaller in absolute value than V_0, intended as a high-pass filter to eliminate secondary electrons and inelastically scattered electrons; E fluorescent screen at a positive voltage of some 5 kV for post-accelerating the diffracted electrons to an energy sufficiently high to excite the fluorescence; the hemispherical geometry results in a high collecting efficiency, at a large solid angle, without any angular deformation of the diffraction pattern; low-energy electrons are very sensitive to any stray fields which must be compensated; (6) glass porthole for observing and recording the diffraction pattern.

of a line of a three-dimensional lattice which are the coordinates of the first lattice point on the line from the origin.

To this surface lattice corresponds a reciprocal lattice which consists of hk points such as the vectors $\mathbf{r}^*(hk)$ connecting them to the origin normal to the set of $[hk]$ surface lattice lines. This definition results in the following relation which is equivalent to the corresponding relation in a three-dimensional lattice (Figure 12.26):

$$\frac{r_g^*(hk)}{n} = \frac{1}{d(hk)} \qquad (12.20)$$

where $d(hk)$ is the equidistance of the lines $[hk]$ and n is an integer.

12.3.3.2 Diffraction Condition

A diffraction condition on a set of surface lattice lines $[hk]$, similar to the Bragg

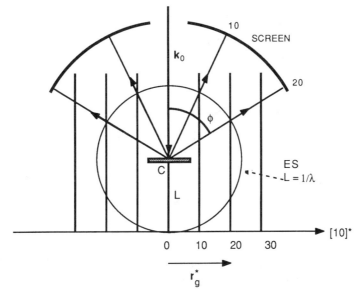

Figure 12.25. Geometrical interpretation of LEED. Section of the reciprocal space by a plane containing the incident beam and the [10]* reciprocal lattice line.

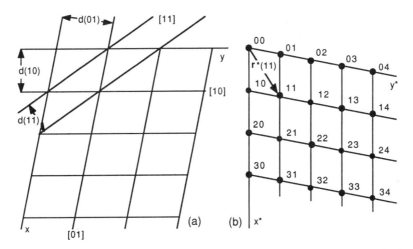

Figure 12.26. Two-dimensional lattices. (a) Direct lattice; (b) reciprocal lattice; the diffraction domains are normal to the plane, passing through the hk lattice points. The relative scales of the lattices are arbitrary.

condition on a set of lattice planes (hkl), can be established as follows, according to Figure 12.25 and Eq. (12.20):

$$d(hk) \sin \phi = n\lambda \qquad (12.21)$$

Knowledge of the wavelength and the measurement of the diffraction angles ϕ on the diffraction pattern leads to the values of the interline spacings $d(hk)$ and consequently to the surface lattice parameters.

A polycrystalline specimen with a fibre axis along the perpendicular to its surface (e.g. graphite), would result in diffraction domains consisting of a set of coaxial cylinders; their intersection with the reflecting sphere would determine circular rings. A polycrystalline material without any preferential orientation would result in a continuous scattering background.

12.3.3.3 Diffracted Intensities. Surface Structure

Due to a finite penetration depth of the slow electrons, the resulting reciprocal space in fact corresponds to a *two-dimensional direct lattice in a three-dimensional space*. There is no periodicity in the third dimension which corresponds to the thickness of the active layer. The related space groups were established by Wood (1964).

The participation of several atomic layers, at various depth levels, results in modulating the diffracted intensity along the fibres of the diffraction domains. The intensities along the diffraction domains can be measured by varying the incident energy E_0 (i.e. varying the radius $1/\lambda$ of the reflecting sphere), or by tilting the crystal. The FT of this intensity function leads theoretically to the structure of the surface layers. The strong interaction requires a dynamic treatment. A quantitative interpretation is made difficult by uncertainties on the intervening parameters, in particular the atomic scattering amplitudes at low energies.

12.3.4 APPLICATION TO CRYSTAL SURFACE STRUCTURE ANALYSIS

The geometrical interpretation of the diffraction pattern leads to the surface lattice of the specimen, either of the intrinsic crystal surface or of the adsorption layer. The intensity interpretation leads theoretically to the symmetry (Figure 12.27) and the structure of the surface. However, it is hampered by strong and ill-defined dynamic effects.

12.3.4.1 Intrinsic Crystal Surface

For this aim the specimen surface must be clean without any 'foreign' atoms. The ideal clean surface can be achieved by various techniques, depending on the specimen:

(a) cleavage in the high vacuum of the diffraction enclosure, limited to certain species;
(b) heating in high vacuum;
(c) ion sputtering, followed by reheating in order to release surface defects and stress;
(d) oxidation or reduction reactions, depending on the species.

The result of certain experiments leads to surface parameters equal to those of the bulk lattice. In other cases the surface parameters are significantly different. The discordance can be due to a real difference between the surface lattice and the three-dimensional lattice. It can be instrumental, related to the *internal potential* of the crystal which, unlike the case of high-energy electron diffraction, is not negligible against the incident energy. Assuming the surface lattice and the bulk lattice to be identical, the value of the potential correction to be introduced in order to fit the respective parameters can then be used as an indirect measurement of the inner potential (Germer et al, 1966).

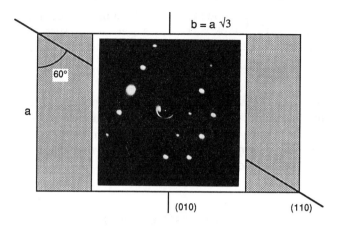

Figure 12.27. Diffraction pattern from 75 eV electrons on a (001) cleavage plate of mica muscovite (space group $C\,2/c$). The pattern shows an apparent surface symmetry plane (110), making an angle of 60° with the (010) symmetry plane of the bulk crystal. It corresponds to the glide plane c. This method, when combined with ion sputtering removing layer after layer, can provide valuable information on polytypism in lamellar crystals, at the scale of the elementary layer. A complementary analysis by Auger spectrometry confirms the localisation of the cleavage at the level of the interlayer potassium atoms (Deville et al, 1967; Deville and Goldsztaub, 1969). The absence of a centre of symmetry in the diffraction pattern shows the failure of Friedel's law (Deville J. P., Lab. Crystallography, Louis Pasteur University, Strasbourg).

An apparently clean crystal surface may lead to a *superlattice* diffraction pattern (Figure 12.14). It may be due to a so-called surface reconstruction. The alternative interpretation of localised ordered adsorption of residual foreign atoms (next section) has led to controversy. The use of complementary techniques may solve the problem (e.g. STM, Figure 22.4).

12.3.4.2 Adsorption Layers

An ordered distribution of foreign atoms on specific sites of the crystal surface also results in a *superlattice* diffraction pattern (Figure 12.14). Thus, LEED can provide valuable data on the structure of adsorption layers which can play a crucial role in determining the surface properties of certain materials, such as chemical reactivity, corrosion resistance and catalytic effect. The method is highly sensitive: a hundredth of one complete atomic layer can thus be detected. The above-mentioned difficulties, however, impose a limit on quantitative interpretation.

12.3.4.3 Attachments

Complementary techniques are often associated with LEED and provide further information:

1. Analysis of desorbed gases, originated from the sample, through a mass spectrometer.
2. Chemical analysis of the sample surface by Auger electron spectrometry emitted inside the low-energy electron diffractograph slightly modified by adding a third grid (see Figure 10.11).

In the currently used LEED facilities, the electron probe is relatively large (0.1–1 mm in diameter), thus integrating the diffraction data over the corresponding area. The recent trend is towards developing *low-energy electron microdiffraction* by means of more sophisticated focusing systems. It permits investigation of surface domains of the order of 100 Å, such as surface defects.

12.4 CONCLUSION

Due to the strong electron–matter interaction and the resulting high absorption:

1. *High-energy electron diffraction* (10–100 keV) is used in the transmission mode with thin specimen (order of 100–1000 Å) or microcrystals, as well as in the grazing reflection mode with bulk specimen for surface studies.
2. *Low-energy electron diffraction* (10–1000 eV) affects few surface layers (order of 10 Å) and can only be performed in the reflection mode.

In any case, surface investigation requires operation in high vacuum, in order to avoid surface contamination.

Thanks to the strong interaction, the diffraction patterns are directly observable on a fluorescent screen.

Parallel-beam high energy electron diffraction is mainly carried out in the form of selected area diffraction (microdiffraction) with an electron microscope. Due to the small diffraction angles, the diffraction pattern represents a plane section of the reciprocal space. The quantitative intensity interpretation for structure analysis is made difficult by multiple diffraction effects. High-energy electron diffraction is of particular use for investigating poorly crystallised materials, down to a nanometre scale.

Convergent beam high-energy electron diffraction, operating with a nanoprobe, provides three-dimensional structural data and is increasingly used for investigating the symmetry or the structure of the intrinsic crystal or of defects on a nanometre scale.

Low-energy electron diffraction is an important tool for studying crystal surfaces. Quantitative intensity interpretation is, however, difficult. It becomes more interesting in the form of microdiffraction and acquires an analytical dimension when associated with Auger spectrometry.

REFERENCE CHAPTERS

Chapters 4, 5, 6, 9, 12, 19. Appendices B, C.

BIBLIOGRAPHY

[1] Andrews K. W., Dyson D. J., Keown S. R. *Interpretation of electron diffraction patterns*. Hilger, London, 1971.

[2] Clark L. J. *Surface crystallography. An introduction to low energy electron diffraction*. John Wiley, New York, Chichester, 1985.

[3] Edington J. W. *Practical electron microscopy in materials science. Monograph 2: Electron diffraction in the electron microscope*. Macmillan, Philips Technical Library, London, 1975.

[4] Gard J. A. (Ed.) *The electron-optical investigation of clays*. Mineralogical Soc., London, 1971.

[5] Haymann P. *Théorie dynamique de la microscopie et diffraction électronique*. Presses Universitaires de France, Paris, 1974.

[6] Heidenreich R. D. *Fundamentals of electron microscopy*. John Wiley, New York, 1964.

[7] Hirsch P. B., Howie A., Nicholson R. B., Pashley D. W., Whelan M. J. *Electron microscopy of thin crystals*. Butterworths, London, 1965.

[8] Hren J. J., Goldstein J. I., Joy D. C. (Ed.) *Introduction to analytical electron microscopy*. Plenum Press, London, 1979.

[9] *International tables for crystallography. Brief teaching edition of vol. A. Space-group symmetry*. (T. Hahn, Ed.). D. Reidel, Kluwer Acad. Publ., Dordrecht, 1987.

[10] JCPDS International Centre for Diffraction Data, Swarthmore, Pa (USA).

[11] Jouffrey B., Bourret A., Colliex C. (Ed.) *Microscopie électronique en science des matériaux*. Editions du CNRS, Paris, 1983.

[12] Loretto M. H. *Electron beam analysis of materials*. Chapman and Hall, London, 1984.

[13] Pinsker Z. G. *Electron diffraction*. Butterworths, London, 1953.

[14] Tanaka M., Terauchi M. *Convergent beam electron diffraction*. Vols 1, 2. Maruzen, Tokyo, 1986.

[15] Vainshtein B. K. *Structure analysis by electron diffraction*. Pergamon, Oxord, 1964.

[16] Zwyagin B. B. *Electron diffraction analysis of clay mineral structures*. Plenum Press, New York, 1967.

Part Four

X-RAY, ELECTRON AND SECONDARY ION SPECTROMETRY APPLIED TO MATERIAL ANALYSIS

Spectrometric techniques are based on the data delivered by characteristic energy transfers between radiations and matter. They provide information about the nature and the concentration of elements in either crystallised or non-crystallised materials. To some degree, they provide additional information about chemical bonds.

Energy transfers during interaction with high-energy radiations (10–1000 keV) can be expressed in two complementary ways: alteration of matter and alteration of radiation, the total energy being preserved. This results in the following complementary effects which are exploitable for spectrometry:

1. **Excitation–de-excitation** processes of atomic core levels leading to characteristic secondary emissions.
2. **Characteristic absorption** processes, with or without energy loss.

Those two types of effects provide data which are theoretically identical and can both be used for analytical purposes, leading to two groups of techniques: **secondary emission spectrometry** and **absorption spectrometry** or **energy loss spectrometry**. They are mostly non-destructive.

The emission of particles of matter through ion sputtering results in a destructive analytical technique by means of **mass spectrometry**.

The data provided by any spectrometric methods surveyed in the following sections are expressed in terms of a spectrum $I(E)$ or $I(\lambda)$. The chemical composition of the analysed material is to be extracted from this spectrum. The method of spectrum processing depends on the basic physical effect.

Spectrometric methods are often associated with diffraction and imaging, thus resulting in structural as well as in chemical analysis.

The particle aspect of radiations dominates in spectrometry.

13
Important Parameters in Spectrometry

Inside the considered energy range, spectrometric methods are mostly based on the effects of excitation and resulting de-excitation (or relaxation) of atomic energy levels, due to the primary radiation.

Some methods are based on the first stage of the process, the *excitation* effect. They are mainly the emission spectrometries of photoelectrons and of secondary electrons, as well as the associated characteristic absorption spectrometries of X-rays and of electrons.

Other methods are based on the second stage of the process, the *relaxation* effect. They are mainly the characteristic secondary emission spectrometries of X-rays and of Auger electrons.

Emission spectrometry is most frequently used. It generally makes possible the selection of the spectrum corresponding to only one characteristic effect of a given element.

Absorption spectrometry or *energy loss spectrometry* is generally more difficult to interpret. Several characteristic interaction effects superimpose on the spectrum, often resulting in a poor signal-to-background ratio and in overlapping.

The main purposes of the spectrometric methods which will be discussed in this chapter are *qualitative analysis* and *quantitative analysis* of the elements in materials.

Qualitative analysis requires us to identify a given element by means of one or several characteristic emission lines or absorption lines.

Quantitative analysis requires us to relate the measured line intensities (detector count rate) to the respective mass concentration or the number of atoms in the volume of analysis considered.

The parameters and the relations between them which are outlined in the present section are valid for most spectrometric techniques. They are directly applicable to emission spectrometry, with certain adaptations to absorption spectrometry and secondary ion mass spectrometry.

13.1 QUALITATIVE ELEMENTAL ANALYSIS

The first step of spectrometric analysis is generally to determine the various elements in the specimen to be analysed. Identifying the *major elements* is mostly straightforward. Analysing the *minor elements* or the *trace elements* is determined by the sensitivity of the method; it may vary over a large range, depending on the physical process involved, on the atomic number and on the instrumental characteristics.

The spectrometric techniques which will be considered in the following chapters involve mainly the core levels of the atoms. They are therefore not very sensitive to the chemical state of the elements, i.e. their bonds. The result is chiefly elemental analysis.

The main parameter in qualitative analysis is the detection limit of elements.

13.1.1 COMPOSITION PARAMETERS OF THE SPECIMEN

For spectrometric purposes, the specimen and its constituent elements to be analysed are characterised by means of the following quantities which will intervene in relations for qualitative analysis.

Specimen. **Mean atomic mass** M (average value of the atomic masses of the elements in the specimen, weighted by their respective atomic concentrations); **density** ρ.

Element A. **Atomic mass** A; **mass fraction** c_A; this expresses the mass ratio of the element A in the specimen, the total mass of the elements taken as unity:

$$\sum_J c_J = c_A + c_B + \ldots = 1$$

Atomic fraction $c_A(at)$; this expresses the ratio of the number of atoms of an element A in the specimen, the total number of atoms taken as unity:

$$\sum_J c_J(at) = c_A(at) + c_B(at) + \ldots = 1$$

The relations, given under the headings below, are currently useful in spectrometry.

13.1.1.1 Number of Atoms of an Element in a Volume Unit

Given the Avogadro number N, the number of atoms per mass unit of the pure element A is N/A; the number $N_A(V)$ of atoms A per volume unit of the specimen is therefore

$$N_A(V) = \frac{N}{A} c_A \rho \qquad (13.1)$$

13.1.1.2 Total Number of Atoms of the Specimen in a Volume Unit

The total number of atoms per mass unit of the specimen is N/M; the total number $N(V)$ of atoms in a volume unit is therefore

$$N(V) = \sum_J N_J(V) = \frac{N}{M} \rho \qquad (13.2)$$

13.1.1.3 Number of Atoms of an Element in a Surface Unit

Consider a specimen layer of thickness t, of mass per surface unit ρt; the number $N_A(S)$ of atoms of element A in a surface unit is written as

$$N_A(S) = \frac{N}{A} c_A \rho t \tag{13.3}$$

13.1.1.4 Mass Fraction and Atomic Fraction

Mass fraction c_A and atomic fraction $c_A(at)$ of an element A are related as follows:

$$c_A(at) = \frac{N_A(V)}{N(V)} = \frac{M}{A} c_A \tag{13.4}$$

13.1.2 DETECTION LIMIT OF AN ELEMENT

The possibility of analysing an element A in a specimen is defined by means of the detection limit of this element. This most important characteristic for qualitative analysis may be defined by various quantities.

13.1.2.1 Definitions

Minimum detectable mass fraction: c_{min}
Minimum detectable number of atoms in the volume of analysis v: N_{min}:

$$N_{min} = \frac{N}{A} c_{min} \rho v \tag{13.5}$$

Minimum detectable mass of an element A in the volume of analysis v: M_{min}:

$$M_{min} = \frac{A}{N} N_{min} = c_{min} \rho v \tag{13.6}$$

For a given analytical method, these detection limits depend on the element to be analysed and on the specimen which contains it.

13.1.2.2 Detection Limits Related to the Count Rate

In emission spectrometry, a given element A is determined by the detector counting when the spectrometer is adjusted on a characteristic line of the element. In absorption spectrometry, the energy range of counting of the spectrum is not so well defined. In any case the significant measurement is the signal-to-noise ratio.

The total counted number N of pulses on a line of an element, during a given time, is the sum of:

(a) the count N_c of the characteristic emission of the element;
(b) the count N_0 of the background; it consists of the continuous radiation due to various interaction effects and of the instrumental noise. The significant measurement for an element is therefore

$$N_c = N - N_0$$

The absolute statistical error to expect on the count N_c is given by Eq. (10.14) as follows, for an error function $p = 0.997$, generally considered as suitable:

$$\Delta N_c = 3(N + N_0)^{1/2}$$

For the element with mass fraction c to be detectable with the foregoing error function, the difference $N - N_0$ must be significant, i.e. greater than the statistical error ΔN_c, resulting in the condition

$$N - N_0 \geq 3(N + N_0)^{1/2}$$

At the detection limit, the signal is just emerging from the background, i.e.

$$N \cong N_0 \Rightarrow N + N_0 \cong 2 N_0$$

The above condition thus becomes

$$\frac{N - N_0}{(N_0)^{1/2}} \geq 3\sqrt{2} \cong 4.2$$

In practice the following detection condition is considered as satisfactory, with an error function of $p \cong 0.96$:

$$\frac{N - N_0}{(N_0)^{1/2}} \geq 3 \tag{13.7}$$

13.1.2.3 Minimum Detectable Mass Fraction

It is more convenient to express the detection limit of an element A by means of its minimum detectable mass fraction c_A rather than by means of the minimum detector count. Within certain limits a linear relation can be written as follows between the mass fraction c and the total count N on a characteristic line:

$$N = N_1 c + N_0 \begin{cases} c = 0 \Rightarrow N = N_0 \\ c = 1 \Rightarrow N = N_1 + N_0 \end{cases}$$

This relation shows that:

1. N_0 expresses the count on a *blank* of the element, i.e. the count which would be measured at the same setting, on the same specimen, but in the absence of

element A. In practice, it is the count on the background extrapolated to the line position.
2. N_1 is the characteristic count on the pure element. Its actual value depends on matrix effects and on instrumental parameters (incident energy and intensity, spectrometer and detector efficiency). For a given specimen, N_1 expresses in some way a quality factor of the spectrometer facility.

Substituting $N - N_0 = N_1 c$ in Eq. (13.7) leads to the following detection criterion of a given mass fraction c:

$$c_{min} = \frac{3\sqrt{N_0}}{N_1} = \frac{3}{\sqrt{N_1}\sqrt{(N_1/N_0)}} \qquad (13.8)$$

N_1/N_0 approximates the signal-to-noise ratio measured on the pure element.

Replacing the respective counts by the count rates (number of pulses per time unit) results in the following expression of the *minimum detectable mass fraction*:

$$c_{min} = \frac{3}{\sqrt{t_c}\sqrt{n_1}\sqrt{(n_1/n_0)}} \qquad (13.9)$$

where t_c is the counting time, and n_1 and n_0 respectively the count rate on the characteristic line of the pure element and on the background.

In practice, the count rates are currently expressed in terms of intensities I_1 and I_0. According to Eq. (13.9):

1. The minimum detectable mass fraction of an element with a given spectrometer facility can be directly deduced from the count rate provided by a pure standard of this element.
2. The detection limit is lowered by:
   (a) increasing the count time;
   (b) increasing the count rate;
   (c) increasing the signal-to-background ratio.

Action on time is limited by instrumental drift, by radiation-related specimen alterations (in particular with electrons and ions) and by the convenient duration of experiments.

Action on the count rate is possible by optimising instrumental conditions (high primary intensity, optimum excitation energy, maximum efficiency of the spectrometer-detector, energy discrimination, electronic differentiation, etc.). Currently observed detection limits are given in Table D-1 and in the following specialised section.

13.2 QUANTITATIVE ELEMENTAL ANALYSIS

The aim of quantitative analysis is to determine the mass fraction of any detectable elements in a specimen. For a given element A, the intensity of a characteristic emission line or the characteristic absorption effect should be measured. The intensity measurement provided by a spectrometer depends on three kinds of parameters: physical parameters characterising the nature and the concentration of the element to be analysed; physical parameters due to any other elements in the specimen (matrix effects); and instrumental parameters.

13.2.1 PARAMETERS CHARACTERISING AN ELEMENT

The characteristic intensity I_A is defined as the number per time unit of photons, electrons and secondary ions, resulting from interaction with the atoms of element A.

The characteristic intensity $I_A(V)$ issued from the volume unit of the specimen is proportional to the following factors:

1. Incident intensity I_0 (number of primary photons, electrons, ions per atom).
2. Number $N_A(V)$ of atoms A per volume unit (Eq. 13.1).
3. Total emission or absorption cross-section σ_A for the given analysing mode. For the methods based on primary excitation effects (XPS, XAS, EELS, etc.), this factor is the excitation cross-section of the related energy level. For the methods based on secondary emissions induced by relaxation (XRF, EPMA, AES, EDX, etc.), this factor is the product of the excitation cross-section with the transition probability in the considered mode leading to the line to be measured. For secondary ion emission, the cross-section is replaced by the sputtering yield of the element in the form of ions.

The characteristic intensity emitted by the volume unit can now be written, according to Eq. (13.1), as

$$I_A(V) = I_0 N_A(V) \sigma_A = I_0 \frac{N}{A} c_A \rho \sigma_A \qquad (13.10)$$

13.2.2 PARAMETERS CHARACTERISING THE MATRIX

Consider an element A to be analysed in a specimen consisting of elements $J(J=A, B, C, \ldots)$. All the elements in the specimen except A ($J \neq A$) are referred to as the matrix. The interaction of radiations with the matrix intervenes in the intensity measurement related to element A. This *matrix effect* may be an attenuating effect or an enhancing effect.

13.2.2.1 Attenuation of the Characteristic Intensity

The matrix attenuation effect on the intensity I_A to be measured may result from interaction of matrix atoms with the primary radiation (*primary matrix effect*) or with the secondary radiation issued from the atoms to be analysed (*secondary matrix effect*). This is the case of X-ray absorption, as well as of electron absorption, backscattering and energy loss.

13.2.2.2 Enhancement of the Characteristic Intensity

The matrix enhancement effect is due to radiation scattered or emitted by matrix atoms which add up to the primary radiation for exciting the atoms to be analysed. This is the case of enhancement by primary or by secondary X-ray fluorescence, of enhancement by electron backscattering and by photoelectrons. In secondary ion emission, the sputtering yield is enhanced through the presence of certain matrix atoms, e.g. oxygen atoms.

13.2.2.3 Matrix Correction

The magnitude of matrix effects depends on the primary radiation and on the radiation to be measured, as well as on the matrix composition. For an analysis to become quantitative, they have to be corrected. The correction process varies according to the technique. It is generally the main factor which determines the accuracy of analysis. Using the factor m_J to express the total matrix effect, the intensity issued from a volume unit of the specimen and leaving the specimen can be expressed as follows:

$$I_A(V) = I'_A(V) \, m_J = I_0 \, N_A(V) \, \sigma_A \, m_J = I_0 \frac{N}{A} c_A \rho \, \sigma_A \, m_J \qquad (13.11)$$

where $I'_A(V)$ represents the characteristic intensity which would be emitted from the volume unit in the absence of matrix effects (Eq. 13.10).

13.2.3 INSTRUMENTAL PARAMETERS

The radiation which characterises the specimen composition is measured in the form of electric pulses by the spectrometer-detector device. Only a fraction of the incoming intensity is actually being measured. The *efficiency* of the measuring system is thus characterised by an instrumental factor τ_A which expresses the ratio of the detected characteristic intensity to the total emitted characteristic intensity. This global instrumental factor can be detailed as follows:

$$\tau_A = \varepsilon_A \frac{\Omega}{4\pi} \qquad (13.12)$$

where ε_A is the efficiency of the spectrometer-detector and Ω the solid detection angle for the characteristic radiation of the element A to be measured.

The instrumental factor τ_A is often called the *transmission factor* of the spectrometer.

13.2.4 MEASURED INTENSITY

The radiation spectrum gathered by the measuring system is expressed by a function $I(E)$. The characteristic intensity of the concentration of an element is usually measured in the form of a count integrated over the width of the characteristic line or of the absorption edge. The actual measuring method depends on the radiation.

The significant measure is the intensity over the background. The latter must therefore be subtracted from the total intensity. In emission spectrometry, the characteristic lines are generally fine and the background is usually expressed by averaging the intensity measured on both sides of the line. In absorption spectrometry the energy range to be integrated for intensity measurements is large and often ill defined and the foregoing method is inadequate; the background is then determined by modelling through an equation which gives the best approximation and which is subtracted from the experimental intensity function. The background correction is now often carried out by means of software packages integrated in the computer programs of spectrum processing.

13.2.4.1 Intensity from a Volume Unit

When making explicit any factors which induce a deviation of the measured intensity from the intensity emitted by the atoms of an element A, the intensity issuing from a volume unit of the specimen and measured by the spectrometer is expressed as follows:

$$I_A(V) = I_0 N_A(V) \sigma_A m_J \tau_A = I_0 \frac{N}{A} c_A \rho \sigma_A m_J \tau_A \qquad (13.13)$$

13.2.4.2 Intensity from a Surface Unit

For surface analysis, it is more convenient to express the intensity issuing from the unit surface of a layer of thickness t or of mass thickness ρt, according to Eq. (13.3), as

$$I_A(S) = I_0 N_A(S) \sigma_A m_J \tau_A = I_0 \frac{N}{A} c_A \rho \sigma_A t m_J \tau_A \qquad (13.14)$$

13.2.4.3 Instrumental Intensity

When using Eq. (13.13) for a real calculation, it is more appropriate to express the incident intensity by means of the primary instrumental intensity I_P incident on the volume of analysis v which is directly accessible to measuring, rather than by means of the incident intensity per atom I_0 which is a theoretical parameter. The number of atoms in a volume v of the specimen is $(N/M)\rho v$ (Eq. 13.2); the incident intensity I_P is therefore expressed by

$$I_P = I_0 \frac{N}{M} \rho v$$

Substituting I_0 from this equation into Eq. (13.13) leads to the following characteristic intensity of a line of element A, issued from volume v and measured by the spectrometer:

$$I_A = \frac{M}{A} I_P c_A \sigma_A m_J \tau_A \qquad (13.15)$$

where M is the average of the atomic masses of the elements in the specimen, weighted by their respective atomic fractions.

When expressing the atomic fraction instead of the mass fraction, this relation simplifies to

$$I_A = I_P c_A(at) \sigma_A m_J \tau_A \qquad (13.16)$$

The foregoing general equations relate the analytical data $N_A(V)$, $N_A(S)$ or c_A to the measured intensity on one side, and to the matrix factors and instrumental factors on the other. If the latter can be evaluated, those equations lead to absolute quantitative analysis. If this is not possible, certain unknown factors are eliminated by comparing with a standard specimen or by limiting the analysis to measuring the concentration ratio of pairs of elements.

The practical analytical processes depend on the physical effect to be exploited and on the kind of specimen. They will be surveyed in more detail in the following specialised sections.

Given the minimum intensity which can be significantly measured, those equations provide in addition an estimate of the detection limits for a specific analytical facility.

REFERENCE CHAPTERS

Chapters 2, 7, 10, 14, 15, 16, 17, 18. Appendix D.

BIBLIOGRAPHY

See references of the specialised Chapters 14 and 15.

14

Elemental Analysis by X-ray Fluorescence

The theoretical bases of X-ray fluorescence were given by Moseley at the beginning of the century; X-ray fluorescence was applied as early as 1932 by Von Hevesy. Thanks to the development of X-ray detection techniques and data processing, it has become a powerful method for material analysis.

Common abbreviations: XRF or XRFA (X-ray fluorescence analysis); WDS or WDX (wavelength-dispersive X-ray spectrometer); EDS or EDX (energy-dispersive X-ray spectrometer).

14.1 BASIC PRINCIPLE AND INSTRUMENTAL LAYOUT

The specimen to be analysed is submitted to a primary X-ray beam. The spectrum of characteristic secondary X-rays (fluorescence radiation) emitted by the specimen contains any data necessary for determining its composition.

Characteristic X-ray wavelengths emitted by an atom are approximately independent of its chemical state; this results in a very simple spectrum, easy to interpret and providing the *elemental composition* in a straightforward way.

In the basic technique the analysing area on the specimen is of the order of several square centimetres, its thickness being several micrometres, resulting in a global analysis. Microanalytical techniques are being developed thanks to the availability of high-intensity and finely focused X-ray beams.

The basic layout, outlined in Figure 14.1, consists of two main elements: the X-ray generation system and the spectrometer-detector system. Computer data processing is increasingly associated with the basic facility.

14.1.1 PRIMARY EXCITATION RADIATION

Conventional X-ray fluorescence facilities make use of X-ray tubes. New possibilities are provided by high-intensity synchrotron sources. Radioactive sources are useful for mobile facilities.

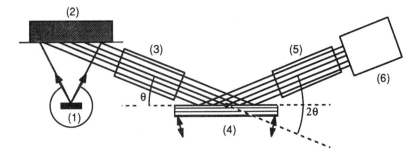

Figure 14.1. Diagram of an X-ray fluorescence analyser with a plane crystal wavelength dispersive spectrometer. (1) X-ray tube; (2) specimen with a lower plane face; (3) inlet Soller slit collimator; (4) plane analysing crystal set to the Bragg angle θ selecting the wavelength to be analysed; (5) exit Soller slit collimator oriented together with the detector (6) at an angle 2θ to collect the selected radiation.

14.1.1.1 X-ray Excitation

The criterion for choosing the target anode and the operating conditions of the X-ray tube is the maximum intensity of fluorescence emitted by the elements to be analysed. It determines the sensitivity limits.

The primary radiation spectrum from an X-ray tube consists of a continuous background (bremsstrahlung) and of the characteristic lines of the target element, both components participating in the excitation process.

Excitation by the Characteristic Lines

The excitation cross-section of the X-level of an element of atomic number Z is maximum when the incident energy is just greater than the excitation threshold W_X. This ideal condition can be approached by using the primary radiation $K\alpha$ of a target element of atomic number $Z+1$ or $Z+2$. For a given target element, this condition can only be approximated for analysing a few elements (e.g. primary radiation Cu–$K\alpha$ for exciting K–Co and K–Fe). An alternative is to use *secondary radiators* made of the corresponding elements.

Excitation by the Continuous Background

The total bremsstrahlung intensity is proportional to the tube voltage and to the atomic number of the target element (see Eq. 7.5), which must therefore both be as high as possible.

The *synchrotron sources* also deliver a broad ranging continuous spectrum, but of much higher intensity.

Instrumental Conditions

The first condition to be met is to use a target which does not contain the element to be analysed. Effectively, a fraction of the primary radiation is elastically scattered by the specimen and superimposes on its characteristic fluorescence radiation. An impurity-free target is essential for analysing trace elements.

High-power X-ray tubes (e.g. 3 kW) are currently used in X-ray fluorescence in order to reach a low detection limit. Specific tubes with front windows and annular cathodes have been developed.

The *continuous background radiation* from a heavy target element (Ag, Mo, Rh, W) is convenient for analysing heavy elements (high fluorescent yield) and major elements (high-intensity output).

The *characteristic line radiation* of target elements selected for the maximum excitation efficiency is used for analysing light elements (low fluorescent yield) and trace elements (e.g. Cr–K for analysing Ti and Ca; Al–K for analysing Mg, Na, F, O). Changing the target, i.e. changing the tube, in order to cover a large range of elements in the optimum conditions is not very convenient. Easily interchangeable *secondary radiators*, made from pure elements meeting the optimum condition, provide a valuable solution. The primary tube radiation excites the fluorescence of the radiator element which in turn excites the elements of the specimen. In good conditions, this method may provide a higher signal-to-noise ratio than direct excitation (e.g. Figure 14.2).

The special tubes with front windows mentioned above, allow the use of ultra-thin windows down to 125 μm and are therefore best suited for light element analysis.

A rhodium tube operated with a high voltage is an often-used compromise for current work: the relatively strong continuous radiation for the heavy elements, the K lines for the medium elements, the L lines for the light elements.

The K-level excitation is used whenever the tube voltage is sufficiently high: $Z=63$ (Eu) is reached with a 50 kV supply, $Z=86$ (Rh) needs at least 100 kV. The L-levels can be excited over the whole Z-range with 50 kV.

A stabilized voltage supply is required for accurate quantitative analysis.

14.1.1.2 Excitation by Radioactivity

Artificial radio elements may provide adequate and light-weight primary radiation sources for mobile X-ray fluorescence facilities. The best-suited sources are long-period elements without γ-ray emission against which the spectrometer is difficult to protect. The β-radiation is the most commonly used, directly or through secondary excitation.

14.1.2 SPECTROMETER

The characteristic X-rays emitted by the elements of the specimen are analysed by means of a spectrometer associated with a system of intensity measurement and data processing.

14.1.2.1 Wavelength-dispersive Spectrometer

The angular wavelength-dispersive spectrometer (WDS or WDX) is the most currently used in conventional facilities. Due to the large irradiated area of the specimen, a plane analysing crystal is associated with a Soller slit collimator (Figure 14.1). Energy resolution is limited by the divergence of the collimator. A better resolution is achieved with a focusing crystal together with a small active analysing area. Real X-ray fluorescence microanalysis can be carried out with a synchrotron source providing a finely collimated (0.1 mm²) high-intensity beam combined with a fully focusing analysing crystal.

14.1.2.2 Energy-dispersive Spectrometer

An energy-dispersive X-ray spectrometer (EDS or EDX) with a Si(Li) diode leads to a set-up without any mechanical motion and to a nearly instantaneous display of the whole spectrum. These advantages are set against a poorer resolution, and a signal-to-noise ratio which is generally smaller. However, when optimized, this type of spectrometer can have a lower sensitivity limit than conventional crystal spectrometers in X-ray fluorescence.

14.2 QUALITATIVE ANALYSIS

The elements are identified by the characteristic lines of the spectrum $I(\lambda)$ issuing from the specimen.

14.2.1 DETECTION LIMITS

The main parameter for qualitative analysis, the detection limit is related to the count statistics. The **minimum detectable mass fraction** of an element is given by Eq. (13.9) which is recalled below:

$$c_{min} = \frac{3}{\sqrt{t_c}\sqrt{n_1}\sqrt{(n_1/n_0)}}$$

where t_c is the counting time, n_1 the count rate and n_1/n_0 the signal-to-noise ratio given by the pure element. This equation highlights the factors which determine the detection limit.

Counting time. This must be sufficiently long, taking into account the instrumental drift and the convenient duration of the measurement.

Signal-to-noise Ratio. Various factors influence this parameter:

1. Excitation energy. It must be optimal, in particular for light elements.
2. Background noise. One cause is the secondary fluorescence of heavy elements in the analysing crystal. This occurs with lead in a PbSt (lead stearate) crystal, with an L-level excitation energy of about 16 keV; this crystal being used for light element analysis, it suffices to decrease the excitation energy below 16 keV in order to eliminate the effect. This is also the case for potassium in a KAP crystal with a K-level excitation energy of 3.6 keV; its secondary fluorescence is a hindrance for analysing minor elements in the range $Z = 8-20$.
3. Weakening of the characteristic radiation to be measured. Absorption increases sharply with the wavelength (Eq. 8.2). In current conditions, an X-ray fluorescence spectrometer can be operated down to $Z = 20$ (Ca) at atmospheric pressure, down

Table 14.1. Detection limit of some elements by means of X-ray fluorescence (from Broll, 1985)

Element	Z	Minimum mass fraction
B	5	3×10^{-3}
C	6	3×10^{-4}
O	8	10^{-2}
F	9	2×10^{-5}
Na	11	10^{-5}
Mg	12	10^{-5}
Al	13	5×10^{-6}
Sc–Mo	21–42	10^{-6}
I–U	53–92	5×10^{-6}

to $Z = 9$ (F) in a primary vacuum. Carbon, boron and beryllium can be detected with a light target element and with very thin tube windows and detector windows (Broll, 1985).

Current Analysing Characteristics.

Minimum detectable mass fraction. This is typically of the order of 1–100 ppm, depending on the atomic number and the instrumental parameters. The values given in Table 14.1 correspond to good laboratory facilities.

Minimum detectable mass. A focusing X-ray fluorescence spectrometer is able to detect a mass of an element of the order of 10^{-8}–10^{-9} g. A limit of 10^{-11} g has been achieved by means of an optimized EDS (Ruch, 1985).

In thin-layer analysis one-hundredth of an atomic layer can be considered as the limit of detection.

With a synchrotron X-ray source, the minimum detectable mass can be as low as 10^{-12} g or even less.

14.2.2 IDENTIFYING UNCERTAINTIES

Elemental identification is based on the characteristic line positions in the spectrum. This results in various causes of ambiguity.

Parasitic Spectrum Lines

1. Weak lines from forbidden transitions and satellites from multiple excitations.
2. Instrumental lines due to the spectrometer–detector system (multiple-order reflections from the analysing crystal, escape peaks from the detector). Multiple-order reflections (e.g. K-lines from heavy elements) can be distinguished from overlapping single-order reflections (e.g. K-lines from lighter elements or L-lines from heavy elements) by decreasing the primary energy below the excitation level of the former or by energy discrimination. As an example, a discrimination window centred on 1.48 keV (Al-Kα_1) would strongly attenuate the overlapping fourth-order peak of Cr-Kβ (5.95 keV $\cong 4 \times 1.48$ keV).
3. Lines due to the primary emission of the anode (target element or impurities), through elastic scattering or through inelastic scattering (Compton peak) in the specimen or in the secondary radiator (see Figure 14.2).

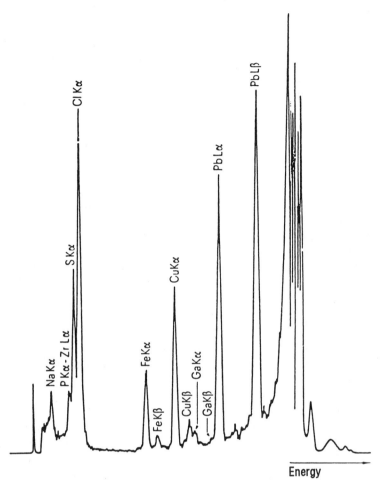

Figure 14.2. X-ray fluorescence spectrum from heavy elements in an organic matrix. Rhodium anode at 40 kV. Zirconium secondary radiator. Si(Li) energy-dispersive spectrometer. The elastic scattering and Compton peaks appear on the right-hand side of the spectrum (from Ruch, 1985).

The parasitic peaks are the most annoying when analysing trace elements. In the foregoing example, discrimination would not work in detecting traces of aluminium in chromium.

Spectrometer Resolution. An insufficient energy resolution of the spectrometer (plane analysing crystal, EDS) favours peak overlapping. Multiple peaks can now be separated into their Gaussian components by means of peak deconvolution thanks to specific computer software. It is, however, difficult to separate peaks with large intensity differences by this method, e.g. from a major element and a trace element.

Chemical Shift. The independence of line positions from chemical bonds is a first approximation. The chemical shift is the larger the lighter the element and the more exterior the excitation level. With a high-resolution spectrometer, this effect can be applied to gather data on the chemical state of the elements.

In any identification problem, it is important to consider all the observable lines of the given element; the relative intensity of the lines is a valuable criterion for identification.

Some of the above-mentioned effects lead to inaccuracies in quantitative analysis.

14.3 QUANTITATIVE ANALYSIS

Consider an element A to be analysed by means of one of its characteristic X-ray fluorescence radiations. The case of the Kα-line, the most frequently used when possible, will be considered below.

The quantities related to the element A in the *specimen* to be analysed will be given as follows: mass fraction c; emitted line intensity I; absorption coefficient μ, etc.

The quantities related to the element A in the *test sample T* will be given as follows: mass fraction $c(T)$; emitted line intensity $I(T)$; absorption coefficient $\mu(T)$ etc.

14.3.1 MATRIX EFFECTS

The element A to be analysed is contained in a specimen consisting of elements J with respective mass fractions c_J ($J = A, B, C \ldots$). The whole lot of elements except element A ($J \neq A$) is called the matrix.

The intensity I of the characteristic A–Kα-line of element A is measured by means of a count rate N_c/t_c over the background. The intensity A–Kα emitted at a point of the specimen, per mass unit and per incident photon, is proportional to the mass fraction of element A. In the absence of any interaction of the incident and emitted radiations with the matrix, the intensity I measured at the detector would likewise be proportional to the mass fraction, and the following ideal relation would be valid:

$$I = \tau c \tag{14.1}$$

In order to eliminate the instrumental factor τ (Eq. 13.12), the intensity from the specimen to be analysed is related to the intensity given in the same conditions by a test sample T containing element A in a known mass fraction, thus leading to the relation

$$\frac{1}{I(T)} = \frac{c}{c(T)} \tag{14.2}$$

The matrix in fact intervenes in three ways:

1. The primary radiation is absorbed on its path to reach the atom to be excited; this effect, called the *primary absorption*, depends on the matrix.
2. The fluorescence radiation A–Kα is absorbed on its path inside the specimen in the direction of the detector; this effect, called the *secondary absorption*, depends on the matrix.

The whole determines the absorption effect which is different in the specimen and in the test sample.

3. If the specimen to be analysed contains one or several elements with atomic numbers greater than A and close to A, the fluorescence radiation of those elements can in turn contribute to exciting the Kα fluorescence of element A (*secondary fluorescence*) and thus enhance the radiation to be measured. This effect is different in the specimen and in the test sample.

In addition to the foregoing fundamental interaction effects, the matrix may intervene by a microabsorption effect due to heterogeneities of the material. This *grain effect* acts differently in the specimen and in the test sample if their texture is different.

As a result, for the ratio of the mass fractions in the specimen and in the test sample to be related to the ratio of the corresponding intensities, the following correction factors must be introduced: *absorption correction* K_a, *secondary fluorescence correction* K_f, *grain (or texture) correction* K_t. Relation (14.2) is then modified to

$$\frac{c}{c(T)} = K_a K_f K_t \frac{I}{I(T)} = K \frac{I}{I(T)} \tag{14.3}$$

14.3.1.1 Absorption Effect

This is generally the main matrix effect. Its formulation is straightforward for a monochromatic primary radiation. Consider:

(a) a specimen with a plane surface;
(b) a layer of thickness dx parallel to the surface, at a depth x;
(c) a parallel monochromatic X-ray beam, of intensity I_0, of incidence angle θ' (Figure 14.3).

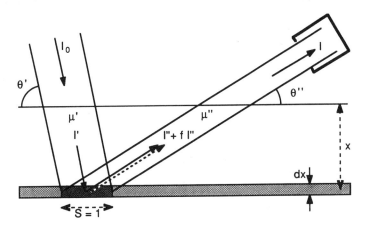

Figure 14.3. Outline of matrix effects in X-ray fluorescence. Absorption effect and secondary fluorescence effect.

Primary Intensity. Let I' be the primary intensity incident on the surface unit of layer dx. It has been submitted to primary absorption (Eq. 8.1) and is written as follows:

$$I' = I_0 \sin\theta' \exp\left[-\frac{\mu'\rho x}{\sin\theta'}\right] = I_0 \sin\theta' \exp(-\chi'\rho x) \quad (14.4)$$

where ρ is the density, μ' the mass absorption coefficient for the primary radiation and $\chi' = \mu'/\sin\theta'$ the primary absorption factor of the specimen (with additivity relations, Eq. 8.3).

Secondary Intensity Emitted in the Spectrum. The secondary intensity dI'' emitted by atoms A per surface unit of layer dx, in the form of a characteristic A-$K\alpha$ radiation, can be written, according to Eq. (13.10),

$$dI'' = I'(N/A)c\rho(\sigma\omega p)\tau dx = I_0 \sin\theta' \, F(A)c\rho \exp(-\chi'\rho x)\, dx \quad (14.5)$$

where $F(A) = (N/A)\sigma\omega p$, the factor depending only on element A, N the Avogadro number, σ the excitation cross-section of level K, ω the fluorescence yield and p the transition probability leading to $K\alpha$ emission.

Secondary Intensity Leaving the Specimen. The intensity dI leaving the specimen in a direction θ'' towards the spectrometer, issued from the surface unit of layer dx becomes

$$dI = \frac{dI''}{\sin\theta''} \exp(-\chi''\rho x) \quad (14.6)$$

where μ'' is the mass absorption coefficient and $\chi'' = \mu''\sin\theta''$ the secondary absorption factor of the specimen for A-$K\alpha$.

Secondary Intensity Measured at the Spectrometer. The intensity measured at the specimen, issued from the surface unit of thickness x results from integrating (14.6)

$$I = K_I \frac{1 - \exp(-\chi\rho x)}{\chi} \quad (14.7)$$

where $\chi = \chi' + \chi''$ is the total absorption factor and K_I the global factor characterizing element A and instrumental conditions. This relation may be compared to the ideal linear relation (14.1).

Result: Mode of Analysis. On examination of the graph representing the measured intensity vs the specimen thickness, two interesting regions are apparent (Figure 14.4): a virtually linear region for a very thin specimen and an asymptotic region for a sufficiently thick specimen.

Thin Specimen. For a small value of $\chi\rho x$, Eq. (14.7) leads to

$$\chi\rho x \ll 1 \Rightarrow \exp(-\chi\rho x) \cong (1-\chi\rho x) \Rightarrow I = K_I c\rho x \quad (14.8)$$

The intensity is proportional to the mass per surface unit of element A. This corresponds to linear condition (14.1). Comparing intensity I from the specimen with intensity $I(A)$ measured in identical conditions from a layer of pure element A, of known mass per surface unit ρt, leads to the mass fraction $c = I/I(A)$.

Experiments have shown that the thin layer approximation is justified for a specimen of thickness t such as

$$\chi \rho t < 0.1 \tag{14.9}$$

As an example, this method can serve to analyse a vacuum-evaporated layer or a fine-grained layer of material deposited on a filtering device. Conversely, the knowledge of $c\rho\chi$ results in a method for determining the thickness of a layer by means of an intensity measurement.

Semi-infinite Specimen. For an infinite thickness, Eq. (14.7) leads to

$$x \to \infty; \quad \exp(-\chi \rho t) \to 0 \Rightarrow I(\infty) = K_I \frac{c}{\chi} \tag{14.10}$$

$$\frac{I}{I(\infty)} = 1 - \exp(-\chi \rho t) \tag{14.11}$$

This leads to the definition of an *efficient thickness* t_e such as $I(t_e)/I(\infty) = 0.999$. Its value is approximated by the following relation:

$$t_e \cong \frac{6.9}{\chi \rho} \tag{14.12}$$

Above this efficient thickness, a specimen can be considered to have an infinite thickness (*semi-infinite specimen*). A further increase in thickness does not result in an intensity increase. A surface layer of thickness t_e only participates in the measured intensity.

Example. Aluminium (Figure 14.4).

Al-$K\alpha$ excited by the primary radiation Cr-$K\alpha$.
Mass absorption coefficients: $\mu' = 129 \text{ cm}^2/\text{g}$; $\mu'' = 324 \text{ cm}^2/\text{g}$.
Absorption factor for $\theta' = 90°$, $\theta'' = 30°$: $\chi \cong 777 \text{ cm}^2/\text{g}$.
Thin layer condition: $\chi \rho t < 0.47 \, \mu\text{m}$.
Efficient thickness: $t_e \cong 33 \, \mu\text{m}$.

Current X-ray fluorescence analysis is most frequently carried out within the semi-infinite specimen conditions. Comparing with a semi-infinite test sample of absorption factor $\chi(T)$, containing element A with a mass function $c(T)$, leads to the following intensity ratio:

$$\frac{I}{I(T)} = \frac{c}{c(T)} \frac{1/\chi}{1/\chi(T)} \tag{14.13}$$

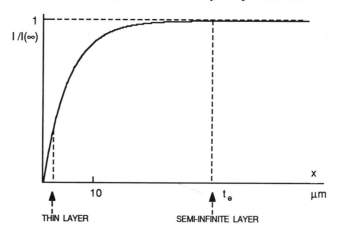

Figure 14.4. Graph representing the function $I(x)/I(\infty)$ according to Eq. (14.11). Aluminium specimen. Al-Kα excited by Cr-Kα.

The absorption effect is represented by factor $1/\chi$ which expresses the deviation from the linear law (14.2), i.e. the deviation from the *apparent mass fraction* given by $I/I(T)$. It can be noticed that when the matrix is more absorbing than the pure element A, the actual concentration is greater than the apparent concentration; the effect is reversed when the matrix is less absorbing than the element to be analysed.

Strictly speaking, the above formulation applies to a primary monochromatic radiation, which is not the case in current X-ray fluorescence set-ups, except with secondary radiators. The reasoning remains approximately valid even for background excitation, because only a narrow range of radiation, just above the excitation level W_K, is really efficient for producing the measured Kα emission. However, it must be taken into account that this filtering effect modifies the energy distribution of the incident beam; its mean energy increases with increasing penetration, the hard X-ray radiations being less absorbed than the soft ones. This causes an accurate absorption correction calculation to be difficult in the general case. The use of test samples with a composition as close as possible to the specimen composition minimises the errors.

14.3.1.2 Secondary Fluorescence Effect

The fluorescence enhancement effect becomes significant only when the specimen contains at least one element B of atomic number just greater than A, thus forming an active pair. The effect becomes important for heavy elements (high fluorescent yield). As an example, analysing Fe in a specimen containing Cu, Ni, Co, results in a substantial reinforcement, the K-radiations of those elements being able to strongly excite the K-level of Fe. The effect is negligible for light element analysis. A preliminary qualitative analysis permits assessment of this possibility.

Let f be the *enhancement rate* of radiation A-Kα (i.e. ratio of the secondary fluorescence intensity to the primary fluorescence intensity). The intensity I'' has therefore to be replaced by $(I'' + fI'') = I''(1 + f)$; the measured intensity is multiplied by the same factor and relations (14.10) and (14.13) are to be replaced by

Elemental Analysis by X-ray Fluorescence

$$I = K_I c \frac{1+f}{\chi} \qquad (14.14)$$

$$\frac{I}{I(T)} = \frac{c}{c(T)} \frac{1/\chi}{1/\chi(T)} \frac{1+f}{1+f(T)} \qquad (14.15)$$

where $f(T)$ is the enhancement rate of radiation A–$K\alpha$ in the test sample. The complete theoretical treatment leading to f is difficult.

14.3.1.3 Grain Effect

Up to now the specimen has been assumed to be homogeneous. The result of measurements on polycrystalline materials shows that intensity increases generally, except for very absorbent materials, when the mean grain size decreases, i.e. with increasing homogeneity. A specimen consisting of well-defined particles may be considered to be homogeneous, with respect to a radiation, when a large number of particles participates to absorb an X-ray photon, thus resulting in an averaging effect. The grain effect is the more significant the more different the composition of the particles, i.e. the more different the individual absorption coefficient μ of a grain with respect to the mean absorption coefficient μ_m of the specimen. A specimen may be approximated to a homogeneous material when the mean diameter d_m of the grains meets the condition

$$d_m |\mu - \mu_m| \ll 1 \qquad (14.16)$$

Due to μ proportional to λ^3, the grain effect is larger for light element analysis than for heavy element analysis.

Minimising the grain effect requires adequate preparation techniques for the specimen as well as for the test sample.

14.3.1.4 Total Correction

In the presence of matrix effects, the mass fraction as a function of the measured intensity can finally be written as follows, according to Eq. (14.15):

$$\frac{c}{c(T)} = \frac{I}{I(T)} \frac{\chi}{\chi(T)} \frac{1/(1+f)}{1/[1+f(T)]}; \quad K_a = \frac{\chi}{\chi(T)}; \quad K_f = \frac{1/(1+f)}{1/[1+f(T)]} \qquad (14.17)$$

This relation replaces the linear relation (14.2).

14.3.2 PRACTICAL METHODS OF QUANTITATIVE ANALYSIS

The practical methods result from the above theoretical considerations.

14.3.2.1 Intensity Measurements

The intensity is measured in the form of a pulse count during a given time, on the chosen characteristic line and on the background, for the specimen as well as for the test sample.

The background correction is relatively low in X-ray fluorescence and can generally be neglected for major heavy elements.

If the element to be analysed is in a different chemical state in the specimen than in the test sample, the respective characteristic wavelengths may be slightly shifted. For accurate measurements, it is therefore of prime importance to reset the spectrometer on the maximum intensity of the line to be measured.

The ratio of respective intensities on the specimen and on the test sample provides the *apparent mass fraction*, after a dead time correction, if necessary (Eq. 10.15).

14.3.2.2 Correction Procedures

In the present section the specimen and the test sample are assumed to be homogeneous. Only the absorption effect and the secondary fluorescence effect are therefore to be corrected. The correction factors can be calculated according to the foregoing theoretical development. Until recently, due to the uncertainty on absorption coefficients and other required physical quantities, empirical correction methods have generally been preferred to mathematical correction. They can be divided into three main categories, respectively based on linear approximation, matrix corrections and internal standards.

In recent years, direct mathematical correction methods have seen new developments, in particular for analysing compact materials for which valuable standards are difficult to prepare.

Linear Approximation

Minimizing the matrix effect permits use of the linear relation (14.2).

Thin Specimen. The thickness limit given by Eq. (14.9), allowing application of the thin-layer approximation (Eq. 14.8), increases when $\chi\rho$ decreases. The specimen thickness may therefore be the greater the lighter the matrix elements (ρ small) and the heavier the elements to be analysed (χ small). This technique may be used for analysing thin vacuum evaporated, sublimated or ion-sputtered layers, of solutions or of powdered material deposited on filters or on organic films. Pure element test samples can be used. The mass per surface unit ρt of the specimen, as well as $\rho t(T)$ of the test sample, is measured by weighting. The grain effect must be taken into account for powder specimen. Due to the small amount of material leading to a low count rate, the method is generally limited to analysing major elements.

Constant Matrix. The linear relation can be applied for analysing small concentrations of elements in specimens whose matrix may then be considered as a constant. The test samples are prepared with a matrix having a similar absorption factor. Linear calibration graphs can be drawn. This technique is useful for analysing trace elements in a series of similar materials. A typical example is the assessment of the Rb/Sr ratio in rocks for geochronological applications; thanks to the neighbouring elements, the intensity ratio is virtually independent of the matrix effect and directly provides the concentration ratio.

Dilution Methods. For a general specimen the constant matrix condition can be approached by diluting the specimen down to a determined concentration in a weakly absorbing matrix. The diluent may be a liquid (e.g. dissolution in water), but is mostly a solid (*dissolution in lithium borate glass*, see preparation methods). Matrix variations are thus attenuated, enabling us to apply the linear relation. Furthermore the grain effect is eliminated by dissolution.

A variant consists in adding to the specimen a given amount of heavy elements (e.g. barium or lanthanum carbonate or oxide), thus forming an *absorption buffer* which minimises the variation of the absorption factor against any variation of the specimen concentration. However, the high absorption decreases the count rate of the elements to be analysed and limits the method to major elements. In addition, the introduced heavy elements may induce strong secondary fluorescence and lead to line overlappings.

In spite of the ensuing sensitivity impairment, the dilution technique is widely used for analysing major elements in various materials, thanks to its simplicity, to its rapidity and to the elimination of any grain effect.

For materials which are in the form of a powder (clays, cements, sublimates, etc.), a simple *powder dilution* may lead to good results. A convenient diluent is boric acid which, in addition, acts as a binder enabling pressing of the specimen into a pellet.

Test Samples Closely Related to the Specimen. When the test sample is similar to the specimen, the matrix effects are minimised and the linear relation applies. A valuable method consists of using the apparent mass fractions proportional to the count rates measured on the specimen (Eq. 14.1) for preparing a test sample. The ratio of intensities from the specimen and from the test sample then provides the actual mass fractions by means of Eq. (14.2).

Matrix Correction

The non-linearity of the intensity vs concentration function is taken into account.

Experimental Calibration Curves. The calibration curves $I(c)$ are drawn by means of a number of standard samples, with matrix effects similar to those of the specimen, for a varying mass fraction of the elements to be analysed. This method is limited in practice to analysing one element, or a pair of elements in related concentration, in a series of specimens, the other elements being approximately constant (e.g. analysis of Se in pyrite FeS_2 or in sulphur; analysis of Fe and Mg in olivine minerals $SiO_4(Mg,Fe)$; Rb and Cs in silicates).

Calculated Calibration Curves. The theoretical $I(c)$ curves related to the matrix effects of a system can be calculated as a function of the composition. A system of n equations has to be set up and solved for analysing a compound with n elements. This method is justified for repetitive analysis of similar materials.

Determining the Absorption Factors

1. By iteration. The apparent concentrations of major elements based on intensity measurements are used to determine the absorption factors which in turn are used to correct the concentrations. The correction factors can be refined by further iteration.

2. Based on major elements. When knowing the major element mass fractions (e.g. those determined by a dilution method), the absorption factor of the specimen can be approximated and then used for correcting the intensities given by trace elements.
3. Direct measurement. The absorption factors are measured by means of a specific device. Corrections are based on Eq. (14.13).

The above correction methods neglect the secondary fluorescence effect. Its possible interference can be predicted through a preliminary qualitative analysis.

Internal Standard Methods

Internal standards are currently used in optical spectroscopy. The intensity of a characteristic line of a given element A is measured vs the intensity of a line S taken as standard. A standard line different from the line to be measured is generally used. However, in the addition method, which may be included in this section, the line to be measured itself is taken as the standard. The secondary fluorescence effect, as well as the absorption effect, is accounted for in this method.

Standard Element Different from the Element to be Analysed. For an internal standard to be valid, its radiation must be excited in identical conditions and affected by the same matrix effects as the element to be analysed, whatever the specimen composition. This condition can only be met approximately by choosing a standard with a wavelength as close as possible to that of the element to be analysed.

Consider the standard to be an element S naturally present in the specimen in a known mass fraction (or added in a given mass fraction). If it meets the above conditions, the ratio of the mass fractions of the element to be analysed and the standard element S is equal to the ratio of the respective intensities, as expressed by

$$\frac{c}{c(S)} = \frac{I}{I(S)} \qquad (14.18)$$

A convenient solution is to use a standard element with an atomic number close to that of the element to be analysed. Emission lines of the same series, i.e. with similar absorption factors, are then compared (e.g. analysis of Rb with Sr as standard; analysis of Cs with I as standard). The standard line can come from a different series, e.g. the L-line of a heavy element close to the K-line of a light element to be analysed. The different emission characteristics must then be taken into account.

The construction of calibration curves is advisable for higher accuracy, in order to correct the remaining matrix differences between the element to be analysed and the standard.

Relation (14.18) fails to apply if the specimen contains a major element B with an absorption edge or a characteristic line located between the two wavelengths to be measured from A and from S. The absorption factor is then significantly different for the two radiations (Figure 14.5).

In order to avoid any additional error due to a grain effect, the distribution of the internal standard element in the specimen must be perfectly homogeneous.

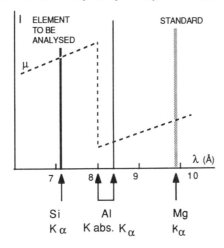

Figure 14.5. The internal standard Mg is incorrect for analysing Si in a specimen containing Al as a major element. Al-Kα between Si-Kα and Mg-Kα excites the secondary K-fluorescence of Mg and does not excite that of Si. The absorption edge K-Al is between the lines Si-Kα and Mg-Kα; the first is therefore much more absorbed than the second. The intensity scale is arbitrary.

Scattered Radiation as Standard. The role of standard can be played by a radiation issued from the primary source and scattered by the specimen (elastically scattered emission line or Compton line), provided that its wavelength is close to that of the element to be analysed and that the above-mentioned conditions are met. Consider I_S and $I_S(T)$, the intensities of the standard line, scattered respectively by the specimen and by a test sample T; the ratio of intensities $I_S/I_S(T)$ can then be considered to give the ratio of the respective absorption factors $\chi(T)/\chi$ of the test sample and of the specimen, according to Eq. (14.10). In the absence of any secondary fluorescence effect, the concentration ratio of Eq. (14.17) approximates to

$$\frac{c}{c(T)} = \frac{I/I_S}{I(T)/I_S(T)} \qquad (14.19)$$

The elastically scattered Lα-line ($\lambda = 1.47$ Å) of a tungsten anode can thus be used as a standard for analysing Cu (1.54 Å) or Zn (1.43 Å) by means of the Kα-lines. The Compton peak of the same anode can be used as a standard for analysing the element tungsten itself by its Kα radiation.

The internal standard method is limited to the analysis of one element or two neighbouring elements. It is suitable for trace elements. One of its advantages is to compensate a possible instrumental drift.

Addition Method. In this technique the element to be analysed itself is used as internal standard. When adding a known mass fraction c' of the element to be analysed itself to the specimen, the measured line intensity increases. Assuming that this addition does not change the absorption factor, the ratio of intensities provides the ratio of mass fractions, thus leading to the determination of c according to the following relation:

$$\frac{I(c)}{I(c+c')} \cong \frac{c}{c+c'} \qquad (14.20)$$

This method could as well be filed under linear approximation. It is valid provided the addition c' is sufficiently small in order not to alter significantly the absorption factor; however, for a reasonable accuracy, the addition c' must be of the same order as the mass fraction c to be measured. These somewhat conflicting conditions limit the addition method to analysis of one minor element per specimen.

Mathematical Correction Methods

The foregoing more or less empirical methods are reasonably adapted to specimens which may be prepared by solid solution (fusion) or by powder mixing. They are less convenient for non-destructive analysis of compact materials, such as alloys, which consist frequently of a great number of various elements leading to strong matrix effects. An entirely mathematical method has been proposed by Sherman (1955), based on determined correction factors; it is valid for limited composition ranges.

In order to analyse n elements a correction algorithm should consist of a system of n equations; according to Lachance and Trail (1966), its general form for absorption correction with a monochromatic primary radiation is given by

$$c_J = \frac{I}{I_0}\left(1 + \sum_{L \neq J} \alpha_{JL} c_L\right) \quad \text{with} \quad \sum_J c_J = 1 \qquad (14.21)$$

where c_J is the mass fraction of the element to be analysed, c_L the mass fraction of the matrix elements $L \neq J$ and α_{JL} the interelemental absorption influence coefficients.

In order to account for interelemental secondary fluorescence effects, Tertian and Claisse (in [6]) have proposed the following algorithm with the corresponding influence coefficients β_{JL}:

$$c_J = \frac{I}{I_0}\left(1 + \sum_{L \neq J} \alpha_{JL} c_L + \sum_{L \neq J} \beta_{JL} c_J c_L\right) \qquad (14.22)$$

More complicated relations are needed to account for a polychromatic primary radiation (Claisse and Quentin, 1967).

The first mathematical methods have made use of interelement coefficients calculated by regression or measured on binary test samples. The *fundamental coefficient* method developed by Broll and Tertian (1983), Broll (1985, 1986), deals in a fundamental manner with the influence coefficients. It is based on the general theory of X-ray fluorescence emission and radiation–matter interaction for the actual composition of the specimen and for the spectrum of primary X-rays actually used. Calculations are performed through iteration. This method needs only one test specimen and provides good results for analysing alloys and other multi-element materials with important matrix effects.

Trace Analysis

Operating Conditions. Analysing trace elements requires optimum conditions for the detection limit to be as low as possible:

1. Anode and excitation voltage chosen for the highest signal-to-noise ratio (e.g. secondary radiator of element $Z+1$ for analysing element Z).

2. Specimen without any dilution or addition which would be liable to decrease the count rate or introduce the trace element to be analysed in the form of impurities.
3. Absence of line overlappings to be checked; overlappings are particularly critical when dealing with weak lines.

Convenient Methods

1. Linear approximation with constant matrix;
2. Calculation of the absorption factors for the major element composition and correction for trace elements;
3. Internal standard methods;
4. Addition method.

14.3.2.3 Accuracy of Analysis

Causes of Error. The following main factors limit the accuracy of quantitative analysis (the means of minimising the errors are given in parentheses):

1. *Instrumental errors.* Drift of the characteristics of the X-ray tube (compensation by making alternate measurements on the specimen and on the test sample or by using internal standards); varying interplanar spacing of the analysing crystal and of the detector gain as a function of temperature (thermostating the spectrometer).
2. Grain size effect of the specimen (liquid or solid solution).
3. Chemical shift different in the specimen and in the test sample (adjustment of the spectrometer on the line maximum); statistical errors of measurement (choice of the best compromise between count time and instrumental drift).
4. Correction errors. These are generally predominant in determining the accuracy of analysis (use of accurate values of correction parameters; choice of the test sample and of the best suited correction method).

Currently Observed Accuracy. The global accuracy of X-ray fluorescence is currently of the order of 1%. It can be better than 0.1% for analysing major elements with rigorous correction methods.

14.4 SPECIMEN PREPARATION

The current diameter of an X-ray fluorescence specimen is of a few centimetres. In the most common semi-infinite specimen analysing mode, its thickness must be greater than the efficient thickness (Eq. 14.12); the required mass is of the order of 1 g. In the thin specimen mode the thickness must meet the condition for thin-layer approximation (Eq. 14.9) with a required mass which can be as low as a few milligrams. Quantitative analysis requires mainly one plane face and an adequate homogeneity; those requirements are the more strict the lighter the elements to be analysed.

Specimen to be Analysed. Solid specimens. Solid specimens are simply to be cut and polished on one face. Polishing should not introduce impurities which could interfere with the elements to be analysed. Diamond polishing is the best method. The grain effect must be taken into account.

Powder specimens. Powder specimens can be used directly, provided the homogeneity condition (14.16) is being met by means of adequate grinding. In the most common X-ray fluorescence facilities where the sample holder is above the X-ray tube (Figure 14.1), the powder is contained in a cylindrical sample holder with a thin organic film (mylar) as the base.

Specimen handling, homogeneity and surface planeness are improved by pressing the powder into a pellet, with or without a binder (light element material such as boric acid, starch, cellulose, graphite, wax). Adding a binder causes the specimen elements to be diluted and must be carefully gauged; it may result in decreasing the signal-to-noise ratio and in introducing foreign elements interfering with the emission lines to be measured; it is to be avoided for trace and light element analysis.

Solid solution. The best-suited preparation method for analysing a large variety of materials is to melt down the specimen with lithium tetraborate ($Li_2B_4O_7$) as a flux. The result of this fusion technique is an easy-to-handle solid specimen of *lithium borate glass* with a perfect homogeneity (Claisse 1956; Rose et al, 1962, 1963, 1965; Welday et al, 1964). The matrix effect is decreased and the grain effect is eliminated; on the other hand the sensitivity is being limited and foreign elements are being introduced. The fusion technique is therefore not well suited for trace and light element analysis.

Test Samples. Test samples can be natural or synthetic materials. They must meet the same requirements as the specimens, in particular concerning their homogeneity. The most common preparation technique is glass fusion. Addition of internal standards is operated in a similar way.

14.5 APPLICATIONS TO MATERIAL ANALYSIS

The main qualities of X-ray fluorescence are its simplicity, its rapidity, its sensitivity, its accuracy and its non-destructiveness (except for dissolution). It has developed into an important technique in science and in industry for elemental analysis of a great variety of materials, crystallised, amorphous or even liquid.

In the form of multichannel spectrometers (up to 28 spectrometers preset on the same number of elements to be analysed), it is widely used as a means of analysing and controlling industrial materials, e.g. metal sheets.

In the form of mobile facilities (with a radioactive source), it can serve for remote-controlled analysis in space probes.

Thanks to recently developed algorithms and powerful calculation means, the mathematical correction modes provide increasingly accurate quantitative analysis for any multi-element materials.

REFERENCE CHAPTERS

Chapters 7, 9, 10, 13. Appendix D.

BIBLIOGRAPHY

[1] Adler I. *X-Ray emission spectrography in geology*. Elsevier, New York, 1966.
[2] Bertin E. P. *Principles and practice of X-ray spectrometric analysis*. Plenum Press, New York, 1975.
[3] Bunshah R. F. (ed.) *Modern analytical techniques for metals and alloys*. John Wiley, New York, 1970.
[4] Jenkins R. *X-ray fluorescence spectrometry*. John Wiley, New York, Chichester, 1989.
[5] Liebhafsky H. A., Pfeiffer H. G., Winslow E. H., Zemany P. D. *X-ray absorption and emission in analytical chemistry*. John Wiley, New York, 1960.
[6] Tertian R., Claisse F. *Principles of quantitative X-ray fluorescence analysis*. John Wiley, New York, Chichester.
[7] Whiston C. *X-ray methods*. John Wiley, New York, Chichester, 1987.

15

Electron Probe Microanalysis

An electron microprobe may be considered as a microfocus X-ray tube in which the pure metal anode has been replaced by a specimen to be analysed. The characteristic X-ray spectrum induced by an electron beam is analysed by means of a spectrometer. The basic principles of this technique have been known since the work of Moseley on X-ray emission. Its quantitative theory and its applications have been worked out by Castaing (1951).

15.1 BASIC PRINCIPLES AND INSTRUMENTATION

Electron probe microanalysis is based on the spectrometry of the characteristic X-rays emitted by the elements of a specimen under the effect of an incident electron beam. Unlike in X-ray fluorescence, the primary beam can be focused on the specimen, thus resulting in a micron-sized probe (*microprobe*) and in microanalysis. In recent facilities associated with electron microscopes, a nanometre-sized probe (*nanoprobe*) can be produced, leading to nanoanalysis.

As well as in X-ray fluorescence, the emission lines are approximately independent from the chemical state, resulting in an *elemental microanalysis*. Qualitative and quantitative analysis can be performed.

The highest accuracy in quantitative analysis on bulk specimens is achieved by means of *dedicated electron microprobes*.

Electron probe microanalysis is increasingly associated with a *scanning electron microscope*, but with lower performances in quantitative analysis. As an attachment to a *transmission electron microscope*, it makes possible analysis of thin specimens at a nanometre scale.

Common abbreviations: EPMA (electron probe microanalysis); SEM (scanning electron microscope); TEM (transmission electron microscope); STEM (scanning transmission electron microscope); EDS or EDX (energy-dispersive X-ray spectrometer); WDS or WDX (wavelength-dispersive X-ray spectrometer).

The basic constituents of any X-ray microanalytical facility are the probe-generating system, the X-ray spectrometer-detector system, the data acquisition and processing system. Accurate positioning of the specimen is performed by means of a goniometric stage (Figure 15.1).

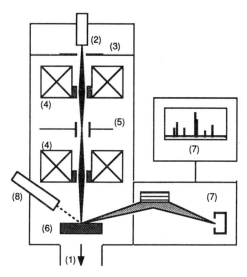

Figure 15.1. Basic diagram of an electron microprobe. The constituents are similar to those of an SEM, with a specificity for optimum X-ray microanalysis. (1) Vacuum enclosure with its pumping system; (2) electron gun, most frequently of the thermionic type and at a negative voltage of 10–50 kV; (3) grounded anode; (4) focusing lenses; (5) beam-scanning device; (6) specimen mounted on a goniometric stage; (7) spectrometer and data acquisition system; (8) light microscope for locating the probe impact on the specimen which appears bright through photoemission. A polarising microscope is useful for petrological thin sections.

15.1.1 ELECTRON PROBE

The probe size on the specimen depends on the electron source and on the electron optical focusing system (see section 9.2). The practical limit is set by the analytical conditions:

1. The primary intensity must be high enough to ensure a sufficient count rate for a good detection limit and accuracy.
2. The electron flux is limited by the specimen resistance to electron bombardment.

Quantitative analysis on a dedicated microprobe is currently performed with a primary intensity of 0.1 μA into a probe diameter of 1 μm. In the microanalytical mode with an SEM, the intensity is generally smaller, of the order of 0.1–5 nA, into a similar probe size. A TEM with a field emission source and a sophisticated condenser–objective focusing system (Figure 9.9) allows analysis of a nanometre-sized area of a thin specimen.

In any case the active specimen area is determined by scattering of incident electrons in the specimen, as well as by the probe diameter.

15.1.2 SPECTROMETER

Thanks to the almost point X-ray source, a *focusing wavelength-dispersive X-ray spectrometer* can be used (see Figure 10.7). Such a spectrometer with high efficiency and resolution is commonly used in dedicated microprobes; these are generally fitted with three or four spectrometers, each being set on a specific energy range for simultaneous measurement on different element lines.

Microanalytical attachments on electron microscopes are commonly fitted with an *energy-dispersive X-ray spectrometer* based on an Si(Li) diode. Such spectrometers are better adapted to the geometry of electron microscopes and easier to operate than wavelength-dispersive spectrometers. Reasonably low detection limits on thin specimens can be reached with optimum set-ups. Ultra-thin or retractable (in high vacuum) detector windows improve the efficiency for light elements.

Common abbreviations: EDX or EDS (energy-dispersive X-ray spectrometer); WDX or WDS (wavelength dispersive X-ray spectrometer).

15.1.3 OPERATING MODES

The electron microprobe may be operated according to two operating modes: stationary probe or scanning probe.

15.1.3.1 Stationary Probe

In qualitative analysis the whole accessible element spectrum is explored by the spectrometer. A WDS requires a somewhat time-consuming angular scanning, whereas an EDS provides an almost instantaneous display of the whole range.

In quantitative analysis of a given element A, the spectrometer is to be set on the characteristic emission line to be measured, e.g. $K\alpha$. The concentration of the element is determined by the count rate after adequate correction.

15.1.3.2 Scanning Probe

A system of deflection coils causes the probe to scan the specimen along a line (one-dimensional scanning) or along a line raster (two-dimensional scanning). With the specimen set on the characteristic line of an element A, the measured intensity determines the distribution of the element along the scanning path.

As a complement to the scanning system, various detectors (for secondary and backscattered electrons and for light photons) are usually attached to the microprobe facility for imaging purposes (see Ch. 20).

The measuring system is connected to a computer which carries out data acquisition and processing. In addition to an automatic operation control of the wavelength-dispersive spectrometers of a microprobe, specific computer software has been developed for various operations:

1. Elemental identification by comparing the experimental spectrum with the theoretical computer-stored emission lines (energies and intensities) of the elements;
2. Deconvolution of overlapping peaks;
3. Peak intensity integration and background subtraction;
4. Matrix corrections in quantitative analysis.

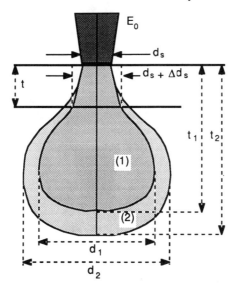

Figure 15.2. Outline of emission volumes of characteristic X-rays from a K-level of an element. (1) Primary K-emission volume ($E > W_K$); (2) primary electron propagation volume ($E > 0$). (Arbitrary forms and volumes, outlined in the case of rather light elements, excited with a high overvoltage ratio.)

15.2 QUALITATIVE MICROANALYSIS

Identifying the spectrum lines leads to determination of the elements in the irradiated area of the specimen, within the detection limits.

15.2.1 SPATIAL RESOLUTION

The spatial resolution of electron probe microanalysis is determined by the emission volume of the X-ray photons to be measured (Figure 15.2).

15.2.1.1 Emission Volume

The emission volume is related to the probe cross-section; however, it is mainly determined by scattering of the primary electrons in the specimen. The resulting spread increases with increasing penetration depth (see Figure 8.3b).

Primary Emission Volume. In order to excite the level X of an element A, the characteristic X-ray photons are emitted within a volume of the specimen where the mean primary electron energy is still greater than the excitation energy W_X. For a probe cross-section d_s, this primary emission volume has a diameter $d_1 > d_s$ and a depth t_1; it is a cube of the order of 1 μm.

Electron Propagation Volume. This volume is the envelope of the zero-energy ends of the primary electron paths, after complete energy loss. It has a diameter $d_2 > d_1$ and a depth $t_2 > t_1$.

Fluorescence Emission Volume. The characteristic X-rays to be measured may likewise result from fluorescence emission induced by primary emitted X-rays from matrix elements heavier than the element to be analysed. Owing to the large X-ray penetration, this fluorescence emission takes place inside a volume of diameter $d_3 > d_2$ and of depth $t_3 > t_2$.

The respective sizes of those scattering volumes depend on various parameters: incident energy, energy of the X-rays to be measured, specimen composition, specimen mean absorption coefficient. The following approximate relation has been established by Castaing (1960) for assessing the excitation penetration depth t_1 related to a level X:

$$t_1 = 0.033 \, (E_0^{1.7} - W_X^{1.7}) \frac{A}{\rho Z} = 0.033 \, W_X^{1.7} (U_X^{1.7} - 1) \frac{A}{\rho Z} \quad (15.1)$$

where $U_X = E_0/W_X$ is the overvoltage ratio, and t_1 and W_X respectively are measured in μm and keV (compound specimens: A/Z given by additivity relations).

15.2.1.2 Emission Surface

The significant quantity for determining the surface distribution of an element is the emission surface of the related characteristic X-rays. It may be assimilated in the diameter d_1 of the primary emission volume.

Thick Specimen

Conventional electron probe microanalysis is mainly performed on a bulk specimen, largely thicker than the depth of the emission volume. The diameter of the primary emission volume may be approximtaed to

$$d_1 \cong d_s + t_1 \quad (15.2)$$

It is of the order of 1 μm in current conditions.

According to Eqs (15.1) and (15.2), space resolution could be improved by optimising various parameters.

Probe Size. Equation (15.2) shows that reducing the probe cross-section d_s of the probe would not significantly improve resolution. On the other hand, the probe intensity is expressed as follows:

$$I_0 = F(B) E_0 d_s^{8/3} \quad (15.3)$$

Factor $F(B)$ is proportional to the brightness B of the source. Reducing d_s while keeping the same brightness would therefore reduce the emission intensity, hence the count rate. Increasing the source brightness (e.g. field emission source) is limited in practice by radiation damage.

Overvoltage Ratio. Reducing the overvoltage ratio $U_X = E_0/W_X$ would generally result in worsening the excitation conditions and hence in decreasing the emission intensity.

Energy. The primary energy E_0 could be reduced, while maintaining an optimal overvoltage ratio, by making use of L or M level lines with smaller excitation energies. However, the counterpart would be an increase of matrix correction errors, due to less accurate correction parameters.

In practice it is convenient to use K-lines up to $Z \leqslant 35$ and L-lines beyond.

Spatial resolution is involved whenever an element has to be located at a micrometre scale or less (e.g. grain limit diffusion, analysis of precipitates, of inclusions, of phase separated solutions). When additional fluorescence interferes, the size of the particles to be analysed should be several times the primary excitation volume.

Thin Specimen

Thin specimen analysis is mainly performed in a STEM. For a thickness t smaller than the penetration depth, the probe spread is less important than with a bulk specimen (Figure 15.2). The probe widening vs t, through elastic scattering in a pure element, has been established as follows by Hren et al (in [5]):

$$\Delta d_s = 6.25 \times 10^5 \frac{Z}{E_0} \left(\frac{\rho}{A}\right)^{1/2} t^{3/2} \qquad (15.4)$$

where Z, A and ρ are, respectively, atomic number (cm), atomic mass (g) and density (eV).

Examples for $E_0 = 100$ keV; $d_s = 200$ Å; $t = 500$ Å
Aluminium: $\Delta d_s = 30$ Å $\Rightarrow d_1 = 230$ Å
Copper: $\Delta d_s = 80$ Å $\Rightarrow d_1 = 280$ Å

These values are in keeping with those calculated by Monte-Carlo modelling. However, they seem to be overestimated for very thin specimens. In the case of aluminium, the analysing area would thus be hardly larger than the probe cross-section. The analysing volume would be of the order of 2×10^{-4} μm^3, which is some 10 000 times smaller than the corresponding analysing volume in conventional electron probe microanalysis on bulk specimens. Fine-focused probes from field emission sources permit even smaller analysing volumes. This result highlights the problems of detection limits in thin-layer X-ray microanalysis.

15.2.2 STATIONARY PROBE OPERATION

Stationary probe operation provides optimum analysing conditions.

15.2.2.1 Detection Limits

The *minimum detectable mass fraction* of an element is determined by the count rate, according to Eq. (13.9):

$$c_{min} = \frac{3}{(t_c)^{1/2}(n_1)^{1/2}(n_1/n_0)^{1/2}} \qquad (15.5)$$

Table 15.1. Optimum primary energy for analysing some elements by means of K-lines in a given specimen (from [10], Obst et al)

Element to be analysed	W_K (keV)	Specimen	Optimum E_0 (keV)
C	0.284	Fe_3C	7.5
C	0.284	Graphite	17.5
N	0.399	BN	10
O	0.532	SiO_2	15
Na	1.072	NaCl	25
Fe	7.114	Fe	17.5
Fe	7.114	Fe_3O_4	13.8
Cu	8.979	Cu	25

where t_c is the count time, and n_1 and n_0 respectively the count rates on the characteristic line of the pure element and on the background.

The higher the count rate of an element the lower the detection limit, i.e. the greater the emission intensity. In order to optimise the detection limits, it is therefore important to set the operating conditions for a maximum emission intensity and a minimum background.

The signal-to-noise ratio is maximum for an optimal overvoltage ratio corresponding to the energy level to be excited. The optimal value is of the order of $U = 3$ for a pure medium element and the K-level. It decreases with increasing absorption of the matrix (Table 15.1).

For analysing heavy elements, the optimal K-excitation energy becomes too high, and the L-lines should be used.

Detection characteristics depend on numerous instrumental factors, on the type of spectrometer and on the type of microprobe facility (dedicated microprobe or microanalysis attachment in an electron microscope).

Identification uncertainties are similar to those in X-ray fluorescence.

15.2.2.2 Thick Specimen Analysis by Microprobe

The emission spectrum is analysed by means of a wavelength-dispersive spectrometer. The probe intensity is of the order of $1\,\mu A$ into a diameter of about $1\,\mu m$.

Current Detection Limits

1. Detected elements: $Z \geqslant 5\text{-B}$;
2. Minimum detectable mass fraction: $c_{min} = 10^{-4} - 10^{-3}$;
3. Minimum detectable mass: $M_{min} = 10^{-15} - 10^{-13}$ g;
4. Minimum number of detected atoms: $N_{min} = 10^7 - 10^9$ (Table D.1).

15.2.2.3 Thick Specimen Analysis by Scanning Electron Microscope

Analyses are usually performed by means of an EDS, with a mean probe intensity of $10^{-10} - 5 \times 10^{-9}$ A in a diameter of $0.1\,\mu m$. The count rate is smaller than with a dedicated microprobe; however, similar detection limits can be achieved when optimising the operating conditions of the Si(Li) diode. Elements from 11-Na upwards are detectable with a reasonable efficiency when working with a conventional beryllium detector window; a retractable window makes it possible to go down to 6-C; ultrathin windows give intermediate results, depending on their thickness.

15.2.2.4 Thin Specimen Analysis

Thin specimen X-ray microanalysis is most commonly carried out with a *transmission electron microscope*, usually fitted with beam scanning. *Scanning transmission electron microscopes*, specifically dedicated to exploit any analytical possibilities of electron beams, provide the best performance. Specimen thickness is currently of the order of some 100 Å. The probe cross-section on the specimen may be as small as a few nanometres, resulting in an analysing volume down to some $10^{-6} \mu m^3$, i.e. 10^6 times smaller than for a conventional microprobe operating on bulk samples. The probe intensity of a TEM is typically of about 0.1 nm, i.e. 10^3 times smaller than in a conventional microprobe.

Thanks to its efficiency which reaches 100% between energies of 2 and 20 keV and to its large detection solid angle when positioned close to the specimen, an EDS based on an Si(Li) diode can achieve an acceptable detection limit. In order to reach the maximum possible signal-to-noise ratio, all instrumental parameters must be optimised for the specific analysing facility. The parasitic emission of X-rays by the various materials surrounding the specimen must be reduced as far as possible, or at least limited to the low-energy range which is not detected by the diode. This is achieved by using light element materials such as beryllium or graphite for the specimen holder, for covering the exit pole pieces and for the detector collimator. A retractable window increases the signal-to-noise ratio in the low-energy range, i.e. for light elements. It requires high vacuum in the specimen–spectrometer enclosure.

Current Detection Limits

1. Detected elements with a Be window: $Z \geqslant 11$-Na; without window: $Z \geqslant 6$-C;
2. Minimum detectable mass fraction: $c_{min} = 5 \times 10^{-3} - 10^{-2}$;
3. Minimum detectable mass: $M_{min} = 10^{-20} - 10^{-18}$ g;
4. Minimum number of detected atoms: $N_{min} = 10^2 - 10^4$ (see Table D.1).

15.2.3 SCANNING MODE

15.2.3.1 Scanning X-ray Microanalysis

With the spectrometer set at a given wavelength of an emission line, the motion of the probe over the specimen results in a display of the concentration variation of the corresponding element. Thus the scanning mode provides data on the element surface distribution.

Linear Scanning

The probe is made to move along a line on the specimen surface. The intensity recording represents a concentration profile of the selected element. The intensity function and the scanning line may be superimposed on the electronic scanning image (Figure 15.3a). Due to the short acquisition time on a specimen point, the count statistic is less favourable than in the stationary mode. If T and L represent respectively the scanning period and the scanning length, d_s the probe diameter, the count time on a point is $t_c = Td_s/L$.

Example: $T(\text{line}) = 10$ s; $d_s = 1$ μm; $L = 100$ μm $\Rightarrow t_c = 0.1$ s.

The minimum detectable mass fraction is proportional to $(t_c)^{-1/2}$ (Eq. 15.5). This example thus results in a detection limit which is about 30 times higher than for a commonly practised count time of 100 s in the stationary mode. It would be illusory to determine in this way the distribution of elements with a mass fraction smaller than 1%.

Plane Scanning. Distribution Maps

In this most common scanning mode the probe is made to scan along a square line raster on the specimen surface. The detector signal monitors the electron beam of a cathode ray tube (CRT) with synchronised scanning. As a result, each detector pulse gives rise to a bright point on the display screen. With the spectrometer set on the characteristic wavelength of a given element, the density of bright pixels on the screen is a function of the element concentration in the corresponding area of the specimen. The ratio of the scan width L' (fixed) of the CRT to the scan width (varying) on the specimen determines a *scan magnification* $G = L'/L$. The number of raster lines and the scanning speed are adjustable. The screen is recorded by means of a camera, the exposure time being a multiple of the scan period. The distribution statistics are improved by cumulating more than one scan period. The result is a *distribution map* of the element on the specimen surface (Figure 15.3).

The concentration limit required to carry out a significant distribution map is obviously even higher than for a concentration profile.

Example: 400 lines; $T(\text{line}) = 1$ s, $T(\text{plane}) = 400$ s; $d_s = 1$ μm $\Rightarrow t_c = 0.01$ s.

As a result, the minimum detectable mass fraction is some 100 times higher than in stationary beam analysis, for commonly practised exposure durations.

A *practical criterion* for recording a significant distribution map is a signal-to-noise ratio of at least five for detecting the given element.

The scanning mode is sometimes used in quantitative analysis for averaging the intensity output from a heterogeneous specimen over an area larger than the probe cross-section. Counting is then integrated over the whole area, resulting in a similar detection limit as in the stationary mode.

The distribution data of several elements may be stored on computer. After assigning a colour to each distribution, they can be displayed on a colour video screen and superposed in order to provide a *multi-element distribution map*. An EDS associated with a multichannel analyser and a computer is best designed for this technique.

15.2.3.2 Miscellaneous Scanning Images

An electron microprobe may in addition provide scanning images, through various electron-induced effects:

(a) emission of secondary electrons and backscattered electrons;

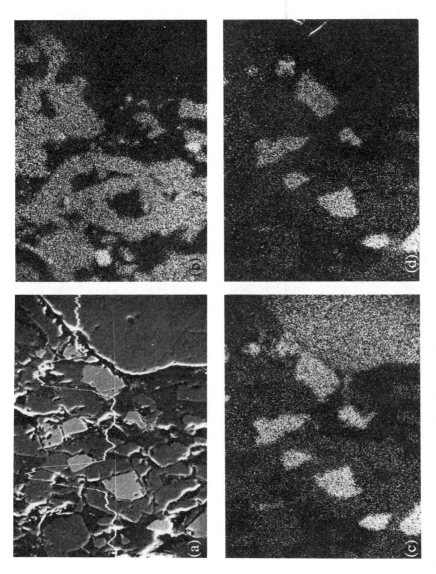

Figure 15.3. Polished surface of a petrological section of ankaratrite, a volcanic rock containing as main minerals: *pyroxene* Ca(Mg, Fe) Si$_2$O$_6$, *titanomagnetite* (Fe, Ti)$_3$O$_4$, *olivine* (Fe, Mg)$_2$SiO$_4$. (a) Secondary electron emission image; superimposed linear scan with intensity-related vertical deflection. The brightest zones correspond to the minerals with highest atomic numbers (see Ch. 20); (b) distribution map by means of Ca-Kα corresponding mainly to pyroxene; (c) distribution map by means of Fe-Kα corresponding to titanomagnetite, to olivine (large particle on the right-hand side) and partly to pyroxene (compare to b); (d) distribution map by means of Ti-Kα corresponding to titanomagnetite and, for a small part, to pyroxene. Dept Electron Microscopy, Lab. Crystallography, Louis Pasteur University, Strasbourg).

314 Structural and Chemical Analysis of Materials

(b) electron absorption;
(c) emission of light photons (cathodoluminescence).

Those imaging modes will be detailed with the SEM (Ch. 20).

15.2.3.3 Kossel Pattern

The divergent characteristic X-rays issued from an almost point sized source inside a single crystal can be diffracted by the lattice planes of the crystal itself, resulting in a so-called Kossel pattern. Such a pattern may be recorded on a film placed before the specimen (back reflection) or after the specimen (transmission). It provides information on the specimen which is similar to that given by a *Kikuchi pattern* in electron diffraction.

Actual Kikuchi patterns may also be observed through backscattered electrons.

15.3 QUANTITATIVE ANALYSIS OF THICK SPECIMEN

The theory of quantitative analysis by electron microprobe is due to Castaing (1951), and development has been continuous since then. The present section is a simplified survey of the basic principles.

Consider c, the mass fraction of an element A, to be analysed in a specimen consisting of elements J in respective mass fractions $c_J (J = A, B, C \ldots)$. The whole set of elements $J \neq A$ form the matrix.

The intensity of a characteristic emission line of element A is measured by means of its count rate. The instance of the $K\alpha$-line, the most commonly used, will be considered hereafter. As with X-ray fluorescence, the instrumental effect is eliminated by comparing the intensity I given by the specimen to the intensity $I(T)$ given in the same conditions by a test sample T which contains A in a known mass fraction.

The relation between the measured intensity ratio $I/I(T)$ and the requested concentration ratio $c/c(T)$ may be established in a fundamental way for electron probe microanalysis. The A–$K\alpha$ intensity, emitted by the mass unit per incident electron, is proportional to the mass fraction of element A. At a first approximation, the stopping power for electrons of a given energy depends solely on the mass of elements concerned (Bethe relations, Eqs 8.4, 8.5 and 8.6). In any case of electron microprobe analysis, the specimen is semi-infinite; it follows that the mass of matter involved by X-ray emission is roughly equal in the specimen and in the test sample.

Emitted Intensities. From the basic theory: the ratio of intensities *emitted* by the atoms of the specimen and of the test sample is equal to the ratio of intensities emitted by the mass unit, i.e. to the ratio of the respective mass fractions of the element to be analysed.

Measured Intensities. The ratio of the corresponding intensities as *measured at the detector*, is therefore expressed by the following relation:

$$\frac{c}{c(T)} = K \frac{I}{I(T)} \tag{15.6}$$

where K is a correction factor depending on the specimen and test sample compositions; it expresses the *matrix effects*.

Thanks to the fundamental intensity–concentration relationship, analysis of a given element may be performed in a straightforward way by reference to a single test sample, e.g. the pure element as far as possible, leading to $c(T) = 1$.

15.3.1 MATRIX EFFECTS

Consider a primary electron beam normal to a flat specimen surface and an atom A located at a depth x inside the primary emission volume (Figure 15.4).

Linear Approximation. Assume that:

1. Electron intensity incident at A is equal in the specimen and in the test sample;
2. Absorption of emitted characteristic X-rays, along the path towards the detector, is equal in the specimen and in the test sample;
3. Primary electron-induced emission is the only source of characteristic X-rays of element A.

As a result, the matrix factor K would be equal to 1 and Eq. (15.6) would simply be written as

$$\frac{c}{c(T)} = \frac{I}{I(T)} \tag{15.7}$$

This linear approximation (the so-called first Castaing approximation) does not provide an accurate analysis, except when the specimen and the test sample have similar compositions.

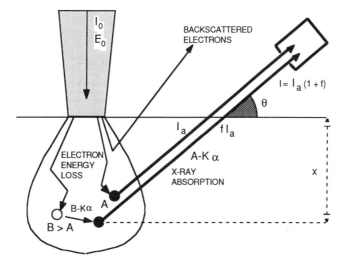

Figure 15.4. Outline of the matrix effects on the measurement of the characteristic $K\alpha$ radiation of an element A in a matrix containing one heavier element B.

Matrix Effects. The actual matrix effects cause the linear approximation to fail. There are three causes of deviation with respect to the three foregoing assumptions.

1. For equal primary beams, the electron intensity incident at atom A is different in the specimen from its value in the test sample, due to interaction effects on the primary beam along its path to the atom. The primary electron intensity and energy decrease depend on the composition, i.e. on the atomic number of the components. The effect must be accounted for by means of a correction factor called the *atomic number correction* K_z.
2. Absorption of the characteristic X-rays to be measured, on their way out of the specimen or the test sample, depends on composition. It is different in the specimen and in the test sample and must be accounted for by means of a correction factor called the *absorption correction* K_a.
3. The characteristic X-rays of element A may also be emitted through X-ray fluorescence caused by electron-induced X-ray emission from elements B heavier than element A. This fluorescence effect is different in the specimen and in the test sample (absent in a pure element test sample) and the resulting enhancement must be accounted for by means of a correction factor called the *fluorescence correction* K_f.

According to Eq. (15.6), the *total matrix correction* is expressed as follows:

$$\frac{c}{c(T)} = K_z K_a K_f \frac{I}{I(T)} \tag{15.8}$$

The correction method acting successively on those three correction factors is accordingly called the *ZAF method*.

The corresponding corrections will now be described in detail.

15.3.1.1 Atomic Number Effect

The primary electrons lose their energy by succeeding atomic interactions along their path through the specimen or the test sample:

1. A fraction of the primary electrons excite the K-level of atoms A, thus contributing to the primary emission intensity I_e of the Kα line to be measured. This intensity is proportional to the number of atoms A per volume unit (Eq. 13.10).
2. A fraction of the primary electrons is backscattered and is therefore lost for analysis. It is expressed by the *backscattering factor* $R(R \leqslant 1)$, the ratio of the number of electrons available for the excitation of atoms A to the total number of primary electrons. It depends on the composition of the specimen and of the test sample.

The intensity A–Kα emitted by the surface unit of a layer of thickness ds can now be written according to Eq. (13.11):

$$(dI)_e = G(A) c R \sigma d(\rho s) \quad \text{with} \quad G(A) = (N/A)\omega p \tag{15.9}$$

where $G(A)$ groups any factors which depend only on element A (independent of the matrix and of the incident energy), and ρs is the mass per surface unit of thickness s.

The electron path s may be approximated to the thickness passed through (Figure 8.3b).

According to Bethe's equation (Eq. 8.4), the mean energy loss of electrons along a path s is a function of the involved mass ρs. It is expressed by the *stopping factor* $S = dE_m/d(\rho s)$. Equation (15.9) is now written as

$$(dI)_e = G(A) c R \frac{\sigma}{S} d(E_m) \tag{15.10}$$

Only electrons with energy $E \geq W_K$ are able to excite the K-level of atoms A. The total primary intensity emitted per electron is therefore given by integrating this equation from E_0 to W_K. Factor $G(A)$ is eliminated by writing the intensity ratio

$$\left[\frac{I}{I(T)}\right]_e = \frac{c}{c(T)} \frac{R}{R(T)} \frac{\int_{E_0}^{W_K} (\sigma/S)\, dE_m}{\int_{E_0}^{W_K} (\sigma/S(T)\, dE_m} \tag{15.11}$$

The expression of S is given by Bethe's equation and the additivity relations. The ratio $S/S(T)$ may be detailed as follows:

$$\frac{S}{S(T)} = \frac{(\overline{Z/A}) \log_e(1.16 E_m/\overline{W}_m)}{(\overline{Z/A})_T \log_e[1.16 E_m/\overline{W}_m(T)]} \tag{15.12}$$

Any variation of factor S is slow with respect to variations of the mean energy E_m which appears in the log and the S factors may therefore be taken out of the integral. The ratio of concentrations and the atomic number correction factor can now be approximated to

$$\frac{c}{c(T)} = \frac{S/R}{S(T)/R(T)} \left[\frac{I}{I(T)}\right]_e \Rightarrow K_z = \frac{S/R}{S(T)/R(T)} \tag{15.13}$$

Factors S and R or ratios S/R may be determined by calculation through Bethe's equation or through various tables or graphs set up as a function of atomic numbers (e.g. Duncumb and Shields, 1963; Bishop, 1965; Thomas, 1964).

Note With increasing atomic number:

1. Z/A decreases (0.46 for 5-B; 0.38 for 92-U), therefore S decreases (Eq. 15.12): for an equal mass, light elements have a higher stopping power than heavy elements.
2. R decreases, due to the atomic scattering cross-section being proportional to Z^2 (Eq. 4.9).

Stopping power and backscattering vary in the same direction, thus resulting in a certain balance which justifies approximations, in particular for high overvoltage ratios. The latter are favourable for atomic number corrections, but are unfavourable for absorption corrections due to their action of increasing the penetration depth.

15.3.1.2 Absorption Effect

The characteristic X-rays to be measured are generated at various depths. They undergo absorption on their path towards the detector. This effect is different in the specimen and in the test sample. It is usually the most important matrix effect.

Consider an emission depth x and a measuring angle θ (Figure 15.4). In the general case of normal incidence, the length of the exit path is $x/\sin\theta$. Let I_a be the intensity leaving the object after undergoing absorption as the only matrix effect. It is related to the intensity emitted per incident electron by the volume unit, along an elementary path ds at the depth x by the following expression (Eq. 8.1):

$$(dI)_a = (dI)_e \exp(-\mu\rho x/\sin\theta) = (dI)_e \exp(-\chi\rho x)$$

where χ is the absorption factor of the specimen for the radiation to be measured and ρ the specimen density (with the additivity relations Eq. 8.3).

The emitted primary intensity is a function of the depth x. It is expressed by the *distribution function* $\Phi(\rho x)$: $(dI)_e = \Phi(\rho x) d(\rho x)$. Replacing $(dI)_e$ in the foregoing relation and integrating leads to the total intensity leaving the specimen, referred to the volume unit per incident electron

$$(I)_a = \int_0^\infty \Phi(\rho x) \exp(-\chi\rho x) d(\rho x)$$

Consider an *absorption function* $f(\chi)$ which expresses the ratio of the leaving intensity after absorption to the emitted intensity

$$f(\chi) = \frac{I_a}{I_e} = \frac{\int_0^\infty \Phi(\rho x) \exp(-\chi\rho x) d(\rho x)}{\int_0^\infty \Phi(\rho x) d(\rho x)} \qquad (15.14)$$

Function $[f(\chi)]_T$ is defined in a similar way for the test sample. The absorption corrected intensity ratio and the absorption correction are now written as

$$\left[\frac{I}{I(T)}\right]_e = \frac{1/f(\chi)}{1/[f(\chi)]_T}\left[\frac{I}{I(T)}\right]_a \Rightarrow K_a = \frac{1/f(\chi)}{1/[f(\chi)]_T} \qquad (15.15)$$

Distribution $\Phi(\rho x)$ has been measured on standard samples by means of tracers (Castaing and Descamps, 1955; Castaing and Henoc, 1966). An analytical form of the absorption function has been put forward by Philibert (1963) as follows:

$$f(\chi) = \frac{1+h}{(1+\chi/\sigma)[1+h(1+\chi/\sigma)]}; \quad h = 1.2\,A/Z^2; \quad \sigma = 4.5 \times 10^5/(E_0^{1.65} - W_X^{1.65})\,(\text{eV}) \qquad (15.16)$$

This is valid for medium and heavy elements and is commonly used for correction procedures. For compiling purposes, the absorption function can be developed into a limited series. Its values have been tabulated as a function of the compositional and instrumental parameters ([1] Adler; [2] Adler and Goldstein; [13] Ziebold). Accurate values for absorption coefficients are of prime importance.

15.3.1.3 Fluorescence Effect

X-rays emitted by matrix atoms can in turn excite the atoms A provided their energy is greater than W_K. The resulting fluorescence of intensity I_f enhances the electron-induced A–Kα radiation, resulting in the definition of an *enhancement rate* $f = I_f/I_e$. The measured intensity at the detector finally becomes $I = I_a (1 + f)$. A similar enhancement rate is defined for the test sample. The exit intensity ratio and fluorescence correction are written as follows:

$$\frac{I}{I(T)} = \left[\frac{I}{I(T)}\right]_a \frac{1+f}{1+f(T)} \Rightarrow K_f = \frac{1/(1+f)}{1/[1+f(T)]} \qquad (15.17)$$

The K-fluorescence of element A may be excited:

(a) by K-radiations of elements B heavier than A;
(b) by the braking background from primary electrons;

The first of the two effects is the more important and easier to formulate.

The enhancement rate has been calculated as follows by Castaing (1951) for a single excited element B whose K-radiation enhances the A–Kα line:

$$f = \frac{I_f}{I_e} = \left[\frac{c\omega}{2}\right]_B \left[\frac{r-1}{r}\right]_A \frac{A}{B} \frac{W_K(A)}{W_K(B)} \mu_B(A)(F_1 + F_2) \qquad (15.18)$$

$$F_1 = \frac{\log_e(1 + \chi/\mu_B)}{\chi}; \qquad F_2 = \frac{\log_e(1 + \sigma/\mu_B)}{\sigma}$$

where A and B are the atomic masses, μ_B and $\mu_B(A)$ respectively the mass absorption coefficients of the specimen and of the pure element A for radiation B–Kα, ω the fluorescence yield, r the ratio of absorption coefficients across the edge K of element A, σ the coefficient from Eq. (15.16) and χ the absorption factor of the specimen for A–Kα.

When using L-lines for analysis of an element A, the fluorescence enhancement effect can be due both to heavier-element L-lines and to lighter-element K-lines.

The fluorescence effect becomes particularly significant for compounds containing pairs of relatively heavy neighbour elements (e.g. analysis of Fe in a matrix containing Cu; analysis of Ti in a matrix containing Fe, etc.). It may be neglected for analysing light elements. The existence of active pairs is easily monitored through a preliminary qualitative analysis.

15.3.2 CORRECTION METHODS

The global correction formula results from combining the foregoing partial corrections (Eqs 15.13 and 15.17):

$$\frac{c}{c(T)} = \frac{I}{I(T)} \frac{S/R}{S(T)/R(T)} \frac{1/f(\chi)}{1/f_T(\chi)} \frac{1/(1+f)}{1/[1+f(T)]} \qquad (15.19)$$

The practice of intensity measurement is the same as in X-ray fluorescence, with the difference that the continuous background must be taken into account for any accurate analysis.

15.3.2.1 ZAF Method

The ZAF correction procedure is based on Eq. (15.19). It is the most commonly practised method for analysing heavy and medium elements. Determining the values of the correction factors S/R, $f(\chi)$, f requires knowing the elemental concentrations, which is precisely the aim of analysis. The problem is solved through an iteration procedure by using the respectively measured intensity ratios of the elements as initial concentrations, then calculating the factors K_z, K_a and possibly K_f which are used for correcting the concentrations and so on, until convergence is reached. Adequate computer programs have been developed and are being constantly refined (e.g. Henoc et al, 1971).

For quantitative analysis, dedicated microprobes are connected to a computer which performs the automatic control of the spectrometer settings, as well as data acquisition and processing. Analytical attachments to electron microscopes are increasingly fitted with similar facilities. However, the correction procedure is most simplified when performed by means of an SEM. Instead of directly comparing with standards, the characteristics of the latter are often stored for a given instrumental primary intensity and energy. Another type of non-standard correction program works by means of iteration procedures based on the theoretical formulation and on memorised element characteristics. The analysis becomes semi-quantitative. The lower accuracy is offset by a greater simplicity and operating speed.

15.3.2.2 Limitation of ZAF

Light Elements

Any causes of error are amplified for analysing light electrons ($Z < 10$) (Ruste, 1979).

The practised overvoltage ratio $U = E_0/W_K$ is generally much higher than the optimal value for light elements, resulting in a low K-level excitation cross-section. Together with a low fluorescence yield, this leads to weak lines with a low signal-to-noise ratio.

Common contamination effects by carbon, oxygen and nitrogen, due to the residual gas molecules in the enclosure, prevent those elements from being analysed.

The efficiency of the specimen–detector system is low for soft X-rays, due mainly to the strong absorption.

Significant chemical shift effects may be another cause of error.

The approximations which have led to the theoretical correction formulae, while valid for analysing medium or heavy elements, fail to apply for light elements, in particular as far as the absorption corrections are concerned; whereas fluorescence corrections are insignificant, absorption corrections become very large. X-rays emitted from light electrons are highly absorbed and are therefore issued from a thin surface layer of the specimen. This consideration has led to a simplified *thin-layer model* as proposed by Duncumb and Melford (1966), permitting a simultaneous atomic number and absorption correction.

Higher analytical accuracies for light elements are achieved by means of a quite different correction technique, based on modelling the most probable individual electron trajectories, hence the name *Monte-Carlo method* (Bishop, 1967); a better accuracy for light element analysis is offset by long calculation times on powerful and costly computers.

In specific instances, more accurate analyses are carried out by using test samples with a composition close to that of the specimen. Determining the concentration of light elements, such as oxygen, is then often limited to normalising to one the sum of mass fractions of the other elements.

Thin Specimens

The ZAF correction formulation has been set up for bulk specimens and test samples. It loses its validity when dealing with thicknesses smaller than the penetration depth of electrons, i.e. of the order of 1 μm. Rather than analysing such thin specimens with a conventional microprobe, they are better dealt with in the specially designed analytical transmission microscope (see section 15.4).

15.3.3 ACCURACY OF ANALYSIS

The best results are achieved with dedicated electron microprobes, for which working parameters have been optimised.

15.3.3.1 Instrumental Errors

Probe Intensity Drift. Due to rather long measuring times, an intensity drift of the order of $\Delta I_0/I_0 = 10^{-3}$ per hour may subsist, in spite of intensity stabilising.

Drift of Characteristics of the Measuring System. Any temperature variation results in altering the analysing crystal setting and the detector amplification, hence the importance of working at constant temperature. Any residual instrumental drift may be offset by alternating measurements on the specimen and on the test sample.

15.3.3.2 Errors due to the Specimen

Positioning and Orientation Error. The specimen setting at a given working distance is very sensitive with a focusing spectrometer, whereas its orientation is critical for absorption correction (intervention of $\chi = \mu/\sin\theta$). Non-planarity of the surface induces local orientation errors or even absorption of the emerging X-rays (surface asperities). The relative error is greater, the smaller the emergence angle, hence the advantage of large detection angles θ, of the order of 45° in dedicated microprobes.

Chemical Shift. Different chemical states of the element to be analysed in the specimen and in the test sample result in a shift of the corresponding emission lines. This chemical shift is the greater the lighter the element and the more external its excited level. For this effect to be annealed, the spectrometer should be reset at maximum intensity before any measurement.

Contamination. The intense electron bombardment on residual air molecules, left by secondary pumping, causes the specimen to be covered by a carbon layer after a short working time. Due to additional absorption, this contamination layer may alter the analysis of light elements and is a severe hindrance for analysing carbon itself. A clean vacuum and an anticontamination device are therefore of prime importance for light element analysis.

Specimen Alteration. Strong electron–object interaction may lead to evaporation of certain elements from the specimen surface. Alkaline elements seem particularly sensitive when occurring in certain materials (e.g. Na and K in silicate glasses). This effect should be taken into account by choosing a suitable test sample and by reducing the primary intensity as far as possible.

Diffraction Effect. When dealing with single crystals, the backscattering factor, which intervenes in the atomic number correction, is direction-sensitive as a result of diffraction, leading to a correlated variation of X-ray emission and hence to analysing errors.

15.3.3.3 Statistical Errors

If any systematic errors are assumed to be eliminated, accuracy is determined by count statistics. As shown by Eq. (10.14), accuracy may be improved by increasing the count time within limits set by electron radiation effects and by instrumental drift.

15.3.3.4 Correction Errors

Uncertainties concerning the values of correction parameters are more significant for correction errors than approximations of the theoretical relations. Parameters of increasing accuracy are currently being measured and compiled for electron probe microanalysis.

15.3.3.5 Currently Observed Accuracy

Major element analysis performed on a dedicated electron microprobe is normally of the order of 1%. A microanalytical attachment to an SEM, equipped with an EDS, is less accurate in the general case (see Table D.1).

15.3.4 SPECIMEN PREPARATION

15.3.4.1 Specimen to be Analysed

Quantitative analysis requires a specimen with at least one flat surface, with a sufficient electrical conductivity and resistance to electron bombardment. Those conditions are less severe for qualitative and semi-quantitative analysis.

Surface Planarity. A planarity to 1 μm is required in order to avoid absorption artefacts in quantitative analysis. This requirement is the more important, the smaller the detection angle θ and the lighter the elements to be analysed. The generally used mechanical polishing process should not introduce foreign elements which could hinder the analysis. The most commonly used abrasive is diamond, rather than alumina or silicon carbide. Specimens in the form of fine-grained materials are embedded in a resin before polishing. Petrological thin sections may be used without covering.

15.3.4.2 Test Samples

In addition to previous specimen requirements, a test sample for quantitative analysis has to meet the following conditions:

(a) easy to prepare with an accurate composition;
(b) homogeneous to less than 1 μm;
(c) stable in vacuum and in the electron beam;
(d) without a significant chemical shift with respect to the specimen to be analysed.

Straightforward corrections are carried out by means of pure elements. Compound specimens are needed when the corresponding element is unstable (e.g. N, O, F, alkalines, Ca, P, etc.). Materials with numerous elements, light electrons as well as heavy elements, like some glasses and ceramics, lead to important matrix effects and chemical shifts. Compound test samples with closely related compositions may then lead to better results. Glasses can be used as test samples for analysing various silicates. However, alkaline glasses are sensitive to electron irradiation. The homogeneity of compound samples must be checked (e.g. phase separations in glasses).

15.3.5 APPLICATIONS TO MATERIAL ANALYSIS

Electron probe microanalysis is a most important, often non-destructive, very sensitive and rapid technique for analysing and controlling any materials at a micrometre scale. Some specific applications may be highlighted:

1. Analysis of micrometre-sized particles, phases, precipitates and inclusions in alloys, ceramics, glasses, minerals and rocks.
2. Analysis of small particles in the atmosphere, e.g. for monitoring pollution.
3. Investigating concentration gradients during grain boundary diffusion, surface corrosion and alteration of materials.
4. Assessing the element distribution in rock minerals has made the electron microprobe into one of the main tools in petrology, as a complement to polarising light microscopy.
5. Identification of mineral species which occur only in microscopic sizes. Numerous new minerals have been discovered thanks to electron probe microanalysis.

15.4 QUANTITATIVE ANALYSIS OF THIN SPECIMENS

A specimen is considered to be thin for electron microprobe analysis when its thickness is smaller than the electron penetration depth. Microanalysis of thin specimens is mostly carried out by means of a TEM or an STEM. The main objective is localising elements at a nanometre scale, rather than very accurate analysis. Due to the small analysing volume involved, analysis is generally semi-quantitative.

The basic principles of quantitative analysis are the same as for bulk specimens. However, thanks to the thickness being smaller than the penetration depth, matrix effects will be reduced and intensity–concentration relationships will be simplified. On the other hand, the use of standards is more difficult.

15.4.1 BASIC FORMULATION. MATRIX CORRECTIONS

Consider an element A with a mass fraction c_A to be measured by means of the intensity I_A of one of its characteristic X-ray emission lines. The K-emission will be considered below.

15.4.1.1 Relationship between Intensities and Concentrations

The intensity of the A–Kα line emitted per surface unit of a thin specimen of thickness t is given as follows from Eq. (15.9):

$$(I_A)_e = I_0 \frac{N}{A} (\sigma \omega p)_A c_A R \rho t \tag{15.20}$$

where σ is the K-level excitation cross-section, ω the fluorescence yield, p the Kα-line emission probability (for element A), $R \leqslant 1$ the backscattering factor and ρ the density (for the specimen).

Before being measured at the detector as an intensity I_A, the characteristic radiation has been affected by *physical factors* due to the matrix (absorption, fluorescence) and by *instrumental factors* (detector acceptance angle, detector efficiency). According to Eqs (13.14) and (15.19) it is written.

$$I_A = (I_A)_e f(\chi)(1+f)\tau_A = I_0 \frac{N}{A} (\sigma \omega p)_A c_A R f(\chi)(1+f)\tau_A \rho t \tag{15.21}$$

where $f(\chi)$ is the absorption function of the specimen, $(1+f)$ the fluorescence enhancement rate by atoms heavier than A and τ_A the transmission factor of the detector–spectrometer system (Eq. 13.12).

15.4.1.2 Thin Specimen Matrix Corrections

A specimen thickness of the order of 100 Å results in reduced matrix effects. When a matrix correction is smaller than the error of measurement to be expected, i.e. some 5%, it may be neglected. When it must, nevertheless, be taken into account, its correction is simplified.

Atomic Number Correction

The atomic number correction is expressed by the backscattering factor R; it is a function of the electron energy loss, which in turn is a function of the mass involved ρt.

Due to the small value of t, this correction may be simplified or even neglected ($R \cong 1$).

The most probable energy loss is given by the Landau relation (Eq. 8.7); if very small ($E \cong E_0$), the ionisation cross-section σ may be assumed to be constant for the considered element A. Backscattering may then usually be neglected, as can be assessed from the graphs which give R as a function of electron energy loss (e.g. Bishop, 1967).

Absorption Correction

Due to the small thickness, the intensity emitted along the path of the primary electrons may be assumed to be constant. Integrating Eq. (15.14) therefore simplifies, leading to the following absorption function:

$$f(\chi) = \frac{1 - \exp(-\chi \rho t)}{\chi \rho t} \qquad (15.22)$$

where $\chi = \mu/\sin\theta$ (μ is the mass absorption coefficient of the specimen for the measured line) and θ the detection angle; normal incidence is assumed (Figure 15.4).

When developed into a limited series, this relation becomes

$$f(\chi) = 1 - \frac{\chi \rho t}{2!} + \frac{(\chi \rho t)^2}{3!} + \ldots \qquad (15.23)$$

For small values of $\chi \rho t$, absorption correction becomes negligible ($f(\chi) \cong 1$). A currently used criterion for neglecting this correction when analysing an element A has been forwarded by Tixier and Philibert (in [11]):

$$\chi_A \rho t < 0.1 \qquad (15.24)$$

where χ_A is the absorption factor of the specimen for the characteristic line of element A. This *thin specimen criterion* is similar to that used in X-ray fluorescence (Eq. 14.9).

For simultaneous analysis of two elements A and B by means of the ratio method, Goldstein (in [5]) proposes the following criterion:

$$\frac{(\chi_B - \chi_A)\rho t}{2} < 0.1 \qquad (15.25)$$

Fluorescence Correction

Enhancement of the A-Kα emission through fluorescence from a heavier element B (Eq. 15.18) may be assumed to be negligible when

$$\mu_B \rho t < 0.1 \qquad (15.26)$$

where μ_B is the mass absorption coefficient of the characteristic line of B which causes the X-ray fluorescence of element A to be analysed ($Z_B > Z_A$).

15.4.1.3 Thin Specimen Approximation

The conditions for the thin specimen approximation to apply are met when matrix corrections are liable to be neglected, i.e. when the above criteria are justified, the most important being the absorption effect. In this case the factors R, $f(\chi)$ and $(1+f)$ are approximated to unity and σ to a constant. The intensity measured at the detector (Eq. 15.21) then simplifies to

$$I_A = I_0 k_A \tau_A c_A \rho t \quad \text{with} \quad k_A = \frac{N(\sigma \omega p)_A}{A} \tag{15.27}$$

15.4.1.4 Concentration Ratio of Two Elements

When dealing with bulk specimens in electron probe microanalysis, the instrumental factors are compensated by comparing with a test sample (pure element or compound test sample). This procedure becomes difficult to apply with thin specimens where thickness intervenes.

The mass fraction of an element could be determined in a direct way, without a test sample, if the physical parameters k_A, ρ, t and the instrumental parameter τ_A were known. The difficulty in determining some of these parameters, in particular thickness, leads to important errors of analysis.

Those problems have led to the most currently used *ratio method* (Cliff and Lorimer, 1975; Champness et al, 1981; Lorimer and Cliff, 1984; Lorimer, 1987). In the thin specimen approximation, the ratio of characteristic line intensities from two elements A and B measured simultaneously is derived from Eq. (15.27). The detector acceptance angles are equal for both elements. Hence the ratio τ_A / τ_B reduces to the detector efficiency ratio $\varepsilon_A / \varepsilon_B$ and the intensity ratio becomes

$$\frac{I_A}{I_B} = \frac{k_A \varepsilon_A}{k_B \varepsilon_B} \frac{c_A}{c_B}$$

The ratio of respective mass fractions is then written

$$\frac{c_A}{c_B} = \frac{k_B \varepsilon_B}{k_A \varepsilon_A} \frac{I_A}{I_B} = k_{AB} \frac{I_A}{I_B} \quad \text{with} \quad k_{AB} = \frac{k_B \varepsilon_B}{k_A \varepsilon_A} = \frac{(\sigma \omega p)_B}{(\sigma \omega p)_A} \frac{A}{B} \tag{15.28}$$

In the thin specimen approximation, the ratio of mass fractions does not depend on thickness, on density and on the detection angle.

Inside the validity range of this approximation, the factor k_{AB} characterizes the pair of elements A and B for a given primary energy (e.g. $E_0 = 100$ keV) and for a given detector. Its values may therefore be tabulated for pairs of elements, thus making the method easy to set up. It must, however, be kept in mind that the factors k_{AB} are the product of physical factors and of instrumental factors; therefore they do not have any fundamental significance.

An EDS with an Si(Li) diode is of particular interest for the ratio method, in so far as it provides a simultaneous intensity measurement of the lines of all elements and thus avoids any instrumental drift.

Table 15.2. Values of some factors k_{AB} measured and calculated for the Kα lines, with respect to reference elements Si and Fe. It can be seen that Fe as reference results in a smaller dispersion of values. From Goldstein in [5] and from Lorimer (1987)

Z	Element	Reference Si (100 keV)		Reference Fe (120 keV)	
		k_A Si exp.	k_A Si calc.	k_A Fe exp.	k_A Fe calc.
12	Mg	1.07	1.25	0.96	1.03
13	Al	1.42	1.12	0.86	0.88
14	Si	1	1	0.76	0.77
20	Ca	1.0	1.02	0.88	0.79
26	Fe	1.27	1.33	1	1

15.4.2 PRACTICAL METHODS OF ANALYSIS

15.4.2.1 Concentration Ratios

Measuring the concentration ratio of a pair of elements A and B requires knowing the characteristic factors k_{AB}. They may be determined with or without a test sample.

Method without a Test Sample

Factors k_{AB} are calculated from the physical and the instrumental parameters of Eq. (15.27). These parameters have been compiled for a certain number of pairs of elements, e.g. by Goldstein, Zaluzec in [5] and by Lorimer (1987). To analyse a given type of material, a constantly present major element is selected as a *reference element* (e.g. Si in silicates). Table 15.2 shows some values of such factors.

Methods with Thin Test Sample

These methods are best suited for analysis by means of a TEM. They consist of determining the correction factors k_{AB} in an experimental way. The same facility must be made use of for measurements carried out both on the specimen and on the test sample. According to the type of test sample, the procedures given under the headings below may be considered.

Pure Element Test Sample. Pure element thin films can be used as test samples, provided they are prepared with a constant and well-known mass thickness ρt (e.g. vacuum-evaporated metal elements, mass thickness measured by weighting). Mass thicknesses are different for two test samples A and B; their intensity ratio is therefore given by Eq. (15.27) as follows:

$$\left[\frac{I_A}{I_B}\right]_T = \frac{k_A \varepsilon_A}{k_B \varepsilon_B}\left[\frac{(\rho t)_A}{(\rho t)_B}\right]_T = \frac{1}{k_{AB}}\left[\frac{(\rho t)_A}{(\rho t)_B}\right]_T$$

For identical instrumental conditions for both the test samples and the specimen, the k_{AB} factors are determined by

$$k_{AB} = \left[\frac{(\rho t)_A}{(\rho t)_B} \frac{I_B}{I_A}\right]_T \qquad (15.29)$$

Compound Thin Test Sample. This is the most commonly used method, inasmuch as it leads to simultaneous analysis of elements from Na upwards, in various materials. It consists of measuring the correction factors k_{AB} for pairs of elements (AB) on a test sample of known composition meeting the thin specimen criteria. The relative concentrations are then provided by Eq. (15.28).

When analysing several elements, the measurements are usually referred to one of them. Whenever possible, it is advisable to choose for this aim a major element with $Z>15$ for a maximum detector efficiency. As an example, Si is most commonly used as a reference element for analysing silicate materials; however, Fe is preferable when present in sufficiently high concentration, as shown in Table 15.2.

Specimen Used as Test Sample. In some cases the specimen itself may serve as the test sample, e.g. for analysing inclusions or altered parts in a matrix of known composition. The matrix then acts as a test sample (see Figure 21.1).

A somewhat more questionable procedure consists of referring the intensity at a point to be measured in the specimen to the average intensity integrated by beam scanning over a large area surrounding the point; the scanned area is then taken as the test sample, assuming its composition is equal to the average specimen composition as measured by another method, e.g. X-ray fluorescence.

Thick Test Sample. Bulk test samples, e.g. with a thickness greater than the electron penetration depth (of the order of 1 µm), are easy to prepare with one flat polished surface. However, any related measurements must undergo the whole correction process (atomic number and absorption). As with the conventional microprobe, one test sample per analysed element is used (pure element or compound). Tixier and Philibert (1969) have set up a relation for thin specimen analysis by means of bulk test samples, based on Bethe's equation and on the common correction formulation. It has the following expression for K-lines:

$$\frac{c_A}{c_B} = \frac{I_A}{I_B} \frac{(W_K/\log_e U_K)_A}{(W_K/\log_e U_K)_B} \left[\frac{[(R/S)f(\chi)]_A}{[(R/S)f(\chi)]_B} \frac{I_B}{I_A}\right]_T \qquad (15.30)$$

where W_K and U_K respectively are the energy and overvoltage ratio of level K and R/S and $f(\chi)$ respectively the atomic number and absorption correction for thick test samples.

15.4.2.2 Absolute Multi-element Analysis

Consider a specimen consisting of n elements J of respective mass fractions $c_J(J=A, B\ldots)$. Knowledge of:

(a) the correction factors $k_{AR}, k_{BR}, \ldots k_{JR}$ related to a major reference element R;
(b) the measured intensities $I_A, I_B, \ldots, I_R, \ldots I_J$;

leads to $n-1$ linear equations, according to Eq. (15.28):

$$c_A = \frac{c_R}{I_R} k_{AR} I_A; \quad c_B = \frac{c_R}{I_R} k_{BR} I_B; \quad \ldots; \quad c_J = \frac{c_R}{I_R} k_{JR} I_J \qquad (15.31)$$

Normalising the sum of mass fractions to 1 provides the nth equation

$$\sum_J c_J = c_A + c_B + \ldots + c_J = \frac{c_R}{I_R} \sum_J k_{JR} I_J = 1 \qquad (15.32)$$

Substituting c_R/I_R from this equation into Eqs (15.31) leads to the absolute values of mass fractions, with their sum normalised to 1:

$$c_A = \frac{k_{AR} I_A}{\sum_J k_{JR} I_J}; \quad c_B = \frac{k_{BR} I_B}{\sum_J k_{JR} I_J}; \quad \ldots; \quad c_J = \frac{k_{JR} I_J}{\sum_J k_{JR} I_J} \qquad (15.33)$$

The practical analysing procedure results from these relations:

1. Measuring simultaneously intensities $I_A, I_B, \ldots I_J$ after background subtraction.
2. Calculating relations (15.31, 15.33) by means of calculated or measured correction factors k_{JR}.

Absolute mass fraction determination by normalising their sum to 1 is an often used analysing procedure, e.g. in earth science. However, it must be handled with care. Any elements of the specimen must be analysed, the light elements included, in order to avoid large errors due to normalising.

15.4.2.3 Accuracy of Analysis

X-ray microanalysis of thin specimens is a very sensitive technique. However, due to the tiny amount of analysed material, its accuracy cannot compare with that of the conventional microprobe performed on bulk specimens.

Systematic Errors

Systematic errors are mainly due to the thin specimen approximations and to the ratio method. In order to avoid gross errors, it is of prime importance to check the validity of the thin specimen approximation for the most sensitive elements (light elements, pairs of neighbouring heavy elements), according to Eqs (15.24)–(15.26). The absorption correction is usually predominant; it is a function of the detection angle θ. This should be as large as possible, but is limited by instrumental requirements in a transmission electron microscope (see Ch. 21). It can be increased by tilting the specimen; however, this increases the path of incident electrons, thus increasing the atomic number correction.

Example. Thickness limit for analysing Mg in a basaltic glass (see Figure 21.1) calculated according to criterion (15.24).

$$\theta = 20° \Rightarrow t < 540 \,\text{Å}$$
$$\theta = 45° \Rightarrow t < 1100 \,\text{Å}$$

Accuracy can be affected by the choice of the reference element for determining the mass fraction ratios. Due to the fact that the resulting error on a ratio is the quadratic sum of respective errors on the elements, the mass fraction of the reference elements should be measured with the highest possible accuracy. It should preferably be a major element with a sufficiently high atomic number.

For multi-element analysis the commonly used method of normalising the sum of mass fractions to 1 may lead to considerable errors if one or several elements have not been detected at all or have been measured with a poor accuracy. This is the case of light elements for which the EDS efficiency is poor. Consider the case of many common materials containing oxygen, e.g. silicate materials such as glass, ceramics, rock; normalising is usually carried out by means of the oxide components. Any uncertainties about the oxidation degree of an element (e.g. Fe, Mn) then results in concentration uncertainties.

The use of a retractable or an ultra-thin spectrometer window increases the efficiency and therefore the accuracy of light element analysis.

Electron energy loss spectrometry (EELS, see Ch. 17) may be useful as a complementary technique for analysing light elements.

High-intensity electron irradiation can cause certain elements to be evaporated, resulting in further errors of analysis.

Statistical Errors

The quadratic deviation of the mass fractions of two elements A and B is the sum of respective quadratic deviation of k_{AB}, I_A, I_B.

The small analysing volume results in a small count rate. Count times on both the specimen and the test sample must accordingly be sufficiently high. Furthermore, it is of prime importance to carry out several intensity measurements on each element to be analysed until the mean count value appears stabilised; 10–20 countings are currently required. In any case, the uncertainty brackets must be objectively estimated for a significant analysis.

Peak interferences are an additional cause of errors if their deconvolution cannot be adequately performed.

Currently Observed Accuracy

Thin specimen analysis on a micrometre scale or even on a nanometre scale is at the present limit of sensitivity for electron intensity measurements. It is actually semi-quantitative. The mean accuracy which can be expected for analysing major elements with $Z > 11$, in a specimen meeting the thin object criteria, is of the order of 5–10%, provided an adequate reference element has been chosen and operations have been carried out according to the above-mentioned rules.

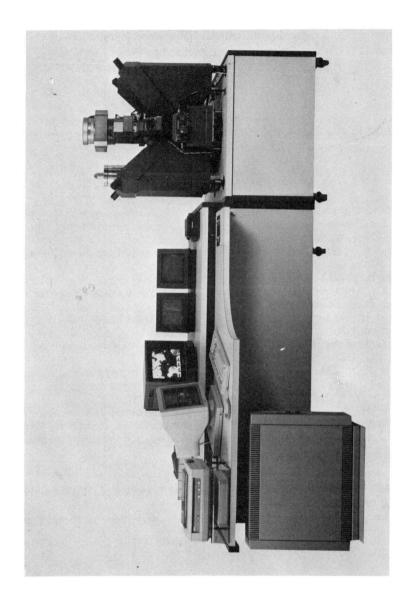

Figure 15.5. Electron microprobe. On the right-hand side is the column with its wavelength dispersive spectrometers. On the left-hand side is the data acquisition and processing system. Document from CAMECA.

15.4.2.4 Applications

Microanalysis of thin specimens is of particular interest when carried out by means of a TEM (STEM whenever possible); it can then be completed by structural data provided by electron diffraction and high-resolution imaging. The best solution would be to combine X-ray microanalysis with electron energy loss spectrometry which is complementary for light element analysis (EELS, Ch. 17).

The various types of materials and objects liable to be investigated by this technique are the same as for transmission electron microscopy.

REFERENCE CHAPTERS

Chapters 7, 8, 9, 10, 13, 14, 21. Appendix D.

BIBLIOGRAPHY

[1] Adler I. *X-Ray emission spectrography in geology.* Elsevier, New York, 1966.
[2] Adler I., Goldstein J. I. *Absorption tables for electron probe microanalysis.* NASA Technical Note D-2984, Washington, 1965.
[3] Bunshah R. F. (Ed.) *Modern analytic techniques for metals and alloys.* John Wiley, New York, 1970.
[4] Hencke B. L. (ed.) *Advances in X-ray analysis.* Plenum Press, New York, 1979.
[5] Hren J. J., Goldstein J. I., Joy D. C. (eds) *Introduction to analytical electron microscopy.* Plenum Press, London, 1979.
[6] Kalos M. H., Whitlock P. A. *Monte Carlo methods.* Vol. 1 *Basics.* John Wiley, New York, Chichester, 1986.
[7] Laves G. *Scanning electron microscopy and X-ray microanalysis.* John Wiley, New York, Chichester, 1987.
[8] Loretto M. H. *Electron beam analysis of materials.* Chapman & Hall, London, 1984.
[9] Maurice F., Meny L., Tixier R. (eds) *Microanalyse par sonde électronique. Spectrométrie de rayons X.* ANRT, Editions de Physique, Orsay, 1987.
[10] Obst K. H., Münchberg W., Malissa H. *Elektronenstrahl Mikroanalyse (ESMA) zur Untersuchung basischer feuerfester Stoffe.* Springer-Verlag, Heidelberg, Vienna, 1972.
[11] Siegel B. M., Beaman D. R. (eds) *Physical aspects of electron microscopy and microbeam analysis.* John Wiley, New York, 1975.
[12] Theisen R., Vollath D. *Tables for X-ray attenuation coefficients.* Stahleisen, Düsseldorf, 1967.
[13] Ziebold T. O. (ed.) *The electron microanalyser and its applications.* MIT, Cambridge, Mass. (USA), 1965.
[14] Zussman J. (ed.) *Physical methods in determinative mineralogy.* Academic Press, London, 1977.

16

Electron Spectrometry for Surface Analysis

Electrons issued from atomic levels and ejected during radiation–matter interaction processes may be used to gather analytical data on materials, provided their energy is characteristic of their origin atom. Electron emission may involve any atomic levels, from core levels to external levels, thus resulting in a rather complicated spectrum when compared to X-ray emission. Its sensitivity to the chemical state of atoms leads to line shifts which in turn provide data on chemical bonds.

The energy of electrons issued from excitation–relaxation processes are mainly in the low-energy range and are therefore strongly absorbed in the specimen. Those which are able to leave the specimen surface to be measured are therefore issued from a thin surface layer which has a thickness of the order of the electron mean free path. As a result, electron spectrometry is mainly a method of surface analysis which has seen an important development in recent decades.

Depending on the primary radiation, the following electron emissions may be observed:

1. **X-ray excitation** generates *photoelectrons* and *Auger electrons.*
2. **Electron excitation** generates *secondary electrons* and *Auger electrons.*

Photoelectrons and secondary electrons issued from primary interactions may in turn induce further excitations of more external atomic levels and thus result in additional *secondary electron* emissions.

Photoelectrons (when excited by monochromatic X-rays) and Auger electrons travel with a characteristic energy which can be exploited through spectrometry for analytical purposes.

Two techniques of electron emission spectrometry are available, **photon-induced** and **electron-induced** electron spectrometry. Exciting photons are mainly X-rays, sometimes ultraviolet.

Common abbreviations. *Photon excitation*:
1. ESCA (electron spectroscopy for chemical analysis) developed by Siegbahn [10]. This best-known technique combines the spectrometry of photoelectrons and of Auger electrons which are in the same energy range.

2. XPS (X-ray photoelectron spectroscopy); UPS (ultraviolet photoelectron spectroscopy).
3. AES-X (Auger electron spectrometry, X-ray excitation).

Electron excitation:
1. AES (Auger electron spectrometry), sometimes with specifying AES-E (electron excitation).
2. SAM (scanning Auger microscope).

Auger spectrometry with electron excitation may be combined with low-energy electron diffraction (LEED).

Secondary electrons are not directly fit for use in spectrometry. They serve for electron imaging in the SEM (see Ch. 20).

16.1 PHOTOELECTRON SPECTROMETRY

16.1.1 BASIC PRINCIPLE

For an incident photon beam to be able to excite the X-level of an element A in a specimen, its energy E_0 must be higher than the binding energy W_X of the level. The result is then ejection of photoelectrons from this level, with a kinetic emission energy $(E_0 - W_X)$. Those photoelectrons are able to leave the specimen to be measured if their energy, when arriving at the surface, is greater than the work function $e\Phi$ (Φ is the potential barrier of the solid).

The kinetic energy E of the outgoing photoelectron can therefore be expressed by the following relation:

$$E \leqslant E_P = E_0 - W_X - e\Phi \qquad (16.1)$$

The maximum energy value E_P is related to photoelectrons issuing from a surface layer of thickness equal to their *escape depth* which can be assimilated in their mean free path l. Those generated at greater depths lose a more or less important fraction of their energy through interactions along their path to the surface. The resulting energy spectrum is therefore dissymmetrical, with a sharp peak at the maximum value E_P, corresponding to the predominance of surface photoelectrons (Figures 7.1 and 16.1). For a primary monochromatic radiation and for a given value of the work function $e\phi$, measuring the energy of these photoelectron peaks leads directly to the energy levels W_X of the atoms in the surface layer of thickness l of the specimen. The mean free path l is of the order of a few nanometres for the commonly observed photoelectron energy range of a few kiloelectronvolts (see Figure 8.2). In addition to identifying the elements, the measurement of the values E_P, when compared to their corresponding values in standards, provides data on chemical bonds.

Photoelectron peaks are most commonly designated by the quantum notation of the corresponding levels, e.g. $3d^{3/2}$ for level M4 (Table 7.1).

The first work on photoelectron spectrometry was carried out by Robinson in 1914 and by de Broglie in 1921, by means of a magnetic analyser combined with a photographic

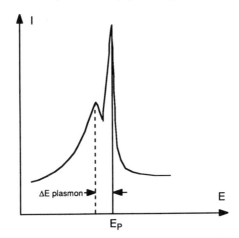

Figure 16.1. Outline of an experimental photoelectron peak related to a given atomic level. The smaller peak on the left-hand side is a plasmon peak due to a characteristic energy loss ΔE (arbitrary scale).

film. However, it was not before the late 1950s that it was practically developed, mainly by Siegbahn, by means of high-resolution electron spectrometry capable of separating the sharp photoelectron peak at E_P with a half-width of less than 1 eV.

In addition to the main photoelectron peak, spectra from conductors or semiconductors often display one or two secondary peaks, some 10 eV on the lower-energy side, corresponding to characteristic energy losses by plasmon excitations (Figure 16.1).

A relative energy resolution of about 10^{-4} is required for recording a significant photoelectron spectrum. It has been made possible by the development of high-resolution electron spectrometers.

16.1.2 INSTRUMENTAL LAYOUT

Common abbreviations: MCA (multichannel analyser); CMA (cylindrical mirror analyser); HSA (hemispheric analyser).

The common layout of an XPS or ESCA facility is outlined in Figure 16.2. Compare with an X-ray fluorescence facility (see Figure 14.1).

Electrons passing through the exit aperture of the spectrometer induce a detector signal which is amplified and recorded as a function of the varying dispersive field, thus leading to the spectrum $I(E)$ of the electrons emitted by the specimen.

Simultaneous detection over a given energy range of the spectrum, e.g. by means of a microchannel plate combined with an MCA, provides a higher sensitivity and a virtually instantaneous display, together with easier spectrum processing.

In order to avoid electron scattering and absorption by air molecules as well as specimen contamination, the whole electron path is in ultra-high vacuum, commonly provided by a turbomolecular and an ionic pump (maximum pressures: at the spectrometer 10^{-6} Torr; at the specimen 10^{-9} Torr). The enclosure is built of stainless steel with metallic gaskets for a clean vacuum.

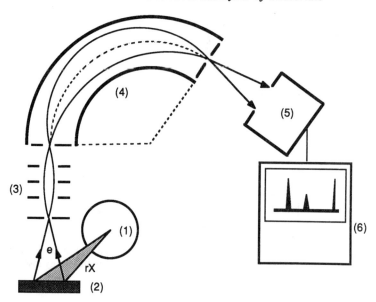

Figure 16.2. Diagram of a common photoelectron spectrometry facility. (1) X-ray tube (with monochromator); (2) specimen; (3) electron-optical focusing system; (4) focusing electron spectrometer (electrostatic prism analyser for linear focusing, hemispherical analyser (HSA) or cylindrical mirror analyser (CMA) for point focusing, see Figure 10.10); (5) electron detector (e.g. channeltron); (6) data acquisition and processing system.

16.1.2.1 Primary Radiation

The most commonly used primary radiation is the characteristic radiation from an *X-ray tube*. The Kα-line is selected either by means of a β-filter or a crystal monochromator, the latter being required for high-resolution spectrometry. The anode choice is guided by the energy levels to be excited in optimum conditions. For various reasons, it is advisable to operate with the lightest possible anode element consistent with the elements to be analysed:

1. The *excitation cross-section* of a given level X is maximum for a close to unit overvoltage ratio $U_X = E_0/W_X$, where E_0 is approximately proportional to the squared atomic number of the anode element.
2. The *characteristic peak over continuous background ratio* is greater for light elements.
3. The relative energy resolution $\delta E/E$ being determined by the spectrometer, the *absolute resolution* is the better, the smaller the energy E.
4. The *natural line width* of light elements is smaller (e.g. Mg-Kα: 0.8 eV; Al-Kα: 0.9 eV; Cu-Kα: 3 eV; Mo-Kα: 7 eV).

The natural width of X-ray lines is still too large for accurate measurements of energy levels which have dispersions as low as 0.1 eV in free molecules, 0.5–1 eV in solids (e.g. graphite 1s-level: 0.7 eV). The incident spectral line width may be reduced by means of a Johansson-type monochromator focused on the specimen surface. A line width as low as 0.2 eV is achieved by using a point focusing hemispheric monochromator (Siegbahn in Shirley [9]).

The most commonly used anode elements are Mg and Al.

A high-intensity *synchrotron X-ray source* together with a focusing monochromator leads to a far lower detection limit and to the possibility of photoelectron surface microanalysis.

16.1.2.2 Spectrometer

Highest resolution and efficiency are achieved with focusing spectrometers:

(a) hemispheric electrostatic analyser (HSA);
(b) cylindrical electrostatic analyser (CMA, Figure 10.10a).

The detector is currently a channeltron with cone-shaped entry.

16.1.2.3 Specimen

Spectrometric data about the specimen are issued from a thin surface layer corresponding to an escape thickness of some 10 Å.

For the intrinsic specimen surface to be analysed, any contamination must be prevented by a clean vacuum of the order of 10^{-9} Torr. An additional surface cleaning, by heating or ion sputtering, is currently performed in an adjoining enclosure (section 16.2).

16.1.2.4 Spectrum Acquisition and Processing

The resulting spectrum $I(E)$ can be directly displayed by a chart recorder. In recent techniques, the digitalised spectrometer signals are memorised in the respective channels of an MCA and displayed on a video screen. This technique is best suited for spectrum processing, e.g. statistical smoothing, peak deconvolution, background subtraction. Spectrum smoothing by means of multiple spectrometer scanning is commonly performed for a maximum peak resolution; thanks to the weak X-ray–matter interaction some 10 spectrometer scans, which may last hours, are currently carried out and integrated with the aim of spectrum line fine structure analyses (Figure 16.3).

The spectrometer background increases towards low photoelectron energies (i.e. high bonding energies). This increase is due to the cumulative effect of photoelectron peak tails on the low-energy side. However, background is relatively low, thanks to X-ray excitation. The resulting signal-to-noise ratio is currently of the order of 10–100, much higher than for Auger spectrometry.

Photoelectron spectra of conducting or semiconducting materials generally display one or several weaker satellite peaks on the low-energy side of the photoelectron peaks. They are due to *plasmon excitation* and to the conduction band (Figure 16.1).

Additional parasitic effects due to *double excitations* may generate subsidiary peaks or alter line profiles; those effects are called the *shake-up* effect (second electron raised from the valence band to the conduction band) and the *shake-off* effect (second electron ejected from the valence band into the continuum).

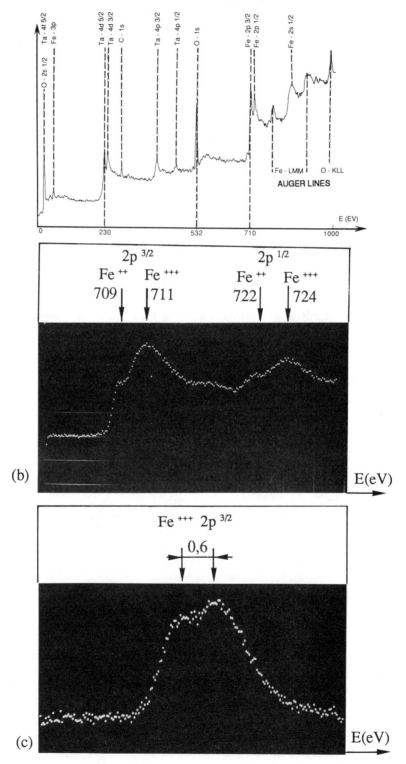

All energies are referred to the Fermi level. The specimen and the spectrometer are at the same electrical potential and thus have equal Fermi levels. As a result, the work function reduces to that of the spectrometer and Eq. (16.1) becomes

$$E_P = E_0 - W_X - e\Phi_{spectro} \qquad (16.2)$$

Non-conducting specimens may be affected by charge effects due to photoelectron ejections, causing surface potentials to be altered. For accurate measurements of bonding energies, it is therefore useful to refer to the photoelectron lines of a standard element, e.g. to the carbon-1s peak (usually present as a contaminant) or an added metal element (Carrière et al 1983).

16.1.3 APPLICATIONS

16.1.3.1 Measurement of Energy Levels

One of the first applications of photoelectron spectrometry has been an accurate measurement of energy levels of elements, leading to a better knowledge of the electronic fine structure of atoms and of molecules. The energy levels of all elements from lithium upwards have been tabulated by Siegbahn (in [10], see Appendix D.2).

16.1.3.2 Qualitative Surface Analysis

The characteristic energy levels of many elements have been tabulated; they may thus be identified by their photoelectron spectra. The whole set of elements from lithium upwards can thus be analysed. In addition to photoelectron lines, the X-ray-induced electron emission spectrum displays Auger lines, notably from light elements. They are usually distinguished from photoelectron peaks by their greater width (due to convolution of three level widths against one level width for photoelectrons), as well as by their position which does not depend on primary energy. In common practice the spectrum energies are referred to the primary energy; changing this energy by changing the anode element therefore leaves the photoelectron peak energies invariable, while shifting the Auger electron peak energies (Figure 16.3).

Currently observed detection limits:

1. *Minimum detectable mass fraction:* $c_{min} \cong 10^{-3}$;
2. *Minimum detectable mass* in an analysing volume of diameter 1 mm and thickness 10 Å (analysing volume of the order of 10^{-9} cm^3): $M_{min} \cong 10^{-11}$ g;
3. *Minimum number* of atoms detectable in the same volume: $N_{min} \cong 10^{10}$–10^{11}.

Figure 16.3. *(opposite)* Photoelectron and Auger electron spectrum from magnetite Fe$_3$O$_4$, with Ta as a standard. ESCA spectrometer with monochromator and Al–Kα source. Resolution 0.5 eV; diameter of analysing area 7 mm. (a) Global spectrum. Note the greater width of Auger lines; (b) part of the spectrum provided by integrating 100 spectra in an MCA, see the split of the 2p$^{1/2}$ and 2p$^{3/2}$ peaks of iron corresponding to Fe^{2+} and Fe^{3+}; (c) fine structure of the Fe^{3+} peak; two peaks separated by 0.6 eV, respectively corresponding to atoms in tetrahedral sites and in octahedral sites, have been resolved by deconvolution (from Baltzinger et al 1983). Lab. Crystallography, Louis Pasteur Univ., Strasbourg).

Table 16.1. Chemical shifts in eV for a given energy level of some elements, according to their oxidation degree (from Bouvy, 1972)

Element and level	Oxidation degree									
	−2	−1	0	1	2	3	4	5	6	7
N (1s)			0	+4.5	+5.1		+8			
S (1s)	−2		0				+4.5		+5.8	
Cl (2p)		0				+3.8		+7.1		+9.5
Cu (1s)			0	+0.7	+4.4					
Eu (3d)					0	+9.6				

One-hundredth of a monolayer of atoms may be detected in good working conditions (see Appendix D.1).

The detection limits can be divided by a factor of 10^2–10^3 when operating with a high-intensity synchrotron X-ray source.

16.1.3.3 Surface Bonding

A photoelectron spectrum includes peaks from external, bond-sensitive energy levels. Photoelectron line positions can thus be directly related to the chemical state of the respective elements by referring to the corresponding positions of pure element lines. Measuring the *chemical shift* provides useful data on solid surface bonding, e.g. oxidation degrees of metals, coordination, etc. In the case of an oxidized metal surface, a fine structure of its photoelectron peak appears on the spectrum; this consists of lines from the pure metal, together with the lines related to its various oxidation states (e.g. spectrum of Figure 16.3 and Table 16.1).

For oxidation degrees and coordinations to be significantly investigated, chemical shifts must be measurable to less than 1 eV. A high-resolution spectrometer with monochromator is required, together with a statistical spectrum smoothing by integrating up to 100 successive spectrum scannings into an MCA.

16.1.3.4 Quantitative Surface Analysis

Quantitative analysis by photoelectron spectrometry is theoretically possible. In practice, it is limited to semi-quantitative analysis, due to the high surface state sensitivity of electron emission.

The intensity measured from a given photoelectron peak of an element A to be analysed is proportional to the following parameters:

1. *Excitation cross-section* σ of the correspondent level of the element.
2. *Escape depth* of the electrons to be measured. It may be approximated to the mean free path l.
3. *Transmission factor* τ of the detector–spectrometer system.

When neglecting the shake-up and shake-off effects which cause a weakening of the characteristic peaks as well as X-ray absorption (negligible for the small depth), the measured intensity per unit area can be written as follows according to Eq. (13.13):

$$I_A = I_0 [N(V)\sigma l \tau]_A \qquad (16.3)$$

where $N(V)$ is the number of atoms A per volume unit.

By limiting the *concentration ratio* of two elements A and B as measured by the intensity ratio of two neighbouring peaks, by assuming that $l_A \cong l_B$ and $\tau_A \cong \tau_B$, certain parameters are eliminated and the ratio of the number of atoms becomes

$$\frac{N_A(V)}{N_B(V)} \cong \frac{I_A}{I_B} \frac{\sigma_B}{\sigma_A} \qquad (16.4)$$

Currently observed accuracy. The accuracy to be expected on surface concentrations of elements is of the order of 10–30%.

16.1.3.5 Applications to Materials

Photoelectron spectrometry is a fundamental method of solid surface analysis. Some examples of applications are as follows:

1. Study of corrosion and alteration effects of materials;
2. Development and monitoring of solid surface protection (painting, enamelling, anodising, nitriding, metal coating, etc.);
3. Investigation on absorption and adsorption processes on solid surfaces and on the related chemical reactions and catalytic effects.

16.2 AUGER ELECTRON SPECTROMETRY

16.2.1 BASIC PRINCIPLE

When excited at a given level, either by photons or by electrons, an atom undergoes relaxation according to two competitive processes, one of them resulting in emission of a characteristic Auger electron. Whereas X-ray emission is predominant for heavy elements, Auger emission is predominant for light elements (see Figure 7.4) and is therefore well designed for their analysis.

The energy E of an Auger electron leaving a solid surface is given by following relation, according to Eq. (7.10):

$$E \leqslant E_A = W_X - W_Y - W'_{Y'} - e\Phi \qquad (16.5)$$

where X is the primary excitation level, Y the origin level of the electronic transition, Y' the origin level of the Auger electron (given that level X is already excited, the energy of level Y' is greater than the first ionisation energy $W_{Y'}$, hence the notation $W'_{Y'}$) and $e\Phi$ the work function with a typical value of some 10 eV.

The Auger electron energy E_A corresponds to electrons emitted from a surface layer of thickness equal to the *escape depth* of Auger electrons which can be assimilated to their mean free path. The emission energy of Auger electrons being of the order of 10–1000 eV (see Table 7.5), it corresponds to an escape depth of a few ångströms (see Figure 8.2). Like photoelectron spectrometry, AES is therefore a method for analysing solid surfaces, i.e. the very first few atomic layers.

Photoelectron energies are characteristic of a single atomic level, whereas Auger electron energies depend on three levels. The resulting spectrum is therefore more complicated and more difficult to interpret.

The half-width of Auger peaks is typically of the order of 10 eV. Koster–Kronig peaks are wider, owing to shorter transition times. An additional widening at the low-energy side is due to Auger electrons emitted from greater depths which have undergone energy losses.

16.2.2 INSTRUMENTAL LAYOUT

As well as photoelectron spectrometry, AES requires to be operated in ultra-high vacuum (10^{-9}–10^{-10} Torr) to prevent any contamination layer which would screen the characteristic spectrum of intrinsic surface elements. Measurements indeed show that a stay of 2 s in a 10^{-6} Torr vacuum results in the build-up of a monolayer of oxygen atoms on a solid surface, whereas the same contamination would require 40 min in a 10^{-9} Torr vacuum.

Auger electron emission may be induced by X-rays (AES-X) or by electrons (AES-E).

16.2.2.1 X-ray Excitation

The X-ray induced Auger electron spectrum superimposes on the photoelectron spectrum as provided by an ESCA facility. In such a composite spectrum (Figure 16.3), the Auger peaks may be distinguished from photoelectron peaks by means of their greater width and by the fact that their energy does not depend on the primary energy (compare Eqs 16.1 and 16.5). The common instrumentation is outlined in Figure 16.1.

16.2.2.2 Electron Excitation

This technique is specific to AES. Its advantage is the possibility of focusing the incident electron beam.

The primary electron energy is commonly of a few kiloelectronvolts corresponding to the optimum excitation cross-section of K-levels in light elements and of more external levels in heavier elements.

Auger peaks emerge weakly from an intense secondary electron background (see Figure 7.1b; Langeron, 1988). Electronic differentiation increases the signal-to-noise ratio.

Retarding Field Spectrometer Combined with LEED

Electrons commonly used for AES and for LEED (see section 12.3) have similar energies. both techniques can therefore be combined. This combined operation (Scheibner and Tharp, 1967; Weber and Peria, 1967) is based on the retarding field (or retarding voltage) spectrometer (see Figure 10.11). This kind of spectrometer produces a cumulative intensity graph (Eq. 10.5). The actual $I(E)$ spectrum is provided by a derivation system based on superimposing a sine-shaped modulation (e.g. frequency 400 Hz) on the signal.

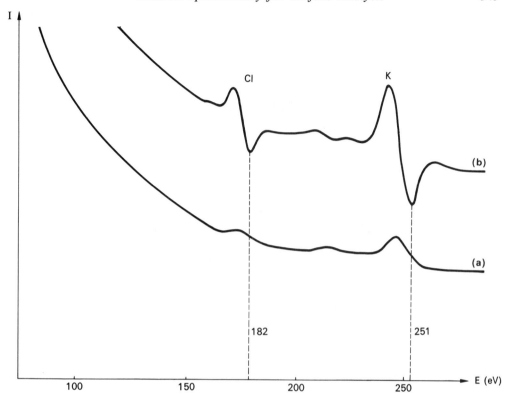

Figure 16.4. Auger spectrum from ultra-vacuum cleaved KCl as displayed by means of a retarding field spectrometer. (a) Spectrum $I(E)$, first derivative; (b) spectrum dI/dE, second derivative; the high-energy negative peak of the second derivative is conventionally used to define the peak positions (from Deville, 1972).

In order to refine the peak positions which are often displayed merely by inflection points of the spectrum, a second differentiation is carried out by detecting its second-order harmonic (Palmberg and Rhodin, 1968; Deville, 1968); Figure 16.4 illustrates the effect of differentiation.

Angular Dispersive Spectrometers

The spectrometers which are used in photoelectron spectrometry are also valid for AES. The most currently used seems to be the *cylindrical mirror electrostatic analyser* (*CMA*, Palmberg et al, 1969; see Figure 10.10a). Thanks to the direct display of the $I(E)$ spectrum, one electronic differentiation is sufficient for a better emergence of the Auger peaks from the background.

Similar electron spectrometers are made use of in the *Auger microprobe* (or *Auger scanning microscope*).

Combining AES with an *ion-sputtering* attachment results in the possibility of carrying out *concentration profiles* as a function of the depth.

Auger electron spectrometry may also be combined with *slow electron energy loss spectrometry* (*SEELS*) which provides data similar to those from transmission

EELS, but on surface layers of bulk specimens instead of thin films (Cazaux, 1985).

Due to long pumping and outgassing procedures, the whole cycle of operations leading to Auger analysis of a specimen is usually long (e.g. several hours); the same is true for photoelectron spectrometry.

Auger Microprobe and Scanning Microscope

Specific electron microprobes for Auger analysis, operating in ultra-high vacuum, have recently been developed. Fitted with a beam deflection system, such a facility actually becomes an Auger scanning microscope (see Figure 16.6). In addition to carrying out fixed beam analysis and depth concentration profiles (with ion sputtering), such systems provide elemental surface distribution maps, in particular of light elements such as oxygen, impossible to obtain by X-ray mapping. A resolution of 1000 Å is currently achieved with a voltage of 10–15 kV.

Using high-energy primary electrons (e.g. 100 keV electrons) and a field emission electron source results in a resolution reaching 100 Å (Cazaux, 1988; Cazaux et al, 1988).

Combining AES with electron energy loss spectrometry (EELS and EXELFS), with X-ray spectrometry (EDS) and with image processing, in a scanning transmission electron microscope (STEM), results in a powerful microanalytical facility for chemical and structural investigations of thin films and material surfaces, commonly called an *analytical electron microscope* (Ch. 21).

16.2.3 SPECIMEN PREPARATION

The conditions of specimen preparation are the same as for LEED and for photoelectron spectrometry.

Specimens must undergo electron bombardment in high vacuum and have a contamination-free surface. Even a monolayer of foreign atoms or molecules on the surface to be analysed would alter the analytical results. A contamination-free surface may be achieved in various ways:

1. *Cleavage or fracture* inside the vacuum enclosure, leading to the analysing of internal atomic layers of solids.
2. *Ion sputtering* of the surface to be analysed. The set-up of an ion source inside the Auger facility provides the additional ability to perform concentration profiles.
3. *Chemical* or *electrolytical surface dissolution* and various chemical reactions, according to the specimen and the aim to be achieved.

The two last-mentioned techniques may introduce foreign atoms and alter the chemical state of the surface. They are therefore to be performed with care.

16.2.4 APPLICATIONS

Among the first published researches by means of AES are those of Lander (1953); they highlighted the importance of the technique for analysing various surfaces of elements and oxides. The actual applications to surface analysis of materials began to develop in the late 1960s. As well as for ESCA, the main applications are in the field of solid surface physics and chemistry. Auger electron spectrometry leads to a better depth profile resolution than photoelectron spectrometry. Thanks to the ability of the primary electron beam to be focused on the specimen surface, AES can be developed into surface microanalysis. Similar to other spectrometric techniques, imaging is developing through SAM.

16.2.4.1 Qualitative Surface Analysis

Identifying Auger peaks is made difficult by their large number due to the three levels involved, by their large width favouring superpositions and by their sensitivity to chemical shifts. It is easier for light elements which have simpler spectra. The uncertainty of peak positions due to an ill-defined surface potential of the specimen (internal potential, electrical charges of non-conductors) is offset by referring to standard element peaks (e.g. absorbed carbon or nitrogen layer, Figure 16.3a).

Commonly observed detection limits. Orders of magnitude for reasonably good instrumental conditions, with light elements (see Table D.1):

1. *Minimum detectable mass fraction:* $c_{min} \cong 10^{-2}$;
2. *Minimum detectable mass* in an analysing volume of diameter 1 μm and thickness 10 Å (analysing volume of the order of 10^{-12} cm^3): $M_{min} \cong 10^{-16}$–10^{-15} g;
3. *Minimum number* of atoms detectable in the same volume: $N_{min} \cong 10^6$–10^7.

Parameters M_{min} and N_{min} are more favourable in the Auger microprobe, owing to a finely focused and intense incident beam. A minimum detectable mass as low as 10^{-19} g has been observed by Cazaux (1988).

Auger electron spectrometry is a sensitive technique for detecting small amounts of light atoms adsorbed on solid surfaces, notably of carbon, oxygen and nitrogen. A hundredth of an atomic monolayer may be detected in good operating conditions.

16.2.4.2 Surface Bonding

Chemical shifts of Auger peaks depend on three levels. Their relationship with chemical surface states is therefore less straightforward than with photoelectron spectrometry. They are typically of a few electronvolts for various degrees of oxidation (Carrière et al, 1985).

16.2.4.3 Quantitative Surface Analysis

The problems of quantitative analysis of surface layers by means of AES are similar to those encountered with photoelectron spectrometry. Auger emission is highly dependent on surface matrix effects, notably on the presence of contamination layers. It is therefore limited to semi-quantitative analysis.

The measured intensity I of an Auger peak, given by an element A to be analysed, is proportional to the following parameters:

1. *Excitation cross-section* σ of the considered level by primary electrons of energy E_0;
2. *Auger emission yield* $(1 - \omega)$, where ω is the fluorescence yield;

3. *Probability p* of the considered transition;
4. *Escape depth*, approximated to the *mean free path l* of the Auger electrons in the specimen;
5. *Transmission factor* τ of the detector–spectrometer system for the radiation considered.

An enhancement effect due to excitation by electrons backscattered in the specimen is expressed by means of the factor $(1+R)$, where R is the backscattering factor (Eq. 15.9). With monocrystalline specimens, an additional orientation-sensitive intensity variation is due to structure-related electron channelling. The resulting intensity as measured per surface unit is expressed as follows, according to Eqs (13.13) and (16.3):

$$I = I_0 N(V) \sigma (1 - \omega) p l \tau (1 + R) \qquad (16.6)$$

where $N(V)$ is the number of atoms A per volume unit.

Comparing with a test sample, containing the element to be analysed in a known surface concentration, permits elimination of factors σ, $(1-\omega)$, p, τ and to determine the surface concentration by the following relation:

$$\frac{N(V)}{N(V)_T} = \frac{I}{I_T} \frac{l_T}{l} \frac{(1+R)_T}{(1+R)} \qquad (16.7)$$

When dealing with spectra emitted from electronic differentiation, the difference between the positive and the negative peak is taken as the intensity measurement (Figure 16.4).

Currently observed accuracy. Depending on the nature and the concentration of the elements to be analysed, the accuracy to be expected is of the order of 10–30%.

16.2.4.4 Applications to Materials

The applications are similar to those of photoelectron spectrometry, i.e. determining the composition and the chemical state of the first few atomic layers of a solid. Combining with ion sputtering leads to depth concentration profiles, e.g. for investigating diffusion or segregation of elements at surfaces or interfaces. Quantitative analysis can be applied, for example, to measure the coverage rate of a surface-adsorbed atomic or molecular layer.

Auger microanalysis combined with Auger imaging enables the surface element distributions to be visualised, in particular of light elements impossible to localise by means of X-ray mapping. Local oxidation of surface elements may thus be imaged (Figure 16.5).

16.3 CONCLUSION

Photoelectron spectrometry and Auger spectrometry have developed into fundamental tools for chemical surface analysis, at a depth of the order of some 10 Å. Their applications range over the whole field of surface physics and chemistry. Their sensitivity allows detection of a hundredth of a monolayer, but their accuracy is limited by important surface matrix effects, in particular by contamination.

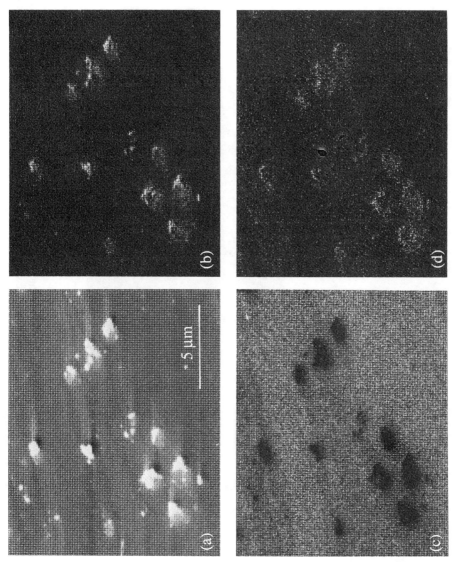

Figure 16.5. Gold particles, vacuum deposited on a silicon surface. (a) Secondary electron emission image; (b) Auger emission image from gold; (c) Auger emission image from silicon; (d) Auger emission image from oxygen, outlining a local gold-induced oxidation of silicon. $G_{dir} = 5000$; acquisition time 240 s (document from RIBER, Lab. Crystallography, IPCM, Louis Pasteur Univ., Strasbourg).

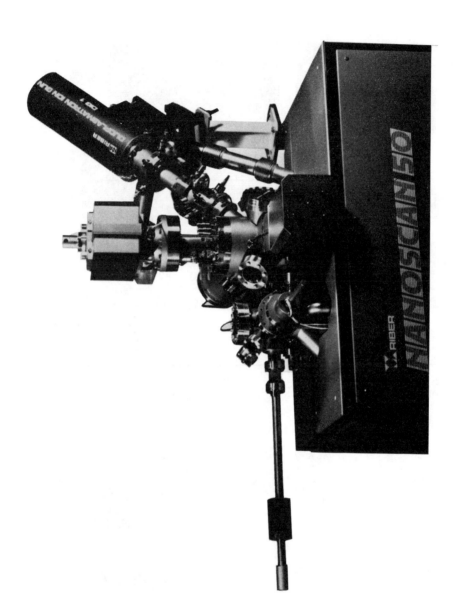

Figure 16.6. General view of the column of an SAM. At the top, the ion pump and the field emission electron source. On the right-hand side the ion gun. At the rear the spectrometer. On the left-hand side the specimen introduction device and the preliminary preparation chamber. At the front the specimen controls (document from RIBER).

The general tendency is a rapid development of microanalysis and of composition imaging. For a sufficient resolution, a finely focused primary beam is required. For a reasonable signal-to-noise ratio, a high-intensity beam is necessary.

Photoelectron spectrometry and Auger electron spectrometry both have their advantages and their drawbacks.

As far as the signal-to-noise ratio is concerned, photoelectron spectrometry is favoured with respect to Auger electron spectrometry; thanks to a much lower emission of secondary electrons, it reaches 10–100 for photoelectrons against 0.1–1 for Auger electrons. An additional advantage of photoelectron spectrometry is the low level of irradiation damage which does not limit the count time, as well as the ease of spectra interpretation. On the other hand, photoelectron microanalysis has been hindered until recently by the lack of intense and finely focused X-ray beams. New developments are predictable with the advent of synchrotron X-ray sources enabling high-intensity X-ray beams to analyse areas as small as 1 μm in diameter. With an additional focusing of primary X-rays by means of Fresnel lenses, a space resolution of down to 500–1000 Å becomes possible.

Auger electron spectrometry has become a microanalytical technique in the form of the Auger microprobe which is being continuously improved. A 500 Å resolution is already available; it may be improved to less than 100 Å when operated with a STEM facility. The combination with electron energy loss spectrometry (chemical analysis by EELS, structural analysis by EXELFS) seems particularly fruitful.

REFERENCE CHAPTERS

Chapters 2, 7, 8, 9, 10, 13. Appendix D.2.

BIBLIOGRAPHY

[1] Agius B., Froment M. et al *Surfaces, interfaces et films minces*. Dunod-Bordas, Paris, 1990.
[2] Briggs D., Seah M. P. *Practical surface analysis by Auger and X-ray photoelectron spectroscopy*. John Wiley, New York, 1983.
[3] Dekeyser W., Fiermans L., Vanderkelen G., Vennik J. (Eds) *Electron emission spectroscopy. Proc. NATO Summer Inst. Gent, 1972*. Reidel, Dordrecht, 1973.
[4] Feldmann L. C., Mayer J. W. *Fundamentals of surface and thin film analysis*. North-Holland, New York, 1986.
[5] Ghosh P. K. *Introduction to photoelectron spectroscopy*. John Wiley, New York, 1983.
[6] Grasserbauer M., Dudek H. J., Ebel M. F. *Angewandte Oberflächenanalyse*. Springer-Verlag, Berlin, 1986.
[7] Ibach H. *Electron spectroscopy for surface analysis. Topics in current physics*, Vol. 4. Springer Verlag, New York, 1977.
[8] Richardson J. H., Peterson R. V. *Systematic materials analysis*, vols 1–2. Academic Press, New York, 1974.
[9] Shirley D. A. (Ed.) Electron spectroscopy. *Proc. Intensity Conf. Asilomar, USA, 1971*. North-Holland, Amsterdam, 1972.
[10] Siegbahn K. *Atomic, molecular and solid-state structure studied by means of electron spectroscopy*. Almquist-Wicksells, Uppsala, 1967.
[11] Thompson M., Baker M. D., Christie A., Tyson J. F. *Auger electron spectroscopy*. John Wiley, New York, Chichester, 1985.

17

X-Ray Absorption Spectrometry and Electron Energy Loss Spectrometry

Any radiation–matter interaction involves absorption effects and energy loss effects of the primary beam. Due to the fact that absorption effects are complementary to secondary emission effects, they theoretically provide similar information. To the characteristic *emission lines* of the secondary emission spectrum correspond characteristic *intensity and energy losses* in the transmitted beam spectrum. These losses are expressed by minima of the transmitted intensity which correspond to maxima of excitation cross-sections of atomic levels in the material concerned. On the transmitted spectrum, they appear as *edges* at the related energies.

The main absorption techniques for material analysis are X-ray absorption spectrometry and electron energy loss spectrometry. As well as the corresponding emission techniques, these absorption techniques are similar as far as the data provided are concerned. Compared to emission spectrometry, they do not depend on a competition between radiative and non-radiative relaxations and therefore have a good efficiency for analysing light elements.

Spectra of both X-ray absorption spectrometry and electron energy loss spectrometry display a fine structure around the absorption edges.

An **emission spectrum** results from only one type of interaction. It is therefore relatively simple, with separate lines corresponding to different elements. The emission lines are usually well defined and their intensity is easily measured in view of quantitative analysis.

An **absorption spectrum**, on the other hand, results from the sum of any effects which produce a decrease of the number and the energy of transmitted photons or electrons. It therefore displays edges on a continuous background which decreases with decreasing energy. For a compound, this gives rise to overlapping and results in an often ill-defined spectrum with a high background. Interpreting such a spectrum is more difficult than for an emission spectrum, notably for relating intensities and concentrations.

Edge positions (absorption or energy loss) lead to qualitative and quantitative elemental analysis.

Fine structures on and around edges provide data on the electronic structure of atoms and on their environment.

X-Ray absorption spectrometry has mainly developed through fine structure analysis for structural purposes. Electron energy loss spectrometry, on the other hand, is mainly applied to elemental microanalysis.

Due to the interdependence of absorption and emission, fine structure effects also exist on emission spectra and may be used for structural investigations. Because of their basic principles, techniques based on fine structure analysis may be considered from the diffractometric as well as from the spectrometric point of view.

The principles of both X-ray absorption spectrometry and electron energy loss spectrometry have been known for a long time; their recent development for material analysis has been linked for the first to the availability of high-intensity polychromatic X-ray sources (synchrotron sources), for the second to the advent of electron microscopes with nanometre-sized electron probes of high brightness.

Common abbreviations. *X-rays*:

1. XAS (X-ray absorption spectrometry);
2. EXAFS (extended X-ray absorption fine structure);
3. SEXAFS (surface EXAFS);
4. XANES (X-ray absorption near edge structure);

Electrons:
1. EELS (electron energy loss spectrometry);
2. EXELFS (extended electron energy loss fine structure);
3. SEXELFS (surface EXELFS).

17.1 X-RAY ABSORPTION SPECTROMETRY (EXAFS)

17.1.1 BASIC PRINCIPLE

X-ray absorption is mainly due to atomic-level ionisations (photoelectric effect or true absorption). When neglecting inelastic scattering (Compton effect), the energy of transmitted photons is equal to their incident energy; intensity losses alone have therefore to be considered.

17.1.1.1 Absorption Spectrum

Consider an incident polychromatic X-ray beam, with an energy range as large as possible, transmitted through a specimen containing elements J ($J = A, B \ldots$). The outgoing spectrum will then display an absorption edge for each excitation energy W_X ($X = K, L \ldots$) of elements J (Figure 17.1a). Measuring the energies corresponding to the positions of these edges would theoretically provide the values of electronic level energies of the constituents, i.e. the same information as photoelectron spectrometry. In a similar way to the latter, X-ray absorption spectrometry used as a 'fingerprint' technique would thus serve for qualitative analysis. Measurement of the height of the edges would lead to quantitative analysis. In fact things are less straightforward, for various reasons:

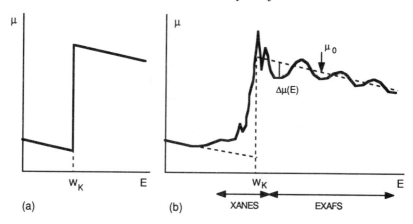

Figure 17.1. Outline of the absorption spectrum of an element in the vicinity of a K-edge, as a function of the incident photon energy (compare to Figure 8.1b where absorption has been plotted against wavelength). (a) Theoretical shape accounting merely for a single K-excitation, without any additional effects; (b) actually observed spectrum shape with fine structure.

1. The primary continuous background intensity from an X-ray tube is rather low, resulting in low count rates and long-lasting experiments. This drawback is avoided by synchrotron sources.
2. Due to the spectrum shape, the effects emitted from different levels of different elements overlap and must be extracted from a high background for element analysis.
3. The absorption spectrum as a function of incident energy in fact does not have the simple theoretical shape, but displays a *fine structure* before, around and after the theoretical edges (Figure 17.1b).

17.1.1.2 Fine Structure

The simplified shape of a K-absorption edge, as outlined by Figure 17.1a, is based on the assumption that the only interaction effect leading to X-ray absorption is the ejection of a K-electron into the continuum. The edge then corresponds to the bonding energy W_K of the ejected electron. Further processes actually intervene, notably raising of the electron to the valence band or the conduction band, and multiple ionisations. In addition, the photoelectron-associated wave undergoes backscattering by neighbouring atoms and returns to interfere with the emitted wave. It is this interference effect which leads to periodic intensity oscillations beyond the absorption edge and which is the most used for structural analysis.

Absorption and emission effects are complementary, in keeping with the energy conservation principle; as a result, this fine structure can similarly be observed in corresponding emission spectra of X-rays, photoelectrons and Auger electrons.

Fine structures around the absorption edges were observed as early as 1930 by Kronig. Their structural meaning has been assessed by Stern et al (1975). The practical development of the related technique has been accelerated by the availability of synchrotron sources.

Observed Effects

The following features can be observed around a given absorption edge (consider a K-edge), as a function of increasing energy (Figure 17.1):

1. *Fine structure before and on the absorption edge*, thus altering its shape with respect to that predicted by the elementary single excitation theory.
2. *Fine structure beyond* the absorption edge, extending typically up to about 1 keV from the edge. It consists of decreasing undulations.

The *EXAFS technique* is concerned with fine structure beyond the absorption edge; it will be the only X-ray absorption technique to be described in some detail in this section. The designation *secondary EXAFS* is used by analogy for correlated fine structure effects in the corresponding emission spectra (X-ray fluorescence, photoelectrons and Auger electrons).

The term *SEXAFS* (*surface EXAFS*) designates the particular fine structure of electron emission peaks (photoelectrons, Auger electrons, secondary electrons) which characterise surface atoms.

The *XANES technique* is concerned with fine structure on and near the absorption edge.

Interpretation of the Extended Fine Structure. EXAFS

For an incident X-ray beam of energy E_0, the dominant absorption effect is core-level excitation resulting in photoelectron emission. The initial photoelectron energy from a K-level of an atom A is given as follows (Eq. 7.1):

$$E = E_0 - W_K \qquad (17.1)$$

When the increasing energy E_0 reaches W_K, the sudden increase of the excitation cross-section results in the K-absorption edge.

A fraction of the wave associated with the outgoing photoelectron is backscattered by surrounding atoms J close to the source atom A. The scattered waves returning to the source atom A, before the end of its excitation state, will interfere with the outgoing wave, their amplitude adding or subtracting according to the phase shift, i.e. the distance of neighbours J. Varying the incident energy E_0 results in variation of the photoelectron-associated wavelength, according to Eq. (17.1), and thus the phase of the backscattered wave. This leads to a periodic amplitude variation of the resulting photoelectron wave as a function of the incident energy.

The absorption probability of an incident X-ray photon by exciting a K-electron depends not only on the initial state of the electron (energy W_K corresponding to the edge), it also depends on its final state represented by its associated wave. Now this wave is the sum of the outgoing wave and of the backscattered waves interfering during the lifetime of the excited state. The resulting absorption spectrum will therefore show in turn those periodic undulations which are actually observed.

Interpreting the EXAFS effect may therefore be carried out by means of a classical *interference function* of two or more spherical waves. The path difference between direct and backscattered waves is determined by the distances of neighbouring atoms to atom A taken as phase reference. The following approximations lead to a simplified model, the *single scattering assumption* (Figure 17.2):

1. *Single central atom* with a single core-level excitation generating a single photoelectron without any secondary interaction with more external levels.
2. *Single backscattering* of the outgoing photoelectron wave by nearest neighbour atoms. Due to the short range of the photoelectron spherical wave and to its amplitude decreasing with the squared distance, multiple scattering may be neglected.
3. *Small atoms* with respect to their distances. The part of the direct photoelectron wave incoming on a scattering atom may therefore be assimilated to a plane wave. This approximation holds for a sufficiently high photoelectron energy (i.e. far from the edge), enabling it to penetrate to the core shells which thus determine the apparent atom size.

Using these approximations results in simplifying the interference function to a limited number of scattered waves interfering with the outgoing wave.

When dealing with elastic scattering of a plane wave by a distribution of scattering elements (see section 3.2), energy (i.e. wavelength) is a constant; the scattered amplitude is plotted against the angle or the scattering vector $\mathbf{R}$ as $F(\mathbf{R})$. In EXAFS, the spherical wave emitted from the source atom inside the specimen undergoes a scattering angle $2\theta = \pi$, resulting in a scattering vector of length $R = 2/\lambda$. Amplitude variations are plotted against energy. This amounts to exploring reciprocal space by means of a reflecting sphere with varying radius.

In order to represent the interference function in a conventional way, the photoelectron energy is to be converted into a *wave vector* of length $k = 1/\lambda$ (*wave number*), given by

$$k = \left(\frac{2m}{h^2}(E_0 - W_K)\right)^{1/2} \quad (17.2)$$

where h is the Planck constant and m the electron mass (rest mass, justified by low energy).

The significant parameter of the observed spectrum is the amplitude of the absorption coefficient oscillations as a function of energy, with respect to the mean absorption coefficient in the absence of any interference. It is expressed by the following normalised absorption function in which the energy E is referred to the excitation energy of the considered level:

$$\chi(E) = \frac{\mu(E) - \mu_0(E)}{\mu_0(E)} \quad (17.3)$$

where $\mu(E)$ and $\mu_0(E)$ respectively are the mass absorption coefficient in the presence and in the absence of interference effects.

The phase of the primary electromagnetic wave on atom A is taken as reference. With the emitted photoelectron wave amplitude as unit, the amplitude of the spherical wave on an atom J at a distance r_J from the reference atom has the form $(1/r_J) \exp 2\pi i k r_J$.

After being scattered by atom J and having travelled a distance $2r_J$ to return to atom A, the amplitude becomes

$$(1/r_J^2) f_J(k) \exp 4\pi i k r_J$$

where f_J is the backscattering amplitude of atom J.

In addition to the phase shift due to the wave path, the electron wave undergoes further phase shifts, respectively, $\phi_{AJ}(k)$ through potential variations along the path there and back and $\pi/2$ through elastic scattering by atom J. The resulting amplitude is then written:

$$(1/r_J^2) f_J(k) \exp i[4\pi k r_J + \phi_{AJ}(k) - \pi/2]$$

The phenomenon is centred on the reference atom A. Only the absolute values of the neighbour distances therefore intervene in phase shifts (Figure 17.2). The surrounding atomic distribution is thus considered as a radial distribution and the resulting amplitude can be expressed by its real part.

Writing the normalised absorption coefficient of Eq. (17.3) as a function of the wave number now results in the following interference function, commonly called the *EXAFS function*:

$$\chi(k) = \sum_J \frac{N_J f_J(k)}{r^2} \sin[4\pi k r_J + \phi_{AJ}(k)] \qquad (17.4)$$

where N_J is the number of atoms J at a distance r_J and $f_J(k)$ the backscattering amplitude of atom J (atomic scattering amplitude for $2\theta = \pi$).

Phase shift $\phi_{AJ}(k)$ includes both the phase shifts at emission from atom A and at the return to atom A. It depends on energy and on the atoms A and J. The atomic backscattering amplitudes $f_J(k)$ also depend on energy.

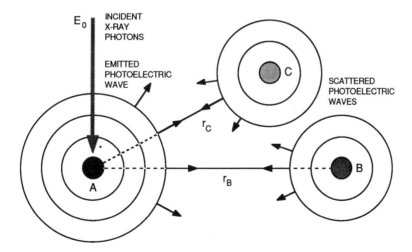

Figure 17.2. Elementary interpretation of EXAFS. The photoelectron wave emitted from the reference atom A is backscattered by two neighbour atoms B and C, respectively at distances r_B and r_C.

Equation (17.4) is a simplified expression which does not take into account the following additional effects:

1. *Temperature and disorder effect* (Debye–Waller factor, Eq. 4.11) which results in the oscillations to be damped out as a function of k.
2. *Coherence losses* due to inelastic scattering and to the limited lifetime of the excited state; this effect is expressed in the form of a mean free path.
3. *Multiple excitation effects*, such as shake-up and shake-off effects (Ch. 16).

To any atom J at a distance r_J corresponds a sine function $\chi(k)$, of spatial frequency $2r_J$ in the k space (reciprocal space) and of phase $[4\pi k r_J + \phi_{AJ}(k)]$. For the whole set of neighbouring atoms, the sum of those sine functions results in the EXAFS function, to the above-mentioned approximations.

The greater the distance r_J of a scattering atom J, the greater the space frequency $2r_J$ of the corresponding oscillations and the lower their amplitude (variation with $1/r^2$). This fact justifies neglecting both the effect of second neighbours and of multiple scattering.

Function (17.4) contains data on the distances of atoms surrounding the reference atom. In a similar way as for elastic scattering, they may be extracted from the spectrum $\chi(k)$ by *Fourier transform* (*FT*, see Eqs 3.6 and 3.7). To any amplitude $\chi(k)$, at a point k (or an energy E) of the EXAFS function, corresponds a sine-shaped distribution function of spatial frequencies k in direct space. The sum of those sine functions over all k values represents the *distribution function of the distances of neighbouring atoms*, i.e. the FT of amplitudes $\chi(k)$.

As a final result, the FT of the EXAFS function provides the radial distribution of nearest atoms around the reference atom.

Fine Structure near to the Absorption Edge

The fine structure observed on the absorption edge and close to it, up to about 50 eV, is not yet fully understood. For a photoelectron energy close to zero ($E \cong W_X$), the small atom approximation no longer holds and multiple scattering becomes significant. The near-to-the-edge fine structure is therefore much more sensitive to charge distributions in atoms. It is notably connected to the electronic structure of the reference atom, to state densities and to multiple excitations. Up to now its interpretation remains semi-quantitative.

17.1.2 PRACTICE OF EXAFS SPECTROMETRY

The following stages of operation may be considered for extracting structural data from a material through EXAFS:

1. Selecting the absorption edge to be investigated, generally a K-edge.
2. Recording the absorption spectrum by measuring the intensity $I(E_0)$ of an X-ray beam of increasing photon energy E_0 after transmission through the specimen. The recording is usually limited to an energy range of about 1 keV beyond the selected absorption edge.

3. Converting the spectrum $I(E_0)$ into a function $\chi(E)$ or $\chi(k)$.
4. Extracting the radial distribution function by FT.

17.1.2.1 Instrumental Layout

Recording an accurate transmission spectrum for EXAFS analysis requires the following attachments:

1. Polychromatic X-ray source of highest possible intensity.
2. Monochromator of sufficient resolution in order to select a linearly varying energy E_0, usually a focusing monochromator.
3. X-ray detector. An additional electron detector is required for secondary EXAFS.
4. Data acquisition and processing system.

X-ray Source

For dealing with the K-absorption edge of a given element, the primary polychromatic X-ray beam must meet the following requirements:

1. Cut-off energy E_{max} greater than the minimum excitation energy W_K of the selected K-level.
2. Intensity high enough for a reasonably good count statistic, for the interference function χ to be well defined.

X-ray Tube. EXAFS may be carried out by a laboratory facility provided that any parameters are optimised. The source is then an X-ray tube with a heavy anode element excited at the highest compatible voltage for the highest energy cut-off. A fine-focus high-power tube (usually with rotating anode) is combined with a focusing monochromator. However, the possibilities of such a set-up are limited by:

(a) the low bremsstrahlung efficiency (see Eq. 7.5), resulting in long acquisition times (1–20 h).
(b) the limited energy range as determined by the practically available voltage.

Synchrotron Source. The best presently available X-ray source is synchrotron radiation which has a high intensity, together with a small beam cross-section and a wide energy range which may reach 200 keV from a storage ring fitted with wigglers. The energy range is then sufficient for exciting K-levels of any element. Spectrum acquisition time drops to a few minutes. EXAFS has in fact developed thanks to the availability of such sources.

Monochromator and Focusing

For plotting the spectrum around a given absorption edge, the incident photon energy must be scanned linearly from the corresponding minimum to the maximum energy. This is carried out by means of a monochromator which is usually placed in front of the specimen; however, a post-specimen layout may be used. An energy resolution of the order of 1 eV is required for extracting the fine structure data from the absorption spectrum.

With an *X-ray tube*, a linear focusing crystal spectrometer (see Figure 10.7) has the highest efficiency and resolution.

With a *synchrotron source*, the low beam cross-section and divergence allows use to be made of a flat crystal monochromator. A set of two parallel crystals provides a monochromatic beam in the same direction as the incident beam (Figure 17.3), resulting in an easier set-up on a synchrotron beam line. A combined rotation–translation motion maintains the transmitted beam in a line with the incident beam; a system of two symmetrical double crystal monochromators provides a similar result without any translation, but attenuates the intensity. An additional focusing system improves both efficiency and resolution. It is mostly based on total reflection of the X-ray beam, e.g. by means of paraboidal mirrors placed in grazing incidence on both sides of the monochromator.

Monochromators and mirrors have to sustain the very intense synchrotron beam. Silicon or germanium single crystals are currently used as monochromators, while silica glass or ceramics are commonly used materials for mirrors (e.g. gold- or platinum-coated silicon carbide).

X-ray energy scanning is performed by rotating the monochromator. To deal with a given K-edge, the energy is made to vary from W_K to about 1 keV beyond. For EXAFS itself, spectrum processing starts at 50 eV beyond the edge in order to avoid any interference of the near-to-the-edge fine structure. The spectrum acquisition system carries out statistical smoothing and background subtraction.

17.1.2.2 Setting up the Interference Function

The measured quantity is intensity $I(E_0)$, whereas the significant one is the interference function $\chi(k)$. The latter is extracted by means of data processing according to the theoretical formulation: conversion of measures $I(E_0)$ into the absorption function $\mu(E)$, then into the normalised absorption function $\chi(E)$ and finally into the interference function $\chi(k)$.

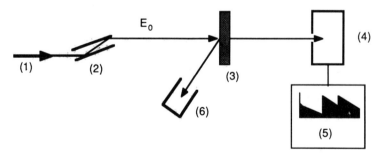

Figure 17.3. Diagram of a common layout for EXAFS. (1) Incident polychromatic X-ray beam (synchrotron beamline); (2) flat double crystal monochromator; (3) specimen; (4) X-ray detector; (5) data acquisition and processing system; (6) electron detector (e.g. channeltron) for global secondary electron emission SEXAFS; an additional electron spectrometer is required to select one emission mode (e.g. photoelectron or Auger electron).

Absorption Function

The intensity–absorption relationship (Eq. 8.1) leads to

$$\mu(E) = \frac{1}{\rho x} \log \frac{I_0}{I} \qquad (17.5)$$

where ρ is the density and x the thickness of the specimen.

The measured absorption function (Figure 17.4a) consists of the cumulative absorption effects of the excited levels of all elements in the specimen, the element to investigate A as well as the matrix elements $J \neq A$. The contribution of the single K-edge effect of reference element A has to be extracted from this composite spectrum. This can be done by adjusting a theoretical function $\mu'_0(E)$ to the experimental continuous absorption spectrum before the edge, extrapolated to energy E, and subtracting it from the global experimental spectrum.

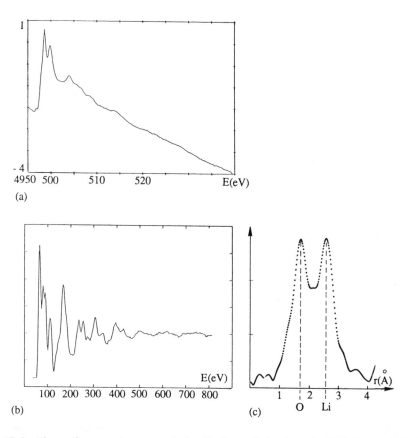

Figure 17.4. Absorption spectra around the K-edge of titanium in $LiTi_2O_4$ at 45 K (ionic superconductor). (a) Experimental $\mu(E)$ spectrum; (b) normalised EXAFS spectrum $\chi(E)$ (50–800 eV); (c) radial distribution function $q(r_j)$ resulting from Fourier inversion (spectra from Kappler J. P., IPCMS–CNRS, Louis Pasteur Univ., Strasbourg).

Interference Function

In order to normalise the absorption function (Eq. 17.3), the function $\mu_0(E)$ expressing absorption of pure element A in the absence of any interference effects (graph of Figure 17.1a) has to be determined. In a similar way as for the continuous matrix background, the procedure consists of least-squares adjusting of an arbitrary function to the mean value (without undulations) of the absorption spectrum beyond the edge. Subtracting and normalising leaves theoretically the sole undulation component $\chi(E)$ with a mean value equal to zero (Figure 17.4b). Equation (17.2) determines the conversion to the interference function $\chi(k)$ with its origin taken at the energy position $(E_0 - W_K)$ of the edge.

17.1.2.3 Extracting Structural Data

The available structural data are extracted from function $\chi(k)$ by means of an FT. The result is theoretically a distribution function of atoms around the reference atom, in a similar way as when derived from elastic scattering (Figure 17.4c).

Various techniques are employed; they are similar to those leading to structure analysis by X-ray diffraction.

Direct Fourier Transform

The knowledge of function $\chi(k)$ at any points k from 0 to ∞ should fundamentally lead to the distribution function around reference atom A by calculating the FT. There are in fact constraints:

1. The interference function is only defined in a restricted range

$$50\,\text{eV} < E < 1000\,\text{eV} \Rightarrow 3\,\text{Å}^{-1} < k < 16\,\text{Å}^{-1}$$

   thus limiting resolution and leading to filtering artefacts. Statistical disorder speeds up convergence and lessens the contribution of high spatial frequencies. Damping out of χ undulations vs energy may be somewhat compensated by multiplying by a factor k^n ($n \cong 3$) in order to favour high values of k.
2. Phases ϕ_{AJ} and atomic scattering amplitudes f_J vary with k. The phase origin is ill defined with respect to the theoretical edge position.
3. Ancillary effects, such as inelastic scattering, are additional causes for deviation from the ideal interference function which would result from purely elastic backscattering.

These drawbacks have led to indirect methods.

Fitting of Experimental and Theoretical Spectra

The best agreement between the experimental spectrum and the spectra calculated from structure models is approached by adjusting atomic distances. Complicated structures with various neighbour atoms result in a large number of parameters to be adjusted and in uncertainties. The problem is simplified when the structure to be investigated is closely related to a known structure which may be taken as a reference structure (e.g. substitution of various elements in a basic structure; crystalline analogues for investigating a glass structure).

Fourier Filtering

Fourier filtering reduces the structure model to a single neighbour element. After a first FT on the experimental spectrum, the inverse transform is carried out on a single peak of the resulting distribution function, i.e. corresponding to a single neighbour. The resulting filtered spectrum is fitted to spectra of single-neighbour models by adjusting their atomic distances.

17.1.2.4 Possibilities and Limitations

One of the main advantages of the EXAFS method is the possibility of selecting a reference element by its absorption edge and subsequently to investigate its surroundings. EXAFS acts therefore as a valuable probe for monitoring the local structure around a selected reference atom. It applies to any kind of material, crystallised or amorphous solid, or even liquid. As an example, it is notably useful for investigating glassy structures.

The *lower atomic number* limitation is determined by the absorption cross-section which becomes low for light elements. Carbon ($W_K = 280$ eV) may therefore be considered as the lower limit.

The *higher atomic number* limitation is determined by the required excitation energy and by the natural level width. The latter increases with increasing atomic number and becomes larger than 1 eV for $Z > 50$ (Xe), i.e. larger than the resolution limit. Heavy elements may be investigated by means of the L-edges which lead to similar results.

According to Eq. (17.4), the *coordination* of an element may be approached by means of the relationship between function $\chi(k)$ and the number of neighbours N_J, provided that all photoelectrons participate in the backscattering effect, which is not the case. In practice, the limit is set by uncertainties on parameters which determine the amplitude of the interference function (atomic scattering amplitude, multiple scattering, etc.). The relationship remains semi-quantitative.

Data on *dynamic disorder* (thermal vibrations) and *static disorder* are liable to be gathered from assessing the dumping-out of the $\chi(k)$ function, expressed by an exponential Debye–Waller factor. Working at various temperatures leads to a separation of the two effects.

Currently Observed Accuracy. Accuracy on atomic distances is of the order of 0.02 Å, similar to that achieved by X-ray diffraction. More accurate measurements of differences in neighbour distances may be achieved through the beat frequency of oscillations.

Variants. Due to the interdependence of excitation and relaxation effects, interference fine structures may also be observed on the emission spectrum $I(E_0)$ induced by incident X-rays of varying energy E_0. Those complementary effects are therefore commonly called *secondary EXAFS*. For this purpose the intensity of a given characteristic emission is measured into a given solid angle and under a given incidence, as a function of the incident photon energy E_0. Two methods may be used:

1. *X-ray fluorescence.*
2. *Electron emission* (photoelectrons, Auger electrons, secondary electron emission induced by photoelectrons). A current method is to cumulate the whole electron emission. Owing to a mean free path limited to a few ångströms, this effect is issued from a thin surface layer. It provides information on the structural surrounding of surface atoms, hence its designation SEXAFS (surface EXAFS).

17.1.3 APPLICATIONS

EXAFS spectrometry is a complement to X-ray diffraction for structure analysis. Its applications for materials result from its characteristics as a probe for exploring local short-range structures around specific atoms, with an 0.02 Å accuracy.

It applies to any type of material, whether crystallised, vitreous or amorphous. It is notably interesting for non-crystalline materials for which conventional diffraction analysis is limited.

The most important data gathered by means of EXAFS concern chemical bonding of the selected reference element with its neighbours:

1. *Bond lengths* through the radial distribution function;
2. *Bond strengths* through determining the disorder factor by way of oscillation dumping-out (Debye–Waller factor);
3. *Number of near neighbours, coordination* through the amplitude of the distribution function peaks.

Examples.

Basic EXAFS techniques—non-crystallised solids. Investigating the surroundings and coordination of glass-lattice forming and modifying elements. It has been shown in silicate glasses that the glass-forming cations are strongly bonded, whereas the glass-modifying cations are linked to the glass lattice by bonds some 30 times weaker. The cation surrounding has been shown to be quite similar, to the local distortions, in glass and in the corresponding crystalline species. The existence of short-range order has thus been confirmed.

Phase transitions. Changes of valence, of coordination, of bond type, during phase transformations; order–disorder transformations in alloys.

Cluster structures. Structures of metal atom clusters have been related to their gas absorption capacity, important in solid catalysis.

SEXAFS variant. The spectrum is related to surface structure:

1. Solid surface structure analysis as a complement to LEED which is limited by dynamic effects.
2. Solid surface absorption and adsorption processes related to catalysis.

17.2 ELECTRON ENERGY LOSS SPECTROMETRY (EELS)

An electron beam is much more absorbed by a specimen than an X-ray beam of similar energy. In addition to intensity losses, electrons undergo energy losses through interactions with specimen atoms. Interaction probability is maximum for energy losses equal to electron-bonding energies in core levels of specimen atoms. As a result the spectrum $I(E)$ measured after transmission of a monokinetic incident beam of energy E_0 will display *energy loss edges* corresponding to any excited level.

Energy loss edge positions provide data on the elemental specimen composition.

Energy loss edge fine structures, similar to that observed by X-ray absorption, provide structural data.

Recall of common abbreviations:

1. EELS (electron energy loss spectrometry);
2. EXELFS (extended electron energy loss fine structure);
3. TEM (transmission electron microscope);
4. STEM (scanning transmission electron microscope).

A high-intensity and fine-focused primary electron beam, as well as a thin specimen (some 100 Å thickness) are required for efficient transmission electron energy loss measurements. EELS as a method of microanalysis has therefore quite naturally developed in connection with transmission electron microscopy (TEM or STEM) operating with primary electron energies in the 100–1000 keV range; for those facilities it has become an important microanalytical attachment which is complementary to thin specimen X-ray microanalysis (see Ch. 21).

The principle of EELS was discovered in the 1940s (Hiller and Baker, 1944). Its practical development began in the 1970s.

17.2.1 BASIC PRINCIPLE

Electrons travelling through matter lose their energy by successive interactions with atoms.

17.2.1.1 Experimental Parameters

Consider a monokinetic incident electron beam, of intensity I_0, energy E_0 and convergence semi-angle α_0. After being transmitted through the specimen, electrons of energy $E<E_0$ are analysed by means of an electron spectrometer of acceptance semi-angle α (Figure 17.5).

The spectrometer measures the intensity of electrons which have sustained a given energy loss ΔE, as transmitted into a solid angle Ω. Referred to the surface unit, this intensity is expressed by

$$I(\Delta E,\Omega) = I_0 N(S)\sigma(\Delta E, \Omega) \qquad (17.6)$$

where I_0 is the incident intensity per specimen atom, $N(S)$ the number of atoms per surface unit of thickness t and $\sigma(\Delta E, \Omega)$ the interaction per atom, integrated into the solid angle of measurement Ω (Eq. 13.14).

When scanning the spectrometer deviation field (magnetic or electrostatic analyser), the spectrum $I(\Delta E)$ is recorded by the detector.

364   *Structural and Chemical Analysis of Materials*

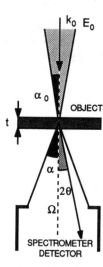

Figure 17.5. Diagram of electron energy loss spectrometry with the main instrumental parameters.

The interaction cross-section may be related to chemical composition and atomic structure of the specimen, in the analysing volume considered.

The outgoing intensity also depends on:

(a) the direction of emergence as defined by the wave vector **k** or by the emergence angle 2θ with respect to the incident direction;
(b) specimen thickness.

17.2.1.2 Observed Spectrum

The general shape of the energy loss spectrum $I(\Delta E)$ is outlined in Figure 17.6. It may be split into three sections, as given under the headings below.

Around Zero-energy-loss Region

This first region corresponds to *elastic and quasi-elastic* interactions ($\Delta E \cong 0$). Given that:

(a) the current electron spectrometer resolution is of the order of a few electronvolts;
(b) the incident electron energy dispersion ranges between 0.3 eV (field emission source) and 1 eV (thermionic source);

it follows that this zero-energy-loss region is characterised by a single peak of a few electronvolts in width. Its high intensity represents more than half of the total intensity integrated over the whole spectrum. This peak is caused by the following effects:

1. electrons transmitted without interaction;
2. elastically scattered electrons;
3. quasi-elastically scattered electrons. This effect, due to excitation of interatomic oscillations and phonon excitations in crystals, corresponds to energy losses ranging

X-Ray Absorption Spectrometry and Electron Energy Loss Spectrometry

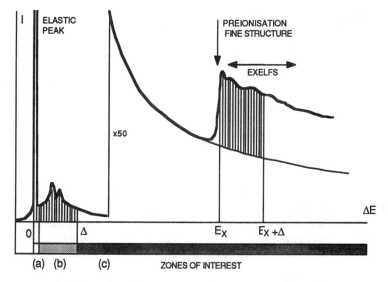

Figure 17.6. General outline of an electron energy loss spectrum, displaying the three significant regions. (a) Elastic and quasi-elastic scattering peak, $E \cong 0$; (b) valence and conduction excitation peaks (plasmon peaks for conductors), $E < 50$ eV; (c) core-level excitation, $E > 50$ eV, displaying one characteristic edge and its fine structure. Hatched areas outline the integration areas on a 50 eV window for the purpose of quantitative analysis. Arbitrary scales.

from about 0.1 eV up to a few electronvolts. It results in heat transfer and possible bond breaking. With available energy resolutions it is not separable from elastic scattering; it merely contributes to widening the peak.

Electrons corresponding to this peak do not provide any analytical data.

Low-energy-loss Region

Energy losses in this region ($\Delta E \leqslant 50$ eV) correspond to *valence-level and conduction-level excitations*.

A *conducting material* (metal, alloy) leads to the occurrence of discrete peaks of decreasing intensity for increasing ΔE. They are due to collective electron plasma oscillations, induced by incident electrons, with energy quanta called *plasmons*. For a sufficient specimen thickness, an incident electron may excite a single plasmon (first peak), two plasmons (second peak), three plasmons (third peak), etc. Excitation probability decreases with the order, resulting in decreasing intensities.

Plasmon peaks may provide valuable information about metallic compounds. Their shift can provide data on concentration variations of a given element in alloys. Measuring the ratio of successive peak order intensities P_0 and P_1 provides the specimen thickness t as follows:

$$\frac{I(P_1)}{I(P_0)} = \frac{t}{l(P)} \qquad (17.7)$$

where $l(P)$ is the mean free path of plasmon excitation.

Table 17.1. Order of magnitude of interaction parameters corresponding to various regions of the energy loss spectrum. Instance of primary 100 keV electrons interacting with light elements such as Al, Si (from data of Joy and Maher in Hren et al [3])

Type of interaction	ΔE (eV)	Integrated cross-section (cm²/at)	Mean free path (μm)	Half-width (mrad)
Elastic	$\cong 0$	10^{-18}	0.1	30
Excitation of conduction and valence levels	$\leqslant 50$	10^{-18}	0.1	10–20
Excitation of core levels	> 50	5×10^{-20}	2	2

An *insulating material* leads to energy losses corresponding to excited individual valence electrons into molecular orbitals (<15 eV) and atomic orbitals (>15 eV).

Interaction cross-section and mean free path are of the same order of magnitude as for elastic scattering (Table 17.1). The result is that cumulated intensities of regions (a) and (b) represent about 95% of the intensity integrated over the whole spectrum $I(\Delta E)$.

High-energy-loss Region

Energy losses in this region ($\Delta E > 50$ eV) correspond to *core-level excitations* which result in characteristic X-ray and Auger emissions. The spectrum consists of a decreasing continuous background, of characteristic edges and of a fine structure around the edges. The peak intensities are about 50 times lower than plasmon peaks, thus requiring a scale amplification of the same order.

This is the most important region: edge positions characterise specimen elements; fine structures around the edges provide data on the electronic and atomic structure of the specimen material.

The shape of the spectrum in this region and the data provided are similar to those of X-ray absorption spectrometry.

Continuous Background. The background is generated by the overlapping tails of absorption edges of the various specimen elements, as well as by valence and plasmon excitations. Due to overlapping of several effects, its accurate mathematical modelling is difficult. For a limited energy range, its decrease is reasonably expressed by an empirical function of the form

$$\frac{dI}{dE} = K_1 E^{-K_2} \qquad (17.8)$$

Parameters K_1 and K_2 are adjusted by least-squares fitting for the best agreement with the observed background. Knowledge of this function enables background subtraction in quantitative analysis.

Characteristic Edges. With decreasing ΔE the successive edges correspond to characteristic energy losses due to exciting the respective levels K, L, M ... of specimen atoms. They express the *primary excitation process* and are complementary to the relaxation process leading to characteristic secondary emissions (X-rays and Auger electrons). Their integrated cross-section is of the order of 100 times lower, in similar conditions, than that of elastic interaction; scattering of the corresponding electrons takes place in a much smaller angular range (Table 17.1).

Exploiting the characteristic edge spectrum for analysis is the purpose of EELS.

Fine Structure around Edges. EXELFS. A fine structure, similar to that in X-ray absorption, is observed around and after the energy loss edges. The structural information provided is similar.

The so-called *extended fine structure* at the K-edge of an element A consists of undulation extending up to some 10 eV at the high-energy side. They are due to the electron wave associated with secondary electrons emitted by the reference atom A and backscattered by neighbouring atoms, interfering with the outgoing wave during the excitation lifetime. By analogy with EXAFS, exploiting the extended energy loss fine structure is usually designated by the abbreviation EXELFS. Methodology and structural data gathered are similar and will not be detailed further.

Table 17.1 summarises the interaction characteristics of the respective regions of the spectrum.

17.2.2 INSTRUMENTAL LAYOUT

17.2.2.1 Spectrometer and Detector

The EELS technique is usually an attachment to a transmission electron microscope with scanning possibilities (TEM–STEM). The probe diameter on the specimen is of some 100 Å. The electron spectrometer is placed under the observation chamber and thus does not alter the imaging function of the microscope (Figure 17.7).

The *spectrometer* is usually a focusing magnetic prism analyser (see Figure 10.10c). Due to prism aberrations the admittance semi-angle is limited to about 10 mrad for a resolution of 1–2 eV. The resolution limit is determined by energy dispersion of incident electrons (about 1 eV for thermionic sources, 0.2 eV for field emission sources). For the aim of EELS analysis, a resolution of some 10 eV is generally sufficient. For fine structure analysis the maximum resolution is required.

The *detector* is usually a channeltron.

The rapid intensity decrease vs increasing energy loss ΔE requires an acquisition electronics of high *dynamic range*, e.g. an MCA (intensity ratio of about 10^5 to be measured between $\Delta E = 0$ and $\Delta E = 2000$ eV).

17.2.2.2 Specimen

For an increasing specimen thickness t, the signal-to-noise ratio increases, passes through a maximum and then decreases, together with edge deformations, due to multiple interactions. The optimum thickness is of the order of the inelastic interaction mean free path, e.g. 1000 Å at 100 keV for light atoms like carbon. The usual electron microscope specimens are therefore appropriate.

17.2.3 QUALITATIVE ANALYSIS. DETECTION LIMITS

Elemental identification is based on the positions of the characteristic energy loss edges in the spectrum acting as a 'fingerprint'.

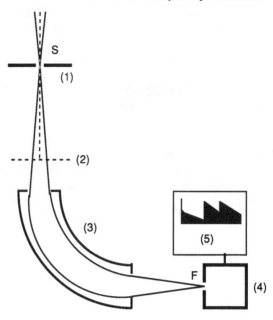

Figure 17.7. Diagram of an EELS attachment mounted on a TEM. (1) Differential vacuum pumping aperture, exit aperture S of the projection lens; (2) imaging screen of the electron microscope; (3) magnetic prism analyser focused on the exit aperture S and on the admittance slit F of the electron detector (4), e.g. channeltron; (5) data acquisition and processing system.

The detection limit is related to the signal-to-noise ratio (Eq. 13.9). The characteristic signal intensity measured at a K- or an L-edge is proportional to the corresponding excitation cross-section. The K-edges are commonly used for light elements, the L23-edge or the M23-edge for heavy elements.

For a given energy loss ΔE, the characteristic scattering angle is $\theta_E = \Delta E/2E_0$ (Eq. 2.9). In order to preserve a good collection efficiency, this angle should be smaller than the spectrometer admittance angle α which is limited to about 10 mrad.

With $E_0 = 100$ keV and $\theta_E \leqslant \alpha/2$, the measurable energy losses are limited to a value of about 1 keV.

The range of elements liable to be analysed is therefore:
K-edge   from 3-Li to 11-Na
L23-edge   from 1-Na to 30-Zn
M23-edge   from 30-Zn upwards

In similar conditions, electrons of 1 MeV energy would permit measurement of the K-edge up to 31-Ga (see Table D.2).

Currently observed detection limits. *Experimental parameters*: probe diameter 100 Å; specimen thickness 1000 Å; analysing volume about 10^{-5} μm³; light element analysis (carbon, nitrogen, elements which are difficult to detect by X-ray microanalysis):

1. *Minimum mass fraction:* $c_{min} \cong 10^{-2}$;
2. *Minimum detectable mass:* $M_{min} \cong 10^{-18}$ g;
3. *Minimum detectable number of atoms:* $N_{min} \cong 10^4$.

Minimum detectable masses as low as 10^{-20} g have been reported. This corresponds to a minimum detectable number of atoms of the order of 100 (Table D.1). This limit largely depends on the background level which itself depends on matrix composition.

17.2.4 QUANTITATIVE ANALYSIS

Quantitative analysis of an element A by means of its characteristic energy loss at a K-edge is based on measuring the corresponding intensity of the edge above the continuous background. Neglecting the matrix effects leads to the following expression of intensity, according to Eq. (13.14):

$$I_A = I_0 \frac{N}{A} c_A \rho t \sigma_A \tau_A \tag{17.9}$$

where $(N/A)c_A \rho t$ is the number of atoms per surface unit of thickness t of the analysing volume, σ_A the excitation cross-section of the level of element A considered and the spectrometer transmission factor.

17.2.4.1 Absolute Quantitative Analysis

Physical and instrumental parameters of Eq. (17.9) must be known for an absolute determination of the mass fraction of element A.

Intensity Measurement. The intensity I_A corresponding to the characteristic K-edge of element A is determined by the following operations (Figure 17.4):

1. Integrating the spectrum intensity from W_K to $(W_K + \Delta)$; the integration window Δ is currently of 50–100 eV.
2. Subtracting the background over the same energy window Δ. It is determined by extrapolating from the empirical Eq. (17.8), least-squares fitted to the experimental background.

The intensity I_0 taken as reference is not the incident intensity, but the value derived by integrating the elastic peak intensity from $\Delta E = 0$ upwards over the same energy window. Making use of this integrated reference value enables multiple interactions to be accounted for somewhat.

Excitation Cross-section. The excitation cross-section σ_A is the integrated value over the admittance angle α and the energy window Δ. It may either be calculated from theory or measured from the pure element A.

Specimen Thickness. Specimen thickness t may be approximated from plasmon peak decrease (Eq. 17.7, limited to conductors) or by means of diffraction contrasts (e.g. through convergent beam electron diffraction, Eq. 12.19).

Specimen transmission factor. Factor τ_A may be determined from physical and geometrical characteristics of the specimen–detector system.

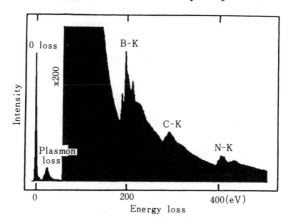

Figure 17.8. Electron energy loss transmission spectrum from a thin film, showing loss edges of light elements and the associated fine structure (document from JEOL).

17.2.4.2 Concentration Ratio of Two Elements

The measurement of the mass fraction ratio c_A/c_B of two elements A and B is often considered as sufficient and is much more straightforward. For identical instrumental conditions, the parameters which do not depend on the elements A and B to be measured are then eliminated. The energy difference corresponding to the two elements is negligible with respect to the incident energy; the spectrometer transmission may therefore be considered as a constant and relation (17.9) becomes

$$\frac{c_A}{c_B} = \frac{I_A}{I_B} \frac{\sigma_B A}{\sigma_A B} \qquad (17.10)$$

This relation can be compared to the corresponding relation set up for thin specimen X-ray microanalysis (Eq. 15.28).

17.2.4.3 Accuracy of Analysis

Due to uncertainties about measurements as well as physical and instrumental parameters, analysis by means of EELS is merely semi-quantitative. Accuracy depends on experimental conditions. Uncertainties ranging between 5 and 30% are currently reported for major element analysis, i.e. of the same order as for thin film X-ray microanalysis on similar analysing volumes.

17.2.5 APPLICATIONS

Electron energy loss spectrometry is complementary to X-ray emission spectrometry performed on thin specimens in an analytical transmission electron microscope. It is notably efficient for analysing light elements concentrated in a tiny volume of a specimen (Figure 17.8), e.g. precipitates, inclusions, segregations, whereas X-ray microanalysis is indicated more for medium and heavy elements.

In addition to the analytical aspect, it provides structural data similar to those of EXAFS through fine structure analysis.

REFERENCE CHAPTERS

Chapters 2, 3, 7, 8, 10, 13, 15, 21, Appendix D.

BIBLIOGRAPHY

[1] Clark L. J., Hester R. E. *Spectroscopy of surfaces*. John Wiley, New York, Chichester, 1988.
[2] Greaves G. N., Fontaine A., Lagarde P., Raoux D., Gurman S. J., Parke S. *Recent developments in condensed matter physics*. Plenum Press, New York, 1981.
[3] Hren J. J., Goldstein J. I., Joy D. C. (eds) *Introduction to analytical electron microscopy*. Plenum Press, London, 1979.
[4] Jouffrey B, Bourret A., Colliex C. (eds) *Microscopie électronique en science des matériaux*. Editions du CNRS, Paris, 1983.
[5] Koningsberger D. C., Prins R. (eds) *X-ray absorption. Principles, applications, techniques of EXAFS, SEXAFS and XANES*. John Wiley, New York, 1988.
[6] Teo B. K. *EXAFS. Basic principles and data analysis*. Springer-Verlag, Berlin, Heidelberg, 1986.
[7] Teo B. K., Joy D. C. (eds) *EXAFS spectroscopy. Techniques and applications*. Plenum Press, New York, 1981.

18

Secondary Ion Mass Spectrometry for Surface Analysis

18.1 BASIC PRINCIPLES AND INSTRUMENTAL LAYOUT

When submitted to a primary ion beam of a few kiloelectronvolts, a solid undergoes various interaction effects (see section 2.5). The most significant effect for analytical purposes is ion-induced *sputtering* of surface atoms. This takes place when the kinetic energy transferred by ions exceeds the bonding energy of certain atoms and when the transmitted momentum has a component normal to the surface. A fraction of the ejected atoms are ionised. These *secondary* ions, negative or positive, have *mass/charge* ratios which characterise the corresponding elements. Analysing them by means of a mass spectrometer leads to the identification of the surface elements and the possible determination of their concentration, without any lower atomic number limit as in the case of X-ray spectrometry.

In addition to surface analysis, this technique may also serve for establishing *concentration profiles* as a function of depth, through ion milling of successive atomic layers.

Combining spectrometry with imaging results in *ion emission microscopy*.

The principle of secondary ion emission spectrometry has been known since the 1930s. Its practical implementation started in the 1960s, notably with the work of Castaing and Slodzian around 1960, a decade after the onset of the electron microprobe (Castaing and Slodzian, 1962; Slodzian, 1964; Slodzian cited in [8]). It has since developed into one of the main techniques of surface analysis and has been the object of numerous papers. In addition to highly specialised publications, general surveys are available (e.g. Stuck and Siffert, 1984; Katz and Newman, 1987) as well as proceedings of recent SIMS symposia [15, 16, 17].

Common abbreviations:

1. SIMS (secondary ion mass spectrometry);
2. SNMS (secondary neutral mass spectrometry).

The designation SIMS covers various facilities. A purely analytical system with fine focused primary beam may be called an *ion probe*, by analogy with the electron probe. A system with imaging is called an *ion emission microscope*.

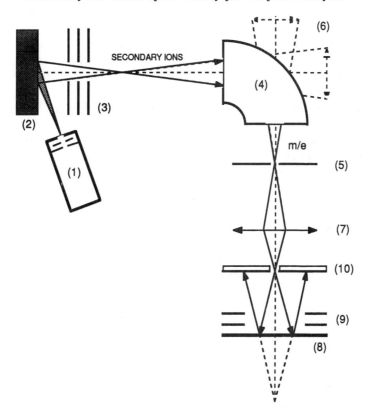

Figure 18.1. Diagram of a SIMS facility with stigmatic imaging (from Castaing and Slodzian, 1962). (a) *Analytical part*: (1) ion source, $E_0 = 3$–10 keV; (2) specimen; (3) system of electrodes for accelerating and focusing the secondary ions on to the entrance slit of the spectrometer; (4) mass spectrometer; (5) spectrometer exit slit selecting secondary ions of mass ratio m/e; entrance of the ion detector. (b) *Imaging part*: (6) virtual image of selected ion distribution given by the objective lens (3); (7) lens system for projecting the magnified ion distribution image on to a conversion electrode (8) generating secondary electrons; (9) lens system for projecting the secondary electron emission image on to the fluorescent screen or film (10). The conversion electrode (8) and its lenses (9) may be replaced by a micro-channelplate.

18.1.1 PURELY ANALYTICAL FACILITY

A simplified diagram of an ion probe with imaging attachment is outlined in Figure 18.1. The analytical part, considered alone, consists of the primary ion source, the specimen, the mass spectrometer and the data acquisition and processing system.

18.1.1.1 Primary Ion Source

Various types of ion sources are utilised (see section 9.4). The chosen ion species depends on operating conditions. Inert gases like *argon* are used when chemical effects are to be avoided; the resulting *sputtering yield* (number of secondary ions per incident ion) is rather low; it varies with the element to be sputtered. Ions such as *oxygen* and *caesium* lead to a higher sputtering yield, relatively independent of the atomic number of the elements; in addition, they have a chemical reactivity which may be useful in some cases; they are used when the lowest possible detection limit is sought.

The optimum primary energy ranges from 3 to 10 keV. The intensity per surface unit depends on the analysing mode, surface analysis or depth profiling. The incidence angle is set for a maximum ion-sputtering yield; depending on the ion species and the specimen, it may vary between grazing and normal incidence.

Depending on the primary beam cross-section, global analysis (millimetre-sized probe) or microanalysis (micrometre-sized probe) will be carried out. Microanalysis requires beam focusing on the specimen through an adequate ion–optical system. In some SIMS facilities, primary ions undergo a preliminary mass filtering by means of a magnetic prism; the specimen is then irradiated by a mono-ionic beam.

18.1.1.2 Specimen

The specimen must be kept free from any surface contamination. In any case the absorption rate of residual gas molecules by the specimen must be lower than the sputtering rate. This requires a high vacuum in the specimen chamber, achieved by means of differential pumping.

18.1.1.3 Mass Spectrometer

Secondary ions extracted by a voltage V of a few kiloelectronvolts are accelerated and focused by a system of electrodes on the entrance slit of the mass spectrometer. This has to select ions of a given mass ratio m/e (see Table 18.1). A *magnetic prism spectrometer*, preceded by a preliminary electrostatic *energy filter*, is generally set up in high-resolution facilities. A reasonable transmission factor, together with a high resolution, is achieved by means of a *double focusing spectrometer* (see Figure 10.12). Currently observed mass resolution ranges from 10^3 to 10^4.

A *quadrupole* may be used in place of a magnetic prism; its lower resolution is balanced by a higher scanning speed.

18.1.1.4 Data Acquisition and Processing

The detection and spectrum recording system is placed at the spectrometer exit slit. Photographic recording of the spectrum $I(m/e)$ provides a good resolution when combined with plane focusing on the film, but is not suited for quantitative measurements.

Ion beam intensity measurements are currently carried out by a conversion electrode, on which ions induce secondary electron emission, which in turn is amplified and measured by means of an electron multiplier, e.g. a channeltron.

As well as for electrons, direct ion intensity measurement is possible simply by an electrode in a Faraday cage, combined with an electrometer (Figure 10.9a).

Electrode alterations by ion sputtering must be taken into account.

Recording and processing of the mass spectrum are carried out in a similar way as for other previously surveyed spectrometric techniques.

18.1.2 ION EMISSION MICROSCOPY

Ions are charged particles which may therefore be focused like electrons by means of lenses and form images. In connection with SIMS, imaging is due to secondary ions, thus resulting in the design of the ion emission microscope.

18.1.2.1 Stigmatic Images

Part (b) of Figure 18.1 outlines the principle of displaying a stigmatic image (Gaussian image). A primary ion probe is focused into a diameter of 300–500 μm on the specimen. At the spectrometer exit, an ion–optical system serves both to select a given ion species and to focus the corresponding mono-ionic image on the conversion electrode. A double magnetic prism with an electrostatic mirror can provide a filtered ion beam in a line with the incident secondary ion beam as emitted by the specimen (Castaing and Slodzian, 1962).

The result is a *distribution image* of the selected ion species at the specimen surface.

The *resolution* of such ion images is of the order of 1 μm, depending on the quality of ion optics.

18.1.2.2 Scanning Images

The basic principle of scanning imaging by means of ion emission is the same as that leading to emission imaging in the SEM.

The specimen surface is scanned by a micrometre-sized primary ion probe, along an area of currently 500×500 μm. With the spectrometer tuned to a given mass ratio, the detector output modulates the electron beam of a synchronised CRT. An *ion distribution image* of the related specimen area is thus displayed, after a completed scan period.

The *ion image resolution* achieved is of the order of 1 μm, similar to that of the stigmatic ion image and of the X-ray emission distribution maps as performed by a SEM. Unlike the case of X-ray emission, ionic imaging resolution does not depend on a scattering volume, due to the fact that ion penetration is limited to a few atomic layers. It virtually depends on the probe size and on ion–optical aberrations.

The primary ion scanning mode generates a flat sputtering crater over the scanned area. This is of particular interest for depth concentration profiling.

18.2 SPECIMEN ANALYSIS

18.2.1 ANALYTICAL MODES

SIMS operates by removing material and is therefore somewhat destructive.

Either static surface analysis or dynamic depth profiling may be performed, depending on the instrumental layout.

18.2.1.1 Surface Analysis

A significant surface analysis requires a low sputtering rate, limited to the very first few atomic layers, during the exposure time. For this *static surface analysing mode*, the incident ion current density must be limited to the order of 1 nA/cm^2, leading to a sputtering rate of about 1 Å/h.

Depending on the sputtered area, the system may be operated according to a *global analysing mode* (millimetre-sized beam diameter or scan length) or a *microanalytical mode* (micrometre-sized beam diameter).

18.2.1.2 Depth Concentration Profiling

Setting up a concentration profile as a function of depth corresponds to a *dynamic analysing mode*. It requires a high sputtering rate, e.g. 1 μm/h, i.e. a high-density ion current of a few milliamperes per square centimetre. A flat sputtering crater is desirable for a significant depth profiling. It is best achieved through the scanning mode.

18.2.2 SPATIAL RESOLUTION

18.2.2.1 Surface Resolution

Surface resolution in the microanalytical mode is mainly determined by the probe size and by ion lens aberrations. A compromise between high resolution and high sensitivity leads to a typical analysing resolution of 1 μm, similar to the resolution of an electron microprobe.

18.2.2.2 Depth Profiling Resolution

Depth resolution is determined by uncertainties about the depth, with respect to the surface, related to an ion concentration measurement carried out at a given moment.

Stationary Analysing Mode

Ion current density across the primary beam has a Gaussian distribution. The resulting sputtering crater therefore has a curved bottom, making a good depth resolution difficult.

Scanning Analysing Mode

Ion beam scanning results in a flat crater bottom if the specimen is quite homogeneous. In order to eliminate analysing errors due to the curved crater edge, the corresponding signals are rejected by means of an aperture or an electronic locking device. A micrometre-sized probe is consistent with a low sputtering rate, between 0.01 and 0.3 Å/s, whereas a high-intensity probe of some 10 μm is used for a high sputtering rate, between 0.3 and 30 Å/s.

The most difficult problem is to relate an ion concentration measurement to the corresponding depth at the specimen. Accurate depth locating is often essential, e.g. for monitoring the implantation depth of doping ions in a semiconductor or for assessing the surface diffusion of an element.

The sputtering depth may be estimated if the sputtering rate is known, e.g. measured on a thin standard specimen of known thickness, ion milled until perforation.

A direct measurement is provided by a mechanical microsensor or by interferometry.

Limiting Parameters of Depth Resolution

Limiting factors may be either fundamental or instrumental.

Detection Limits. For a significant signal to be measured, a minimum specimen volume has to be sputtered. Depth resolution for analysing an element A is therefore linked to the concentration and the detection limit of this element. In commonly encountered instrumental conditions and with a homogeneous concentration of 1 ppm ($c = 10^{-6}$), the volume to be sputtered is of the order of 10^{-12}–10^{-10} cm^3 (see qualitative analysis). With an efficient scanning area of $100 \times 100\ \mu$m, i.e. 10^{-4} cm^2, the thickness to be sputtered would thus be 1–100 Å.

Ion-induced Surface Unevenness. A perfectly flat surface becomes somewhat rough through ion irradiation. In a polycrystalline material, the sputtering rate varies according to composition and orientation of the grains. Roughness increases with increasing depth, resulting in an increasingly altered resolution.

Beam Heterogeneity. In spite of edge effect suppression, beam intensity is not quite constant during scanning, thus resulting in depth variations which increase with depth.

Errors on Depth Measurement. Converting the time scale into a depth scale induces errors which are usually greater than any other errors. In the best case, direct depth measurements result in an accuracy of the order of 100 Å. This is the actual depth resolution limit.

18.2.3 QUALITATIVE ANALYSIS

18.2.3.1 Observed Emission Spectrum

Specimen atoms ejected through primary ion irradiation are mainly neutral particles which are not detected. A fraction of about 1% are ions. Their characteristic mass ratio m/e leads to their identification by means of a mass spectrometer.

Emission Parameters

Two quantities characterise secondary ion emission:

(a) the *sputtering yield S*, the number of ejected atoms per incident ion;
(b) the *ionisation degree α*, the fraction of atoms ejected as ions with respect to the total number of sputtered atoms.

Any of these parameters may be defined with respect to the global amount of sputtered atoms, with respect to a given element, or with respect to a given ion species.

The sputtering yield of an element in the form of ions of a given sign equals the total sputtering yield of that element (neutral atoms + ions) multiplied by the corresponding ionisation degree. In stabilised conditions, it is the ionisation degree which varies from one element to another (see also quantitative analysis).

Ion Species

There are various types of ions:

(a) *monoatomic ions with single charge*, positive or negative (A^+, A^-), usually more than 95% of the total number of ions;
(b) *monoatomic ions with multiple charge* (A^{n+}, A^{n-});
(c) *polyatomic ions* (A_n^+, A_n^-);
(d) *compound ions* ($A_nB_m^+$, $A_nB_m^-$);
(e) *natural isotope ions*.

Multiple charge ions, polyatomic ions and compound ions represent a very small fraction of ejected atoms, usually less than 1%. Natural isotope ions occur with their characteristic ratios (isotopic abundance).

The relative amounts of the various ions depend on primary ions, as well as on specimen composition.

Experimental Spectrum

The complete mass spectrum from a compound specimen generally consists of a large number of peaks, often with very close values of mass ratios. An insufficient mass resolution can result in overlappings, i.e. identification uncertainties. As an example, a peak with mass ratio 56 could correspond either to the 56-Fe or to the 28-Si ion (Table 18.1); their separation requires a mass resolution of at least 3000 (Stuck and Siffert, 1984).

Uncertainties may generally be removed by observing the natural isotopes, in their characteristic relative abundances. In the previous example, the mass ratios 54 and 57, corresponding to the major isotopes of iron, 54-Fe (5.8%) and 57-Fe (2.2%), are not

Table 18.1. Major positive ions observed by sputtering silicon by means of primary oxygen ions. Ion intensities are referred to the intensity of the single charge monomer set to 100 (from Whatley and Davidson cited in [13]

Mass ratio m/e	Relative intensity	Ion species
14	0.32	28-Si^{2+}
14.5	0.02	29-Si^{2+}
15	0.01	30-Si^{2+}
28	100	28-Si^+
29	5.1	29-Si^+
30	3.3	30-Si^+
44	6.3	28-SiO^+
45	0.32	29-SiO^+
46	0.21	30-SiO^+
56	0.19	28-Si_2^+
72	0.83	28-Si_2O^+
88	0.25	28-$Si_2O_2^+$
104	0.02	28-$Si_2O_3^+$

found in the spectrum, whereas the silicon isotopes 29-Si and 30-Si are observed in their natural abundance.

Compound ions can have numerous values of mass ratios and may therefore be difficult to identify.

Ion identification by means of their mass ratios is made easier by tables of nuclides providing the isotopes of all elements (e.g. [11, 18]).

18.2.3.2 Detection Limits

Any elements, as well as their isotopes, may be analysed by SIMS, whatever their atomic number. For a given element A to be analysed, the minimum detectable mass fraction depends on various fundamental and instrumental parameters, which may vary by large amounts. The values of those parameters given below in parentheses are approximate:

1. *Primary ion intensity* on the specimen ($I_P \cong 10^{-11}$–10^{-5} A);
2. *Sputtering yield* of element A corresponding to the ions to be measured ($S_A \cong 10^{-5}$–1);
3. *Composition* of the specimen (matrix effects);
4. *Volume of sputtered matter* during measurement (related to sputtering yield and primary ion current);
5. *Minimum measurable secondary ion intensity* ($I_{min} \cong 10^{-18}$–10^{-14} A);
6. *Transmission factor* of the spectrometer-detector system ($\tau_A \cong 10^{-4}$–10^{-2}).

It is therefore not possible to give standard detection limits.

For given instrumental conditions, the detection limits of an element A may be approximated from Eq. (13.15), in which the interaction cross-section is replaced by the ion-sputtering yield. For an analysing volume receiving an incident intensity I_P and emitting a secondary intensity I_{min} equal to the minimum significantly measurable value, the *minimum detectable mass fraction* of that element can be written as follows, when neglecting matrix effects:

$$c_A(min) = \frac{A}{M} \frac{I_{min}}{I_P S_A \tau_A} \tag{18.1}$$

Expressing concentration in terms of the atomic fraction, this relation becomes (Eq. 13.16)

$$c_A(atom.min) = \frac{I_{min}}{I_P S_A \tau_A} \tag{18.2}$$

where A and M respectively are the atomic mass of element A and mean atomic mass of the specimen.

18.2.3.3 Minimum Volume to be Sputtered

SIMS analysis involves removing matter. The minimum required volume to be sputtered, for significantly detecting an element, may be approximated when the determinant parameters are known. Consider N_c the *minimum number of ions* of an element A to be detected (e.g. ions A^+) to characterise that element; this number is related to the total number N_{sputt} of atoms to be sputtered by the following equation:

$$N_c = N_{sputt} c_A \alpha_A \tau_A$$

where α_A is the ionization degree of element A in the form of ions to be detected.

The volume v occupied by these atoms to be sputtered is

$$v = \frac{M}{N} \frac{N_c}{c_A \alpha_A \tau_A} \tag{18.3}$$

Example of detection limits. Analysis of aluminium in silicon:

1. Specimen parameters: $A = 27$ g; $M = 28$ g; $\varrho = 2.3$ g/cm³; $S_A = 2 \times 10^{-2}$; $\alpha_A = 10^{-2}$;
2. Instrumental parameters: $I_P = 10^{-6}$ A; $I_{min} = 10^{-17}$ A; $\tau_A = 10^{-3}$; $N_c = 100$ (100 ions/s correspond to 1.6×10^{-7} A).

> *Minimum detectable mass fraction*: $c_A(min) = 5 \times 10^{-7}$ (0.5 ppm).
> *Minimum detectable number of atoms*: $N_A(V) \cong 10^{17}$ at/cm³.
> *Volume to be sputtered*: $v = 4 \times 10^{-10}$ cm³.

For a scanning area of 300×300 μm $\cong 10^{-3}$ cm², this represents a thickness of 40 Å to be sputtered.

Depending on the SIMS facility and on the specimen, the typical minimum detectable mass fraction ranges from a fraction of 1 ppm to about 100 ppm.

18.2.3.4 Choice of Primary Ions

The ion-sputtering yield of an element may vary by a factor as large as 10^{-4}, depending on the primary ions. Relative proportions of positive and negative secondary ions likewise depend on primary ions and on the elements to be analysed.

The best choice of primary ions corresponding to a given specimen composition is therefore essential for determining the detection limits.

Various theoretical models have been proposed for interpreting the observed variations. The following general rules are in keeping with experiments:

1. Positive ion emission is favoured by a high electropositivity of the elements to be analysed as well as by a high electronegativity of the elements used as primary ions.

   This statement may be explained by an increase of the work function of surface electrons, due to absorption of electronegative primary ions which have a greater affinity for electrons than for specimen atoms. This leads to a decrease in the probability of exciting electrons across the potential barrier, i.e. decreasing the number of electrons intended to neutralise sputtered positive ions or to generate negative secondary ions.

   As an example, using primary oxygen ions results in increasing the positive ion yield by a factor of $10-10^4$, whereas the negative ion yield is merely increased by a factor of 1–10.

2. Negative ion emission is favoured by a high electronegativity of the elements to be analysed as well as by a high electropositivity of the elements used as primary ions.

   This may be explained conversely by a decrease of the surface electron work function, due to absorption of electropositive primary ions which have less affinity for electrons than for the specimen atoms. More electrons can then cross the potential barrier to neutralise sputtered positive secondary ions or to form negative secondary ions.

   As an example, using primary caesium ions results in increasing the negative ion yield by a factor of 10–10^4, whereas the positive ion yield remains unchanged.

As a result of these rules, reactive primary ions may improve the detection limit of certain elements. The most commonly used reactive ions are *oxygen* (O^- or O_2^+) and *caesium* (Cs^+). A discharge ion source is required for generating oxygen ions, whereas caesium is generated by means of a surface ionisation of a liquid metal source (see Figure 9.10).

A hot cathode source can merely produce ions from inert gases. The ion-sputtering yield of electropositive elements may then be increased by blowing oxygen on to the specimen surface, thus leading to a chemical effect similar to oxygen bombardment. The sputtering rate must be kept low to leave enough time for the oxygen to be absorbed on the surface, thus limiting sensitivity. On the other hand, introducing oxygen into the specimen chamber may have drawbacks.

Post-ionisation is a further technique for increasing the ionisation degree. It is currently applied in *secondary neutral mass spectrometry* (SNMS). It may be carried out by means of a laser beam grazing the specimen surface which is submitted to the primary ion bombardment. The tuned laser and multiphoton ionisation techniques lead to an ionisation degree of nearly 100%. Post-ionisation may similarly be achieved through an incident beam of electrons or of an electron plasma.

A quasi-total ionisation of the generated secondary ions results in a detection limit which may be as low as 10^{-8}. A further advantage of total ionisation is to make the measured secondary ion intensity from an element virtually proportional to its concentration in the specimen, which is essential for quantitative analysis (Eq. 18.5).

18.2.4 QUANTITATIVE ANALYSIS

18.2.4.1 Analytical Parameters

Measuring the number of secondary ions of a given element A, emitted by a specimen in a given time, leads theoretically to the mass fraction of this element, provided all parameters of the intensity–concentration relation are known. This relation can be derived from Eq. (13.15) when replacing the interaction cross-section σ by the sputtering yield S related to the detected ions:

$$c_A = \frac{A}{M} \frac{I_A}{I_P S_A \tau_A m_J} \qquad (18.4)$$

Instrumental parameters may be either measured or eliminated by comparison with a test sample of known composition.

Sputtering Yield

The sputtering yield of a pure element can be measured. However, in a compound specimen it will be affected by *matrix effects* expressed by the factor m_J. It varies as a function of the specimen elements and their chemical bonds. As an example, oxygen (commonly present in surface oxidation layers), as well as other electronegative elements, may induce an enhancement of the sputtering yield of various elements, an effect which is generally the stronger, the more electropositive the elements concerned. Thus, in a gold–copper alloy, the sputtering yield of copper is more sensitive to the presence of oxygen than that of gold.

In a polycrystalline material, the sputtering yield depends on *grain orientation*. In addition, the sputtering yield varies with the practical analytical mode and is different in surface analysis and in depth analysis. During ion irradiation of a specimen, a *transitional stage* corresponds to sputtering of the first few atomic layers, followed by a *steady state* of deep-layer sputtering. In the first stage, the sputtering yield varies greatly from one element to another, thus resulting in an enrichment of low-yield elements on the surface. This alteration of surface composition will progressively offset differences between sputtering yields. Once the steady state is reached, the sputtering yield is virtually equal for all elements, i.e. ion milling becomes relatively steady. As a result for analysis:

1. *Ion irradiation of a homogeneous material, by means of a constant ion intensity across the whole scan area, leads to a reasonably uniform thinning.*
2. *During the steady state the respective proportions of sputtered elements (neutral and ionised atoms) equals approximately the respective proportions of atoms in the specimen, i.e. the elemental composition.*

Ionisation Degree

Secondary ion intensity as measured at the spectrometer corresponds in fact to the ionised fraction of sputtered atoms, determined by the ionisation degree. It is mainly the variation of this factor from one element to another which induces the observed variation of the sputtering yield. The problem of quantitative analysis of a given element, during the steady state, therefore reduces mainly to determining the relative amounts of neutral atoms and ions in the cloud of sputtered particles, in relation to the specimen composition.

With the knowledge of the ionisation degrees of two elements A and B, their concentration ratio would be related as follows to the ratio of respective ion intensities:

$$\frac{c_A}{c_B} = \frac{\alpha_B}{\alpha_A} \frac{I_A}{I_B} \tag{18.5}$$

This linear relation is valid for elements with low concentrations and negligible interaction.

A theoretical approach for determining the ionisation degrees of elements sputtered from a compound is based on *plasma thermodynamics* (Andersen and Hinthorne, 1973). This theory states that ionic irradiation of a solid generates at its surface a plasma of excited atoms and ions which is in a thermodynamic equilibrium. An ionisation equation at equilibrium may thus be set up and the relative amounts of neutral atoms and various ions calculated, at a given temperature and electron density, from the knowledge of physical constants (partition function of the various particles, ionisation energies). Now, with the adequate values of ionisation degrees, measuring in steady conditions the relative number of the various secondary ions emitted leads directly to specimen composition.

A quite different approach consists of eliminating the uncertainty about ionisation degrees by making them equal to one through *post-ionisation* (see qualitative analysis).

18.2.4.2 Analytical Practice

The spectrometer is to be operated in view of optimum results in quantitative analysis:

1. Oxygen as primary ion source (O^+) is favourable for quantitative analysis. When compared to argon, it leads to smaller variations of the sputtering yield for different elements. In addition, the secondary compound ions induced by oxygen are mainly oxide ions, for which the ionisation constants are well known.
2. For a quantitative analysis to be significant, the measured ion intensities must equal the initial proportions of emitted secondary ions. For this aim the efficiency of the spectrometer–detector system must remain constant for any measured ions.
3. The energy range of the spectrometer must be large enough in order to collect all secondary ions of a given mass ratio, despite their energy dispersion which is of the order of 100 eV. A double-focusing spectrometer (focusing with respect to a direction dispersion as well as to an energy dispersion, Fig. 10.12) provides an optimum compromise between resolution and energy range.

18.2.4.3 Accuracy of Analysis

In spite of optimum instrumental conditions, accuracy is generally limited to some 20%, due to the number of parameters involved and to uncertainties about their values. It is therefore semi-quantitative.

18.3 SPECIMEN PREPARATION

The specimen to be analysed may be a bulk material, a thin film or particles deposited on a support. In the two last-mentioned cases, an even approximate quantitative analysis may be impossible, due to an insufficient thickness to reach the steady sputtering state.

For significant results in depth profiling as well as in quantitative analysis, the specimen must meet the requirements given under the headings below.

Flat Analysing Surface. In the absence of a natural plane surface (cleavages), a flat surface is to be produced by conventional cutting and polishing techniques similar to those used for electron microprobe analysis (see 15.3.4).

Clean Surface. For surface analysis, any existing contamination layer has to be removed by preliminary sputtering before a significant analysis. As noted in previous sections, ion milling is similarly used as an attachment for specimen cleaning in other techniques of surface analysis.

Vacuum Resistance. The specimen should not contain volatile components liable to evaporate in the ultra-vacuum of the specimen chamber, thus excluding most organic materials.

Electrical Conductivity. SIMS involves charged particles. The specimen surface must therefore conduct electricity and be connected to the grounded specimen holder in order to drain charges, thus avoiding altering secondary ion emission and ion trajectories towards the spectrometer. Due to ion sputtering, this condition is difficult to meet for an insulating material. A vacuum-deposited carbon layer of some 100 Å thickness results in the surface becoming virtually equipotential; it is then locally perforated by the primary beam. This technique is obviously not adequate for element distribution mapping; a metallic grid (e.g. aluminium) of about 0.1 mm mesh is then vacuum deposited on the specimen surface, thick enough to endure ion milling during the operation. Small particles may be embedded in a conducting resin (copper or silver charged resin).

The total ion sputtering yield is smaller than 1. An incident positive ion beam therefore generates a local charge of the same sign. However, in addition to secondary ions, it also generates secondary electrons resulting in an increased negative charge deficit, i.e. an increase in excess positive charge. It follows that with an incident negative ion beam, the charge effect would be partially compensated. A negative primary ion beam (e.g. O^-) therefore seems to be advantageous for insulating materials. However, its advantage is somewhat offset by the resulting ion intensity which is rather low, due to the conventional ion sources favouring the generation of positive ions.

An electron beam of adequate energy and intensity, focused on the analysing area of the specimen, represents another often used and efficient method for charge compensation.

The acuteness of the charge problem for analysing insulating specimens may vary somewhat with the design of the SIMS facility (e.g. specimen bias, ion extraction voltage, ion-induced secondary electron emission by apertures).

18.4 APPLICATIONS

Secondary ion mass spectrometry is a valuable method of surface analysis. Its main qualities with respect to other surface analysing methods are:

(a) access to the third dimension, resulting in depth profiling and three-dimensional elemental analysis;
(b) isotope analysis;
(c) microanalysis of light electrons, without atomic number limit;
(d) surface composition imaging.

Some typical examples of applications are surveyed below.

Isotope Analysis. SIMS characterises ion masses and is therefore suited to isotope analysis. *Isotopic abundances* may be easily determined by switching the spectrometer to the various mass ratios of the isotopes of an element, thus avoiding any error due to instrumental drift.

Determining the isotopic abundances of trace elements leads to material dating (radiochronology). However, accuracy of analysis by means of SIMS is generally insufficient for geochemical requirements.

Composition Profiles. The possibility of concentration depth profiling is very useful in material science:

1. Monitoring *surface diffusion* or *implantation* of dopants in semiconductors is one of the most important applications of SIMS, since it is presently the only available

Figure 18.2. General view of a SIMS facility (CAMECA document).

method of measuring and locating small amounts of dopants and impurities which are significant for semiconductor operation.
2. Monitoring *surface treatments* of materials carried out by chemical reactions, by surface diffusion, by implantation (e.g. steel nitriding, chemical glass hardening).
3. Study of *surface absorption* on solids (e.g. contamination of catalysers).
4. *Surface corrosion* studies (e.g. alloys).
5. *Depth analysis* of defects, inclusions, grain boundaries in a material, by progressive ion sputtering.

Light Element Microanalysis. In its microanalytical or imaging form, SIMS is interesting for its ability to analyse and to locate light elements in small amounts, e.g. in tiny aggregates. It is notably the case for *hydrogen* which plays an important role in metals, in semiconductors (e.g. in polycrystalline silicon diodes for solar energy conversion). It allows detection and location of light elements like *lithium* or *boron* in glass, ceramics and rock.

Organic Materials and Polymers. The primary ion irradiation induces the emission of molecules and of compound secondary ions (clusters) which may provide data about the molecular structure of the material.

REFERENCE CHAPTERS

Chapters 2, 9, 10, 13, Appendix D.

BIBLIOGRAPHY

[1] Anciello O., Kelly R. *Beam modification of materials.* Vol. 1, *Ion bombardment modification of surfaces.* Elsevier, Amsterdam, 1984.
[2] Behrish R. (ed.) *Sputtering by particle bombardment. Topics in applied physics.* Springer-Verlag, Berlin, 1983.
[3] Benninghoven A., Rudenauer F. G., Werner H. W. (eds.) *Secondary ion mass spectrometry.* John Wiley, Chichester, 1987.
[4] Blaise G. *Material characterization using ion beams.* Plenum Press, London, 1978.
[5] Clark L. J., Hester R. E. *Spectroscopy of surfaces.* John Wiley, New York, Chichester, 1988.
[6] Davis R., Frearson M. *Mass spectrometry.* John Wiley, New York, Chichester, 1987.
[7] Feldmann L. C., Mayer J. W. *Fundamentals of surface and thin film analysis.* North-Holland, New York, 1986.
[8] Grant W. A., Procter R. P. M., Whitton J. L. (eds.) *Surface modification of metals by ion beams.* Elsevier Sequoia, Lausanne, 1988.
[9] Grasserbauer M., Dudek H. J., Ebel M. F. *Angewandte Oberflächenanalyse.* Springer-Verlag, Berlin, 1986.
[10] Jouffrey B. (ed.) *Méthodes et techniques nouvelles d'observation en métallurgie physique.* Soc. Franç. Microsc. Electron., Paris, 1972.
[11] Lederer C. M., Shirley V. S. *Table of isotopes.* John Wiley, New York, Chichester, 1978.
[12] Oechsner H. (ed.) *Thin film and depth profile analysis. Topics in applied physics,* Vol. 37. Springer-Verlag, Berlin, 1984.
[13] Richardson J. H., Petersen R. V. (eds.) *Systematic materials analysis*, Vol. 4. Academic Press, New York, 1978.
[14] Ryssel H, Ingolf R. *Ion implantation.* John Wiley, New York, Chichester, 1986.

[15] Slodzian G., Huber A. M., Benninghoven A., Werner H. W. Secondary ion mass spectrometry. *Proc. 6th Intern. Conf. on SIMS, Versailles 1987*. John Wiley, New York, Chichester, 1988.
[16] Storms H., Benninghoven A. (eds.) Secondary ion mass spectrometry. *Proc. 7th Intern. Conf. on SIMS*. John Wiley, New York, Chichester, 1988.
[17] *Proceedings of SIMS conferences*. Springer-Verlag, Berlin.
[18] *Handbook of chemistry and physics*. CRC, Cleveland.

Part Five

TECHNIQUES OF ELECTRON MICROSCOPY

During recent years probably the most significant developments in scientific instrumentation have taken place in the field of computer data acquisition and processing, as well as in the field of imaging. Virtually any information on materials given by radiation interactions may now be displayed as images. Electrons are best designed for imaging purposes, thanks to their easy generation, acceleration and focusing. As a result, techniques of electron microscopy have been the first to be developed.

The present electron microscopes may be classified into three categories:

1. **Electron microscopes based on stigmatic imaging (Gaussian images).** As in conventional light microscopes, images are generated in a simultaneous way by lenses. The laws of classical optics apply to electrons, leading to electron lenses which have characteristics similar to those of light lenses. Such microscopes are generally operated by transmission, thus leading to the common designation *transmission electron microscope*. However, this may now result in ambiguity, due to the advent of other types of transmission microscope.

   Continuously improved during more than 50 years, the top models have reached a quasi-atomic resolution.
2. **Scanning electron microscopes.** Similarly to television techniques, images are generated in a sequential way by processing the various data provided by an electron beam scanning the specimen surface. Such microscopes may operate either through emission, reflection or transmission. In the emission mode, images may be generated by any electron-induced secondary radiation, such as secondary electrons, Auger electrons, X-rays, light photons, infrared photons, ions and sound waves. It is this technique of microscopy which has most developed in all fields of material science during the last 30 years. It is mostly combined with spectrometric techniques, as described in Part Four.
3. **Tunnelling electron microscopes.** This most recently developed technique does not make use of incident electrons, transmitted or reflected, or of electron-induced

emissions. Imaging is performed by means of the electron tunnel current between the specimen and a tungsten tip; this may be considered as a probe for exploring the electrical potential of the electron cloud surrounding a solid surface. It thus generates a surface image. This rapidly progressing technique complements the existing surface investigation methods.

19

Transmission Electron Microscopy

The present chapter deals with electron microscopy based on stigmatic (or Gaussian) imaging by means of electronic lenses.

It is the oldest technique of electron microscopy, first developed in 1935 in Berlin by Ruska, von Borries and Knoll. It won Ruska a late Nobel Prize in 1986.

Common abbreviations.

General abbreviation:

1. TEM (transmission electron microscope).

Specific abbreviations:

1. CTEM (conventional TEM);
2. HVEM (high-voltage electron microscope);
3. HREM (high-resolution electron microscope).

The most common TEM (sometimes called conventional TEM) is operated with so-called high-energy electrons (10–120 keV). Specific characteristics of the high-voltage electron microscope (1–3 MeV) are summarised at the end of this chapter.

The imaging theory will be dealt with in an elementary way, based on the kinematic formulation; it is usually sufficient for qualitative or semi-quantitative interpretations. A more quantitative approach, based on dynamic theory, may be found in specialised treatises.

Imaging and diffraction are closely related in a TEM. This fact is of crucial importance for studying crystallised materials.

The main quality of any microscope is its *resolution* (or resolving power). In a lens-imaging microscope, resolution is limited by diffraction effects on the entrance pupil of the objective lens. Consider a thin, plane and amorphous specimen, submitted to an incoherent radiation. Resolution is defined as the smallest distance d between two object points resolved by the objective lens; it is related to the *wavelength* λ, the *refractive index* n of the medium between specimen and objective lens and to the *angular aperture* α of the objective lens (Figure 19.1) by the following well-known relation of conventional optics:

$$d = \frac{0.61 \lambda}{n \sin \alpha} \tag{19.1}$$

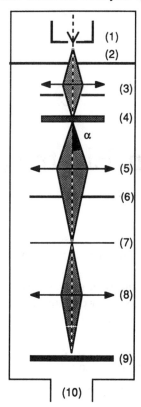

Figure 19.1. Basic components of a TEM. (1) Electron source at a voltage V_0; (2) grounded anode; (3) condenser system; (4) thin specimen; (5) objective lens; (6) objective back focal plane, with its diaphragm limiting the angular aperture to α; (7) objective image plane (Gaussian plane), conjugate of the object plane with respect to the objective lens; (8) system of projection lenses; (9) fluorescent screen or photographic film; (10) vacuum pumps.

Improving the resolution means reducing the resolution distance d. It may be achieved in three ways:

(a) reducing wavelength λ;
(b) increasing aperture α;
(c) increasing refractive index n.

Photon Microscopes. The manufacture of high aperture immersion objectives is well developed for light microscopes. Resolution is mainly limited by wavelength.

Order of magnitude of resolution limits

$$\lambda \cong 5000 \text{ Å} = 0.5 \ \mu\text{m}; \quad \alpha \cong 1; \quad n \cong 2 \Rightarrow d \cong 0.15 \ \mu\text{m} = 1500 \text{ Å}.$$

UV microscopes may go down to 0.1 μm. Lenses for shorter electromagnetic wavelengths (X-rays, γ-rays) are not available at present.

Particle Wave Microscopes. Stigmatic images may be generated by means of the wave associated with any moving particle, as well as by electromagnetic waves. The electron wave is the most straightforward to use, for several reasons: it is easy to generate, to focalise and to observe on a fluorescent screen or on a photographic film; furthermore, electron–specimen interaction is lower than for any other particles. As a result it is the electron microscope which has been mainly developed. However, proton microscopes and ion microscopes have been built.

High-aperture objectives are not available for electrons, because it is at present impossible to correct spherical aberration. Aperture is therefore limited to about 10^{-2} rad. The refractive index of vacuum is equal to 1. The wavelength is a function of the acceleration voltage (Eq. 1.6) which therefore determines the resolution limit.

Order of magnitude of resolution limits:

$E_0 = 100$ keV; $\quad \lambda = 0.037$ Å; $\quad \alpha = 5 \times 10^{-3}$ rad $\Rightarrow d \cong 4.5$ Å.

$E_0 = 1000$ keV; $\quad \lambda = 0.0087$ Å; $\quad \alpha = 5 \times 10^{-3}$ rad $\Rightarrow d \cong 1$ Å.

The resolution which is actually achievable depends on various factors related to the instrument and to the specimen. The ultimate resolution reached at present is of the order of 2 Å.

19.1 BASIC COMPONENTS AND CHARACTERISTICS

The basic components of a TEM are similar to those of a light microscope, with characteristics corresponding to the respective nature of waves.

19.1.1 ELECTRON SOURCE

Electrons are generated by an electron gun, then accelerated through a positive stabilised voltage V_0 (see Figure 9.5). A secondary vacuum of at least 10^{-5}-10^{-6} Torr is required for the most common thermionic source with tungsten filament. A high vacuum better than 10^{-8} Torr is necessary for a LaB_6 source and for a field emission source.

A conventional thermionic electron gun associated with a double condenser system (see 9.2.3), operated in the focusing mode on the specimen, generates a probe of the order of 1 μm in diameter and 1 μA in intensity. Additional condenser lenses (e.g. condenser-objective, Figure 9.9) lead to a minimum beam diameter of some 1 nm (nanoprobe) with an intensity of some 1 nA, notably combined with a LaB_6 source or a field emission source.

The *electron flux density* on the specimen is approximately proportional to the illumination aperture angle (Eq. 9.7). In normal operating conditions, the filament current is set to saturation; the related intensity maximum depends on the filament position relative to the Wehnelt; a greater brightness is paid for by a shorter filament lifetime.

The *lowest illumination coherence* is achieved with the electron beam focused on to the specimen (cross-over image projected on the specimen), thus resulting in the maximum beam aperture consistent with acceptable irradiation effects. This focusing mode reduces diffraction effects and is suitable for imaging with high-point resolution.

The *highest illumination coherence* is achieved by overfocusing the beam (maximum excitation of condenser 2), resulting in a reduced beam aperture. This mode increases the diffraction effects and is therefore suitable for periodic imaging (interference microscopy) as well as for electron diffraction.

19.1.2 ELECTRON LENSES. ABERRATIONS AND RESOLUTION

19.1.2.1 General Characteristics

Lenses used in electron microscopes are mainly magnetic lenses with stabilised input (see Figure 9.6a). Electron lenses have characteristics which are quite similar to those of conventional lenses, with parameters such as the focal length and the aberrations. Focal length is given by Eq. (9.3) as follows:

$$\frac{1}{f} = \frac{k}{E_0} \int_{gap} H_z^2 \, dz$$

where k is a constant and H_z the axial component of the magnetic field in the pole piece gap.

Convergence $1/f$ increases continuously with increasing lens current, resulting in a zoom effect. A common objective lens focal length is 2–3 mm. Electrons moving through a magnetic field follow a helicoidal path. This results in a rotation of the electron propagation plane which leads to image rotation **when chang**ing the setting of a lens.

A TEM includes three lens systems, as described below.

Condenser System. Its setting varies the specimen illumination mode. It usually consists of two or three lenses.

Objective. Its setting results in focusing on the specimen. Its characteristics determine the resolution.

Post-objective Lens System (Diffraction Lenses and Projection Lenses). The aim of the projection lenses is to transfer the first image given by the objective lens on the screen or the film. Setting the various lenses results in varying the final magnification, switching from the image mode to the diffraction mode and varying the diffraction length.

Recent electron microscopes may have up to nine lenses to ensure ample working versatility in the various operating modes: two to four condenser lenses, one objective lens, two to four lenses for the post-objective system (diffraction lens, intermediate lens, one or two final projection lenses).

Optical qualities are determined by lens aberrations. Point resolution in particular is limited by objective lens aberrations.

19.1.2.2 Point Resolution

The significant resolution of a microscope is its ability to separate in its image two object points as close as possible. It is determined by three effects of the objective lens which do not anneal on the optical axis:

1. Diffraction at the aperture diaphragm (diffraction aberration);
2. Spherical aberration;
3. Chromatic aberration.

The first limiting effect is the only one to intervene in a perfect objective lens. Any of these effects results in the image of a point becoming a *confusion disc*.

The following analysis assumes the incident electron wave to be incoherent, i.e. the condenser to be focused on the specimen (see Figure 9.9b).

Point Imaging by a Perfect Objective Lens

Image of a Single Point. The image of a point given by a lens with limited aperture is the diffraction pattern generated by its circular aperture diaphragm; it is expressed by the *Airy function*, the FT of the disc function corresponding to the aperture (Figures 19.2a and b). Limited to only one dimension (the diameter), this function is similar to the size function of which the FT is expressed by Eq. (3.27) and illustrated by Figure 3.4.

The Airy pattern is most easily observed on the image of a star provided by a telescope. The radius d_0 of the first dark ring in the image plane is approximately equal to the half-width of the central maximum. In the small-angle approximation its value is $d_0 = 0.61 \lambda g/\alpha$, where g is the objective magnification and α its aperture.

Image of Two Close Points. In incoherent illumination, intensities add up. According to the *Rayleigh criterion*, for resolving just two points, the central maximum of the Airy function of one point should fall on the first dark ring of the other, i.e. the distance d' of the central maxima should be greater than or at least equal to the radius d_0 of the first minimum (Figure 19.2c). Referring this value to the object plane leads to the resolution distance as given by Eq. (19.1). In vacuum and for a small aperture angle, the resolution distance is expressed by

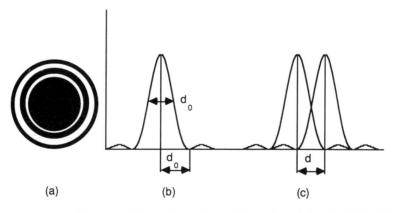

Figure 19.2. Aperture diffraction effect and resolution. (a) Outline of the Airy diffraction pattern, corresponding to the image of a point; (b) intensity variation along a diameter of the Airy pattern; (c) overlapping of the images of two points at the resolution limit.

$$d = \frac{0.61\lambda}{\alpha} \qquad (19.2)$$

The image of the whole object is the superposition of the images of its individual points. It can be expressed as the convolution product of the stigmatic image by the Airy diffraction pattern (Appendix C).

In coherent illumination, an image consists of the sum of amplitudes of the respective point images, thus leading to much more pronounced diffraction effects which may result in image contrast artefacts and in incorrect interpretations.

Resolution Consistent with Aberrations

Assuming astigmatism to be easily corrected, resolution is limited by spherical and chromatic aberrations.

Rayleigh Rule. An aberration may be considered not to alter the resolution, determined by aperture diffraction, as long as the diameter $2r'$ of the aberration confusion disc remains smaller than the half-width of the central maxima of the Airy pattern (which may be considered as the diffraction confusion disc). Given that this half-width is itself approximated to the image resolution distance d', the rule is expressed accordingly as $2r' \leq d'$.

When referred to the object plane, Rayleigh's rule becomes

$$2r \leq d \qquad (19.3)$$

where $r = r'/g$, $d = d'/g$ and g is the objective magnification.

Spherical Aberration. Rayleigh and Scherzer Resolution. The aperture diffraction effect may be assimilated in an aberration, as well as any other resolution limitation effects, with its central maximum acting as a confusion disc. The respective radii of the diffraction confusion disc $r_d = d/2$ (Eq. 19.2) and of the spherical confusion disc $r_s = C_s\alpha^3$ (Eq. 9.4) vary in opposite ways as a function of the objective aperture α. The radius of the confusion disc of the total instrumental aberration (diffraction aberration + spherical aberration) is the quadratic sum of the respective radii. Its value r_{tot} passes through a minimum which determines the optimum objective aperture. As shown by Figure 19.3, this minimum is close to the intersection $r_s = r_d$ of the respective functions, thus justifying Rayleigh's rule.

Condition $r_s = r_d$ leads to the definition of the **Rayleigh aperture and resolution**, according to Eqs (9.4) and (19.2):

$$\alpha = \left[\frac{0.61\lambda}{2C_s}\right]^{1/4} \Rightarrow d = 0.8\lambda^{3/4} C_s^{1/4} \qquad (19.4)$$

where C_s is the spherical aberration coefficient.

Defocusing of D_s, according to Eq. (9.5), leads to the more favourable definition of the **Scherzer aperture and resolution**

$$\alpha = \frac{4}{3}\left[\frac{0.61\lambda}{2C_s}\right]^{1/4} \Rightarrow d = 0.6\lambda^{3/4} C_s^{1/4} \qquad (19.5)$$

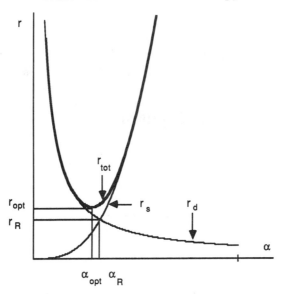

Figure 19.3. Variation of confusion discs with aperture α: diffraction disc $r_d = 0.6\lambda/2\alpha$; spherical aberration disc $r_s = C_s\alpha^3$; total instrumental aberration $r_{tot} = (r_d^2 + r_s^2)^{1/2}$.

The Scherzer resolution defined in this way may be called the *optical point resolution*. It expresses the limit resolution corresponding to a given wavelength and to a given spherical aberration.

Example. Microscope Philips CM 12-EM 420 operating at 100 keV. Instrumental characteristics:

(a) $f = 0.27$ cm;
(b) $C_s = C_c = 0.2$ cm;
(c) $D_s = -100$ nm (Scherzer defocusing).

Optical characteristics (at Scherzer defocusing):

(a) optimum aperture: $\alpha = 6.5 \times 10^{-3}$ rad (diameter of the objective diaphragm about 35 μm);
(b) optical point resolution: $d = 0.34$ nm $= 3.4$ Å.

Chromatic Aberration. For the resolution limit imposed by spherical aberration to be preserved in spite of chromatic aberration (Eq. 9.6), the condition expressed by Eq. (19.3) must be met, thus resulting in

$$2r_c \leq d \Rightarrow 2C_c\alpha \frac{\Delta E}{E_0} \leq d \Rightarrow \frac{\Delta E}{E_0} \leq \frac{d}{2C_c\alpha} \qquad (19.6)$$

where C_c is the chromatic aberration coefficient.

398   *Structural and Chemical Analysis of Materials*

In the above example, the resolution $d = 3.4$ Å would be preserved if relative energy loss $\Delta E/E_0$ remains smaller than 1.3×10^{-5}. According to calculations (see Eq. 8.6), in the instance of an aluminium foil and 100 keV electrons, this would correspond to a limit thickness of about 20 Å (ΔE being assimilated in the mean energy loss ΔE_m). Experiments show this thickness limit to be too severe, notably for thick specimens. The confusion disc significant for resolution is in fact smaller than stated above, for various reasons forwarded by Cosslett (1969):

1. The intensity distribution across the confusion disc of radius r_c' is not a constant, but follows a Gaussian function with a half-width of about $r_c'/2$; the efficient confusion disc diameter becomes then $r_c'/2$ instead of $2r_c'$, i.e. four times smaller.
2. In very thin foils, the image results mainly from a significant fraction of electrons which have not been subjected to any energy loss. Chromatic aberration is then mainly determined by the difference ΔE_p between the incident energy and the most probable transmitted energy E_p (Figure 8.3).
3. In thick foils, all electrons undergo energy losses. Imaging is then mainly due to electrons with the most probable transmitted energy E_p and chromatic aberration is determined by the half-width δE of the energy distribution function; this width is smaller than ΔE_p for thick specimens (see Figure 8.3 and Table 8.3).

The limit values of relative energy losses and the related resolution limits are then given by the following relations:

$$\frac{\Delta E_p}{E_0} \leqslant \frac{2d}{C_c \alpha} \Rightarrow d \geqslant C_c \alpha \frac{\Delta E_p}{2E_0} \quad \text{(very thin foils)}$$

$$\frac{\delta E}{E_0} \leqslant \frac{2d}{C_c \alpha} \Rightarrow d \geqslant C_c \alpha \frac{\delta E_p}{2E_0} \quad \text{(thick foils)}$$

(19.7)

The thickness limits of Table 19.1 have been calculated according to these relations and the energy losses of Table 8.3. It may be seen from Table 19.1 that chromatic aberration does not impair point resolution of the above-mentioned electron microscope for observing 100 Å thick foils.

To Summarize. The optical point resolution of an electron microscope is determined by the objective lens aperture and spherical aberration. In order to preserve it the specimen should be very thin. An often used rule of thumb states that the specimen

Table 19.1. Orders of magnitude of resolution limits due to chromatic aberration according to Eqs (19.7) for: $V_0 = 100$ keV; $C_c = 0.2$ cm; $\alpha = 6.5 \times 10^{-3}$ rad

Thickness (t)	Resolution distance d (Å)		
	C	Al	Au
Thin foil (100 Å)	1.5	1.2	3.4
Thick foil (1 μm)	150	180	1000

thickness should not exceed 10 times the expected resolution. According to the above analysis, a factor of 30 seems more realistic.

The *point resolution* is defined in incoherent illumination conditions; it depends on optical characteristics and should not be mistaken for the *lattice resolution* (or line resolution) of a crystalline specimen. The latter depends mainly on electronic and mechanical stability of the microscope and may be much more favourable (see section 19.4).

19.1.2.3 Resolution Tests

As has been stated above, a point resolution test on an amorphous specimen is the only significant way of assessing the optical qualities of an electron microscope. A thin layer of an amorphous Pt–Ir alloy, vacuum deposited on a carbon foil, is one of the most commonly used test preparations.

A perfect focusing and astigmatism correction of the objective lens is required for resolution testing. It is carried out by means of specially prepared support foils with circular holes of the order of 0.1 μm in diameter (see section 19.5.1). Objective lens defocusing results in Fresnel diffraction fringes appearing at the edge of the holes, fringes which are similar to those commonly observed with light on the edge of an opaque screen. A coherent illumination provided by condenser overfocusing improves the fringe contrast.

Objective Lens Focusing. Fresnel fringes disappear at exact focusing on the object plane. The fringe contrast varies with the sign of defocus: *underfocusing* results in a bright fringe at the edge (Figures 19.4a and 19.31), whereas *overfocusing* leads to a first dark fringe at the edge (Figure 19.4b).

Astigmatism Correction. A perfectly corrected objective lens causes the fringe to have a constant brightness along the edge (case of Figures 19.4a and b). Astigmatism of a focused objective lens results in overfocusing along one direction, underfocusing along the perpendicular direction (case of Figure 19.4c), thus determining the direction of astigmatism. Its correction is then carried out by means of a *stigmator* device which generates an elliptic magnetic field of adjustable magnitude and direction.

Focusing and astigmatism correction may be carried out without specific preparation, by observing Fresnel fringes in the specimen, e.g. at the edge of opaque grains (see Figure 19.31). Astigmatism varies during operating time, e.g. through contamination of the objective lens aperture due to electron impacts. Visual adjustment of the stigmator is generally insufficient for operating at high resolution. The best method is to carry out a *focal series*: several photographs are taken while changing the objective lens focusing step by step, between an underfocusing setting and an overfocusing setting. On recent microscopes, focal series as well as Scherzer defocusing are carried out automatically.

19.1.2.4 Depth of Field and Depth of Focus

Depth of Field

The depth of field D is the distance of extreme specimen positions on both sides of the theoretical object plane Π of the objective lens, without impairing resolution

Figure 19.4. Fresnel fringes at the edge of a circular hole in a carbon film. (a) Objective lens without astigmatism; 1 μm underfocusing; (b) objective lens without astigmatism; 1 μm overfocusing; (c) objective lens with a strong astigmatism (Ehret G., Dept Electron Microscopy, Lab. Crystallography, Louis Pasteur University, Strasbourg).

Transmission Electron Microscopy

(Figure 19.5). A specimen shift $D/2$, with respect to the object plane Π, results in a confusion disc of radius $r = \alpha D/2$, where α is the objective aperture. For the resolution to be unimpaired, the diameter of this disc should be smaller than or equal to the resolution distance d; the limit value corresponds to $\alpha D = d$, leading to

$$D = \frac{d}{\alpha} \qquad (19.8)$$

In the previous example ($d = 3.4$ Å; $\alpha = 6.5 \times 10^{-3}$ rad), the depth of field would be $D = 520$ Å, i.e. greater than the specimen thickness consistent with the resolution to be achieved. As a result, the specimen is focused on its whole thickness. The observed image integrates the whole specimen thickness. Locating a given specimen detail is possible through a stereoscopic method: two photographs, taken with relative tilting of about $+6°$ and $-6°$ with respect to the optical axis, provide a stereoscopic pair which can easily be observed by means of a stereoscope. This three-dimensional technique may be very useful, e.g. for depth-locating defects or other details in a crystal. It is of particular interest when carried out on a high-voltage microscope which permits the use of much thicker specimens.

Depth of Focus

The depth of focus D' is defined in the final image plane, in a similar manner as the depth of field in the object plane. It expresses the maximum distance between extreme positions of the observation plane (photographic film or fluorescent screen), on both sides of the theoretical image plane Π', without impairing resolution. A shift $D'/2$ of the observation plane with respect to the image plane Π' results in a confusion disc of radius $r' = \alpha' D'/2$, where α' is the convergence angle of the image beam. For the resolution to be preserved, the diameter of the disc should be smaller than or equal to the image resolution distance d'. The limit value corresponds to $\alpha' D' = d'$, i.e. $D' = d'/\alpha'$. Now $d' = dG$ and $\alpha' = \alpha/G$, where G is the final magnification of the microscope; taking into account Eq. (19.8), the depth of focus then becomes

$$D' = DG^2 \qquad (19.9)$$

Thus the depth of focus is G^2 times the depth of field. In the previous example with $G = 10^5$ it would measure $D' = 520$ m. Compared to the microscope dimensions, depth

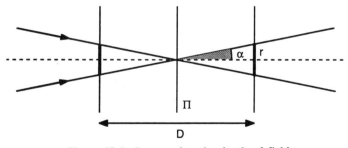

Figure 19.5. Interpreting the depth of field.

of focus is virtually infinite at high magnifications. As a result, positioning of the fluorescent screen and the photographic cameras is not critical.

19.1.2.5 Specimen Stage

The specimen holder has to meet various and severe requirements:

1. Positioning of the specimen in the pole piece gap must be carried out with great adequacy and stability, such that any displacement during exposure time (mechanical vibrations, thermal drift) remains small with respect to the expected resolution.
2. For the investigation of crystalline materials, the ability of extended orientation along various crystal zone axes requires a eucentric goniometric stage, with tilt rotation or double tilt motions. Specimen tilt is usually limited to some 60° for mechanical reasons.

Side-entry stages are most commonly used. They are easy to operate with an *airlock* enabling specimen introduction without vacuum breaking, as well as with a cryoshield for reducing specimen contamination, based on a liquid-nitrogen-cooled metal enclosure surrounding the specimen. Furthermore they are well designed for X-ray analysis by means of an EDS attachment. *Top-entry stages* are less sensitive to vibration and to thermal drift (thanks to their circular symmetry) and are therefore often used in high-resolution microscopes, despite some drawbacks (limited tilting, unsuitable for analytical attachments, difficult to accommodate with airlock and anti-contamination). Recently developed symmetrical side-entry stages have performance similar to the best top-entry stages (Hewitt et al, 1989).

19.1.2.6 Observing and Recording

Visual observation takes place through a fluorescent screen. Photographic recording makes use of specially designed films with fine-grained and thin emulsion, permitting extensive enlargements as required for high resolution. Direct magnification (G_{dir}) is usually limited to some 300 000–800 000, depending on the microscope; it is often insufficient for eye-resolving the limit image resolution distance d'. Considering the eye to be able to separate 0.3 mm, the final magnification has to reach 10^6 for a 3 Å object resolution. It is generally advisable to operate at a maximum direct magnification of some 10^5 followed by photographic enlargement, in order to limit exposure time, i.e. limit specimen drift.

19.1.2.7 Vacuum System

The minimum operating vacuum in the microscope enclosure is of the order of 10^{-5}–10^{-6} Torr, currently provided by a cascade system of rotary pump and oil diffusion pump. The present trend is towards localised higher and cleaner vacuum of at least 10^{-8} Torr, provided by ion pumps, in the specimen chamber for reducing contamination and in the electron source chamber for operating LaB_6 sources or field emission sources. High-vacuum parts are separated in a dynamic way from low-vacuum parts by apertures of some 200 μm (differential pumping). A clean vacuum up to

10^{-6} Torr may be provided by turbomolecular pumps which have the advantage of operating from atmospheric pressure and thus avoid any hydrocarbon contamination from the vacuum system. Their vibrations are eliminated by magnetic bearings. The conventional rubber O-rings are replaced by metallic gaskets, mostly indium, or by stainless steel bellows.

19.1.2.8 Attachments and Developments

Various complements may widen the working field of an electron microscope.

Heating and Cooling Specimen Stage. A heating stage (usually up to 1000 °C) and a cooling stage (200 K by liquid nitrogen, 4 K by liquid helium) lead to the possibility of analysing solid state transitions as a function of temperature. Combining with a chemical reaction chamber (e.g. oxidation, sulphurisation) may be interesting. Reactions are best observed by means of the diffraction pattern.

Strain Stage. Observing stress-induced deformation effects of a specimen is useful in material science, e.g. for studying generation and motion of glide planes, of dislocations in crystals, by means of their image.

Image Intensifying. Image intensifying (e.g. by means of a microchannel plate detector) enables us to operate at electron flux densities which may be some 100 times lower and, as a result, can considerably reduce electron bombardment. Adjusting and preliminary observations are then carried out through a video monitor screen. In addition, such a device permits image recording on a videotape recorder. An intensifying device is interesting, notably for studying electron bombardment sensitive materials, such as hydrated minerals and organic materials.

Scanning Attachment. A beam scanning device, notably when combined with a nanometre-sized electron probe, provides a TEM with some of the possibilities of a scanning transmission microscope (STEM), in particular for implementing analytical techniques which are quoted below.

Spectrometers. An X-ray spectrometer (EDS) permits analysis of major elements in a specimen area of some 10 nm.

An electron spectrometer mounted at the beam exit, after transmission through the specimen, makes possible elemental analysis in a nanometre area by means of characteristic electron energy losses (EELS).

Energy Filtering. A set of magnetic or electrostatic prism analysers enables us to select electrons which have undergone a given energy loss (Castaing and Henry, 1962). When set on the incident electron energy, filtering thus eliminates inelastic scattering and resolution limitations through chromatic aberration; it enables us to operate with thicker specimens and improves the image contrast. Set on a given characteristic energy loss, it leads to a specific form of composition mapping. When combined with a scanning attachment, those operations may similarly be performed by the EELS spectrometer.

19.1.2.9 Summary: Main Characteristics of a TEM

Spherical aberration requires a small objective lens aperture, of the order of 10^{-2} rad, thus limiting the optical point resolution.

Electron energy losses require a very thin specimen (e.g. some 100 Å), in order to preserve the optical point resolution.

Depth of field is greater than the normal specimen thickness.

Depth of focus is virtually infinite with respect to the dimensions of the microscope.

19.2 IMAGE CONTRAST. GEOMETRICAL INTERPRETATION

19.2.1 GENERATION OF CONTRAST

In order to be observable, an image must display a contrast, i.e. a difference in brightness between two adjacent image points. Contrast arises from the specimen as well as from the instrumental characteristics:

1. *Specimen effect*: Intensity and direction distribution of electrons at the specimen exit face.
2. *Instrumental effect*: Imaging of the specimen exit face by the objective lens.

For both of these effects, imaging processes differ in the electron microscope from those in the light microscope.

19.2.1.1 Specimen Effect

The specimen acts on the transmitted beam through absorption and scattering.

The *absorption effect* is prominent for image contrast in light microscopy. From one point to another of the specimen, transmitted light differs mainly in intensity.

The *scattering effect* is prominent for image contrast in electron microscopy. From one point to another of the specimen, transmitted electron waves differ mainly in direction.

19.2.1.2 Instrumental Effect

A light microscope has a large objective lens aperture α. The resulting image displays the intensity distribution at the specimen exit face.

An electron microscope has a small objective lens aperture α, usually less than 10^{-2} rad. Only electrons scattered through an angle smaller than α may contribute to imaging, while electrons scattered at higher angles are stopped by the objective lens aperture. Contrast is therefore generated by subtracting scattered or diffracted electrons, resulting in a scattering or diffraction contrast.

19.2.2 SCATTERING AND DIFFRACTION CONTRAST

19.2.2.1 Amorphous Scattering Object

Consider a thin specimen and a parallel incident electron beam. Objective lens aperture α is materialised by a circular diaphragm located in the back focal plane. Consider in the specimen a small scattering particle AB in a non-scattering matrix. As outlined by Figure 19.6(a), only electrons scattered into an angle smaller than α pass through the aperture. The result in the image plane is a steady brightness, except at the image A'B' corresponding to the scattering particle AB. The image of this particle thus appears dark on a bright background and is therefore called a *bright field image*.

Fully elastic scattering leads to a perfect point correlation between object and image. Inelastically scattered electrons undergo energy losses and thus are not focused on the theoretical image plane, leading to a loss in image sharpness and contrast.

The smaller the objective lens aperture, the stronger the image contrast; it is therefore commonly called the *contrast aperture*. It is interchangeable. In a current conventional electron microscope with a focal length of 3 mm, an aperture diaphragm of 35 μm corresponds to an angular aperture of 6×10^{-3} rad, a reasonable compromise between aberration and resolution.

19.2.2.2 Crystalline Diffracting Object. Bright Field and Dark Field Image

Diffraction contrast imaging due to a crystalline particle AB is similarly outlined in Figure 19.6(b). Consider the particle to be at the first-order Bragg orientation for one of its sets of lattice planes (hkl). The resulting parallel diffracted beam emerges at an angle 2θ with respect to the optical axis. In the absence of any objective diaphragm, it converges at a point T(hkl) of the back focal plane, whereas the non-scattered transmitted beam converges at the back focus F'. Due to both the small specimen thickness and the small wavelength, an extended electron diffraction pattern is actually located in the back focal plane (see section 12.1); it displays a plane section of the reciprocal lattice with the transmitted spot T(000) and a whole set of diffraction spots T(hkl). In the small-angle approximation, the objective back focal plane may be assimilated in the reflecting sphere (Ewald sphere), of radius the focal length f, leading to the *diffraction constant* $K = f\lambda$.

In the imaging mode, the objective lens aperture α is usually smaller than the diffraction angle 2θ and therefore stops the diffracted beams.

Example. Cleavage platelet (001) of periclase MgO (cubic F, $a = 4.2$ Å); 100 keV ($\lambda = 0.037$ Å) electrons. The smallest diffraction angle corresponds to the first possible reflection on (100), i.e. the 200 reflection (see section 5.3.3). In the small-angle approximation

$$\theta = n\lambda/2d(\text{hkl}) \Rightarrow 2\theta(200) = 17 \times 6.10^{-3} \text{ rad}$$

The diffraction angle is seen to be significantly greater than the aperture related to the optical point resolution.

This results in two imaging modes, according to the aperture position in the back focal plane, the bright field mode and the dark field mode.

Bright Field Image

With the aperture centred on the transmitted beam, only the corresponding electrons contribute to the image. Any crystallised parts of the specimen appear dark on a bright background, thus resulting in the bright field image (Figure 19.6b). A crystalline particle

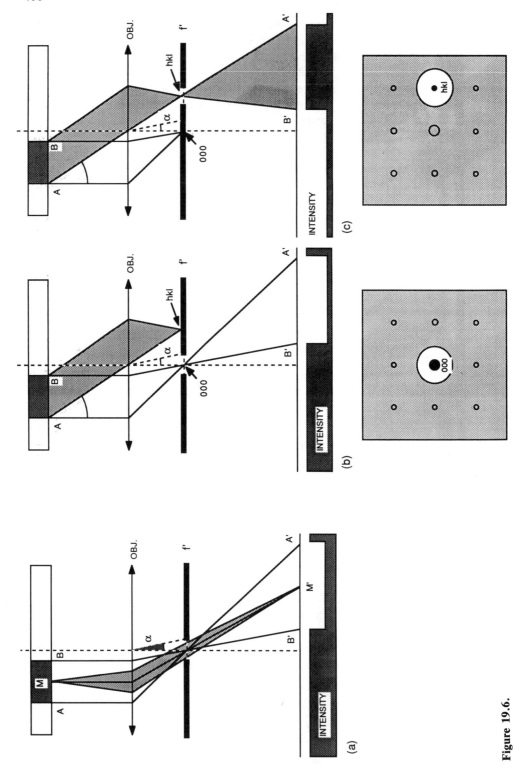

Figure 19.6.

appears darker the closer its orientation to a prominent zone axis. Any image point appears the brighter the smaller its scattering power.

Dark Field Image

With the aperture centred on a given diffracted beam *hkl*, only the corresponding electrons contribute to the image. Any crystallised parts of the specimen appear bright on a dark background, thus resulting in the dark field image (Figure 19.6c). In the image of a polycrystalline specimen, only the crystallites corresponding to the selected reflection *hkl* appear bright. A crystalline particle appears the brighter the closer its orientation to a prominent zone axis.

Any possible reflection *hkl* may give rise to a corresponding dark field image, provided that the related Bragg orientation may be achieved by means of the motions of a goniometric specimen stage. In practice a given diffracted beam is made to pass through the objective lens aperture by adequate tilting 2θ of the incident electron beam, by means of a set of deflection coils. The operation is monitored in the diffraction mode with the objective lens aperture diaphragm inserted.

A dark field image of an amorphous specimen may similarly be observed by placing the objective diaphragm on a scattering ring. The image intensity is generally low in this case.

Multiple Dark Field Imaging

Dark field images corresponding to all reflections of a diffraction pattern may be simultaneously observed. For this aim, the microscope is operated in the microdiffraction mode (without objective diaphragm, with selection diaphragm, see section 12.1.2); intermediate lens excitation is then decreased for defocusing the diffraction pattern (focusing below the back focal plane). As a result, all diffraction spots are widened and become a small circular domain displaying the corresponding dark field image related to the selected area. Due to defocusing, those images are not sharp. They may, however, be useful, e.g. for indexing Bragg fringes (see section 19.3.2) or for orienting specific contrast details with respect to crystallographic axes which are materialised by the displayed reciprocal lattice plane. Such images may be made sharp by subsequent objective focusing; however, they then no longer correspond to the initial selection area.

Selective Dark Field Imaging

When a small amount of crystallites is distributed in an amorphous or poorly crystallised matrix, the resulting diffraction pattern is usually too weak to appear above the scattering background. This is due to the diffraction effect of the few crystals to be spread over the whole scattering pattern. On the other hand, the contrast effect in the image is located at the actual crystal positions and is therefore generally easier to observe. To operate

Figure 19.6. *(opposite)* Geometrical interpretation of image contrast. (a) Amorphous scattering particle; (b) crystalline diffracting particle; bright field image generated with only the transmitted beam; (c) crystalline diffracting particle; dark field image generated with only a diffracted beam *hkl*. The respective positions of the objective lens aperture in the back focal plane are outlined at the bottom of (b) and (c).

in this mode, a small objective diaphragm, of a few micrometres in diameter, is moved radially in the back focal plane, i.e. in the reciprocal plane of constant $K=f\lambda$, where f is the focal length. Local illuminations of crystallites with interplanar spacings such as $d(hkl)=K/R$ can then be observed in the dark field image any time the diaphragm is located on a diffraction ring of radius R. Measuring the values of R thus provides the values of $d(hkl)$ with an accuracy which may be sufficient to characterise the crystalline phase.

This very sensitive method was initially developed for investigating carbon crystallising by Oberlin and Terrière (1972). It may be useful for studying any onsetting crystallisation, e.g. nucleation of clay minerals by alteration of silicate minerals or glasses (Badaut and Risacher, 1979; Crovisier et al, 1983).

19.2.2.3 Summary

Post-objective lenses may be set either to focus on the image or to focus on the corresponding diffraction pattern.

Imaging Mode. The post-objective lens system is focused on the objective image plane (Gaussian plane). A diaphragm in the back focal plane is intended to set the objective lens aperture at the optimum value. According to its position with respect to the diffraction pattern (reciprocal lattice plane), either a bright field image or a dark field image can be observed. The image contrast is mainly controlled by scattering or diffraction.

Diffraction Mode. The post-objective lens system is focused on the back focal plane (Fourier plane) and the objective aperture diaphragm is removed. The object diffraction area is selected by means of a diaphragm inserted in the objective lens image plane. The observed extended diffraction pattern represents a plane section of the reciprocal lattice.

19.3 CRYSTAL IMAGE CONTRAST. KINEMATIC AND DYNAMIC INTERPRETATION

The following theoretical development is based on the kinematic approximation; it leads to a qualitative interpretation of the image contrast provided by a very thin crystal. The dynamic theory is required for a quantitative interpretation. The two-wave dynamic approximation may be a reasonable compromise in some cases.

In this section, the resolution distance d is assumed to be greater than the crystal lattice parameters; structure resolution therefore does not intervene in the image contrast. This amounts to assuming the internal crystal potential $V(xyz)$ to have a constant value. Structure resolution will be dealt with in section 19.4.

As has been previously stated, the image generated by the objective lens of an electron microscope corresponds to the distribution of wave vectors $\mathbf{k}$ (in direction and intensity) at the object exit plane. For a straightforward formulation, only the waves leaving the specimen in a same given direction will be considered. This involves forming the dark field image with a given reflection hkl. Only outgoing waves with wave vectors $\mathbf{k}(hkl)$ will then contribute to imaging (case of Figure 19.6c).

In order to simplify formulation, a reflection *hkl* will usually be designated as reflection **g** (corresponding to the reciprocal vector $\mathbf{r}_g^*$ and to the diffraction deviation **s**. This conventional notation is commonly used in electron diffraction (Eq. 6.2).

19.3.1 COLUMN APPROXIMATION

Consider a point P of the exit face of a single crystal platelet. Let **s** be the diffraction deviation for an *hkl* reflection.

For electrons to be transmitted without significant energy losses, the thickness of the platelet must be very small with respect to its lateral dimensions which may be macroscopic. The resulting reciprocal space consists of fine rod-shaped diffraction domains normal to the platelet (see Figure 12.1a). With the specimen assumed to be at normal incidence, the geometrical conditions are such that the diffraction domains are parallel to the incident direction, resulting in $\mathbf{s} \parallel \mathbf{k}_0$ and $\mathbf{r}_g^* \perp \mathbf{k}_0$ (Figure 19.7a). Their intersection with the reflecting sphere results in very fine diffraction spots and in a very small divergence of the diffracted beam. Thus any electrons leaving the specimen at P are issued from a very fine column of matter passing through P and parallel to $\mathbf{k}(hkl)$ (Figure 19.7b). Due to the small diffraction angle, the direction of the column may be approximated to the incident direction $\mathbf{k}_0$ (Figure 19.7c), leading to the so-called *column approximation*.

As a result, the contrast at an image point P' is determined by interaction of incident electrons with a fine column of matter OP surmounting the corresponding point P of the object exit face.

Width of the Column. To assess an order of magnitude of the width of that column, it may be assumed that electrons leaving the specimen at P are issued from scattering by points located inside a cone of apex P and of axis **k**, thus delimiting a few Fresnel zones on the incident wave plane (see Figures 6.2 and 19.7b). Consider the instance of a thin single gold crystal ($a = 4.08$ Å) of thickness $t = 100$ Å and 100 keV electrons ($\lambda = 0.037$ Å). Limiting the participation to the first 10 Fresnel zones, the radius of the cone intercept at the entrance face would be $r \cong (10\lambda t)^{1/2} \cong 6$ Å (Eq. 6.6). For reflection 200 ($2\theta \cong 0.02$ rad), the shift O'P of the column with respect to OO' would thus be 2 Å, i.e. smaller than the radius of the cone. For a prominent zone axis [*uvw*], it may therefore be assumed that the column consists virtually of a few atomic rows surmounting point P.

Conventions. For a simplified formulation, the following conventions will be applied in this section.

1. *Lattice.* Thin platelet of a crystal with orthogonal axis, of parameters a, b, c.
2. *Orientation.* Incident direction along [100], assumed to be normal to the platelet (zone axis [100], i.e. reflections $0kl$). This results in $\mathbf{s} \parallel \mathbf{k}_0 \parallel [100]$. A column OP is thus oriented along Ox and corresponds to a thickness $t = Ua$ (Figure 19.7c).

19.3.2 IMAGE OF A PERFECT CRYSTAL

It is recalled that a perfect crystal consists of the three-dimensional periodic repetition of a unit cell. The *crystal structure* (i.e. unit cell content) is invariant by any lattice translations (Eq. 5.1)

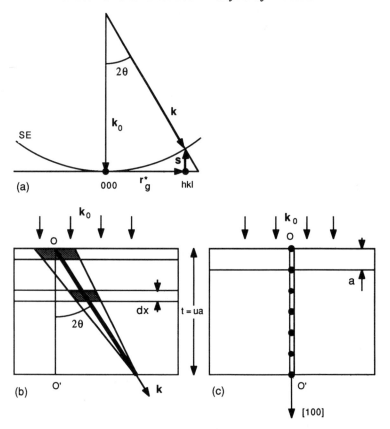

Figure 19.7. Column approximation. [001] zone of a thin parallel-faced crystal platelet. (a) and (b) Outline with exaggerated angle 2θ; (c) column assimilated to be normal to the platelet, consisting, in practice, of a row of atoms along [001], surmounting point P.

$$\mathbf{r}_n = u\mathbf{a} + v\mathbf{b} + w\mathbf{c}$$

where u, v, w are integers and $\mathbf{a}$, $\mathbf{b}$, $\mathbf{c}$ basic lattice vectors.

For a dark field image generated by a **g** reflection, amplitude and intensity at a point P are expressed in the kinematic approximation by the following equations (Eqs 6.10 and 6.11):

$$G(\mathbf{g}) = \frac{i\pi}{t_g} \int_0^t \exp(2\pi i\, xs)\, dx \qquad I(\mathbf{g}) = \frac{\pi^2}{t_g^2} \frac{\sin^2 \pi st}{(\pi s)^2}$$

where t_g is the extinction distance for reflection **g** and s the absolute value of the diffraction deviation.

These equations remain valid for a point P on the exit face of a crystal platelet. Intensity at a point of the exit face of the specimen, i.e. at the related point of the image, thus depends on two variables: *thickness t* of the specimen and *diffraction deviation s* (i.e. specimen orientation).

19.3.2.1 Thickness Fringes

Consider an undistorted crystal platelet (s constant), of varying thickness t. According to Eq. (19.10), intensity at a point P may be written

$$I_P = K \sin^2 \pi s t \quad (K \text{ constant}) \tag{19.10}$$

The outgoing intensity is thus a sine function of thickness t, of period $1/s$ along the normal to the platelet. As a result the image of a wedge-shaped crystal plate will display sinusoidal fringes along thickness contours (Figure 19.8). These *thickness fringes* are commonly observable at crystal edges (Figure 19.9).

19.3.2.2 Bend Fringes

Consider a parallel-faced crystal (t constant) with a varying incidence with respect to the electron beam. Intensity at P as given by Eq. (19.10) is then a periodic function of s with main maxima and secondary maxima (interference function, Figure 5.7). The hkl dark field image of a distorted stationary crystal will thus display a bright fringe along the crystal points where the crystal is at Bragg incidence for reflection hkl, resulting in *bend contours*. For a very thin crystal (Figure 19.10), this main fringe (corresponding to the main maximum hkl) is surrounded by secondary fringes (corresponding to the secondary maxima of the interference function). Those *bend fringes* (or *Bragg fringes*) are commonly observed (Figure 19.11). They are generally an indication of good crystallinity.

19.3.2.3 Bright Field Complementarity. Failure of Kinematic Theory

The following discrepancies may be noted when confronting observed images with their kinematic interpretation:

1. The transmitted intensity $I(0)$ should be constant and equal to the incident intensity I_0. In fact bright field images display a contrast which is complementary to that of dark field images.
2. The period $1/s$ of thickness fringes should become infinite for $s = 0$. In fact they are still observed in exact diffraction conditions.

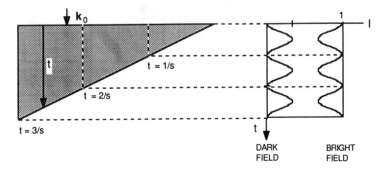

Figure 19.8. Kinematic interpretation of thickness fringes given by a wedge-shaped crystal.

Figure 19.9. Thickness fringes as observed in the image of cubic MgO crystals prepared by collecting the combustion smoke of a magnesium ribbon on a support foil. The two crystals with fringes are cubes deposited on a small {110} face; in the part delimited by {100} cube faces, their thickness thus increases from the edge to the centre. The crystal at the top is deposited on a {100} cube face; its thickness, i.e. its brightness, is therefore constant ($V_0 = 100$ keV; $G_{dir} = 10^5$). (a) Dark field image with 220; (b) bright field image. (Ehret G., Dept Electron Microscopy, Lab. Crystallography, Louis Pasteur University, Strasbourg).

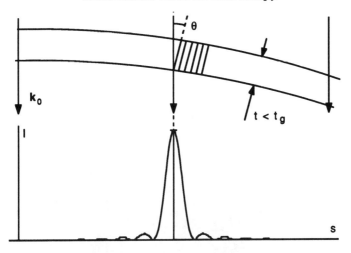

Figure 19.10. Kinematic interpretation of bend fringes.

3. The observed contrast differs from that predicted by the kinematic theory, in intensity and in position, when the thickness reaches a certain value. Thickness fringe amplitudes are damped out with increasing thickness. Their profile becomes dissymmetric.

These observations highlight the failure of the kinematic theory for electron–matter interactions, for which the basic weak interaction hypothesis no longer holds. The dynamic theory is required for a full interpretation consistent with observation. The two-wave dynamic approximation (one transmitted wave and one strong diffracted wave) is often sufficient and easy to apply. It provides diffracted and transmitted intensities through the following relations (Eq. 6.16):

$$I(\mathbf{g}) = \frac{\pi^2}{t_g^2} \frac{\sin^2 \pi s' x}{(\pi s')^2} \qquad s' = \left[s^2 + \frac{1}{t_g^2}\right]^{1/2}$$
$$I(0) = 1 - I(\mathbf{g}) \tag{19.11}$$

The kinematic diffraction deviation s is then replaced by a dynamic deviation parameter s'. The period of intensity variation as a function of x thus becomes $1/s'$ instead of $1/s$.

Far from exact diffraction conditions, it may be assumed that $s \gg 1/t_g$, thus leading to $s' \cong s$: The kinematic theory may then apply.

When approaching the exact diffraction conditions, i.e. for $s \neq 0$ (but remaining low), the variation period $1/s'$ increases; an effective extinction distance t'_g may be defined as follows:

$$t'_g = \frac{1}{s'} = \frac{t_g}{(1 + s^2 t_g^2)^{1/2}} \tag{19.12}$$

At *exact diffraction conditions* $s = 0$, the deviation parameter becomes $s' = 1/t_g$; the intensity period $1/s'$ becomes equal to the extinction distance, thus leading to the real signification of t_g.

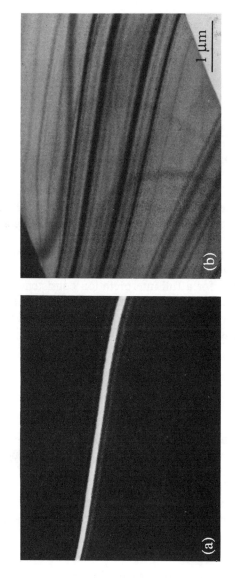

Figure 19.11. Bend fringes in the image of a thin {001} cleavage of mica muscovite ($V_0 = 100$ keV; $G_{dir} = 12\,500$). (a) Dark field image with 200; (b) bright field image. (Ehret G., Dept Electron Microscopy, Lab. Crystallography, Louis Pasteur University, Strasbourg).

Equations (19.11) then simply become

$$I(\mathbf{g}) = \sin^2 \frac{\pi x}{t_g}$$

$$I(0) = \cos^2 \frac{\pi x}{t_g}$$

(19.13)

This formulation expresses well the complementarity between diffracted and transmitted intensities, i.e. between dark field images and bright field images. At the Bragg incidence the diffracted and the transmitted wave oscillate in phase opposition as a function of depth x (so-called *Pendellösung*, i.e. pendular solution). However, this simplified formulation does not account for the observed damping-out of oscillations with increasing depth, nor for the profile dissymmetry of bend fringes in bright field imaging. Those effects are taken into account by introducing *absorption*.

19.3.2.4 Conclusion

The kinematic interpretation of electron diffraction contrast is valid only for very thin specimens and for weakly excited reflections (i.e. s large). The extinction distance t_g is commonly used as a thickness limit criterion, i.e. the kinematic theory is considered to hold if the thickness is smaller than t_g/π (see Table 6.1). For common cubic F metals, the corresponding thickness limits are of the order of 50–200 Å for imaging with strong reflections.

Note. Thickness Assessing. Bright field images display dark bend fringes corresponding to subtracting diffracted rays at any points of the specimen which are in Bragg orientation. In the two-wave case (one transmitted and one diffracted beam hkl), this image is complementary to the corresponding hkl dark field image. In the general multi-reflection case, a bright field image represents the superposition of all the complementary images of corresponding dark fields. For a given set of lattice planes in a crystal plate bent along a cylinder, the result is a system of parallel dark main fringes related to successive reflection orders. For a small number U of periods, there are $U-2$ secondary fringes between any two main fringes. This may be useful in assessing the thickness $t = Ua$ of a crystal (see Figures 5.7 and 19.11b).

19.3.3 IMAGE OF A REAL CRYSTAL

All interpretations in the present section are based on the kinematic approximation and therefore remain qualitative.

19.3.3.1 Effect of Lattice Distortions

Displacement Vector

The occurrence of a lattice defect inside a column OP of a crystal plate results in a phase shift at P, i.e. in a change of contrast.

Consider a wave scattered by a unit cell of a perfect crystal, located at r_n, into a direction defined by $R = g = r_g^* + s$ (reflection g); according to Eq. (5.31) its phase is given by

$$\phi_n = 2\pi r_n s$$

A geometrical lattice defect results in a local deviation from periodicity, i.e. in a shift of lattice points. The perfect lattice position vector r_n is then replaced by r_n' such that

$$r_n' = r_n + \rho_n \tag{19.14}$$

Where ρ_n is the *displacement vector* of the corresponding lattice point $N(uvw)$ with respect to its position in the perfect crystal (Figure 19.12).

Defect-induced Contrast

The amplitude diffracted by a crystal part containing a lattice defect, into a direction determined by a reflection g, may be expressed by the following relation which results from replacing r_n by r_n' in Eq. (5.31):

$$G'(g) = F(g) \sum_{r_n} \exp[2\pi i (r_n + \rho_n)(r_g^* + s)] \tag{19.15}$$

When calculating the product in square brackets, the term $r_n r_g^* = m$ may be omitted (m integer $\Rightarrow \exp 2\pi i\, r_n r_g^* = 1$), and the term $\rho_n s$ may be neglected (product of two vectors which are both small with respect to r_n and r_g^* respectively). The diffracted amplitude then becomes

$$G'(g) = F(g) \sum_{r_n} \exp(2\pi i r_n s) \exp 2\pi i (\rho_n r_g^*) \qquad \delta\phi_n = 2\pi \rho_n r_g^* \tag{19.16}$$

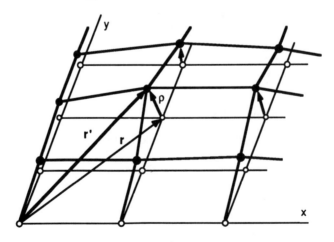

Figure 19.12. Diagram outlining the definition of the displacement vector ρ of a lattice point. The distorted lattice is represented in heavy lines.

where $\delta\phi_n$ is the additional phase shift of a unit cell at $\mathbf{r}_n$ with respect to the perfect crystal lattice.

Consider a thin crystal platelet of thickness t, with the same orientation as previously stated ($\mathbf{k}_0 \parallel [100]$). A defect inside a column results in a shift $\boldsymbol{\rho}_x$ along $[100]$; amplitude and phase of the outgoing wave, related to a dark field image hkl, may now be written according to Eqs (19.10) and (19.16):

$$G'(\mathbf{g}) = \frac{i\pi}{t_g}\int_0^t \exp(2\pi i x s)(\exp 2\pi i \boldsymbol{\rho}_x \mathbf{r}_g^*) dx \qquad \phi' = \phi + \delta\phi = 2\pi s t + 2\pi \boldsymbol{\rho}\mathbf{r}_g^* \qquad (19.17)$$

Extinction of Defect Contrast

For a general displacement vector $\boldsymbol{\rho}$, the defect-induced phase shift $\delta\phi$ becomes zero, i.e. the additional contrast at the column exit disappears, when $\boldsymbol{\rho}\mathbf{r}_g^* = m$ integer.

1. For $m \neq 0$ and $\boldsymbol{\rho}$ a general vector, the above condition refers both to orientation and modulus of the two vectors; it may therefore only be met by chance. On the other hand, it may be met if $\boldsymbol{\rho}$ is a particular vector (i.e. lattice vector or fraction of a lattice vector).
2. For $m = 0$, the condition of contrast extinction refers only to orientation; it may therefore be met experimentally for whatever the value of $\boldsymbol{\rho}$.

The *extinction condition of defect contrast* due to a lattice defect may therefore be expressed as follows for a general displacement vector $\boldsymbol{\rho}$ along the column:

$$\boldsymbol{\rho}\mathbf{r}_g^* = 0 \Rightarrow \boldsymbol{\rho} \perp \mathbf{r}_g^* \Rightarrow \boldsymbol{\rho} \parallel (hkl) \qquad (19.18)$$

A lattice defect does not appear in a dark field image hkl when its related displacement vector $\boldsymbol{\rho}$ is parallel to the reflecting planes (hkl).

This criterion is crucial for determining the displacement vector, i.e. to characterise a lattice defect by electron microscopy. Determining the displacement vector involves looking for at least two reflections, emitted from different sets of lattice planes, which lead to extinction of the defect-induced contrast. The displacement vector is then parallel to the intersection of those planes.

The effects due to the most common lattice defects will be briefly surveyed below.

19.3.3.2 Stacking Faults

The defect plane separates the crystal into two perfect parts (1) and (2), shifted through a vector $\boldsymbol{\rho}$; the shift may be parallel or normal to the defect plane. For a column P_1P_2 crossing the defect plane, the diffracted waves undergo a single phase shift at the interface which amounts to $\Delta\phi = 2\pi\boldsymbol{\rho}\mathbf{r}_g^*$ (Figure 19.13a). The defect-induced contrast anneals for $\boldsymbol{\rho} = \mathbf{r}_n$ (lattice vector) and for $\boldsymbol{\rho}\mathbf{r}_g^* = m$ (integer), in particular for $\boldsymbol{\rho} \parallel (hkl)$. In the general case of a plane defect inclined to the electron beam, the result is a set of sine fringes similar to those induced by a wedge consisting only of part (1), with complementary effects in bright field and in dark field.

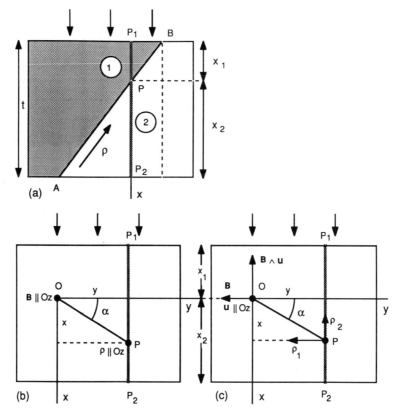

Figure 19.13. Interpreting the defect-induced contrast by means of the column approximation (column P_1P_2) in normal incidence on a parallel-faced crystal. Orthogonal reference axes $Oxyz$, with $Ox \parallel \mathbf{k}_0$; xOy in the figure plane; Oz normal to the figure plane. (a) Column crossing an inclined stacking fault AB; instance of the displacement vector parallel to the fault; (b) column passing close to a screw dislocation along Oz ($\mathbf{B} \parallel Oz \Rightarrow \rho \parallel Oz$); (c) column passing close to a wedge dislocation along Oz ($\mathbf{B} \parallel Oy \Rightarrow \rho \parallel Oy$); xOz slide plane.

A similar contrast effect is induced by other plane defects, e.g. a very thin planar precipitate, a twin plane, a grain limit, provided the two individual parts are phase coherent (see Figure 19.14a). A dissymmetry of the set of fringes is actually observed above a certain thickness, due to absorption and to dynamic effects.

19.3.3.3 Dislocations

A dislocation is surrounded by a strain field. Along a column P_1P_2 passing close to the dislocation, it therefore induces a phase shift which is a continuous function of depth x.

Screw Dislocation

In order to simplify formulation, consider a screw dislocation parallel to the surface of the crystal platelet. The Burgers vector $\mathbf{B}$ of such a dislocation is parallel to it (orientations are defined in Figure 19.3b). Along a closed circuit, i.e. through a rotation

2π around the dislocation, the displacement vector amounts to $\rho = \mathbf{B}$, by definition of the Burgers vector. At a point $P(x, y)$ of a column close to the dislocation, the corresponding rotation amounts to $\alpha = \tan^{-1}(x/y)$, i.e. to a displacement vector

$$\rho = \frac{\mathbf{B}\alpha}{2\pi} = \frac{\mathbf{B}}{2\pi} \tan^{-1}\left(\frac{x}{y}\right)$$

The dislocation-induced phase shift at a point $P(x, y)$ is therefore

$$\delta\phi = 2\pi\rho\mathbf{r}_g^* = \mathbf{Br}_g^* \tan^{-1}\left(\frac{x}{y}\right) \qquad (19.19)$$

A perfect dislocation, i.e. when the Burgers vector is a lattice vector, leads to $\mathbf{Br}_g^* = m$, with m an integer, order of the dislocation. The total induced phase shift at the exit face results from integrating along the column. A column passing through the dislocation core would result in a rotation π, i.e. in a phase shift $m\pi$.

The dislocation contrast becomes invisible for $\mathbf{Br}_g^* = 0$, i.e. when the Burgers vector, hence the dislocation, is parallel to the reflecting planes (hkl) generating the dark field image.

This invisibility criterion enables us to determine $\mathbf{B}$. The most probable Burgers vectors of the crystal species considered are compiled in a table together with the reflections hkl meeting the invisibility criterion. When the corresponding dark field images of two reflections show the extinction of a given dislocation, this is parallel to the intersection of those planes. The choice of hkl reflections is limited by the orientation possibilities of the specimen stage. A goniometric stage is compulsory.

According to Eq. (19.17), the total phase at a point $P(x, y)$ of the column amounts to

$$\phi' = \phi + \delta\phi = 2\pi xs + m \tan^{-1}\left(\frac{x}{y}\right) \qquad (19.20)$$

Depending on the respective signs of the two terms, the phases ϕ and $\delta\phi$ are added or subtracted. The variable x being in both terms, the sign of ϕ' depends on the sign of the product ysm:

1. The sign of y reverses on both sides of the dislocation; the contrast is therefore dissymmetrical on both sides of the dislocation.
2. The sign of s reverses when the dislocation image crosses a main bend contour ($s = 0$); the contrast is shifted on both sides of the fringe.
3. Reversing the sign of m corresponds to two dislocations with opposite Burgers vectors, thus forming a dipole. Their contrast is symmetrical with respect to one another.

For a dislocation inclined to the surface by an angle ψ, x is replaced by $x \cos \psi$ in the phase relationship. The resulting contrast is weaker.

Edge Dislocation

Consider again a dislocation parallel to the surface of the platelet. The Burgers vector is now normal to the dislocation (Figure 19.13c). Let $\mathbf{u}$ be the unit vector along the dislocation. The displacement vector may be expressed as the sum of two components

$$\boldsymbol{\rho} = \boldsymbol{\rho}_1 + \boldsymbol{\rho}_2 \quad \text{with} \quad \begin{array}{l} \boldsymbol{\rho}_1 \parallel \mathbf{B} \quad \text{(parallel to the glide plane)} \\ \boldsymbol{\rho}_2 \parallel (\mathbf{B} \wedge \mathbf{u}) \quad \text{(perpendicular to the glide plane)} \end{array}$$

The displacement vector at a point (P(x, y)) is provided by the Burgers relation which may be written in the following simplified form:

$$\boldsymbol{\rho} = \boldsymbol{\rho}_1 + \boldsymbol{\rho}_2 = \frac{\mathbf{B}}{2\pi} F_1 + \frac{\mathbf{B} \wedge \mathbf{u}}{2\pi} F_2 \tag{19.21}$$

(F_1 and F_2 are functions of x, y and of the Poisson ratio.)

The dislocation-induced phase at P then becomes

$$\delta\phi = 2\pi \boldsymbol{\rho} \mathbf{r}_g^* = \mathbf{B} \mathbf{r}_g^* F_1 + (\mathbf{B} \wedge \mathbf{u}) \mathbf{r}_g^* F_2 \tag{19.22}$$

For $\mathbf{B} \mathbf{r}_g^* = 0$, there remains a residual symmetrical contrast, due to F_2 being independent of the sign of y. A total extinction occurs for $\mathbf{r}_g^*$ simultaneously perpendicular to $\mathbf{B}$ and to $(\mathbf{B} \wedge \mathbf{u})$, i.e. when the dislocation is normal to the reflecting planes (hkl).

General Dislocation

A general dislocation consists of a screw component and an edge component. The contrast invisibility criterion is similar to the edge dislocation case. Along a dislocation loop the angle of $\mathbf{B}$ to the dislocation varies, i.e. the contrast varies. In a dark field image such as $\mathbf{B} \mathbf{r}_g^* = 0$, a residual contrast varies along the loop, with zero minima at two opposite points such as $(\mathbf{B} \wedge \mathbf{u}) \mathbf{r}_g^* = 0$.

Summary. A single dislocation may be visualised by means of the electron microscope. The strain is located in the close vicinity of the dislocation; it induces a contrast along the projection of the dislocation on the specimen exit face. The result is an 'image' of the dislocation which is displayed as a more or less fine line, either bright or dark, with a dissymmetrical contrast (Figure 19.14).

19.3.3.4 Inclusions and Precipitates

Consider a solid consisting of a single crystal matrix containing dispersed small particles (e.g. precipitates or inclusions of some 10 Å size). Various types of particles may be considered, depending on the matrix–particle structure relationship:

1. *Fully coherent particles* with a three-dimensional continuity of the lattice in the matrix and the particles, with or without strain.
2. *Partially coherent particles* with only one- or two-dimensional continuity.
3. *Incoherent particles*.

Coherence is due to a more or less close structural relationship between both the matrix and the particles (e.g. Guinier–Preston zones in alloys, topotaxial phase transformations).

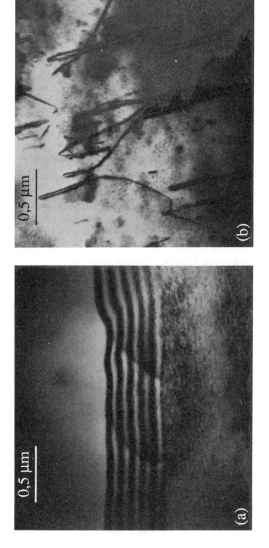

Figure 19.14. Lattice defect images from calcite ($CaCO_3$). Bright field image, $V_0 = 100$ kV. (a) Boundary of a mechanical twin inclined to the incident beam, with dislocations; (b) dislocations in a {100} cleavage plane of a crystal submitted to compression along the threefold axis {111} (Braillon P., Dept. Physics, Univ. Claude Bernard, Lyon).

Two causes of contrast may be considered, due to the particle itself and due to matrix deformation in its vicinity:

1. *Contrast due to the particle.* An incoherent particle produces a diffraction contrast due to a difference in orientation or in structure factor, with respect to the matrix. The contrast of a coherent particle is due to a mutual orientation leading to moiré fringes.
2. *Contrast due to matrix deformation.* This is determined by lattice point displacements $\boldsymbol{\rho}$.

For an easy formulation, consider a spherical particle of radius r_0 inducing a radial displacement ρ in the matrix, depending solely on the distance r to the centre of the particle, according to the relation

$$\rho = \epsilon r_0^3 / r^2 \qquad (19.23)$$

where ϵ is a parameter expressing the lattice misfit of the two phases and their mechanical characteristics.

The corresponding strain contrast observed in a dark field image hkl is annealed for $\boldsymbol{\rho} \perp \mathbf{r}_g^*$; this occurs at any point of the equatorial plane of the sphere which is normal to the reciprocal vector $\mathbf{r}_g^*$, i.e. parallel to the reflecting planes (hkl). Given that $\mathbf{r}_g^*$ is approximately parallel to the image plane, the result in the image is a line of zero contrast along the particle diameter normal to $\mathbf{r}_g^*$ (Figure 19.15).

19.3.3.5 Mutual Orientation of Two Phases. Moiré Pattern

Moiré fringes can be observed in an electron microscope when two overlapping crystals have a mutual orientation such that a strong reflection of each of them may pass simultaneously through the objective aperture. The generation of moiré patterns is therefore related to double diffraction (see 12.1.3). The waves emitted from the two reflections are then able to interfere and to display in the image plane a set of fringes normal to the line connecting the two diffraction spots (without accounting for image rotation). These moiré patterns amount to two-slit interference fringes and are similar to those observed in two-wave lattice imaging (section 19.4.3).

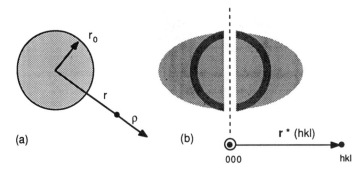

Figure 19.15. Interpretation of the matrix contrast caused by a small spherical particle. (a) Object definition; (b) outline of the image contrast; it in fact varies continuously.

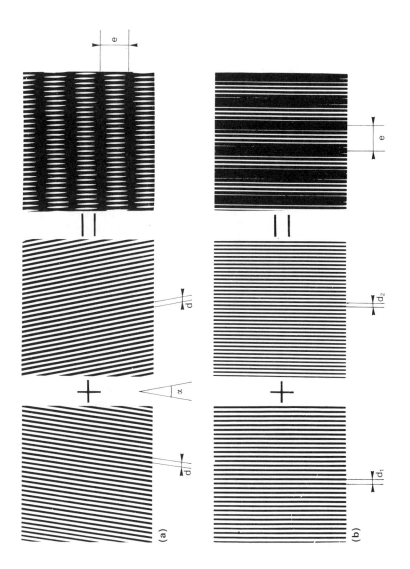

Figure 19.16. Diagram featuring the generation of moiré patterns by overlapping one-dimensional lattices (from Menter, 1958). (a) Rotation moiré; (b) parallel moiré.

Rotation moiré patterns are caused by two overlapping crystals of the same species, with a mutual disorientation angle α.

Parallel moiré patterns result from two different crystal species with parallel lattices.

Macroscopic moiré patterns are commonly observed when superposing fine sieves or silk fabrics (Figure 19.16).

Consider a column OP passing through two crystals. The phase shift at the interface causes a contrast at point P of the exit face. For an easy interpretation, it is convenient to consider the displacement vector $\boldsymbol{\rho}^*$ in the reciprocal lattice ($\mathbf{r}'^* = \mathbf{r}^* + \boldsymbol{\rho}^*$), rather than the displacement vector $\boldsymbol{\rho}$ in the direct lattice. The product $\mathbf{r}\,\mathbf{r}^*$ remaining a constant, it may be written as follows:

$$\mathbf{r}\,\mathbf{r}^* = \mathbf{r}'\,\mathbf{r}'^* = (\mathbf{r}+\boldsymbol{\rho})(\mathbf{r}^*+\boldsymbol{\rho}^*) \cong \mathbf{r}\,\mathbf{r}^* + \mathbf{r}\,\boldsymbol{\rho}^* + \boldsymbol{\rho}\,\mathbf{r}^*$$

This results in

$$\mathbf{r}\,\boldsymbol{\rho}^* = -\boldsymbol{\rho}\,\mathbf{r}^* \Rightarrow \delta\phi = 2\pi\boldsymbol{\rho}\,\mathbf{r}_g^* = -2\pi\,\mathbf{r}\,\boldsymbol{\rho}^* \qquad (19.24)$$

Consider ρ^* to be the distance between two neighbouring points G_1 and G_2 of the respective reciprocal lattices of the two crystals (1) and (2) (i.e. two diffraction spots T_1 and T_2, to the diffraction constant $L\lambda$). The locus of equal intensity on the specimen exit face is the locus of equal phase $\delta\phi$, i.e. the locus of the end point of vector $\mathbf{r}$ such that $\mathbf{r}\,\boldsymbol{\rho}^* = $ constant. It consists of parallel and equidistant lines, normal to $\boldsymbol{\rho}^*$; their distance e is determined by a 2π phase shift, i.e. $e = 1/\rho^*$ (Figure 19.17). For an actual measurement of ρ^* on a diffraction pattern of constant $L\lambda$, the spacing referred to the object plane becomes $e = L\lambda/\rho^*$; in the final image plane, at a magnification G, the fringe spacing is Ge.

Rotation Moiré Pattern

Two thin crystals of the same species are rotated with respect to each other through a small angle α. The distance of their respective hkl reflections in reciprocal space is

$$\rho^* = r_g^* \alpha = n\,n^*(hkl)\alpha = n\alpha/d(hkl)$$

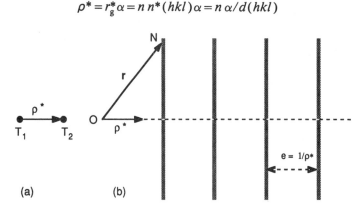

Figure 19.17. Interpretation of moiré fringes. (a) Diffraction pattern in the back focal plane; (b) fringes at the object exit face.

where n is the order of reflection selected for imaging and $n^*(hkl)$ the inter-point spacing of the reciprocal lattice line $[hkl]^*$.

Fringe spacing in the object plane becomes

$$e = \frac{d(hkl)}{n\alpha} \qquad (19.25)$$

Such moiré patterns are commonly observed with easily cleavable lamellar crystals (Figure 19.18a). A bright field image may display several overlapping fringe systems if the diffraction pattern includes several strong doubled reflections (the case of certain portions of Figure 19.18a).

Parallel Moiré Pattern

Consider two different crystal phases with mutual orientation such as a prominent set of planes $(h_1k_1l_1)$ of interplanar distance d_1 from phase (1), parallel to a prominent set of planes $(h_2k_2l_2)$ of interplanar distance d_2 from phase (2). Consider the n_1th lattice point G_1 of $[h_1k_1l_1]^*$ to be close to the n_2th lattice point G_2 of $[h_2k_2l_2]^*$; the distance ρ^* between G_1 and G_2 is given by

$$\rho^* = r_2^* - r_1^* = \frac{n_2}{d_2} - \frac{n_1}{d_1} = \frac{n_2d_1 - n_1d_2}{d_1d_2}$$

Fringe spacing in the object plane becomes

$$e = \frac{1}{\rho^*} = \frac{d_1d_2}{n_2d_1 - n_1d_2} \qquad (19.26)$$

This kind of moiré pattern is notably observed in cases of epitaxial and topotaxial growth (Figure 19.18b).

Examples. Rotation moiré. Two overlapping cleavage foils of mica are twisted by 1°. A bright field image is generated by the doubled reflection 020, related to an interplanar spacing of $d(010) = 9$ Å. According to Eq. (19.25), the image displays a moiré pattern of fringe spacing e, referred to the object plane:

$$e = 9 \times 180/2\pi = 257 \text{ Å}$$

Parallel moiré. Epitaxial growth of silver (cubic F, $a = 4.08$ Å) by vacuum evaporation on to a (001) mica cleavage foil. The mutual orientation is such that the silver planes (111) and $(1\bar{1}0)$ are respectively parallel to the mica planes (001) and (010). The resulting diffraction pattern displays reflections from the [111]-Ag zone and the [001]-mica zone. Strong reflections 220-Ag and 060-mica are close. Fringe spacing given by Eq. (19.26) is then

$$d_1 = d(1\bar{1}0)\text{-Ag} = 2.88 \text{ Å}; \quad d_2 = d(010)\text{-mica} = 9 \text{ Å} \Rightarrow e \cong 36 \text{ Å}$$

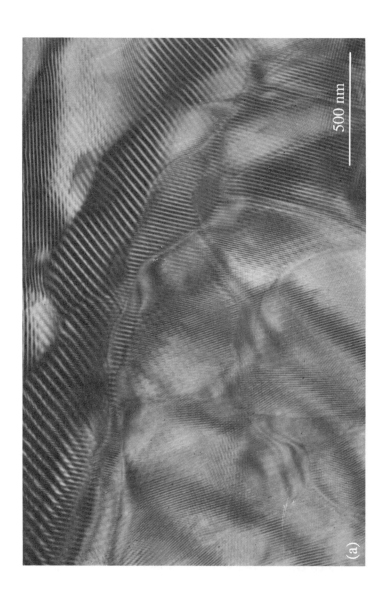

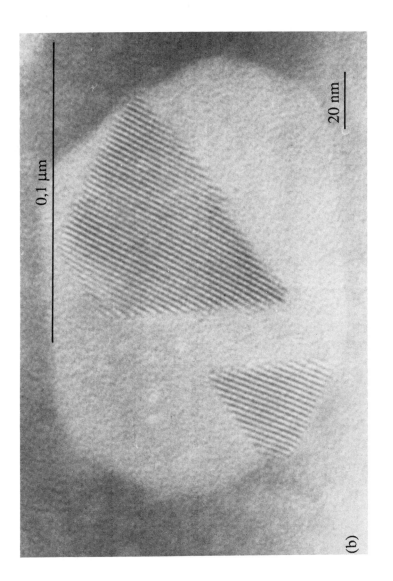

Figure 19.18. Moiré fringes. Bright field images; $V_0 = 100$ kV. (a) Rotation pattern through overlapping of thin, slightly disoriented {001} foils of mica muscovite; $G_{dir} = 28\,000$; (b) parallel pattern generated by topotaxial growth of a spinel-type phase in mica muscovite, at 1000 °C; fringe spacing $e \cong 19$ Å due to interferences between neighbouring strong reflections 060-mica and $4\bar{4}0$-spinel; $G_{dir} = 160\,000$ (G. Ehret, Dept Electron Microscopy, Lab. Crystallography, Louis Pasteur University, Strasbourg).

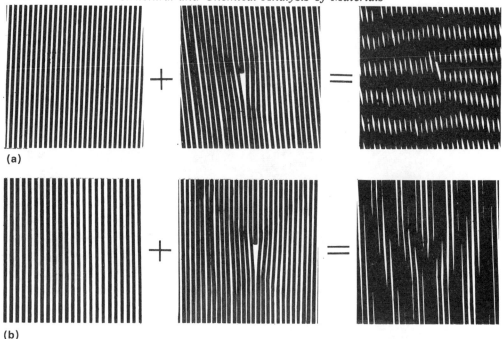

Figure 19.19. Interpretation of the visualisation of an edge dislocation by means of overlapping one-dimensional lattices (from Menter, 1958). (a) Rotation moiré; (b) parallel moiré.

Moiré patterns may display possible lattice defects in one of the phases, e.g. dislocations. They provide in some way a magnified image of the defect, observable at low resolution (Figure 19.19).

19.4 STRUCTURE RESOLUTION. HIGH-RESOLUTION ELECTRON MICROSCOPY

When interpreting the diffraction contrast of crystal images in the previous section, it has been assumed that the internal potential $V(xyz)$ of the crystal, i.e. its scattering power, was constant, to the resolution limit of the microscope.

When the resolution distance becomes smaller than crystal parameters (or atomic distances in an amorphous solid), an additional structure-related contrast appears, leading to the so-called *high-resolution electron microscopy* or *nanoscopy*. This field opened up in the 1970s thanks to the advent of a new generation of electron microscopes with near-ångström resolution; it has been favoured by a better theoretical comprehension of electron imaging, as well as by the development of high-performance computing and image processing. The frontier of atomic resolution has now been reached.

For an easy comprehension of the imaging process, consider successively the path of the electron wave from the object to the objective lens, to its back focal plane and to the image plane (Figure 19.20).

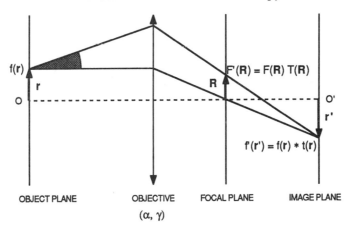

Figure 19.20. Ray diagram outlining the basic principle of electron microscope imaging. Transfer of the object function $f(\mathbf{r})$ by an objective lens with aperture α, with aberration- and defocusing-phase shift γ; these characteristics are expressed by the phase contrast transfer function $T(R)$.

19.4.1 STRUCTURE IMAGING

19.4.1.1 Wave Transmission through the Object. Object Function

The object function $f(\mathbf{r})$ expresses the distribution of wave amplitudes (magnitude and phase) at any point $\mathbf{r}(xy)$ of the object exit face. It thus corresponds to the projection on plane xy of the crystal structure as 'seen' by electrons.

Referred to a unit incident amplitude, function $f(\mathbf{r})$ may be represented as the *transparency function* defined as follows:

$$f(\mathbf{r}) = [1 - s(\mathbf{r})] \exp i \, \phi(\mathbf{r}) \qquad (19.27)$$

where $s(\mathbf{r})$ is the *absorption* or magnitude component and $\phi(\mathbf{r})$ the *phase* component.

In the *weakly scattering object* approximation (s and $\phi \ll 1$), this function may be approximated to its first-order development:

$$f(\mathbf{r}) = 1 - s(\mathbf{r}) + i \, \phi(\mathbf{r}) \qquad (19.28)$$

19.4.1.2 Wave Function in the Back Focal Plane of a Perfect Objective Lens

The back focal plane may be approximated to the reflecting sphere. To any scattering direction $\mathbf{k}$ at the object exit plane corresponds a point $\mathbf{R}(XY)$ in the back focal plane, such that $\mathbf{R} = \mathbf{k} - \mathbf{k}_0$ (neglecting the scale factor λf, i.e. the diffraction constant).

The amplitude $F(\mathbf{R})$ at a point $\mathbf{R}$ is expressed by the Fourier transform (FT) of the object function $f(\mathbf{r})$ (Appendix C):

$$F(\mathbf{R}) = \mathrm{FT}\{f(\mathbf{r})\} = \Delta(\mathbf{R}) - S(\mathbf{R}) + i \, \Phi(\mathbf{R}) \qquad (19.29)$$

where $\Delta = \mathrm{FT}(1)$, the Dirac function, expresses the central peak of the transmitted wave ($\Delta = 0$ for $R \neq 0$), $S(\mathbf{R}) = \mathrm{FT}\{s(\mathbf{r})\}$ and $\Phi(\mathbf{R}) = \mathrm{FT}\{\phi(\mathbf{r})\}$, the components of the scattered amplitude.

Fourier transforms will be denoted from now on as follows: $S = \mathrm{FT}\{s\}$ and $\Phi = \mathrm{FT}\{\phi\}$.

Amplitudes scattered by a crystalline object are localised at small diffraction domains surrounding reciprocal lattice points. The diffracted amplitude is then expressed by a sum of discrete terms, respectively corresponding to the *hkl* reflections, representing the *spatial frequency spectrum* of the object. Spatial frequency $R = r^*(hkl) = 1/d(hkl)$ is defined by analogy with the time-related frequency $\nu = 1/T$ of a propagating wave.

19.4.1.3 Effect of Objective Lens Defects. Transfer Function

The objective lens acts as a filter with respect to the object wave. Its effect in the back focal plane is expressed by the so-called *phase contrast transfer function*, which has the form of an amplitude–phase function. Its amplitude is determined by the objective lens aperture; its phase is determined by spherical aberration and defocusing.

Objective Lens Aperture

Aperture limitation is materialised by an objective lens diaphragm of radius R_0 in the back focal plane; it is expressed by the aperture function $A(R)$ defined as follows:

$$A = 1 \text{ if } R < R_0$$
$$A = 0 \text{ if } R > R_0$$

Spherical Aberration and Defocusing

The phase shift γ_s in the back focal plane, at a distance R from the origin, induced by spherical aberration, is written

$$\gamma_s(R) = \frac{\pi}{2} C_s \lambda^3 R^4 \qquad (19.30)$$

Any defocusing D with respect to Gaussian focusing (either accidental or intentional) results in an additional phase shift γ_D at a distance R in the back focal plane, expressed as follows:

$$\gamma_D(R) = -\pi \lambda D R^2 \qquad (19.31)$$

Astigmatism (correctable) and chromatic aberration (negligible in the weak scattering object hypothesis) are not taken into consideration.

The total phase shift at a distance R of the origin in the back focal plane can now be written:

$$\gamma = \gamma_s + \gamma_D = \pi \lambda \left(\frac{1}{2} C_s \lambda^2 R^4 - D R^2 \right) \qquad (19.32)$$

Phase Contrast Transfer Function

The global effect of the objective lens may be expressed in the back focal plane by a phase contrast transfer function $T(R)$. Due to the revolution symmetry of the objective lens the transfer function depends solely on the radial distance R and may be written

$$T(R) = A(R) \exp[-i\gamma(R)] \qquad (19.33)$$

Wave Function in the Back Focal Plane of a Real Objective Lens

The actual wave function $F'(\mathbf{R})$ may be expressed as the product of the ideal wave function $F(\mathbf{R})$ by the transfer function $T(R)$. This may be written, according to Eqs (19.29) and (19.33),

$$F'(\mathbf{R}) = F(\mathbf{R}) T(R) = (\Delta' - SA \cos\gamma + \Phi A \sin\gamma) + i(SA \sin\gamma + \Phi A \cos\gamma) \qquad (19.34)$$

with $\Delta' = 0$ for $R \neq 0$.

In the weakly scattering object approximation, the amplitude of the transmitted wave expressed by Δ' is large with respect to the amplitude of the scattered waves expressed by the other terms. In bright field imaging, i.e. by means of an axial aperture including the transmitted wave, the imaginary part of function $F'(\mathbf{R})$ may be neglected with respect to the real part which contains Δ'.

As a result, the back focal plane wave amplitude can be approximated to

$$F'(\mathbf{R}) = \Delta' - SA \cos\gamma + \Phi A \sin\gamma \qquad (19.35)$$

19.4.1.4 Wave Function in the Image Plane

When omitting objective lens magnification and image rotation, the amplitude $f'(\mathbf{r})$ at a point $\mathbf{r}(xy)$ of the image plane is the FT of the amplitudes $F'(\mathbf{R})$ scattered at any points $\mathbf{R}(XY)$ of the back focal plane:

$$f'(\mathbf{r}) = \text{FT}\{F'(\mathbf{R})\} = \text{FT}\{F(\mathbf{R}) T(R)\} = f(\mathbf{r}) * t(r) \qquad (19.36)$$

It may be seen that the image function is the convolution product of the object function (identical to the ideal image function) by a confusion function which is the FT of the transfer function. The stigmatic image of an object point is replaced by the confusion function $t(r)$ (see Appendix C).

In the weakly scattering object approximation, the function may now be written, according to Eq. (19.35),

$$f'(\mathbf{r}) = 1 - \text{FT}\{SA \cos\gamma\} + \text{FT}\{\Phi A \sin\gamma\} \qquad (19.37)$$

The *contrast* in amplitude due to scattered waves is expressed by subtracting the transmitted amplitude (normalised to one):

$$c'(\mathbf{r}) = f'(\mathbf{r}) - 1 = \text{FT}\{SA \cos\gamma\} + \text{FT}\{\Phi A \sin\gamma\} \qquad (19.38)$$

19.4.1.5 Structure Image

Two types of contrast arise from Eq. (19.38), determined inside the objective lens aperture ($A = 1$): absorption contrast and phase contrast.

Absorption Contrast

$$\sin \gamma \cong 0 \Rightarrow \cos \gamma \cong 1$$

This condition being approximated in the main part of the aperture-limited objective back focal plane, the first term is predominant, resulting in

$$c'(\mathbf{r}) \cong \mathrm{FT}\{S(\mathbf{R})\} = s(\mathbf{r}) \qquad (19.39)$$

Absorption contrast is predominant. Figure 19.21 shows that this case occurs for small apertures, resulting in resolution lower than 100 Å, i.e. far from structure resolution.

Phase Contrast

$$\sin \gamma \cong \pm 1 \Rightarrow \cos \gamma \cong 0$$

This condition being approximated in the main part of the aperture-limited objective back focal plane, the second term is predominant, resulting in

$$c'(\mathbf{r}) \cong \mathrm{FT}\{\Phi(\mathbf{R})\} = \phi(\mathbf{r}) \qquad (19.40)$$

Phase contrast is predominant. The *phase object* approximation is fulfilled.

Now the phase shift $\phi(\mathbf{r})$ of the wave at a point $\mathbf{r}(xy)$ of the exit face of a crystalline platelet, for an incident energy E_0 and a wavelength λ, is proportional to the crystal potential projection $V(\mathbf{r})$, according to

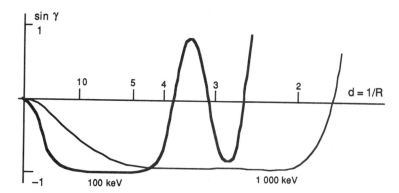

Figure 19.21. Outline of typical transfer functions of objective lenses. The abscissa expresses the axial distance in the back focal plane, i.e. the spatial frequency R; however, for practical purposes, it has been graduated in terms of the ångström value of the corresponding spatial period $d = 1/R$. It may be noted that the zone of interpretable structure contrast extends down to 2 Å at 1000 keV, whereas it is limited to some 4 Å at 100 keV.

$$\phi(\mathbf{r}) = \frac{\pi V(\mathbf{r})}{\lambda E_0}$$

Hence the contrast at a given point is proportional to the potential projection at this point:

$$c'(\mathbf{r}) \cong \frac{\pi}{\lambda E_0} V(\mathbf{r}) \qquad (19.41)$$

This may be compared to Eq. (3.16) expressing the scattering power.

As a result, the contrast may be directly interpreted in terms of potential projection, i.e. of structure projection.

This is the field of the *interpretable structure image*.

As shown in Figure 19.21, the phase condition $\sin \gamma \cong \pm 1$ is approximately met for large apertures, in an angular range limited by the first zero of the phase contrast transfer function. This maximum aperture corresponds to the optical resolution limit of the objective lens, i.e. to the Scherzer resolution. It is therefore often called the *interpretable structure point resolution*.

For larger apertures the phase $\gamma(R)$ undergoes oscillations about zero, leading to contrast artefacts which make direct interpretation hazardous.

In a non-crystallised solid (e.g. glass platelet or carbon support foil), atom projections along a column parallel to $\mathbf{k}_0$ overlap in an aleatory way, making contrast interpretation difficult.

In a crystal oriented along a prominent zone axis, a large number of atoms are aligned along dense lattice rows parallel to $\mathbf{k}_0$. Contrast interpretation in terms of crystal structure is then much easier.

19.4.2 STRUCTURE IMAGE OF A THIN CRYSTAL PLATELET

The potential projection of a perfect crystal is a periodic function which may be expressed by a two-dimensional Fourier series.

Considering the instance of the [001]-zone of a centrosymmetrical crystal, the potential projection $V(xy)$ at a point $\mathbf{r}(xy)$ of the object exit face may be written as follows (Eq. 11.30):

$$\begin{aligned} V(xy) &= \sum_h \sum_k V(hk0)\cos[2\pi(hx+ky)] \\ &= V(000) + V(100)\cos 2\pi x + V(200)\cos 4\pi x + \ldots \\ &\quad + V(hk0)\cos[2\pi(hx+ky)] \end{aligned} \qquad (19.42)$$

This object *potential projection function* represents a sum of sinusoidal potential distributions, of amplitude $V(hk0)$, of spatial period $d(hk0)$, parallel to $(hk0)$ planes. The series is unlimited. Its first term expresses the mean internal crystal potential.

The image of a phase object displayed by a perfect objective lens with unlimited aperture would faithfully reproduce the object potential projection. It would thus be expressed by the same unlimited Fourier series, in which the potential terms $V(hk0)$ would be replaced by the diffraction amplitudes $F(hk0)$ which are proportional to the $V(hk0)$ terms, as shown by Eq. (3.16).

The aperture limitation of a real objective lens results in a limitation of the series: only the reflections $hk0$ passing through the aperture take part in the imaging process. As a result the *image function* may be written as the following limited Fourier series:

$$f'(xy) = \sum_h \sum_k F(hk0) \cos[2\pi(hx+ky)]$$
$$= F(000) + F(100)\cos 2\pi x + F(200)\cos 4\pi x + \ldots \quad (19.43)$$
$$+ F(hk0)\cos[2\pi(hx+ky)]$$

This series is limited to spatial frequencies such as

$$R(hk0) < R_0$$

The resulting limitation of period resolution is

$$d(hk0) > \frac{1}{R_0}$$

Each term of series (19.43) expresses a sinusoidal function. To any $hk0$ reflection in the back focal plane, passing through the objective lens aperture, will therefore correspond a set of parallel sinusoidal fringes:

(a) normal to $\mathbf{R}(hk0)$ (not taking into account image rotation);
(b) of amplitude $F(hk0)$ proportional to $V(hk0)$;
(c) of spatial frequency $gd(hk0)$, where g is the objective lens magnification.

This set of fringes superposes over a continuous background of amplitude $F(000)$ proportional to $V(000)$, the mean internal crystal potential.

The result is similar to a *Fourier–Bragg projection* as calculated by X-ray diffraction (see Figures 11.28 and 11.29). X-ray structure projection is operated in successive stages: intensity measurements, phase determinations, Fourier series calculation. Electron microscope structure imaging, on the other hand, is operated optically, i.e. electron waves focus into image points with their attached phases and thus directly display the structure projection. Unfortunately the series is aperture limited and phases are altered by objective lens aberrations.

Defects in a real crystal lead to scattering outside of discrete $hk0$ reflections. The FT of resulting scattered amplitudes consequently modify the image.

19.4.3 DIRECT INTERPRETATION OF STRUCTURE IMAGES

19.4.3.1 Operating Mode

Optimum conditions for directly interpretable structure imaging are achieved at Scherzer defocusing (Eq. 9.5), with an aperture corresponding to the first zero of the phase contrast transfer function. As shown in Figure 19.21, the resulting point resolution is of the order of 3.5 Å at 100 keV; it improves at higher electron energies, thus highlighting one of the advantages of high-voltage electron microscopy.

The number of terms of the Fourier series (Eq. 19.43) which take part in imaging is determined by the number of reflections passing through the objective lens aperture.

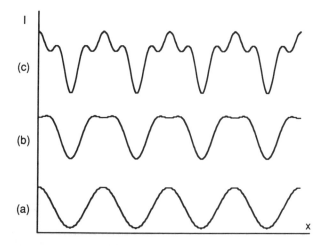

Figure 19.22. One-dimensional Fourier series as calculated from $00l$ reflection amplitudes of mica, with 10 Å interlayer distances (compare with Figures 19.24 and 11.30). (a) Two terms: $F(000) + F(100) \cos 2\pi z$; sinusoidal amplitude variation in the objective lens image plane; (b) three terms: $F(000) + F(001) \cos 2\pi z + F(002) \cos 4\pi z$; (c) four terms: $F(000) + F(001) \cos 2\pi z + F(002) \cos 4\pi z + F(003) \cos 6\pi z$.

19.4.3.2 Two-beam Imaging. Lattice Resolution

To display the interplanar period of a set of lattice planes (hkl), the Fourier series must consist of at least the first two terms, i.e. interference must take place between at least two waves, the transmitted wave 000 and a diffracted wave, e.g. 100. In this specific instance the series (19.43) is limited to

$$f'(xy) = F(000) + F(100) \cos 2\pi x \qquad (19.44)$$

With the first term being constant and the second sinusoidal, the objective lens image plane will display a set of sine-shaped fringes, of intensity proportional to $[f'(xy)]^2$, parallel to the (100) lattice planes (to the image rotation), of period $gd(100)$. This fringe system corresponds to a classical two-slit interference effect, with the diffraction spots T(000) and T(100) in the back focal plane acting as sources. It may be similarly compared to decomposing a structure projection into its sine components (i.e. Fourier analysis by photosummation, Figures 11.28 and 19.22a).

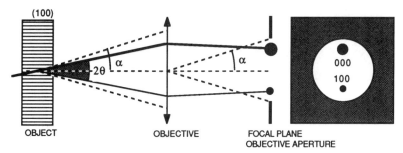

Figure 19.23. Diagram outlining the principle of imaging with two symmetrical beams, the transmitted beam and one diffracted beam.

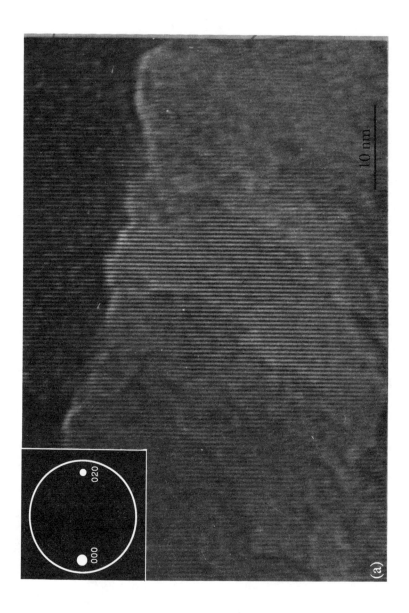

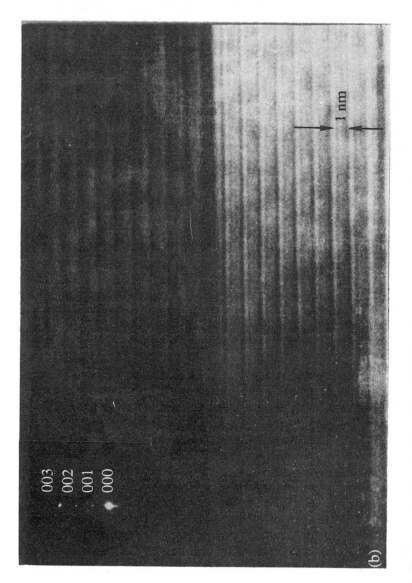

Figure 19.24. One-dimensional resolution. $G_{dir} = 400\,000$; $V_0 = 100$ kV; objective lens aperture 20 μm, approximating the instrumental resolution at Gaussian focusing. (a) (001) cleavage foil of mica muscovite in near to normal incidence; symmetrical imaging with the transmitted beam 000 and the reflection 020; resolution of the interplanar period $d(010) = 4.5$ Å; (b) ultramicrotomic section of a mica crystallite normal to the (001) layers, with interlayer spacing 10 Å, after preferred orientation embedding in epoxy resin. Imaging with the tilted transmitted beam and three orders of reflection 001, 002, 003. Compare with Figure 19.22(c). One-dimensional resolution of the order of 3.3 Å. The insets display the aperture-limited diffraction patterns in the objective back focal plane (Ehret G., Dept Electron Microscopy, Lab. Crystallography, Louis Pasteur University, Strasbourg).

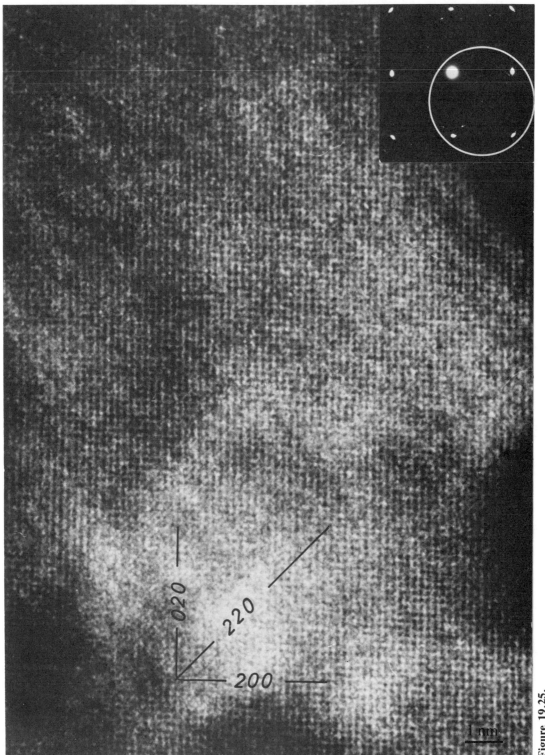

Figure 19.25.

Lattice Resolution Conditions. For the two beams to pass through the objective lens aperture (i.e. $\theta \leqslant \alpha$, Figure 19.23), the small-angle approximation of the Bragg equation requires that

$$\frac{n\lambda}{2d(hkl)} \leqslant \alpha \Rightarrow d(hkl) \geqslant \frac{0.5\lambda}{\alpha} \qquad (19.45)$$

The period resolution condition is slightly different from the point resolution condition (Eq. 19.2), due to the disc-limiting function of point resolution being replaced by a one-dimensional limiting function. In fact the resolution conditions are even more favourable if the two imaging beams are set symmetrical with respect to the optical axis, as outlined in Figure 19.23. Both beams are then inclined at the same angle θ to the axis, resulting in equal axial defocusing D due to spherical aberration:

$$D = C_s \alpha^2 g^2 = C_s \theta^2 g^2 \qquad (19.46)$$

Example

$$200\text{-MgO}; \ \theta = 8.8 \times 10^{-2} \text{ rad}; \ g = 100; \ C_s = 0.16 \text{ cm} \Rightarrow D = 12.5 \ \mu\text{m}$$

When correspondingly underfocusing the objective lens, the spherical aberration is compensated for both beams. The interplanar period of any given set of lattice planes could thus theoretically be resolved, whatever the optical objective lens qualities, by simply making use of sufficiently large objective lens apertures (Figure 19.24a). However, other causes of limitation must be accounted for, such as electron energy losses, and electronic and mechanical instabilities. Nevertheless, lattice resolutions of the order of 1 Å may thus be achieved.

19.4.3.3 Multi-beam Imaging. Structure Resolution

More than two beams must take part in imaging if structure details are to be displayed. In a similar way as for X-ray structure analysis, the greater the number of terms of the Fourier series participating in imaging (Eq. 19.43), the better the resolution achieved. Compensating for spherical aberration remains possible when first-order reflections and the transmitted beam are displayed symmetrically in a regular polygon inside the objective lens aperture. This is the case for certain cubic crystals along a [100] zone; a two-dimensional lattice resolution may then be observed, with two perpendicular sets of sinusoidal fringes, as shown in Figure 19.25.

In any other case of multi-beam imaging, correcting the spherical aberration is no longer possible for all beams involved. However, as previously stated, the phase relationship remains acceptable at Scherzer defocusing, with an aperture limited to the first zero of the phase contrast transfer function (Figure 19.21), leading to the optical

---

Figure 19.25. *(opposite)* Two-dimensional lattice resolution of a thin, monocrystalline gold film (cubic F, $a = 4.06$ Å) observed along a [001] direction. The inset shows the position of the objective lens aperture. Resolution of interplanar distances: $a/2 = 2.03$ Å of lattice planes (100) and (010); $a/2\sqrt{2} = 1.44$ Å of lattice planes (110) ($V_0 = 100$ kV; $G_{dir} = 210\,000$) (SIEMENS document).

point resolution, which in structure imaging is often called *interpretable structure resolution*: the resulting structure image then displays the projection of the potential distribution of the specimen and may be interpreted in terms of its structure.

A crystal orientation along a prominent [uvw] zone leads to a diffraction pattern displaying a high-density reciprocal lattice plane (uvw)*, i.e. consisting of closely packed hkl lattice points. A more or less high number of reflections can then pass through the objective lens aperture, i.e. participate in imaging. The bright parts of the resulting structure image then correspond generally to a low-phase shift of the exit wave, i.e. to a low-integrated potential value along the direction of the projection. They may, for instance, correspond to structure channels between atom rows (see Figure 19.26). Even when those structure imaging conditions are met, direct image interpretation must be handled with care.

19.4.4 INDIRECT INTERPRETATION OF STRUCTURE IMAGES

An objective lens aperture larger than that determined by the above-mentioned interpretable point resolution transmits a greater number of diffracted beams. As a result, the Fourier series (19.43) then comprises terms corresponding to space periods $d(hk0)$ smaller than the optical point resolution distance d. However, the image is then affected by significant phase shifts γ of the transfer function, as outlined by Figure 19.21. Contrast artefacts then result in difficulties for direct image interpretation which should be replaced by indirect methods through image processing or through image modelling.

19.4.4.1 Image Processing

In the weakly scattering object approximation, the object function $f(\mathbf{r})$ may theoretically be reconstructed through deconvolution of the image function $f'(\mathbf{r})$ which is altered by aberrations and defocusing (Eq. 19.36). The FT of the contrast function (Eq. 19.38) is used for this aim:

$$C'(\mathbf{R}) = \text{FT}\{c'(\mathbf{r})\} = SA \cos \gamma + \Phi A \sin \gamma \qquad (19.47)$$

It may be seen that, in the approximation considered, the spatial frequency spectrum $C'(\mathbf{R})$ of the image contrast is a linear combination of the respective spatial frequency spectra of the absorption and of the phase components of the object function (Eq. 19.27), modulated by the transfer function components $A \cos \gamma$ and $A \sin \gamma$. This linear approximation is the basis of image processing, aiming at extracting S and Φ from Eq. (19.47) when the transfer function components are known. The latter parameters may be determined by means of the optical FT of the image of an amorphous carbon or silicon film (e.g. in [7], Hawkes; [19], Saxton).

19.4.4.2 Image Modelling

Electron wave propagation through a crystalline object (perfect or defect structure) may be simulated on the bases of a structure model or a structure hypothesis. Presently used methods are based on various approaches of the electron diffraction theory. The so-called *multi-slice method* considers the crystal as consisting of a stacking of elementary layers which successively diffract the electron waves (Cowley and Moodie, 1957; Goodman and Moodie, 1974). An alternative approach is based on the *Bloch wave formalism* of the dynamic theory (Metherel, 1973). At any object exit point **r**, the calculated diffracted amplitude $F(\mathbf{R})$ is multiplied by the transfer function $T(R)$ determined for the given instrumental conditions (spherical aberration, aperture, defocus). The squared FT of this product leads to the theoretical intensity distribution at any point **r**, i.e. to the theoretical image. This is then compared with the experimental image. Various computer software packages have been developed and are being steadily improved (e.g. O'Keefe et al, 1978; Stadelmann, 1987; Stadelmann, in [15]). Thanks to the steady improvement in computer performance, such modelling programs may now be easily carried out on microcomputers (e.g. 256×256 beam multi-slice calculation on a microcomputer with hard disc in a few minutes) and the simulated image displayed by means of a simple matrix printer. This imaging simulation seems at present the most fruitful way to atomic resolution by means of electron microscopy (Figure 19.26). The sequence of Figures 19.24–19.26 illustrates the improvement of resolution achieved by using an increasing number of reflections for imaging.

19.4.5 APPLICATIONS

Whereas X-ray diffraction analysis leads to a crystal structure averaged over millions of unit cells, high-resolution electron microscopy permits structure analysis at the scale of one unit cell. Due to atom overlappings caused by the finite specimen thickness, easily interpretable crystal structure images by transmission electron microscopy can only be observed at prominent zone orientations. The contrast then results from the projections of atom rows parallel to the incident electron beam. Nanometre-thin amorphous foils may result in the visualisation of individual atoms.

Examples of Applications

1. Structure of crystal nuclei and precipitates in the very first stages of growth;
2. Structure of crystal defects: dislocation cores, point defects;
3. Phase transitions at the unit cell scale;
4. Structure of poorly crystallised materials, e.g. clay particles, gels, mixed layers and polytypes of lamellar particles;
5. Structure of grain boundaries;
6. Solid state reaction mechanisms at a unit cell scale;
7. Structure of organic and biological molecules.

The best results in structure imaging are presently achieved by means of medium-voltage electron microscopes (300–600 kV) which provide a good compromise between theoretical and practical resolution (Figure 19.21).

19.5 SPECIMEN

19.5.1 PREPARATION TECHNIQUES

The same preparation methods apply for any observations carried out by transmitted electrons, imaging as well as diffraction.

442 Structural and Chemical Analysis of Materials

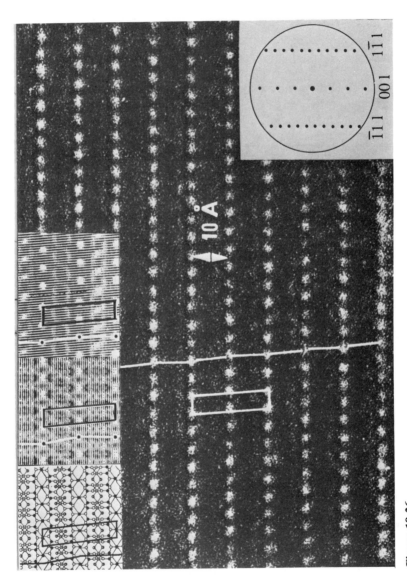

Figure 19.26.

The first requirement is the very small thickness, i.e. at 100 keV a maximum of some 100 Å for current low-resolution work, some 10 Å only for high-resolution studies. High-voltage electron microscopy is less demanding on specimen thinness (see section 19.6).

19.5.1.1 Specimen Support

Before observing a specimen, it has to be mounted on a support which has to meet the following requirements:

(a) electron transparency;
(b) resistance to electron bombardment in vacuum;
(c) no addition of any artefact to the specimen image or diffraction pattern.

It must be taken into account that electron energy losses in the specimen support are impairing resolution as well as those in the specimen itself, whereas the support slide of a light microscope does not have any effect on resolution. The most common support is a thin amorphous foil made of light elements, such as a carbon film or an organic film (formvar or collodion film). This foil is itself supported by a metal grid, with a normalised diameter of about 3 mm, made of electrolytic copper for current applications. Various other metals are used for specific applications (e.g. beryllium or carbon for X-ray analysis, platinum or palladium for heating or reaction stages).

Collodion Film by the Solution Drop Method

The principle is outlined in Figures 19.27(a)–(c). The metal grids are placed on a support (e.g. metal sieve) at the bottom of a glass vessel filled with filtered distilled water. A drop of the corresponding solution (e.g. 1–3% collodion in amyl or butyl acetate) is deposited on the water surface; it spreads into a thin floating layer which forms a thin film after evaporation of the solvent. The first film is conveniently removed in order to clear the water surface of possible dust particles. After a renewed operation, the water is slowly evacuated through a tap or through siphoning. The film is thus deposited on the grids. After dust-free drying, the resulting collodion film is somewhat sensitive to electron bombardment.

Formvar Film by the Dipping and Flotation Method

This method (Figures 19.27d–e) is better suited for formvar films. A clean glass slide is dipped in a formvar solution (e.g. 1.5–3% of the formvar powder in dichloro-1,2-ethane,

---

Figure 19.26. *(opposite)* Structure image of mica muscovite projected along zone axis [1$\bar{1}$0]. The insets at the top display, from left to right: structure model projection; calculated electron density projection; simulated image calculated by the multi-slice method, for $E_0 = 100$ keV; $C_s = 0.18$ cm; $D = -120$ nm. The inset at the bottom right-hand side displays the diffraction pattern which has taken part in imaging, limited by a 35 μm objective lens aperture. Comparing the experimental image (on which the projection of a unit cell and the stacking sequence along [001] have been visualised) with the simulated image leads to an interpretation of the bright dots as structure channels between the potassium atoms in the interlayers. The stacking sequence is typical of a $2M_1$ polytype (Amouric et al, 1981, CRMC[2]–CNRS, Marseille).

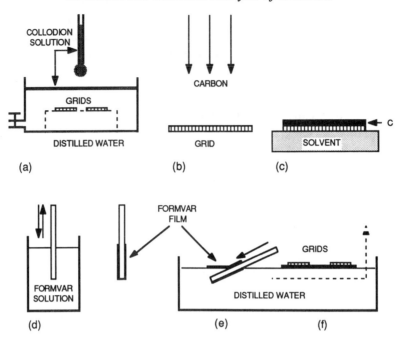

Figure 19.27. Principle of organic support film preparation. (a)–(c) Deposition of a collodion film by the solution drop method; (d)–(f) deposition of a formvar film by the dipping and flotation method.

according to the film thickness to be achieved). The formvar-coated slide is left to dry for a few minutes. The edges of the slide are scraped free of formvar by means of a razor blade. The coated slide is then dipped at low incidence into distilled water and the film is made to unstick through surface tension and to float on the water surface. Removing the film is sometimes difficult; it may be facilitated by a slight moistening (e.g. breath moistening). Support grids are deposited on the floating film, retrieved by means of a metal sieve and left to dry for some 24 h.

Carbon Film

Amorphous carbon is currently the best material known for preparing a support film which is very thin as well as electron-beam resistant. It is best suited for high-resolution electron microscopy. Vacuum-evaporated carbon, produced by resistance heating at the contact points of two pointed carbon arc rods, or cathode sputtered carbon, is deposited in the form of a thin amorphous film either on an organic foil or on a clean glass slide. Carbon-reinforced organic foils may be directly used for current operations. A purely carbon film, suited for high-resolution operation, may be prepared by dissolving the underlying organic film, e.g. by placing the grids on filter paper floating on the specific solvent; dissolution then takes place by capillarity. The carbon film may be directly deposited on a glass slide; it is then collected and deposited on metal grids in a similar way as the formvar film in the previous section; its stripping from the glass slide is, however, more difficult. By adjusting the evaporation time, reasonably resistant carbon films as thin as a few nanometres can be prepared by this technique.

Microgrids

Despite the very small thickness of carbon films as prepared by the above technique, the film may be a hindrance for high-resolution structure imaging. Operating without any support film in the observed specimen area is possible with the microgrid technique (Fukami et al, 1972). After hydrophobic surface treatment and cooling to $-5\,°C$, a clean glass slide is exposed to the air; water soon condenses on its surface in the form of tiny droplets of $0.05-10\,\mu m$ in diameter. A thin layer of a polymer solution (e.g. triafol in ethyl acetate) is then deposited on the slide (at room temperature or after dipping the slide in liquid nitrogen which freezes the droplets). When removed after drying, the plastic film is covered with small holes corresponding to the droplets. The size and the form of the holes depend on the details of the operation. The hole-covered films may be reinforced by carbon or metal vacuum coating. Thicker and purely metallic microgrids are prepared by electrolytic metal coating (silver or copper) and subsequent dissolution of the organic film.

Holed Foils

Organic films with circular holes are also useful for correcting astigmatism, for exact focusing and for testing the point resolution (see section 19.1.2). Various techniques, similar to the microgrid technique above, are currently being employed. The holes may simply be generated in a fresh formvar film by electron bombardment. Adding glycerol to the formvar solution also results in a holed film after the usual preparation process. A more complicated variant consists in dipping a clean glass slide into a chloroform solution of formvar (1%) and hexane (0.05%). Dried in a chloroform atmosphere and then dipped into paraffin, it should result in a film with circular holes of about 200 Å in diameter.

19.5.1.2 Specimen Preparation

Various thin specimen preparation techniques are employed for electron microscopy, depending on the kind of material and on the aim in mind. The most common techniques are outlined in this section. More detailed descriptions are to be found in specialised treatises, e.g. [18], [20].

Grinding

This most simple, but somewhat rough and more or less destructive method, is limited to certain rather soft or brittle materials which are easy to reduce to powder, e.g. clays. It is liable to generate defects and stress effects. Grinding is usually followed by suspension in a fluid, a possible ultrasonic dispersion, differential sedimentation or centrifugation for collecting the required mean grain size. A drop of the final suspension is deposited on a support grid and left to dry.

Cleavage

Some crystallised materials, i.e. those with a layered structure, may be cleaved into thin platelets adequate for electron microscopy (e.g. mica, graphite). Repeated cleavage is made easier by the use of adhesive tape. Specimen orientation is limited to the cleavage plane directions.

Ultramicrotome Cutting

Ultramicrotomy is the most frequently used technique for preparing biological specimens; it may be applied to mineral materials consisting of small, but not too hard, particles. Its main disadvantage is to introduce specimen deformations. It is, however, useful for observing thin sections along any chosen direction, e.g. lamellar crystals cut normally to their layers or fibrous crystals cut normally to their fibre axis. The particles are dispersed in a resin (epoxy or methacrylate) and then cut by means of a diamond knife along a given direction. Specific techniques have thus been developed for studying the rolled sheet structure of chrysolite (Yada, 1967), the globular structure of halloysite (Vernet and Gauthier, 1962), the layer structure of various phyllosilicates (Eckhardt, 1961; Eberhart and Triki, 1972). The specimens leading to the micrographs of Figures 19.24 and 19.26 were prepared in this way. This technique is equally useful for investigating alteration or corrosion layers on solid surfaces, e.g. on glasses (Crovisier and Eberhart, 1985; Ehret et al, 1986; see Figure 21.3).

Thinning of Bulk Materials

Thin preparations of bulk materials are generally carried out in two stages, preliminary thinning followed by the final thinning to the required thickness.

Preliminary Thinning. The initial specimen is cut into a millimetre-thin slice by conventional techniques (e.g. sawing by diamond disc, spark cutting). Mechanical abrasion and polishing lead to a thickness of some $10\,\mu\text{m}$. Chemical dissolution may sometimes be used for preliminary thinning.

The specimen is cut into small pieces corresponding to the size of the support grid.

Final Thinning. Various methods are available, depending on the material:

1. *Chemical thinning* (chemical polishing) is achieved by means of a jet of reactant directed on to the specimen surface (single jet on one face or double jet on both faces). Nature, concentration and temperature of the reagent are chosen as a function of the specimen. Chemical thinning applies to conducting as well as to non-conducting materials. It is difficult to carry out for silicate materials (glasses, ceramics) requiring hydrofluoric acid. The dissolution rate is typically of the order of $50–500\,\mu\text{m/min}$.
2. *Electrolytic thinning* (electropolishing) is limited to conducting materials. It works at lower temperatures and concentrations than chemical thinning and is therefore easier to control. In this technique, the specimen forms the soluble anode of the electrolytic device. The thinning rate is of the order of $5–50\,\mu\text{m/min}$.
3. *Ion milling.* The specimen surface is locally eroded by means of an ion beam (e.g. argon, nitrogen) of a few kiloelectronvolts. A flat and symmetrical crater is achieved

by means of grazing incidence on the rotating specimen. Two opposed ion guns permit thinning on both faces. Ion sputtering is very slow (order of 0.1 μm/min) and depends on various parameters, notably on composition and orientation of crystal grains. It may be carried out inside the microscope. It applies to any material, but is notably useful for silicate materials which are difficult to thin by the chemical method.

With all thinning techniques, dissolution or sputtering is conducted until the specimen is locally perforated, and may be monitored by a light beam and a photoelectric cell.

Any thinning methods of bulk specimens pose homogeneity problems when applied to polycrystalline materials.

Direct Preparation of Thin Films

Thin films of certain materials, notably metals and alloys, may be prepared by vacuum evaporation, provided that they do not decompose. Deposition on to an amorphous substrate (e.g. formvar or carbon film) usually results in a polycrystalline or an amorphous film. Epitaxial growth on a convenient crystalline substrate may lead to monocrystalline layers (e.g. vacuum evaporation of cubic all-faces-centred metals on a {100} cleavage of NaCl or a {001} cleavage of mica). The specimen film is deposited on a support grid after being stripped from the substrate or after dissolving the substrate.

19.5.1.3 Surface Topography of Bulk Specimen. Replicas

The replica technique is suitable for observing the surface of a solid specimen, without destroying it. It is based on transferring the surface topography of any bulk specimen to a thin film observable by a TEM. The successive operations are outlined in Figure 19.28.

Direct Replica (Negative Replica)

A drop of a resin solution (e.g. formvar) is deposited on the specimen and made to spread over its surface by holding it vertically. After drying or polymerising, the film is stripped from the specimen and cut into adequate pieces to be placed on the support grids. Resolution reaches some 100 Å.

A higher resolution (30–50 Å) is achieved by means of a carbon replica. Stripping the carbon film from a porous surface without dissolving the specimen may be a tricky operation. A formvar film deposited on the carbon film facilitates its removal; it is subsequently dissolved. With a metallic specimen, the use of an electrolytic set-up with an acid solution and the specimen as anode results in a hydrogen release which unsticks the carbon film (Figure 19.28a).

Two-stage Replica (Positive Replica)

A thick layer of formvar, softened by its solvent, is deposited on the specimen and strongly pressed against its surface for a perfect imprinting of its topography. Easily stripped off after drying, it is then carbon coated on the significant side and subsequently dissolved to leave the positive carbon replica (Figure 19.28b).

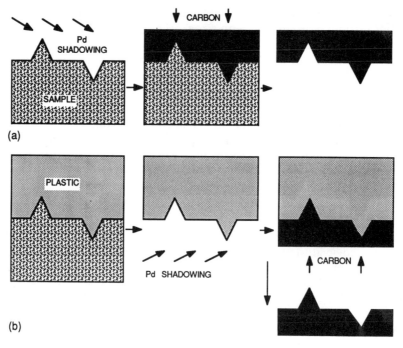

Figure 19.28. Preparation of carbon replicas. (a) Direct (or negative) replica; (b) two-stage (or positive) replica.

The image contrast of replicas is poor; it may be enhanced by shadow-casting. For this aim, a heavy metal (e.g. Au, Pt, Pd) is vacuum evaporated at grazing incidence. In the direct carbon replica technique, shadowing may be carried out on the specimen before replicating; the metal coating is then stripped off together with the carbon film. In the techniques using formvar, the replica itself is generally shadowed after stripping.

The use of replicas is limited to surfaces without too pronounced a surface relief. They are notably unsuitable for porous and crumbly materials. In addition, the replica technique has lost a great part of its importance since the onset of scanning electron microscopy which provides direct imaging of bulk materials, whatever their surface topography and virtually without any preparation.

Extraction Replica

The replicating technique may be used for tearing off fine particles from the surface of a bulk material for transmission investigation. This technique is very useful for studying thin surface layers on solids, e.g. formed by corrosion, by alteration or by deposition. Combined with ultramicrotome cutting after extraction, it makes it possible to investigate such layers along a normal to the surface, by microscopy, diffraction and microanalysis (see Figure 21.3).

19.5.2 EFFECT OF ELECTRON BOMBARDMENT

In order to be observable, an object has to be illuminated. Whereas the resulting irradiation effects are negligible in a light photon microscope, they become significant in an electron microscope, due to the strong electron–matter interaction. They must be taken into account in image interpreting. In addition to the predominant thermal effect, chemical and structural effects are currently observed.

19.5.2.1 Basic Considerations

An important part of specimen irradiation effects may be interpreted through elastic interaction. The maximum amount of energy ΔE transferred to an atom during an electron impact is expressed by Eqs (2.6) and (2.7). The non-relativistic approximation is valid for electron energies in conventional electron microscopy, up to 100 keV.

Atom Displacement Energy

Removing an atom from its potential well requires the provision of a minimum displacement energy E_D; this varies between 15 and 50 eV, with an average value of some 25 eV, depending on the elements and their bonding. In crystals it depends on the incident direction. According to the basic theory of elastic collision, only light atoms (H to N) are liable to be displaced by the impact of 100 keV electrons (Figure 19.29). Some values of the minimum incident energy, deduced from the relativistic relation (2.7), are displayed in Table 19.2, assuming a displacement energy of 25 eV for any element. It shows that at 1.2 MeV all atom species may be displaced.

Energy Transfer below Displacement Energy

The following irradiation effects are observed (see section 2.3):

1. *Thermal effect* due to irradiation induced atomic vibrations, corresponding to energy transfers of the order of some 0.1 eV.

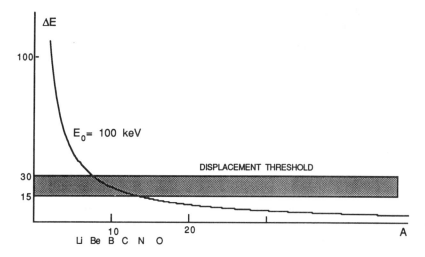

Figure 19.29. Maximum energy transferred to atoms by 100 keV incident electrons as a function of atomic mass, calculated from Eq. (2.6) and compared to the average displacement energy (in grey tint).

2. *Bond breaking.* The breaking of chemical bonds requires a minimum energy of about 3–5 eV. According to Eq. (2.6), this energy may be reached for elements up to atomic mass 72 (31-Ga), in a single collision with incident 100 keV electrons. However, even at lower energies, multiple collisions due to the high-intensity electron beam of electron microscopes may result in bond breakings, notably in organic materials.

Energy Transfer above Displacement Energy

Atom displacements in a solid result in *Frenkel defects* (vacancy + interstitial atom). Multiple collisions may occur when $\Delta E > 2E_D$. Displaced species depend on transferred energies and on chemical bonds. Through thermal diffusion, the radiation-induced defects may either recombine or segregate into clusters or dislocation loops. These become visible even at low resolution, whereas imaging of isolated point defects requires near atomic resolution.

19.5.2.2 Irradiation Effects in Electron Microscopes

These are predominantly a hindrance. The instrumental layout and the operating conditions are correspondingly set to minimise irradiation effects. In some cases, however, they may be useful for inducing local defects or chemical reactions and studying their evolution by means of imaging or diffraction.

Thermal Effect

Heat transfer depends on various factors:

1. Beam intensity. This is adjusted by the condenser lens setting.
2. Beam size on the specimen. At equal intensity, a fine electron probe produces a higher temperature gradient, i.e. a higher heat flow, and therefore results in a lower specimen temperature at the beam centre.
3. Heat conduction of the specimen and the support. A specimen particle close to the copper grid edge is subjected to a lower temperature than a particle at the centre of a mesh.
4. Specimen temperature. This may be reduced by specimen cooling through liquid nitrogen or even helium.
5. Specimen thickness. A thin specimen is less electron absorbing, i.e. less heated than a thick specimen.
6. Excitation cross-section of thermal vibrations.
7. Emissivity of the specimen.

Thermal effects are currently observed in an electron microscope. Metal particles may even be molten on a support film in extreme conditions. In current conditions, the temperature effect causes most hindrance for observing heat-sensitive materials (e.g. hydrated or hydroxylated materials such as clays, zeolites, biological and organic materials). The lifetime of sensitive materials is often shorter than the time required for observing, focusing and exposing. This is notably the case for high-resolution studies. An image intensifier may be of great use for reducing electron intensity during adjusting operations.

Table 19.2. Minimum electron energy in keV required to displace atoms (from Jouffrey and Rocher in [12])

Element	Atomic mass	E_0 min. (eV)	Element	Atomic mass	E_0 min. (eV)
C	12.0	130	Ag	107.9	723
Al	27.0	248	Ta	180.9	1030
Si	28.1	257	W	183.9	1042
Fe	55.8	444	Au	197.2	1092
Cu	63.6	494	U	238.1	1233
Mo	96.0	664			

Chemical Effects

Due to electrons, the beam induces mainly reducing reactions.

The most visible reduction effect is the photographic effect permitting image recording. Intense electron bombardment makes organic specimen support foils fragile, finally resulting in their breaking.

Specimen *contamination* is a nuisance which has been observed since the onset of electron microscopy. Electron-beam induced alteration of the image and the diffraction pattern is due to the build-up of a carbon layer on the specimen surface, due to residual hydrocarbons in the microscope enclosure; these result mainly from machining oils, diffusion pumps, organic O-rings. Hydrocarbon molecules are ionised by the electron beam; positive ions generated in the vicinity of the specimen are then attracted by its negatively charged surface where they are reduced to carbon.

The contamination effect is limited by a *cryoshield*, a small metal enclosure surrounding the specimen, cooled by liquid nitrogen. In recent high-performance electron microscopes, differential pumping of the specimen chamber, by means of turbomolecular or ionic pumps, results in reducing further contamination.

Atomic Displacements

As shown by Table 19.2, displacement effects for any atom species become significant for electron energies above 1 MeV, i.e. in the high-voltage electron microscope. In the conventional electron microscope operating at 100 keV, this effect is limited to very light elements, such as boron or beryllium (Fig. 19.29).

19.6 HIGH-VOLTAGE ELECTRON MICROSCOPY

The most common conventional electron microscopes are limited to a voltage of 100 or 120 kV. Medium-voltage facilities, operating with voltages up to 300–600 kV, are increasingly used for high resolution. So-called high-voltage electron microscopy (HVEM) starts at 1 MV. The first 1 MV HVEM was developed at Toulouse in 1960, followed by a 3.5 MV facility 10 years later (Dupouy et al, 1970). Several high-voltage facilities are now available in specialised research centres.

The design of an HVEM faces technical problems, such as high-voltage insulating, high lens excitation currents (leading to the use of superconductors), protection against

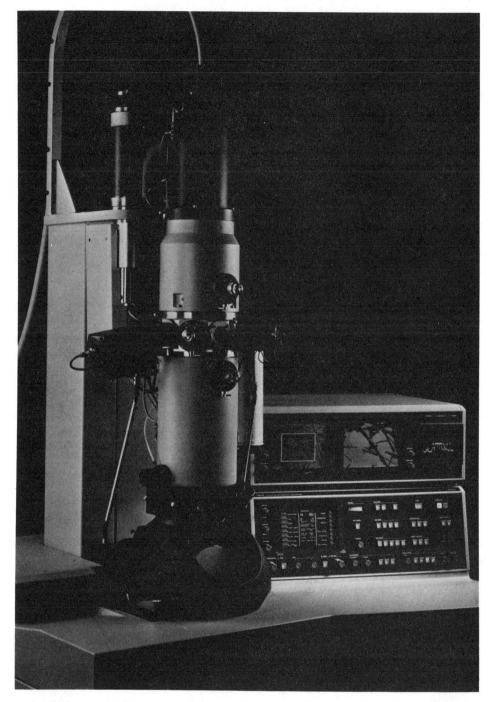

Figure 19.30. General view of a high-resolution TEM–STEM facility. On the left-hand side the microscope column; on the right-hand side the control unit with its commands and its video display screens (Philips Industry document).

19.6.1 RESOLUTION

The point resolution at Gaussian focusing is given by Eq. (19.4) as follows:

$$d = 0.8 \lambda^{3/4} C_s^{1/4}$$

The positive effect of reducing the electron wavelength at high energies is seen to prevail over the negative effect of an increase of spherical aberration, due to the objective lens excitation which is less favourable at high energies.

Wavelengths associated with high-energy electrons are to be calculated by means of the relativity-corrected relation (1.6). They are very short: $\lambda = 87 \times 10^{-4}$ Å at 1 MeV; $\lambda = 36 \times 10^{-4}$ Å at 3 MeV.

The positive effect of high voltages is similarly highlighted by the shift of the first zero of the phase contrast transfer function towards higher space frequencies (Figure 19.21). The resulting interpretable point resolution may be better than 2 Å.

The actual resolution limit is determined by electron energy losses in the specimen. For a given specimen thickness, the mean energy loss is the lower the higher the electron energy (see Table 8.3). According to Eq. (19.6), the resulting limit resolution distance decreases with increasing incident energy, which is in favour of the HVEM. To reach a similar resolution, the latter enables a much thicker specimen to be used. As an example, the resolution observed with a 1 μm thick aluminium film would be about 15 Å at 1 MeV, compared with about 80 Å at 100 keV. Observing conditions are therefore closer to those required for investigating structural characteristics in bulk specimens.

19.6.2 IRRADIATION EFFECTS

As shown by Table 19.2, the atomic displacement threshold is reached for most elements at 1 MeV, for all elements at 3 MeV. The high-energy and intensity electron beam therefore generates a large number of defects (see section 19.5.2). According to the scope which is sought, this effect may be positive or negative. On the one hand, it enables the generation and evolution of crystal defects to be followed directly as a function of operating parameters; on the other hand, it makes it difficult to study the existing intrinsic defects of a material, therefore losing a part of the advantage of operating on thick specimens, closer to bulk conditions.

Thermal effects are less important than in a conventional microscope.

19.6.3 APPLICATIONS

The main quality of an HVEM is the possibility of operating with specimen thicknesses some 3–10 times larger than with a conventional microscope. Observing conditions are thus closer to those of the bulk material. A whole cell structure may thus be observed in a biological material. Carrying out stereoscopic pairs is therefore of particular interest; it provides three-dimensional information over a few micrometres, e.g. on the depth location of defects in a crystal.

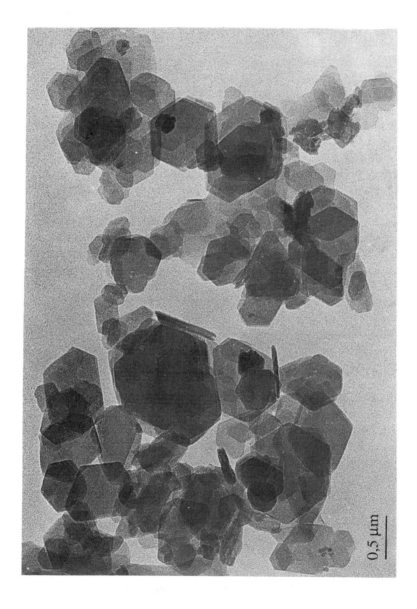

Figure 19.31. Low-resolution electron microscopy. Kaolinite particles. Preparation carried out by grinding the bulk material, water suspension, separation of the fraction smaller than 0.2 μm by differential centrifugation, sedimentation on to a carbon film. Slight under-focusing results in a bright fringe around the particles. $V_0 = 100$ kV; $G_{dir} = 50\,000$ (Ehret G., Dept Electron Microscopy, Lab. Crystallography, Louis Pasteur University, Strasbourg).

Figure 19.32. High-resolution electron microscopy. Thin film of monocrystalline chlorine copper phthalocyanide. Scherzer defocus imaging with 220 *hk*0 reflections, without objective lens diaphragm. The final image has been processed by integrating over aleatory translations along the lattice periods **b** and **c**. Structure interpretation with respect to the structure model (bottom left inset) has been checked by comparing with computed images calculated at various defocusing. Medium-voltage electron microscope ($V_0 = 500$ kV; $G_{dir} = 200\,000$; scale pointer 1 nm). JEOL document, from Uyeda et al, 1981.

Certain advantages of the HVEM are found in the STEM at much lower voltages, i.e. at much lower cost. A reasonable compromise seems to be provided by the medium-voltage electron microscope (300–600 kV) with scanning attachment.

19.7 CONCLUSION

The TEM allows the local structure of solids to be investigated, at a nanometre scale with low-resolution microscopy, and at an ångström scale with high-resolution microscopy.

It provides notably valuable data on defects, deformations, grain boundaries, inclusions, precipitates and any new phases at the very first growth stages.

It is of particular interest for studying any poorly crystallised materials at the single particle scale, e.g. clay minerals (Figure 19.31), zeolites, microfibres, solid particles suspended in the atmosphere, carbons, small particles of catalysers, etc.

Highly developed optical and electronic characteristics, together with a good understanding of the theoretical bases of electron imaging and the availability of high-power computer processing and modelling techniques, permits at present the electron microscope to have access to near-atomic structure resolution.

Whereas X-ray diffraction analysis of a crystal reproduces a mean structure, averaged over a macroscopic volume of matter, high-resolution electron microscopy provides local structural data and their variation at a unit cell scale, e.g. around lattice defects. Structure projections may be carried out (e.g. organic structure, Figure 19.32).

The main drawback of the TEM technique is the requirement for very thin specimens, often difficult to prepare, with properties which do not necessarily correspond to the bulk material. This requirement is, however, less severe with the HVEM.

Combining electron imaging and electron diffraction permits extraction of any possible structural data for investigating crystalline materials.

With a scanning attachment, an energy-dispersive spectrometer and an electron loss spectrometer, the TEM becomes an analytical electron microscope (Ch. 21).

REFERENCE CHAPTERS

Chapters 2, 3, 5, 6, 8, 9, 10, 12, 21. Appendix C.

BIBLIOGRAPHY

[1] Duffieux P. M. *The Fourier transform and its applications to optics*. 2nd edn. John Wiley, New York, Chichester, 1982.
[2] Eberhart J. P. High resolution electron microscopy applied to clay minerals. In Fripiat J. J. (ed.) *Advanced technique for clay mineral analysis*. Elsevier, Amsterdam, 1981.
[3] Edington J. W. Practical electron microscopy in materials science. Monograph 2: *Electron diffraction in the electron microscope*. Macmillan, Philips Technical Library, London, 1975.
[4] Gard J. A. (Ed.) *The electron–optical investigation of clays*. Mineralogical Soc., London, 1971.
[5] Glaser W. *Grundlagen der Elektronenoptik*. Springer-Verlag, Berlin, 1952.
[6] Grivet P. *Electron optics*. Pergamon Press, Oxford, 1965.

[7] Hawkes P. W. (Ed.) *Computer processing of electron microscope images*. Springer, Berlin, 1980.
[8] Heidenreich R. D. *Fundamentals of electron microscopy*. John Wiley, New York, 1964.
[9] Hirsch P. B., Howie A., Nicholson R. B., Pashley D. W., Whelan M. J. *Electron microscopy of thin crystals*. Butterworths, London, 1965.
[10] Hornbogen E. *Durchstrahlungs-Elektronenmikroskopie fester Stoffe*. Chemi GmbH, Weinheim (RFA), 1971.
[11] Hren J. J., Goldstein J. I., Joy D. C. (eds) *Introduction to analytical electron microscopy*. Plenum Press, London, 1979.
[12] Jouffrey B. (ed.) *Méthodes et techniques nouvelles d'observation en métallurgie physique*. Soc. Franç. Microsc. Electron., Paris, 1972.
[13] Jouffrey B., Bourret A., Colliex C. (eds) *Microscopie électronique en science des matériaux*. Editions du CNRS, Paris, 1983.
[14] Kihlborg L. (ed.) *Direct imaging of atoms in crystals and molecules*. Royal Swedish Acad. of Science, Stockholm, 1979.
[15] Krakow W., O'Keefe (eds) *Computer simulation of electron microscope diffraction and images*. The Minerals, Metals & Materials Soc., 1989.
[16] Loretto M. H. *Electron beam analysis of materials*. Chapman & Hall, London, 1984.
[17] Reimer L. *Transmission electron microscopy. Physics of image formation and microanalysis*. Springer Series in Optical Science, Vol. 36. Springer-Verlag, Berlin, 1984.
[18] Reimer L. *Elektronenmikroskopische Untersuchungs- und Präparationsmethoden*. Springer-Verlag, Berlin, 1959.
[19] Saxton W. O. *Computer techniques for image processing in electron microscopy*. Academic Press, London, 1978.
[20] Schimmel G. (ed.) *Methodensammlung der Elektronenmikroskopie*. Wissenschaftl. Verlag mbH, Stuttgart, 1973.
[21] Steward E. G. *Fourier optics. An introduction*. 2nd edn. John Wiley, New York, Chichester, 1987.
[22] Sudo T., Shimoda S., Yotsumoto H., Aita S. *Electron micrographs of clay minerals*. Elsevier, Amsterdam, 1981.
[23] Wenk H. R. (ed.) *Electron microscopy in mineralogy*. Springer-Verlag, Berlin, 1976.
[24] Willaime C. (ed.) *Initiation à la microscopie électronique à transmission. Minéralogie. Science de la matière*. Soc. Franç. de Minér. et Crist., Paris, 1988.

20

Scanning Electron Microscopy

The basic principle of scanning electron microscopy is quite different from that of conventional transmission electron microscopy. There is no optical imaging process by means of lenses. Its operation is similar to TV imaging.

Common abbreviations. *Basic facility:*

1. SEM (scanning electron microscopy);

Variants:

1. STEM (scanning transmission electron microscopy);
2. SAM (scanning Auger microscopy).

Attachments:

1. EDS or EDX (energy-dispersive X-ray spectrometry);
2. EBIC (electron-beam induced current).

20.1 BASIC PRINCIPLE

The basic layout of an SEM is similar to that of the electron microprobe (Figure 15.1).

In a microprobe, emphasis is laid on the analytical aspect, notably on accuracy of quantitative analysis.

In an MEB, the imaging possibilities are predominant, thanks to a very finely focused electron probe, an elaborate scanning system and various detectors.

When scanned by the electron probe, the specimen displays various interaction effects: electron scattering and diffraction; secondary electron and Auger electron emission; photon emission; electron absorption and energy loss; phonon and plasmon excitation; generation of electric and magnetic fields, etc. Each of these effects may be used for imaging, provided that a suitable measuring device is available for converting the object effect into an electrical signal for processing.

The amplified output of the selected detector serves to control the intensity of the electron beam of a CRT with synchronised scanning. Illumination at a given point of the screen is thus directly related to the intensity of the selected interaction effect at

Scanning Electron Microscopy

the corresponding point of the specimen. After completion of at least one scan period, the recorded data result in an image.

This imaging mode has several advantages over conventional stigmatic imaging:

1. Imaging is not limited by lens focusing possibilities. X-ray emission images may thus be displayed.
2. Resolution is not limited by lens aberrations. As a result, relatively thick specimens may be observed in the STEM mode with a reasonable resolution.
3. The scanning mode is well designed for image digitalising and subsequent image processing.

The main drawback of the system is its resolution limitation caused by the probe size and the various spread effects due to scattering.

Current Characteristics in the Basic Imaging Mode. The following characteristics are typical for a current SEM, operating in the basic secondary electron emission mode:

(a) **resolution** of the order of 30–100 Å;
(b) **magnification** varying from some 10 to 100 000 (or more);
(c) ability of observing **bulk specimens**, of centimetric or even decimetric size, with a high **depth of field**, resulting in an image with three-dimensional effect.

The basic principle of the SEM was assessed as early as 1935 by Knoll, i.e. at about the same time and by the same research group as the conventional TEM. Whereas the latter soon developed, the SEM had to wait until the 1960s for practical implementation. Its development has been closely linked to the progress of electronics and video techniques.

20.2 INSTRUMENTAL LAYOUT

As outlined in Figure 20.1, an SEM facility includes the following main components (along the incident electron path): electron source, scanning device, specimen on its stage, imaging system, data processing.

20.2.1 ELECTRON PROBE

The electron probe generating system includes the *electron source* at a negative voltage V_0 of some 10 keV, a *magnetic condenser system* (currently two lenses) generating a reduced image of the cross-over, and a *magnetic objective lens* projecting the cross-over image on to the specimen surface.

The lower limit of the probe diameter d_s is determined by objective lens aberrations (see Figure 19.3) and by the minimum electron intensity on the specimen required for a sufficient signal-to-noise ratio. For a given brightness, the probe intensity (Eq. 15.3) is proportional to $d_s^{8/3}$, i.e. it increases rapidly with the probe size.

A conventional tungsten filament source, operating with a 20–40 kV voltage, may lead to an intensity of the order of 10^{-12}–10^{-11} A (1–10 pA) into a probe of 20–30 Å.

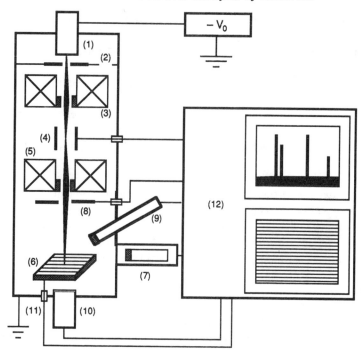

Figure 20.1. Basic diagram of an SEM. (1) Electron source at negative high voltage $(-V_0)$; (2) grounded anode; (3) condenser lenses; (4) beam deflection and scanning system; (5) objective lens; (6) specimen on a goniometric stage; (7) electron detector (scintillation counter, PM); (8) backscattered electron detector (semiconductor); (9) X-ray spectrometer (EDS); (10) transmitted electrons detector; (11) absorbed current measurement; (12) control unit, data acquisition and processing, display screen.

Compared to a dedicated electron microprobe with a common probe size of about 1 μm, the electron density is about the same, i.e. about 10–100 A/cm². The resulting electron intensity at the specimen may be considered as the lower limit for imaging with a reasonable signal-to-noise ratio.

A further decrease of the probe size without decreasing its intensity requires a higher source brightness, i.e. different types of electron sources (see section 9.2). A thermionic cathode with *LaB$_6$ filament* increases source brightness by a factor of 10, whereas a field emission cathode results in an increase by a factor of 10^4–10^5. Probe diameters as small as a few ångströms can thus be achieved (Crewe, 1966; Crewe et al, 1969, 1970).

The required *vacuum* at the cathode depends on the type of source. A conventional tungsten cathode may be operated in a secondary vacuum of some 10^{-5} Torr, as provided by a rotary pump and an oil diffusion pump in cascade. An LaB$_6$ cathode requires 10^{-8} Torr, produced by a turbomolecular or an ionic pump. A high vacuum of 10^{-10}–10^{-11} Torr as generated by an ionic pump is demanded for operating a field emission gun in stable conditions. A system of differential pumping (see Ch. 19) allows achievement of adequate vacuum levels in the various parts of the SEM. A high-vacuum specimen chamber enables Auger electron imaging to be carried out, whereas a low-vacuum specimen chamber is useful for the study of biological materials without pre-treating and of non-conducting materials without metal or carbon coating.

20.2.2 SCANNING SYSTEM

Beam scanning is achieved by means of a deflection system, usually consisting of magnetic coils operated by a large-range scan generator.

Image *magnification* is determined by the ratio of the image scanning length (instrumentally fixed) to the object scanning length which is adjustable.

Two video tubes are currently used.

A large-sized screen with high-remanence phosphor provides the *visual image display* with various scan speeds. In addition to slow scanning (period of the order of 1–10 s or more, depending on the available signal-to-noise ratio) for high magnification observing and focusing, most SEM facilities possess high-speed scanning, including a standard TV mode, for easy observation at low magnification in the secondary electron mode. The TV mode allows video monitoring and recording.

Image recording is performed through a smaller CRT (order of 10×7 cm), with low remanence, actinic light output and high resolution (e.g. 2500 lines). For a maximum signal-to-noise ratio, the scan speed is low, with a period of the order of 1 min. Recording is carried out by means of a conventional camera (Polaroid for quick image control), with an exposure time amounting to one or several scan periods.

Thanks to a digital data output, any information (images as well as spectra) may be easily computer-processed and stored on disc. This can be later redisplayed on video screen or by means of a printer (e.g. screen copy on thermal paper).

20.2.3 SPECIMEN

The specimen to be observed is mounted on an eucentric goniometric stage providing the necessary motions (x–y translation, tilt, rotation). For most observation modes (emission and reflection), the specimen is bulk. Its size is currently a few centimetres, but may reach up to 10 cm or more in certain SEMs (Fig. 20.2).

Transmission imaging is possible with thin specimens similar to those used in TEMs. However, the transmission mode is better dealt with in dedicated STEM facilities (Ch. 21).

In some microscopes the pumping time after specimen change is reduced by means of an object airlock designed for standard specimen holders.

20.2.4 IMAGE PROCESSING

As previously mentioned, any electron–matter interaction effect may be converted into a scanning image, provided that it can be expressed as an electrical signal.

Pixel-by-pixel imaging is well designed for digitalising and processing. Processing may be carried out on-line or on the stored data. Various processing operations are currently carried out as follows:

1. *Contrast processing*, e.g. switching from positive to negative contrast, linearising, gamma control, differentiation.
2. *Contrast quantification* by means of contour mapping and colour mapping.
3. *Image integration* over a certain number of scan periods for enhancing the signal-to-noise ratio. This mode is notably useful for low-level signals, e.g. for backscattered electrons.

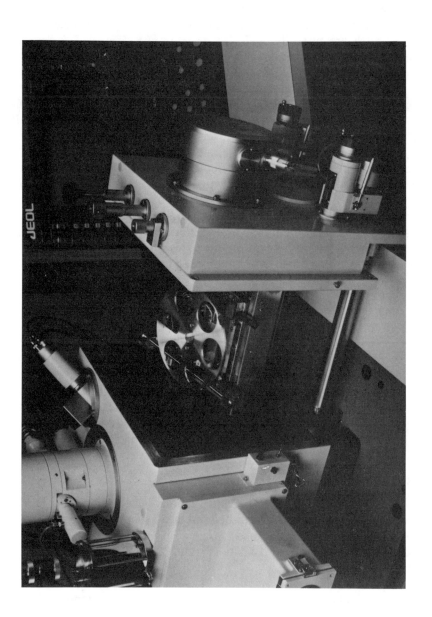

Figure 20.2. General view of a large-size SEM specimen holder (JEOL document).

4. Various *filtering* modes.
5. *Image analysis*: measurement of distances, areas, phase fractions, by means of different grey levels.

20.3 EMISSION AND REFLECTION IMAGES

20.3.1 IMAGING BY SECONDARY ELECTRON EMISSION AND BACKSCATTERED ELECTRONS

Imaging through emission of secondary electrons is the basic mode of common SEMs. Imaging by backscattered electrons is generally combined with the secondary electron mode.

20.3.1.1 Imaging Radiations

Secondary Electrons

The primary electrons undergo aleatory energy losses before exciting specimen atoms. The resulting secondary electrons in turn undergo aleatory energy losses along their paths to the specimen surface and to the detector. Secondary excitation leads to additional secondary electrons with various energies. The overall result is a wide spectrum of secondary electrons with a maximum at low energies around 50 eV, which cannot directly be used for characterising specimen elements. On the other hand, their emission cross-section is high, thus resulting in a high signal-to-noise ratio which is favourable for imaging.

Backscattered Electrons

Elastic and quasi-elastic scattering at an angle $2\theta > 90°$ results in so-called backscattered electrons with high energies, close to the primary energy. The backscattering cross-section is far smaller than the secondary emission cross-section.

20.3.1.2 Detection

In current SEMs, electron detection is mostly carried out by means of a scintillation counter combined with a PM (Figure 10.9b). According to the respective voltages V_c at the collector grid and V_d at the scintillator, two operating modes may be implemented.

Secondary Electron Mode

$$V_c \cong +200 \text{ V}; \qquad V_d \cong +10 \text{ kV}$$

1. *Secondary electrons* have low energies ($E_s < 200$ eV). They are therefore easily deviated and attracted by the collector field. Most of them pass through the grid and are then subjected to the high scintillator field. They are accelerated to an energy eV_d which is sufficient to generate an efficient scintillation pulse. Secondary electrons

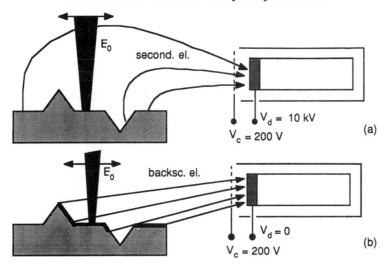

Figure 20.3. Ray path diagram of electrons reaching the detector. (a) Secondary electrons ($E_s \ll E_0$); (b) backscattered electrons ($E_r \cong E_0$).

leaving the specimen surface in any direction are thus collected, resulting in a high signal-to-noise ratio (Figure 20.3a).

2. *Backscattered electrons* have high energies ($E_r \gg eV_c$). They are therefore hardly affected by the collector field. Only backscattered electrons with trajectories passing straight through the detector aperture reach the scintillator and generate a signal. Due to the small detection solid angle, their signal-to-noise ratio is low.

In this mode the sum (secondary electrons + backscattered electrons) is in fact detected. Due to the large predominance of secondary electrons, it is commonly called the **secondary electron mode**.

Backscattered Electron Mode

Scintillation Detector

$$V_c \cong +200 \text{ V}; \qquad V_d = 0$$

In this operating mode the secondary electrons are captured by the collector grid. Only the backscattered electrons reach the scintillator thanks to their high energy $E_r \cong E_0$. As in the secondary electron mode, their signal-to-noise ratio is low (Figure 20.3b).

Semiconductor detector. A far higher signal-to-noise ratio is achieved by a disc-shaped semiconductor device with surface junction, positioned above the specimen for a high collection angle (Figures 10.9c and 20.1). The primary beam passes through a hole. The detector is usually divided into two or four sectors, leading to various contrast modes by combining the respective signals.

Scanning Electron Microscopy 465

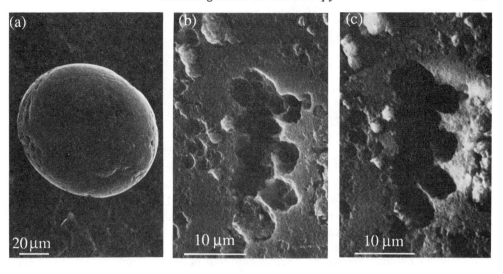

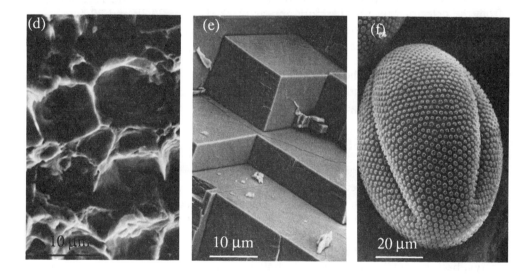

Figure 20.4. Examples of topographical contrast. The scintillation counter is located at the right-hand side with respect to the images. $V_0 = 30$ kV. (a) Inclination contrast; metallic spherule; $G_{dir} = 400$; (b) shadowing contrast, secondary electron image; pores on the surface of a retinite (vitreous volcanic rock); $G_{dir} = 2200$; (c) shadowing contrast, like (b), but backscattered electron image from the scintillation counter; (d) edge contrast; bronze fracture; $G_{dir} = 2500$; (e) inclination and shadowing contrast; {100} cleavages of galena PbS; $G_{dir} = 1500$; (f) spike contrast; linen flower pollen; $G_{dir} = 600$ (Dept Electron Microscopy, Lab. Crystallography, Louis Pasteur University, Strasbourg).

20.3.1.3 Image Contrast

Electron-induced images display two types of contrast: topographic contrast and composition contrast.

Topographic Contrast

Several effects add up to generate a contrast related to the surface topography of the specimen: inclination, shadowing and edge effects.

Inclination Effect. The yield of secondary electrons and backscattered electrons increases with increasing inclination of the primary beam to the specimen surface. Effectively, at grazing incidence, secondary electrons and backscattered electrons are generated close to the surface, within their escape depth. The edges of a cylindrical wire or a sphere thus appear brighter than their centre (Figure 20.4a).

Shadowing Contrast. Secondary electrons emitted in all directions are collected even if generated at specimen points 'unseen' by the detector (Figure 20.3a); however in this case, the signal is lower, resulting in 'unseen' parts of the specimen appearing darker than parts which are directly 'seen' from the detector aperture (Figure 20.4b). This shadowing contrast is much sharper for backscattered electrons which travel in straight lines. In the backscattered electron mode the parts of the specimen which are not 'seen' from the detector aperture therefore appear black (Figures 20.3b and 20.4c).

When imaging backscattering through a semiconductor, a shadowing contrast is achieved by subtracting the signals originating from opposite sectors of the detector.

Edge or Spike Contrast. The output of secondary electrons and backscattered electrons increases at sharp edges and at spikes of the specimen surface. Those features appear as bright parts of the image (Figure 20.4d).

Composition Contrast

The secondary electron emission yield I_s/I_0 varies as a function of specimen composition. Due to their low energy, detected secondary electrons are emitted from a thin surface layer of a few ångströms thickness. This layer is usually not related to the bulk composition (e.g. metal coating, adsorption, contamination). As a result the contrast displayed in the secondary electron mode is not related to the specimen composition.

The backscattering yield I_r/I_0 increases with the atomic number. Due to their high energy, backscattered electrons from the bulk specimen can easily pass though a possible foreign surface layer. The backscattered electron image therefore displays a composition contrast. To observe a purely composition contrast, unaltered by topographical effects, the specimen surface has to be perfectly plane (Figure 20.5a). Making use of this composition effect for analysis would theoretically be possible. Quantitative assessment of the grey levels of the backscattered electron image may provide a rough value of the average elemental composition of a polished specimen, after adequate calibration. It may be noted that additional secondary electron emission is induced by backscattered electrons; resulting in that secondary electrons themselves convey compositional data (Figure 20.5b).

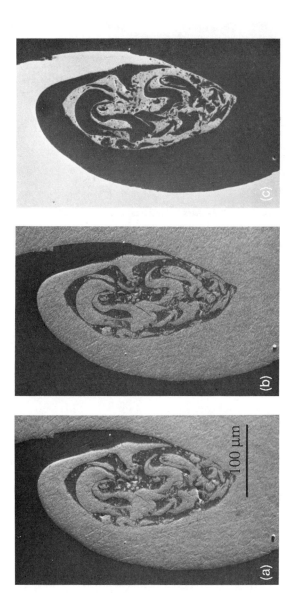

Figure 20.5. Scanning micrographs outlining the composition contrast. Propagation wave at the interface of two explosion-welded plates, one of copper (right-hand side), the other of titanium (left-hand side). Polished section normal to the interface. $G_{dir} = 160$. (a) Backscattered electron image; copper ($Z = 29$) appears brighter than titanium ($Z = 22$); (b) secondary electron image, similar to the backscattered electron image in the absence of a significant topographical contrast; (c) absorbed electron image; the contrast is inverted; any topographical contrast (polishing scratches) is absent (Dept Electron Microscopy, Lab. Crystallography, Louis Pasteur University, Strasbourg).

468 *Structural and Chemical Analysis of Materials*

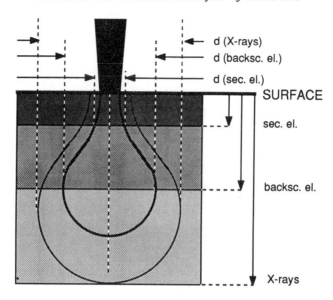

Figure 20.6. Diagram illustrating penetration depths and spatial resolution for various imaging modes by means of a bulk specimen with rather light elements.

A disc-shaped semiconductor detector with two sectors instead of the scintillation detector leads to a higher signal output and permits a switch from a topographical contrast (signal subtraction) to a composition contrast (signal addition).

The global contrast of the electron image may be altered by parasitic secondary electrons and backscattered electrons emitted from the impact of primary electrons on components of the specimen chamber.

20.3.1.4 Resolution

In the electron emission and backscattering modes, resolution depends on the specimen volume from which the detected electrons are emitted. For a given probe size, it varies with the specimen composition, the primary electron energy and the detected electron energy, as outlined in Figure 20.6. The best secondary electron image resolution is achieved with heavy elements.

Secondary Electrons

Due to their low energy, secondary electrons taking part in imaging come from a thin surface layer, of some 10 Å thickness. As shown in Figure 20.6, the emission diameter determining resolution is only slightly larger than the probe diameter. With a conventional thermionic electron source providing a 30 Å probe diameter, resolution is of the order of 40 Å with a heavy element specimen, but reaching 100 Å and more with light elements. In the latter case, metal coating can play an important role in optimizing resolution (see section 20.6.2). In optimum conditions, a field emission source permits a resolution of 10–20 Å.

Scanning Electron Microscopy

Efficient Magnification. When generating a secondary electron scan image at a magnification G, the diameter d_s' of the probe, referred to the image plane, acts as a confusion disc. In optimum conditions, i.e. without scattering-induced probe widening, the image will display a blurring of $d_s' = Gd_s$. In order to extract all available data from the image, the diameter of the image confusion disc should equal the eye separation distance which is about $\epsilon \cong 0.2$ mm. It corresponds to the efficient magnification G_e:

$$G_e = \frac{\epsilon}{d_s} = \frac{2 \times 10^{-2}}{d_s} \quad (\epsilon \text{ and } d_s \text{ in cm}) \tag{20.1}$$

There is nothing to gain but image blurring in increasing magnification further.

Minimum Magnification. Image resolution is limited by the scan interline distance of the photographic CRT. To achieve the limit resolution equal to the probe diameter, the minimum magnification G_m should be such that the image-referred probe diameter Gd_s is larger than the interline:

$$G_m = \frac{L'}{N_b d_s} \tag{20.2}$$

where L' is the image height and N_b the scan line number.

Photographic enlargement then leads to the expected resolution without displaying the scan lines.

Example. Probe diameter 30 Å; $L' = 10$ cm; $N_b = 2500$ ⇒ $G_e = 70\,000$; $G_m = 1300$.

Backscattered Electrons

Due to a greater penetration depth (order of 100–1000 Å), the backscattered electrons taking part in imaging come from a larger specimen volume than secondary electrons (Figure 20.6). The expected resolution is usually of the order of 100–500 Å, depending on the specimen. Backscattered electron imaging is especially useful for providing composition images of polished specimens.

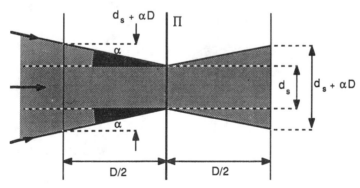

Figure 20.7. Geometrical interpretation of the depth of field of the secondary electron image provided by an SEM.

20.3.1.5 Depth of Field

The low angular beam aperture α results in a high depth of field. As shown in Figure 20.7, a shift $D/2$ of the object plane along the incident direction, on both sides of the focusing plane Π, adds a blurring αD to the probe diameter which thus becomes $(d_s + \alpha D)$. The total blurring in the image at magnification G must be smaller than the eye separation distance ϵ, i.e. the object blurring must be limited to ϵ/G, hence the relation

$$d_s + \alpha D \leqslant \frac{\epsilon}{G} \Rightarrow D \leqslant \frac{\epsilon/G - d_s}{\alpha} \tag{20.3}$$

Example. Depth of field as compared to the object scan length: Secondary electron image; $\epsilon = 0.2$ mm; $d_s = 100$ Å; $\alpha = 2 \times 10^{-3}$ rad; $L' = 10$ cm (square image). Object scan length: $L = L'/G$.

$$G = 100 \Rightarrow L = 1000 \, \mu m \quad D \cong 1000 \, \mu m$$
$$G = 1000 \Rightarrow L = 100 \, \mu m \quad D \cong 100 \, \mu m$$
$$G = 10\,000 \Rightarrow L = 10 \, \mu m \quad D \cong 8 \, \mu m$$

As a result in the above example, the depth of field is of the same order of magnitude as the scan length, i.e. a sharp image would be achieved from any points of a cube with its faces equal to the scan area.

Figure 20.8. Stereoscopic pair showing the dendritic growth of rutile TiO_2 in a vitreous matrix. $V_0 = 27$ kV; $G_{dir} = 3000$ (Dept Electron Microscopy, Lab. Crystallography, Louis Pasteur University, Strasbourg).

20.3.1.6 Basic Operating Mode

The secondary electron mode provides topographical scan images with the highest possible resolution and depth of field. It is the most commonly used imaging mode, even the only mode on simplified SEMs. Due to the depth of field, combined with the shadowing effect, a SEM provides a view of bulk specimens with a *three-dimensional effect* which may be quite impressive, notably with biological specimens (Figure 20.4f). For this reason, one of the first available SEMs was called *Stereoscan*. The space effect may easily be enhanced by recording stereoscopic pairs, carried out by tilting the specimen through an angle of about $\pm 5°$ around an axis normal to the incident beam (Figure 20.8).

For data interpretation, however, the fact that a fraction of backscattered electrons take part in imaging should be taken into account. Imaging with only backscattered electrons can be useful for displaying compositional variations and for detecting low relief effects thanks to the strong shadowing effect.

20.3.2 IMAGING BY X-RAY EMISSION

The X-ray spectrometer attachment of a SEM is usually of the energy-dispersive type (EDS).

The X-ray emission image is performed by means of the characteristic $K\alpha$ or $L\alpha$ emission of a selected element, resulting in a *distribution map* of that element in a surface layer of about 1 μm thickness (for more details, see section 15.2.3 and Figure 15.3). For a distribution map to be significant the specimen face has to be plane. Any topographical effect alters the outcome.

A SEM may be operated with stationary probe, leading to the possibility of quantitative analysis. In fact a SEM facility is not as well designed for accurate analysis as a dedicated microprobe. It is usually fitted with a single spectrometer of the energy-dispersive type. The probe current is lower and less stabilised (e.g. order of 5 nA against 1 μA in a microprobe). A plane specimen face, polished to less than 1 μm, is required. Simplified matrix correction programs (see section 15.3.2) have been developed for X-ray microanalysis with an EDS-equipped SEM. A standardless iteration method provides a rapid semi-quantitative analysis. Another method is based on a catalogue of computer-stored element standards.

20.3.3 IMAGING BY AUGER ELECTRON EMISSION

Electron-beam induced Auger electrons are emitted simultaneously with characteristic X-rays. Their yield is notably high for light elements, where X-ray analysis becomes inefficient. Due to their low energy, they provide analytical data of the order of an ångstrom thick surface layer. As a result, their use in an SEM for displaying a surface distribution map of an element requires a clean high vacuum of some 10^{-10} Torr. This requirement has led to the development of specific Auger scanning microscopes (SAM, section 16.2).

20.3.4 IMAGING BY LIGHT EMISSION. CATHODOLUMINESCENCE

Electron irradiation of a material results in light photon emission, as well as in X-ray photon emission. The emission ranges from ultraviolet to infrared. It is notably related to defect structures in minerals, oxides and semiconductors.

Semiconductors and non-conducting solids have a forbidden energy band between the valence band and the conduction band. The gap is of the order of 1 eV for the first and of several electron volts for the second. An incident radiation of adequate energy causes electrons to rise from the valence band to the conduction band, thus generating electron–hole pairs. The electron return to the fundamental state may occur through non-radiative transitions with heat release or through radiative transitions with light photon emission.

Photon-induced light emission is called *photoluminescence*. It has wide applications, e.g. ultraviolet fluorescence, phosphors for X-ray fluorescent screens and X-ray scintillation counters.

Electron-induced luminescence is called *cathodoluminescence*. It has similarly practical applications, e.g. phosphor screens for CRTs, and for electron scintillation counters. For an energy gap ΔW, the required electron excitation energy is about $3\Delta W$.

In a SEM specimen, each impacting electron may generate thousands of electron–hole pairs. Radiative excitation–relaxation processes are helped by the occurrence of point defects and impurity atoms which result in local lowering of the energy barrier. This effect justifies the addition of dopants in materials used as phosphors, to enhance their luminescence.

The emitted photon intensity is generally rather low. In order to achieve a sufficient signal-to-noise ratio, the incident electron intensity has to be of the same order as for significant X-ray microanalysis in a microprobe, i.e. some 0.1–$1\,\mu$A into a probe of diameter $1\,\mu$m. The scan speed is low, for a better noise integration. Possible specimen irradiation effects must therefore be taken into account.

The light photon signal may be globally detected by means of a photomultiplier or analysed by means of a light spectrometer. In the latter case, defects may be characterised by their spectrum.

The space resolution achieved is of the order of 1–$10\,\mu$m.

As with X-ray analysis, cathodoluminescence may be carried out in the stationary probe mode or in the scanning mode. Scanning leads to a distribution map of luminescent centres which displays the distribution of point defects, dislocations, impurities, grain boundaries, etc. Interpretation is facilitated by combining with X-ray microanalysis.

The technique of cathodoluminescence is widely applied for investigating various defects in materials (Goni and Remond, 1969); it is of particular interest for studying semiconductors (Dussac, 1985, in [5]).

20.4 SCANNING TRANSMISSION IMAGING

When using a thin specimen, scanning images may be provided by detecting the transmitted electrons, resulting in a contrast similar to that produced by a conventional TEM.

The transmission mode is usually available as an attachment to common SEMs. The normal specimen holder is then replaced by a device fitted for the use of standard TEM grids and which enables the transmitted beam to pass through it in order to reach the detector placed under the stage. Thanks to a limited scattering by thin specimens (see Figure 20.6), resolution is of the order of the probe diameter, i.e. 30–40 Å with a conventional thermionic electron source. This resolution is far from high resolution as

20.5 MISCELLANEOUS IMAGING MODES

20.5.1 IMAGING BY ABSORBED CURRENT

The intensity of the current absorbed by a bulk specimen is measured by means of a pico-amperemeter inserted between the insulated specimen holder and the ground. The absorbed current intensity I_a is given by

$$I_a = I_0 - (I_s + I_r + I_t) \qquad (20.4)$$

where I_s, I_r, I_t respectively are the secondary emission, backscattering and transmitted intensities.

In the instance of a bulk specimen, $I_t = 0$, leading to

$$I_a = I_0 - (I_s + I_r) \qquad (20.5)$$

As a result, the scan image displayed by means of the absorbed current signal is complementary to the (secondary electrons + backscattered electrons) image, but without topographical contrast (Figure 20.5c). Resolution is similar to that of backscattered electron images. An absorption signal complementary to the only backscattering signal may be achieved by preventing secondary electrons from leaving the specimen and avoiding parasitic scattering. The backscattering factor may thus be assessed. Applying this method to elemental analysis has been proposed by Philibert and Weinryb (1962) and Heinrich (1965).

20.5.2 IMAGING BY POTENTIAL CONTRAST

A potential variation at the specimen surface induces a variation of the secondary electron emission. When operating with a semiconductor, applying a voltage difference by means of electrodes results in a display of a potential distribution contrast through the secondary electron image. This effect is largely applied in the field of electronics for investigating and monitoring semiconductor devices.

20.5.3 IMAGING BY MAGNETIC FIELD CONTRAST

Due to low energies, the trajectories of outgoing secondary electrons are very sensitive to magnetic fields on the specimen surface, however weak they are. The secondary electron image therefore provides data on the distribution of such fields, e.g. of magnetic domains in ferromagnetic crystals. The magnetic fields generated by currents in an electronic component are similarly displayed.

20.5.4 ANGULAR SCANNING ON A CRYSTAL

The backscattered intensity from a crystal is determined by diffraction on its lattice planes. As a result, it is a function of the crystal orientation with respect to the incident beam. Consider a stationary crystal and an electron beam which is made to tilt about its impact point by means of an angular scan synchronised with the linear scan of the CRT. The result is a line pattern similar to the Kikuchi pattern (see 12.1.3). Its applications are the same, i.e. investigations on crystal orientation, lattice deformations (Vicario and Pitaval, 1972).

20.5.5 ELECTRON-BEAM INDUCED CURRENT

This mode, commonly abbreviated to EBIC, applies only to semiconductors. In such a material, an incident electron may generate thousands of electron–hole pairs (see cathodoluminescence, section 20.3.4). During its scan path on a specimen subjected to a voltage difference through an external device, the incident beam thus induces a current. The corresponding amplified signal results in a scan image. The image contrast is related to any factors which influence generation, diffusion and recombination of the charge carriers; it provides valuable data on the characteristics of a semiconductor, such as junctions, diffusion length, charge carrier life time and defects. Together with the potential contrast mode, EBIC has become of crucial importance for studying microelectronic components (Bresse, 1981).

20.5.6 ELECTROACOUSTIC IMAGING

For electroacoustic imaging, the incident beam is pulsed at a megahertz frequency. The resulting periodic local heating of the specimen induces dilatation which is detected by means of a piezoelectric transducer, tuned on the pulse frequency, fixed under the specimen. The transducer signal modulates the CRT beam.

The electroacoustic image contrast is induced by any local variations of the factors which determine the diffusion of heat in the specimen. These variations result from crystal defects, composition deviations, grain boundaries, cracks, etc. Determining the phase shift of the electroacoustic waves as a function of depth for a given pulse frequency leads to depth localising of defects (Balk, 1988; Bresse in [5]). It may be noticed that similar results are achieved by means of a primary photon beam (e.g. laser beam), leading to the technique of *photoacoustic microscopy.*

20.6 SPECIMEN

This part is only concerned with bulk specimens; thin specimens for the transmission mode are similar to those commonly used for TEM facilities.

Specimen size is currently a few centimetres; some SEMs may accommodate objects up to some 10 cm or more.

20.6.1 ELECTRON AND VACUUM RESISTANCE

The SEM specimen has to bear a strong electron irradiation in vacuum. The prominent irradiation effect (see section 19.5.2) is the thermal effect. It poses a problem notably for biological and organic samples which have a high water content, as well as for hydrated mineral specimens such as zeolites and clays.

Specimens with high water content must firstly be subjected to fixation procedures intended to eliminate water without destroying their structure. Commonly used techniques are *lyophilisation* (Figure 20.4f) and *critical point sublimation*. These techniques are detailed in specialised treatises (e.g. in [8], Reimer and Pfefferkorn).

Lyophilisation. The specimen is rapidly cooled to a low temperature of about $-150\,°C$. The frozen water is then vacuum sublimated.

Critical Point Sublimation. Specimen water is first progressively replaced by other liquids with a more accessible critical point. Commonly used fluids are: a succession of alcohol-refrigerating fluids like chlorofluorocarbons (Frigen 13, $CClF_3$: $P_c = 40$ bars; $T_c = 29\,°C$); a succession of acetone–carbon dioxide (CO_2: $P_c = 73$ bars; $T_c = 31\,°C$). The water-replacing fluid is then evaporated above the critical point in a small autoclave.

An example of the application of this technique to materials is the study of the microscopic texture in clay materials with a high water content, e.g. in soils (Tessier, 1984).

20.6.2 SURFACE CONDUCTIVITY

The specimen must have a sufficient surface conductivity in order to avoid the build-up of local electrical charges which would alter electron emission, i.e. the image (Figure 20.9a).

Conducting materials do not need any preparation. The specimen is simply fixed on to the grounded holder in such a way as to ensure a good electrical contact; this is commonly done by pasting the specimen by means of a special conducting lacquer (silver lacquer).

In the general case of non-conducting materials, the surface is coated with a conductor by means of vacuum evaporation or cathode sputtering.

20.6.2.1 Vacuum Evaporation

For mere topographic imaging, the specimen is mostly gold coated. For additional X-ray microanalysis, gold is replaced by less absorbing, i.e. light elements, usually carbon.

Topographical shadowing effects may lead to local absences of coating, i.e. to local charges. For a homogeneous coating of a specimen with a pronounced surface relief, vacuum evaporation should therefore be carried out at varying incidence, by means of a tilt–rotation motion of the specimen.

20.6.2.2 Cathode Sputtering

A large angular dispersion of incident-coating atoms is provided in a much simpler way by cathode sputtering. The coating device consists of an enclosure at a primary vacuum

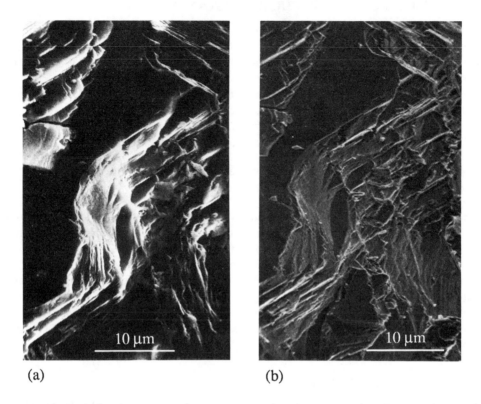

Figure 20.9. Effect of the acceleration voltage on electrical charge build-up at the surface of an insulating material without metal coating (cleavage $KAlSi_3O_3$). $G_{dir} = 2000$. (a) $V_0 = 20$ kV; charges appear as abnormally bright zones; (b) $V_0 = 5$ kV, no charging. (Louis Pasteur University, Strasbourg.)

pressure (0.05–0.2 Torr) of argon (or possibly air), generated by a rotary pump. An electrical discharge is induced between the coating material which acts as the cathode and an anode supporting the specimen. Positive ions are accelerated towards the cathode where they induce sputtering of coating atoms, in a similar way as in an ion-milling device or in SIMS. The sputtered atoms undergo multiple scattering effects with gas atoms, resulting in their omnidirectional incidence on the specimen, thus avoiding any shadowing effects.

In the instance of a 2.5 kV voltage and a 50 mA intensity, a 350 Å gold layer is deposited in about 1 min. Various metals (e.g. Au, Pt, AuPd, Ag, Ni), as well as carbon, may thus be sputtered, by simply changing the cathode plate.

In any case, the coating layer should not alter the resolution by blurring fine topographical details. The optimum gold coating thickness is of the order of 100 Å, corresponding to a bluish tint on a test glass slide.

In view of quantitative X-ray analysis, the specimen must be polished on one face as for operating with an electron microprobe.

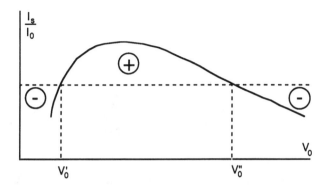

Figure 20.10. Variation of the secondary electron emission yield as a function of the accelerating voltage.

20.6.2.3 Low-voltage Operating

Imaging of insulating materials without any coating may be carried out in given operating conditions. In order to highlight the effect of significant parameters, consider the graph outlining the variation of the secondary emission yield I_s/I_0 as a function of the accelerating voltage V_0. It may be seen from Figure 20.10 that $I_s/I_0 = 1$ for two different values of V_0. The yield is positive between those values, the electron deficit results in a positive specimen charge. The yield is negative outside those values, the electron excess results in a negative specimen charge.

Avoiding any specimen charging amounts to operating at one of the two voltages leading to a secondary electron emission yield equal to 1. These voltages depend on the specimen composition. The lower voltage V_0' is not in fact practicable. The higher voltage V_0'', depending on the specimen, is commonly of the order of 1 kV and may be practised with certain SEMs. In effect, recent facilities can be operated at voltages lower than 1 kV. Due to the electron gun operating in far from optimum conditions, the resulting resolution is usually lower. However, this drawback is balanced by the possibility of imaging non-conducting materials without any surface coating (Figure 20.9b).

20.7 APPLICATIONS

Thanks to its capability to provide images of bulk specimens, the SEM has a very large field of application.

Its main practical advantages are: easy observation and straightforward interpretation of the images of virtually any kind of solid specimen; great variety of imaging modes; ease of pixel-by-pixel image processing; analytical possibilities. As a result, the SEM has developed as a basic instrument in laboratories dealing with material research and control.

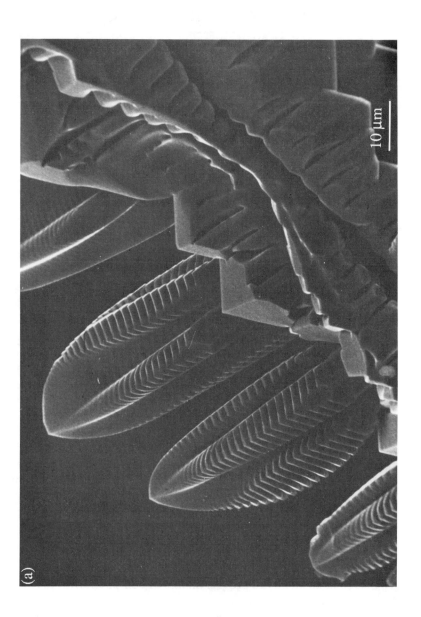

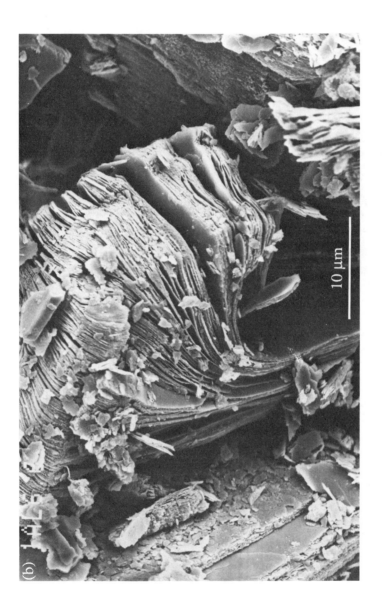

Figure 20.11. SEM images of materials in the secondary electron mode. (a) Dendritic growth of tungsten crystals. $V_0 = 20$ kV; $G_{dir} = 800$; (b) kaolinite lamellas, alteration products formed in a rock cavity. $V_0 = 10$ kV; $G_{dir} = 1500$ (Dept Electron Microscopy, Lab. Crystallography, Louis Pasteur University, Strasbourg).

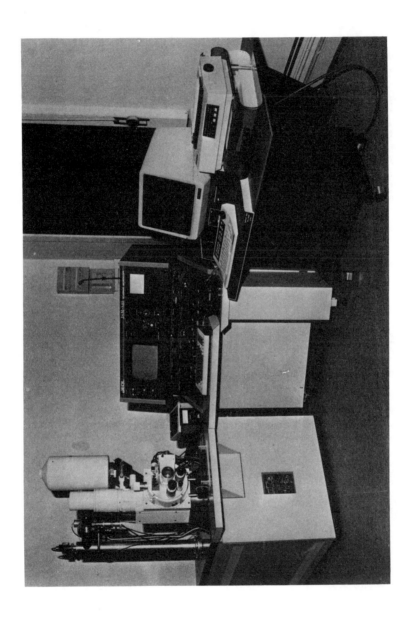

Figure 20.12. General view of a SEM with an attachment for X-ray microanalysis (EDS). From left to right: microscope column with the liquid nitrogen cryostat of the Si(Li) diode; control and display unit of the microscope; data acquisition, displaying and processing system of the X-ray spectrometer (Dept Electron Microscopy, Lab. Crystallography, Louis Pasteur University, Strasbourg).

When limited to the imaging aspect, the following typical applications may be noted:

(a) microscopic texture of materials (Figure 20.11a);
(b) fracture and wearing facies of materials (e.g. mechanical parts);
(c) surface topography of catalysers;
(d) surface corrosion and alteration of materials;
(e) microscopic facies of crystals and particles in rocks, in ceramics (Figure 20.11b);
(f) study and control of microelectronic components;
(g) identification of microfossils in sediments.

The possibility of utilising stereoscopic pairs, notably by using screen fractioning, is illustrated by Figure 20.8. In addition to their spectacular aspect, such pairs may be exploited for three-dimensional measurements.

An attachment of X-ray microanalysis (Figure 20.12) gives the SEM an additional interest, with analytical performance which is now close to that of a dedicated electron microprobe.

REFERENCE CHAPTERS

Chapters 2, 7, 9, 10, 15, 16, 19, 21.

BIBLIOGRAPHY

[1] Goldstein J. I., Newbury D. E., Echlin P., Joy D. C., Fiori C., Lifshin E. *Scanning electron microscopy and X-ray microanalysis*. Plenum Press, New York, 1981.
[2] Hearle J. W., Sparrow J. T., Cross P. M. *The use of scanning electron microscopy*. Pergamon Press, New York, 1972.
[3] Holt D. B., Muir M. D., Grant P. R., Boswarva I. M. (eds.) *Quantitative scanning electron microscopy*. Academic Press, London, 1974.
[4] Laves G. *Scanning electron microscopy and X-ray microanalysis*. John Wiley, New York, Chichester, 1987.
[5] Maurice F., Meny L., Tixier R. (eds.) *Microanalyse et microscopie électronique à balayage*. Editions de Physique, Orsay, 1985.
[6] Oatley C. W. *The scanning electron microscope*. Cambridge University Press, London, 1972.
[7] Reimer L. *Scanning electron microscopy. Physics of image formation and analysis*. Springer-Verlag, Berlin, 1985.
[8] Reimer L., Pfefferkorn G. *Raster Elektronen Mikroskopie*. Springer-Verlag, Berlin, 1973.
[9] Scanning electron microscopy. *Proc. Ann. Symp. I.I.T. Resolution. Inst., Chicago*. Annual since 1968.

21

Scanning Transmission Electron Microscopy. Analytical Electron Microscopy

The considerable development of the conventional transmission electron microscope (TEM), since its creation in the 1930s, has led to a continuous improvement of its resolution, resulting in the present high-resolution microscope with close to atomic resolution. However, in addition to structural information provided by images, the electron–matter interaction effects contain chemical data. These are conveyed in the form of characteristic emissions and in the form of characteristic electron energy losses. They are lost in a conventional TEM operated in the imaging and the diffraction mode only.

In the case of the **scanning electron microscope**, the development of the analytical aspect has been concomitant with the development of imaging, thanks to the relatively large interaction volume and to the design of the specimen chamber, well adapted for accommodating spectrometers. However, when working with thin specimens, a conventional SEM with a transmission attachment has a poor image resolution, of the order of 50 Å, determined by the probe size.

In the case of the conventional **transmission electron microscope**, the design is far less favourable to analysis, due to the small interaction volume of the thin specimen and to the lack of space for installing a spectrometer in an optimum position.

The way towards efficient **microanalysis in the transmission electron microscope** has been cleared thanks to the advent of specifically designed facilities, with a finely focused and high-intensity electron probe, with beam scan, with a specimen chamber accommodating spectrometers, with high-performance spectrometers, and with data acquisition and processing systems. The result has been a new generation of analytical TEMs combining high-resolution imaging, selective diffraction at a nanometre scale and analysis at a similar scale.

Microanalysis at a nanometre scale (sometimes called *nanoanalysis*) is carried out by two complementary techniques:

(a) *X-ray emission microanalysis* by means of *energy-dispersive spectrometry* (EDS);

(b) *microanalysis by characteristic electron energy loss spectrometry* (EELS).

Microscopes operating in high vacuum may in addition carry out *Auger spectrometry*.

At the present state of technology, the analytical electron microscope is the most powerful instrument available for exploring the crystallographical and chemical structure of solid matter at a close to atomic scale.

Common abbreviations.

Imaging mode: STEM (scanning transmission electron microscope). Normally reserved for the electron microscope specifically designed for scanning transmission imaging with the best possible resolution (i.e. dedicated STEM), it is, however, often applied similarly to the conventional TEM with scanning attachment.

Analysing mode:

1. EMMA (electron microscope microanalysis);
2. AEM (analytical electron microscope).

These two abbreviations are less common. The second is ambiguous, due to its use for the Auger electron microscope. However, the latter is now more frequently called SAM (scanning Auger microscope).

3. EDS or EDX (energy-dispersive X-ray spectrometry);
4. EELS (electron energy loss spectrometry);
5. EXELFS (extended electron loss fine structure).

21.1 SCANNING TRANSMISSION ELECTRON IMAGING

21.1.1 BASICS OF IMAGING

The general principle of imaging is the same as that of the conventional scanning electron microscope (see Figure 20.1). The transmitted fraction of the electron beam through the thin specimen is collected by an electron detector located under the specimen holder. The image is similarly generated by beam scanning and displayed on a CRT.

21.1.1.1 Reciprocity

The resulting image may be interpreted by analogy to that provided by a conventional TEM. This analogy is highlighted by the *reciprocity relation* (Cowley, 1969), as outlined by Figure 21.1. It makes contrast interpretation easier. It is valid in the limit of the weakly scattering object approximation (see section 19.4.1) and at medium resolution. It supposes the characteristics of the microscope constituents to be similar, e.g. objective lens transfer function, convergence angle on the object (TEM) and detector acceptance angle (STEM).

21.1.1.2 Resolution

Beam spreading through scattering in the thin specimen is negligible (see Figure 15.2), i.e. resolution is approximately determined by the probe diameter. To reach a resolution similar to that of a conventional high-resolution TEM, the probe diameter on the

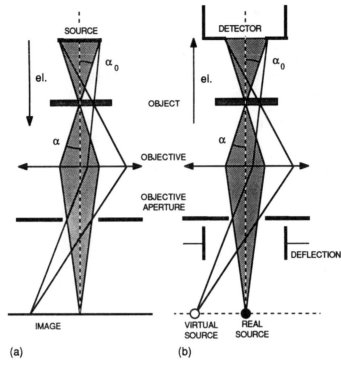

Figure 21.1. Outline of the principle of reciprocity in ray paths. (a) Conventional TEM; (b) STEM.

specimen must be as small as a few ångströms. This aim is reached in dedicated STEMs by means of a field emission source and a careful beam focusing. The technique is derived from that implemented by Crewe et al (1969, 1970) which led for the first time to scanning images with close to atomic resolution. The resolution of conventional TEMs operating in the scanning nanoprobe mode is usually lower.

21.1.2 INSTRUMENTAL LAYOUT AND CHARACTERISTICS

As far as the technical design and the characteristics are concerned, it is convenient to distinguish between the dedicated STEM and the TEM with scanning attachment.

21.1.2.1 Dedicated Scanning Transmission Electron Microscope

The actual STEM has a number of specific features designed to optimise its imaging performance.

Components

The main components are detailed below and illustrated in Figure 21.2.

Electron Source. A field emission gun is compulsory for generating an electron probe smaller than 10 Å. The high brightness of such a source (i.e. of the order of 10^9 A/cm^2 at 100 keV) is similarly important for selected area analysis.

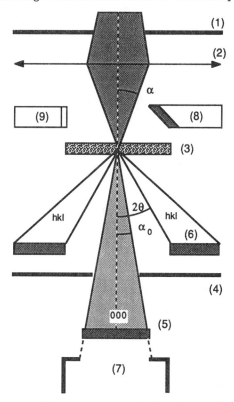

Figure 21.2. Basic outline of a dedicated STEM, limited to the parts located behind the objective lens aperture. (1) Objective lens aperture diaphragm determining the convergence angle α at the specimen; (2) objective lens; (3) specimen; (4) transmission detector aperture diaphragm determining the acceptance angle α_0; (5) axial electron detector for bright field imaging; (6) annular detector for dark field imaging; (7) EELS spectrometer; (8) EDS spectrometer; (9) detector for emitted and backscattered electrons. In the most common facilities, electron propagation takes place in a direction opposite to that of this figure, i.e. electrons travel upwards.

The beam intensity on the specimen is of the order of some 0.1–1 nA. for a probe diameter of 5 Å, it corresponds to an electron flux of the order of 10^5 A/cm^2, i.e. some 1000 times the commonly used electron flux in a microprobe or a conventional SEM in the analytical mode.

For good working conditions, a field emission electron gun requires a high vacuum of at least 10^{-10} Torr, in order to prevent the tip from being contaminated. As well as for electron spectrometers, the whole vacuum enclosure and pumping system is therefore designed for high-vacuum operation.

Scanning System. The beam deflection device is similar to that of a conventional SEM and permits linear scanning as well as angular scanning about a specimen point.

Objective Lens. A combined condenser–objective lens system (see Figure 9.9c), with high excitation current, leads to focusing the beam on the specimen with an intensity of typically 0.1–1 nA. Its final diameter depends on the spherical aberration of the objective lens. In a similar way as in the TEM, the optimum aperture is a compromise between diffraction aberration at the entry pupil and spherical aberration; it is of the same order, i.e. $\alpha \cong 7 \times 10^{-3}$ rad for a focal length of 3 mm. The minimum probe diameter on the specimen is then about 4–5 Å.

Detectors and Spectrometers. Electron detectors for transmission imaging are positioned behind the specimen (semiconductor, microchannel plate or channeltron, depending on the required acceptance angle).

An additional lateral electron detector in front of the specimen may be used for secondary electron and backscattered electron imaging.

Various spectrometers operate in the analytical mode.

Data Acquisition and Processing. The sequential imaging mode, pixel by pixel, makes it easy to digitalise and then to process the image.

Operating Modes

The most common STEM operating mode uses an axial and an annular electron detector (Figure 21.2), resulting in two contrast modes.

Bright Field Imaging. The bright field image is generated by means of the *axial detector* which only collects the electrons transmitted and scattered inside the detector acceptance angle α_0. The resulting signal corresponds to a large fraction of inelastically scattered electrons which are concentrated into a limited angular range (see section 2.3.2 and Table 17.1). When using a small aperture and a thin specimen, the resulting image contrast is similar to the bright field diffraction contrast as observed with a conventional TEM.

Dark Field Imaging. The dark field image is generated by means of an *annular detector* which collects electrons scattered and diffracted into an angle greater than the axial detector acceptance angle. The resulting signal corresponds predominantly to diffracted and to elastically scattered electrons, which are dispersed in a large angular range (see section 2.3.1 and Table 17.1).

Atomic Number Contrast. The elastic scattering cross-section varies according to Z^2/V_0^2, whereas the inelastic scattering cross-section varies according to Z/V_0^2. As a result, the ratio of respective intensities as measured at the annular detector and at the axial detector is approximately equal to the atomic number Z. Imaging with this intensity ratio (so-called *ratiocontrast* imaging) displays differences in atomic number. This mode has been used by Crewe et al (1970) for imaging heavy isolated atoms in an organic matrix.

Image Filtering. The use of an energy filter, e.g. consisting of a magnetic analyser, leads to images only generated by the electrons transmitted into an energy window ΔE. The same analyser may be used for EELS.

Convergent Beam Electron Diffraction. When operating in the common imaging mode with an electron probe focused on the specimen, the convergent beam diffraction conditions are met at any specimen point selected by the beam (compare Figures 12.21 and 21.2). The diffraction pattern can be visualised and recorded on a phosphor screen placed behind the specimen. An alternative way is to explore the pattern with the axial detector by means of an auxiliary scanning attachment or by angular scanning of the incident beam.

Interference Images. With sufficiently large beam convergence and detector acceptance angles, the detector may collect the transmitted beam 000 as well as a diffracted *hkl* beam. The result is an interference image (or *lattice image*) displaying a sinusoidal resolution of the corresponding lattice planes (*hkl*), similar to the lattice fringes provided by two-beam imaging in the conventional TEM (Figure 19.23).

Parallel Beam Electron Diffraction. Overfocusing the condenser lens (e.g. focusing on the objective lens aperture plane) results in illuminating the specimen with a virtually parallel electron beam, but with a low intensity. The corresponding diffraction pattern may then be recorded as mentioned above. Fresnel fringes similar to those observed with a TEM may also be displayed in this mode.

21.1.2.2 Transmission Electron Microscope with Scanning Attachment

Conventional TEMs of the present generation, with the possibility of the nanoprobe mode, are often fitted with a beam-scanning device. As a result they acquire some of the characteristics of a dedicated STEM, but with limitations:

1. The optical system is less favourable to an optimum fitting of detectors and spectrometers.
2. The electron source is usually of the thermionic type. Even with an LaB_6 cathode, its brightness is lower than that of the field emission gun of a dedicated STEM; the probe is generally larger than 1 nm and it has a higher energy dispersion. The common result is a resolution of a few nanometres in the scanning mode and a lower signal-to-noise ratio in the analytical applications. The increasing use of field emission sources even in conventional TEMs obviously results in higher performance.
3. The specimen chamber vacuum is not sufficiently high for carrying out Auger analysis.

21.1.3 COMPARING STEM AND CONVENTIONAL TEM

The STEM has certain advantages with respect to a conventional TEM, notably:

1. A simpler optical system.
2. The absence of chromatic aberration, permitting thicker specimens. At a far lower cost and many fewer installing difficulties, a STEM thus has some of the advantages of a high-voltage TEM.
3. The ease of energy filtering with a transmitted beam energy loss spectrometer.
4. The readiness of image processing thanks to the possibility of digital data acquisition and storage.
5. The higher sensitivity of performing selective dark fields by means of an annular detector with varying radius (see section 19.2.2).

The main drawback of a STEM is its resolution limited to the probe diameter. In this field, its performance is lower than that of a high-resolution TEM, notably with medium voltage, which now reaches a resolution of 1–2 Å.

21.2 ANALYTICAL ELECTRON MICROSCOPE

21.2.1 AVAILABLE SPECTROMETRIC METHODS

In the commonly used energy range, the electron–specimen interaction provides analytical data mainly by means of the following effects:

(a) *characteristic X-ray emission*;
(b) *light photon emission*;
(c) *characteristic electron emission (Auger electrons)*;
(d) *characteristic electron energy losses*.

The first three effects are commonly detected and measured by reflection, whereas the fourth is carried out by transmission.

The dominant characteristic of analysis with a STEM is the excessively small analysing volume, of the order of 1 nm^3. This requires optimisation of any analytical parameters in order to reach a sufficient signal-to-noise ratio, i.e. an acceptable detection limit. In emission spectrometry notably, any parasitic radiation induced by the primary beam on parts surrounding the specimen must be eliminated, e.g. by using beryllium specimen holders and by carbon coating of sensitive parts. Position and characteristics of detectors and of spectrometers must similarly be optimised. Furthermore, a retractable window is advisable for X-ray analysis by means of an Si(Li) diode, notably for detecting light elements.

When performing analysis by means of a conventional TEM fitted with a spectrometer, it is compulsory to operate with an LaB$_6$ cathode, or even better with a field emission source. In any case, the dedicated STEM facilities have higher analytical performance. In addition, their ultra-vacuum enables surface analysis to be carried out by means of Auger electrons.

As a result, the following previously detailed analytical techniques may actually be combined with imaging when using a STEM (the related reference sections are noted):

(a) elemental microanalysis (or nanoanalysis) by EDS (15.4);
(b) elemental microanalysis (or nanoanalysis) by EELS and structural analysis by EXELFS (17.2);
(c) cathodoluminescence (20.3.4);
(d) EBIC (20.5.5);
(e) microanalysis by Auger electrons (with a high-vacuum specimen chamber; 16.2).

Any analytical data may be displayed as a *spectrum* or as a *distribution map*.

21.2.2 COMPARING X-RAY EMISSION SPECTROMETRY AND ELECTRON ENERGY LOSS SPECTROMETRY

The basic analysing techniques of an analytical electron microscope are X-ray emission spectrometry (EDS) and electron energy loss spectrometry (EELS). It therefore seems interesting to compare the performances of both techniques, implemented with a STEM, as a function of the specimen to be analysed.

When omitting the matrix corrections, the measured respective intensities of a K-emission line of an element A (Eq. 15.20) and of the characteristic energy loss at the K-edge of the same element (Eq. 17.9) are expressed by the following relations:

$$I_A = I_0 \frac{N}{A} c_A \rho t (\sigma \omega p)_A \tau_A \qquad (21.1)$$

$$I_A = I_0 \frac{N}{A} c_A \rho t \sigma_A \tau_A \qquad (21.2)$$

where $(N/A) c_A \rho t$ is the number of atoms per surface unit of the irradiated volume, of thickness t, σ_A the excitation cross-section of the considered energy level of element A, ω the corresponding fluorescence yield, p the probability of the considered transition, and τ_A the instrumental transmission factor. It may be seen that both expressions are similar to the factor ωp.

The measured intensities depend on the respective values of the above factors in the respective techniques:

1. The fluorescence yield ω is low for light elements and tends to 1 for heavy elements (see Figure 7.4). The probability p of transition, corresponding to the line to be measured (e.g. KL2 or KL3 for Kα), is accordingly less than 1. The relaxation factor ωp is therefore unfavourable for X-ray analysis, notably for light elements.
2. The instrumental factor τ_A characterises the collection efficiency of the detector-spectrometer system. The acceptance angle of the electron spectrometer is smaller than that of the Si(Li) X-ray spectrometer which is positioned close to the specimen. However, the inelastic scattering angle of electrons, corresponding to core-level excitation, is itself very small (Table 17.2), whereas X-ray emission takes place in all directions. The result is an actual collection efficiency reaching 20-50% in EELS, against some 1% in EDS.
3. The signal-to-noise ratio does not appear in Eqs (21.1) and (21.2); it nevertheless is determinant for the detection limit. It is far more favourable for X-ray analysis than for EELS, notably for heavy elements.

One technical detail has to be mentioned: the EELS analyser is positioned under the specimen chamber and thus does not in the least disturb imaging. This is not necessarily the case for the X-ray spectrometer which may require a modification of objective lens pole pieces for maximum efficiency.

It is generally considered that both techniques are similar in their performance for $Z \cong 12$-20. EELS performs better for lighter elements, whereas X-ray microanalysis is better for heavier elements.

The complementarity of both techniques is illustrated by Figure 21.4 which represents the respective spectra simultaneously displayed from the same analysing area.

21.2.3 AUGER SPECTROMETRY AS PERFORMED WITH STEM

Ultra vacuum in an STEM permits Auger analysis. The main difference with respect to conventional Auger spectrometry is the high primary electron energy commonly used in an STEM. For analysing light elements, operating conditions are therefore far from optimum excitation (see section 7.4). This results in a low excitation cross-section. What is in fact determinant for the detection limit is the signal-to-noise ratio; this has been shown to increase between 30 and 100 keV. It similarly increases with the specimen surface inclination to the incident beam.

The main advantage of STEM Auger spectrometry is in fact the small probe size which allows a real surface microanalysis to be performed. With a 10 Å probe, the spatial Auger analysing resolution is estimated to reach about 80 Å, whereas the resolution of a conventional low-energy Auger scanning microscope is presently limited to 500–1000 Å (Cazaux, 1988).

However, it must be taken into account that the detection limit is directly related to the count rate, i.e. to the beam intensity (Eq. 13.9). Now the beam intensity of an STEM is typically only some 0.1–1 nA.

On the other hand, for an efficient probe diameter of 5 Å, the beam intensity leads to an electron flux of the order of 10^5 A/cm^2 (about 10^8 electrons/s per Å^2), i.e. some 1000 times the electron flux in a microprobe. This high value underlines the problem of irradiation damage in an STEM which is liable to alter rapidly the specimen in the nanoprobe mode.

21.2.4 ELEMENTAL DISTRIBUTION MAPS

For a facility based on imaging, the generation of images by means of characteristic signals from a selected element has a particular importance. The best sensitivity in element localising is presently performed by EELS. For the K absorption edge of a given element, imaging is carried out by means of the intensity integrated over a certain energy range over the edge (Figure 17.6), from which the intensity of a similar range of the background is subtracted pixel by pixel on either side of the edge. The operating conditions being optimised, clusters of some 10 atoms may be displayed in this way.

21.3 SPECIMEN PREPARATION AND APPLICATIONS

Specimen preparation methods are similar to those commonly used for conventional electron microscopy.

Thanks to the absence of chromatic aberration, a specimen of greater thickness may be used, similar to that common in medium-voltage or even high-voltage microscopy.

Due to the scanning mode, the irradiation time at a specimen point during observation is short, thus limiting the heating effects. High-vacuum operation further reduces contamination.

The field of applications is similar to that of the conventional TEM, with the following specifications:

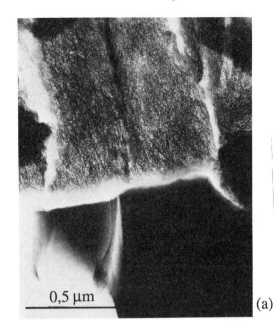

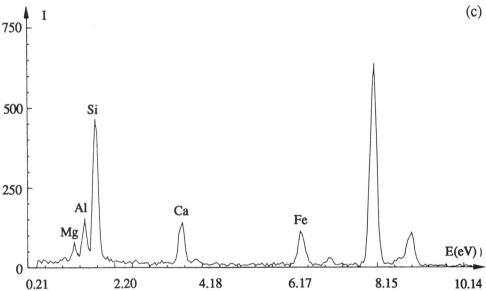

Figure 21.3. Example of a morphological and a chemical investigation carried out simultaneously by means of a TEM with scanning and EDS attachments. Section across the surface layer build up on sea-water altered glass of basaltic composition. Extraction replica, epoxy embedding and ultramicrotomic section normal to the surface. Specimen thickness about 600 Å; $V_0 = 100$ keV; diameter of analysing area 200 Å. (a) Electron micrography showing the non-altered glass (bottom), surmounted by the alteration layer consisting mainly of poorly crystallised phyllosilicates; (b) approximate elemental composition as determined by X-ray microanalysis; (c) corresponding X-ray emission spectrum. Na and Ti are at the detection limit; their peaks do not emerge from background noise (from Crovisier and Eberhart (1985), Ehret et al (1986). Lab. Crystallography, Louis Pasteur Univ., Strasbourg).

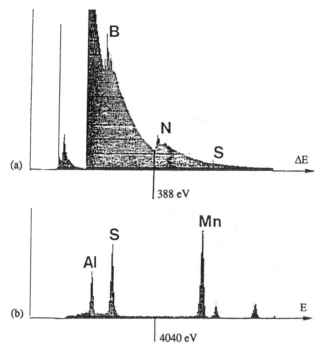

Figure 21.4. Comparing spectra simultaneously recorded by EELS (a) and EDS (b). Thin inclusion extracted by replica from aluminium; $V_0 = 300$ keV. It may be noted that EELS provides high intensities from light elements, while EDS is more suitable for heavier elements, from Al upwards (spectra from Satoh, Tokyo, Philips Industry document).

1. Possibility of operating with relatively thick specimens.
2. Easier observation of radiation-sensitive materials, e.g. organic specimens.
3. Imaging of clusters of a few heavy atoms in a matrix of light elements, thanks to the ratio detection mode (elastic scattering/inelastic scattering).
4. Simultaneous investigation of structure (by high-resolution imaging) and composition of thin solid layers (by EDS or EELS), at a nanometre scale (e.g. corrosion layers, Figure 21.3). Correlation between composition and defects at a close to atomic scale (Baronnet and Onrubia, 1988).
5. Study of poorly crystallised materials by means of annular dark field imaging.

REFERENCE CHAPTERS

Chapters 2, 7, 8, 9, 10, 13, 15, 17, 19, 20.

BIBLIOGRAPHY

[1] Colliex C. (ed.) *Symposium STEM*. Strasbourg 1985. *J. Microsc. Spectrosc. Electron.*, **10**, 1985.
[2] Hren J. J., Goldstein J. I., Joy D. C. (eds) *Introduction to analytical electron microscopy*. Plenum Press, London, 1979.
[3] Jouffrey B., Bourret A., Colliex C. (eds) *Microscopie électronique en science des matériaux*. Editions du CNRS, Paris, 1983.

22

Scanning Tunnelling Microscopy

The most recently developed technique of electron microscopy is based on the tunnelling effect, well known since the origin of quantum mechanics. This effect is similarly applied in the field emission sources as well as in field electron emission microscopy and in field ion emission microscopy developed by E. Muller in the 1950s. The tunnelling electron microscope is only a few years old (Binnig et al, 1982a,b, 1983; Binnig and Rohrer, 1983, 1987) and won its inventors the Nobel Prize for Physics in 1986. It may be noted that they shared the prize with E. Ruska, inventor of the conventional electron microscope some 50 years earlier.

In contrast to conventional electron microscopes, tunnelling techniques do not need external electrons. They operate just with the electrons emitted from the specimen itself.

Common abbreviations:

1. FEEM (field electron emission microscopy);
2. FIM (field ion microscopy);
3. STM (scanning tunnelling microscopy);

Subsidiary techniques:

1. STIPE (scanning tunnelling inverse photoemission);
2. STOM (scanning tunnelling optical microscopy);
3. AFM (atomic force microscopy).

22.1 PRINCIPLE OF THE TUNNEL EFFECT

22.1.1 WORK FUNCTION

According to classical solid physics, electrons are confined inside a solid by a *potential barrier* Φ at their surface. This barrier is due to the fact that any electron leaving the solid would generate at its surface a residual charge $+e$ resulting in a drawback force. When referring to the Fermi level, the work $e\Phi$ required to bring an electron from the solid to infinity is called the *work function*. The drawback force decreases rapidly with increasing distance from the surface. The whole work is virtually completed at a distance

Table 22.1. Approximate work functions of some metals and of LaB$_6$. The values depend on the crystalline state and, for single crystals, on the Miller indices of the surface planes (e.g. for tungsten it amounts to 4.39 eV for {111} planes, 4.52 eV for {100} planes and 4.69 eV for {112} planes)

Material	Work function (eV)
K	1.6
Cr	4.6
Fe	4.5
Ni	4.6
Mo	4.3
Cs	1.8
Ba	2.0
Pt	5.3
Ta	4.2
W	4.5
LaB$_6$	2.6

of a few atomic radii, i.e. a few ångströms, thus justifying the classical representation of a potential barrier of height Φ. In the instance of metals, the typical work function is of the order of a few electronvolts (Table 22.1).

For an electron to be able to leave the surface, it has to gather a minimum energy $e\Phi$. This energy may be provided by excitation through an incident radiation (e.g. photoelectron, Eq. 16.1; secondary electron; Auger electron, Eq. 16.5).

22.1.2 FIELD EFFECT

The potential barrier is decreased by $\Delta\Phi$ as the result of an external electric potential which tends to extract electrons from the surface. In addition its thickness at the Fermi level is reduced (Figure 22.1). This is the so-called Schottky effect. At high temperature, a moderate electric field is sufficient to provide electrons with the required kinetic energy to pass over the potential barrier in great numbers, resulting in thermionic emission (Richardson–Dushman equation, Eq. 9.1).

22.1.3 TUNNEL EFFECT

As shown by experiments, a certain number of electrons can leave a solid with energies largely lower than the extraction energy corresponding to the work function $e\Phi$. It appears as if electrons were crossing the barrier through a tunnel, hence the name *tunnel effect*. The crossing probability increases with increasing external extraction field.

Tunnelling is well interpreted by quantum mechanics. The electron wave function in a conducting solid, the solution of *Schrödinger's* equation, does not abruptly fall to zero at the surface as predicted by the classical theory. As a result, in the absence of any electric field, the plasma of conduction electrons overflows the surface with a density decreasing rapidly with distance. The probability of the presence of an electron outside the surface decreases with distance according to an exponential law, falling virtually to zero at a few atomic distances. At a distance z in vacuum, it is expressed by the following relation, the solution of *Schrödinger's* equation for an electron of energy smaller than eΦ:

$$\Psi = A \exp\left[-2\pi z \frac{(2m_0 e\Phi)^{1/2}}{h}\right] = A \exp\left[-\frac{2\pi z}{\lambda}\right] \quad (22.1)$$

where A is a constant and λ the wavelength associated with an electron of energy $e\Phi$ (Eq. 1.5).

The width z_t of the potential barrier at the Fermi level is commonly called the *tunnelling distance*. As outlined by Figure 22.1, for an initial barrier height Φ_0 and an external field E, the tunnelling distance is approximated by

$$z_t \cong \frac{\Phi_0}{E} \quad (22.2)$$

Tunnelling intensity is proportional to the squared wave function. As shown by Eq. (22.1), it becomes significant for $z_t \cong \lambda$, i.e. for high field values, as illustrated by the example given below, calculated from Eq. (22.1).

Example

$$e\Phi = 4\,\text{eV} \quad \lambda = 12.26/2 \cong 6\,\text{Å} \quad z_t \cong 6\,\text{Å} \quad E \cong 10^8\,\text{V/cm}$$

This is the domain of field emission. With a low external field, tunnelling intensity is low, but not zero.

Consider two electrodes in vacuum, at a distance s, subjected to a voltage difference V. The tunnelling intensity through the vacuum gap as derived from Eq. (22.1) may be expressed by the following relation:

$$I = f(V) \exp\left(-\frac{4\pi(2m_0 e\Phi)^{1/2}}{h} s\right) \quad (22.3)$$

where $f(V)$ is a function of the density of states in the electrodes.

$$\frac{4\pi(2m_0)^{1/2}}{h} = 1.02\,\text{Å}^{-1}(\text{eV})^{-1/2}$$

Substituting numerical values gives (Eq. 1.7).

$$\frac{4\pi(2m_0)^{1/2}}{h} = 1.02\,\text{Å}^{-1}(\text{eV})^{-1/2}$$

Intensity may thus be approximated as follows:

$$I = f(V) \exp\left[-(e\Phi)^{1/2} s\right] \quad (V \text{ and } \Phi \text{ in V; } s \text{ in Å}) \quad (22.4)$$

It may be seen from this relation that for a distance s of a few ångströms, a distance increase of 1 Å divides intensity by a factor of 10. It is this exponential variation of the tunnelling current with distance that is the key for operating the tunnelling microscope.

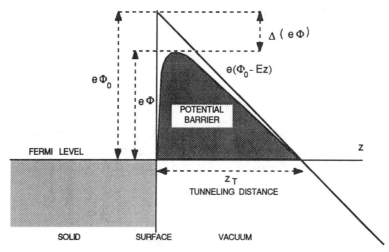

Figure 22.1. Diagram outlining the variation of the potential energy of an electron as a function of its distance z to the surface. Reduction of the work function and of the tunnelling distance induced by an external electric field.

22.2 APPLICATIONS OF THE TUNNEL EFFECT

22.2.1 FIELD EMISSION ELECTRON SOURCES

As stated before, applying an electric field E tending to extract electrons from a conductor decreases the tunnelling distance, i.e. increases the tunnelling intensity. Applying Eq. (22.2) has shown that high intensities were reached for field values of the order of 10^8 V/cm. Intensity is then expressed by the *Fowler–Nordheim* law

$$I = BE^2 \exp\left(-\frac{b\Phi^{3/2}}{E}\right) \quad (B \text{ and } b \text{ constants}) \qquad (22.5)$$

An electric field sufficient for generating a high-intensity electron beam at room temperature may be reached locally at the surface of a sharp metal tip subjected to a moderate voltage. The resulting *field emission electron source* consists actually of a fine tungsten pin acting as cathode, and of an anode at a positive voltage of a few kilovolts (see section 9.2.2).

Absorption of extraneous molecules on the metal surface markedly changes the work function, i.e. the tunnelling current generating electron emission. For a stable operation of such an electron source, the cathode has to remain clean of any contamination, i.e. be operated in ultra-high vacuum.

22.2.2 FIELD EMISSION MICROSCOPY

The principle of field-induced electron emission by a metal tip is applied in the field emission microscopes.

22.2.2.1 Field Electron Emission Microscopy

The conducting specimen acting as the cathode is cut into a very fine tip of terminal radius r. An hemispheric fluorescent screen of radius R acts as the anode. The result on the screen is an electronic field emission image of magnification $G = R/r$ (e.g. $r = 100$ Å, $R = 10$ cm, $G = 10^7$). The image represents a distribution map of the work function at the apex of the tip. When imaging results from a terminal single crystal, the image displays a contrast due to the various limiting lattice planes by means of the differences of respective work functions (see text of Table 22.1). The corresponding facility is called *field electron emission microscope* (FEEM).

22.2.2.2 Field Ion Microscopy

With a similar set-up as above, the pin-shaped specimen at a positive potential, the phosphor screen at a negative potential and helium at low pressure as a medium between the electrodes, the result is *field ion emission microscopy* (FIM), with a magnification $G = R/r$ as well.

Gas atoms in the pin surface vicinity are subjected to a high electric field. This results in ionising the atoms through a tunnelling effect on their electrons and in accelerating the generated positive He-ions towards the negative phosphor screen. The field effect is enhanced in the close vicinity of specimen surface atoms; the ionising probability, i.e. the ion emission current, thus varies at an atomic scale. The resulting image therefore displays the distribution of atoms at the specimen tip surface. The temperature effect is reduced by liquid helium cooling of the specimen.

Implementing this atomic scale surface microscopy is, however, limited to conducting specimens liable to be cut into a fine pin. Due to the small tip radius, the observed surface structure does not necessarily reflect the actual intrinsic surface structure. Barring fundamental research, this technique has therefore not been developed for investigating materials.

22.2.2.3 Tunnel Effect Microscopy

The foregoing limitations are averted by tunnel effect microscopy which may be considered in some ways as inverted FEEM: the pin serves to extract electrons and thus acts as a probe with respect to the flat specimen. The surface emission image is provided by a scanning device, resulting in the technique of scanning tunnelling microscopy.

22.3 SCANNING TUNNELLING MICROSCOPY

This most recent technique, proposed by Binnig and Rohrer in 1982a,b, is in its full development phase. The present section provides merely a survey of the basic layout and the applications. More details are given by review papers, e.g. Binnig and Rohrer, 1983, 1985, 1987; Hansma and Tersoff, 1987; Klein et al, 1987.

The overall principle is similar to that of field emission microscopy, with an inverse set-up averting the drawbacks of the latter. Due to the scanning mode image display, it is currently abbreviated to STM (*scanning tunnelling microscopy*). Related spectrometric techniques may provide additional data.

A very sharp metal pin acting as the anode is brought as close as a few ångströms to the flat specimen surface acting as the cathode. Biasing voltage ranges from a few millivolts to a few volts. The tip of the pin enters the electron cloud surrounding the specimen surface, causing electrons to tunnel across the gap. It thus acts as a local probe of the electron density. Thanks to the sharpness of the tip, the tunnelling current may become appreciable along a narrow column between it and the specimen surface, and may accordingly be measured by means of an associated electron device. Scanning the tip along a parallel to the mean specimen surface results in a tunnelling current that varies in an exponential way with the local distance, i.e. with surface topography (Eq. 22.3 and Figure 22.2).

Considering the uncertainties on the tip position to be negligible, resolution depends on the following parameters.

Normal Resolution

Resolution along the normal to the surface depends on the accuracy of intensity measurement. The tunnelling intensity varies by a factor of 10 for a distance variation of 1 Å (Eq. 22.4); as a result normal resolution may be better than 1 Å.

Lateral Resolution

Resolution parallel to the surface is determined by the terminal diameter of the tip of the pin. It may reach a few ångströms.

22.3.1 INSTRUMENTAL LAYOUT

The basic operating principle is outlined in Figure 22.2. The practical set-up includes the following constituents:

(a) sharp metal pin anode acting as the electron extracting probe;

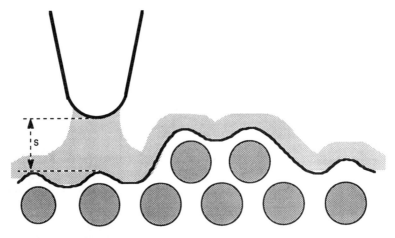

Figure 22.2. Diagram outlining the principle of electron tunnelling between a pin (positive voltage of a few volts) and a solid surface (conductor or semiconductor), the latter being surrounded by the electrons having tunnelled through the potential barrier.

(b) three-dimensional pin displacement attachment, with ångström accuracy;
(c) electronic system for monitoring the pin movements, and for data acquisition, processing and display;
(d) specimen to be observed, with its nanometre-scale displacement device;
(e) vibration damping system.

The whole mechanical part of an STM may be as small as a few centimetres.

22.3.1.1 Probe Pin

The tunnelling probe is generally a tungsten pin, usually at a positive voltage of 2–5 V with respect to the specimen surface. The resulting lateral image resolution is of the order of its terminal tip diameter. For atomic resolution, the pin must therefore be terminated by a single atom. Platinum–iridium pins are increasingly used as well. Several techniques are possible to provide a nanometre-scale (or less) tip diameter: mechanical cutting, high-temperature evaporation, electrochemical etching, ion sputtering, field emission. In the last mentioned method, under the very high local field, a specimen surface atom may be torn away and fixed on the pin surface. Out of all these methods, ion bombardment seems at present to be the least aleatory one (Binh, 1988).

Generating an atomic-scale pin tip is presently the most difficult step towards atomic surface structure resolution. Controlling the state of the tip is best carried out by means of a FIM. In the ideal instance of a single atom tip, due to the exponential current variation vs specimen–tip distance, the tunnelling intensity is virtually determined by this atom alone; the participation of any other, more remote, atoms may then be neglected, leading to atomic resolution and point source interpretation.

22.3.1.2 Pin Displacement System. Control and Data Acquisition

The probe pin has first to be displaced along a normal to the specimen surface in order to bring it as close as a few ångströms, without touching (z movement).

It has then to be translated along successive parallels to the surface (xy movements).

All pin movements must be carried out with an accuracy and a reproducibility better than the resolution to be expected, i.e. to less than 1 Å for atomic resolution. Such accuracy is achieved by means of a ceramic *piezoelectric translator system*, e.g. with three piezo tubes respectively along x, y and z (Figure 22.3a).

In the most usual constant current mode, the z-movement is controlled by an electronic feedback in such a way that the tunnelling current remains constant during the xy translations. The feedback voltage is used as the tunnelling signal, which is accordingly processed, recorded and visualised as a function of the xy coordinates, resulting in a topographic scanning image of the specimen surface.

According to the processing mode, the surface function $z = f(x, y)$ is displayed either as a three-dimensional image (Figures 22.3b and 22.4), a grey level image or a colour-coded image.

Image magnification is determined by the ratio of image scan length to specimen scan length. It commonly reaches 10^8.

Working tunnelling intensities are currently of the order of 1 nA.

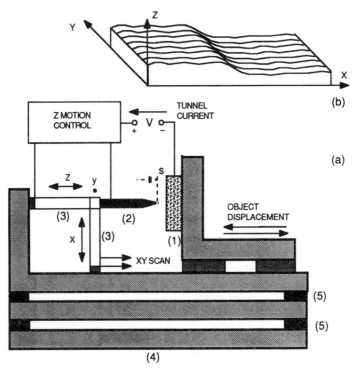

Figure 22.3. Diagram of a scanning tunnelling microscope. (a) Basic constituents: (1) specimen on its mount, positioned by way of a piezoelectric step-by-step transporter; (2) tunnelling pin positioned at a distance s of the specimen surface (a few ångströms) with a positive bias of the order of the volt; (3) ceramic piezoelectric translator tubes along x, y, z for ångström-scale pin displacements; (4) metal plates; (5) viton dampers for vibration damping. (b) Three-dimensional topographic display carried out by xy scanning; deflection is controlled through the constant tunnelling current feedback signal (after Binnig and Rohrer, 1987).

22.3.1.3 Specimen

The specimen must be a conductor, thus limiting applications. Monomolecular layers of organic materials, deposited on a conducting substrate, may also be analysed.

On the other hand, the STM does not require a vacuum for operating. Observations may be carried out in air, in a gas, in a non-polar liquid, as well as in a vacuum. To study a given area, the specimen is positioned by means of a transporter, mostly based on piezoelectric displacement by steps of some 10 nm to 1 μm.

22.3.1.4 Vibration Damping

Such a lightweight system operating at ångström resolution is sensitive to the slightest vibration. Various damping systems are used, mostly based on an accurate combination of absorbing and elastic materials. Low-frequency vibrations may be damped by induction of Foucault currents produced by magnets in copper plates.

The present trend is to produce scanning tunnelling microscopy layouts which are increasingly simplified with respect to the first prototypes.

Figure 22.4. Three-dimensional relief image with atomic resolution displayed by means of STM. Gold single crystal (110) surface showing areas without reconstruction, as well as reconstructed areas with superstructures 2×1 and 3×1. Lateral image dimensions 190×190 Å. Normal amplitude 0.4 Å. (2×1 atoms) (Gimsewski J. K., Schlittler R., IBM Research Division, Zurich Research Lab., Rueschlikon).

22.3.2 POSSIBILITIES AND DEVELOPMENTS

Consider the constant tunnelling current mode.

For a specimen surface with constant work function, the pin tip would follow the surface topography.

For a perfectly flat surface, the pin tip would display a surface barrier equipotential. The actual image displays a combination of these effects. It may be regarded as a *relief map* representing a locus of equal density of states at the specimen surface. At atomic resolution the image may be expressed as the convolution product of the electron density profile of the pin by the electron density profile of the specimen surface.

22.3.2.1 Resolution

Normal Resolution

$d(z) \cong 0.05$ Å.

This is limited by operating stability and by z displacement accuracy.

Lateral Resolution

$d(xy) \cong 2-6$ Å.

This depends on the following parameters:

(a) terminal radius of the pin;
(b) working distance s;
(c) work function;
(d) extension of surface state wave function.

22.3.2.2 Spectrometric Developments

STM may provide additional spectrometric data, e.g. by means of the *inverse photoelectric effect*. In this technique, electrons are injected through the probe pin into the specimen and induce local optical transitions resulting in photoemission in the visible to ultraviolet range. Operation is then similar to field emission, with the advantage of a nanometric electron beam and electron energies ranging between zero and some 100 eV. The connection of a photon analyser thus results in auxiliary techniques called STIPE (scanning tunnelling inverse photoemission) and STOM (scanning tunnelling optical microscopy) which combine the structural and the spectrometric aspects (Coombs et al, 1988).

Tunnelling electrons interact with vibrational modes of molecules on the specimen surface. This additional spectrometric aspect may be developed for characterizing surface atoms and molecules (Smith et al, 1987).

22.3.2.3 Miscellaneous Developments

A new approach for studying non-conducting specimen surfaces, based on STM, is the so-called AFM (atomic force microscopy, Binnig et al, 1986). Similarly to the tunnelling pin, it makes use of a sharp tip connected to a thin flexible platelet acting as a spring. When scanned over the specimen surface, the tip follows the surface profile under the sole action of atomic forces. The z displacements of the tip are measured by means of the tunnelling current between the spring platelet and the tunnelling pin of an associated conventional STM device, in which the spring takes the place of the specimen. In the layout referred to, a diamond microtip was fixed at the end of a 25 μm thin gold lamella measuring 0.8×0.25 mm.

The z displacement of the tip may be measured by means of the tunnelling current at an accuracy of 10^{-4} Å, resulting in the measurement of forces as low as 10^{-18}–10^{-16} N. Atomic forces vary in the range of some 10^{-7} N (ionic and covalent bonds) to some 10^{-12} N (residual bonds); they may therefore be displayed by such a nanomechanical layout.

A further possibility, which may have important developments in the future, is the use of a facility based on STM for carrying out solid *surface lithography* at a nanometre scale (Schneir and Hansma, 1987).

22.3.3 APPLICATIONS TO MATERIAL RESEARCH

STM and its variants are developing into a major technique for surface analysis at atomic scale, as a complement to other available methods, such as LEED, ESCA, AES and SEXAFS.

22.3.3.1 Basic Tunnelling Mode

In the basic tunnelling mode, its applications are presently limited to surfaces of conducting materials and to thin layers deposited on a conducting substrate. Up to the present, investigations have mainly been centred on metals and semiconductors (Binnig and Rohrer, 1985; Hochella et al, 1989; Klein et al, 1987). Biological and organic materials are now being increasingly investigated. Some examples of the studies are given below.

Solid Surface Analyses at an Atomic Scale

Surface structure analysis by means of LEED has led to ambiguities due to the difficulty of quantitative intensity interpretations. One of the first results achieved with STM has been to confirm silicon surface reconstruction through a (7×7) superstructure (Binnig and Rohrer, 1983).

Crystal Surface Topography

The observation of surface topographies at a nanometre scale (e.g. steps, edges, kinks, undulations) provides valuable data for crystallogenetic interpretations. Furthermore, surface topography plays a decisive role in mineral solubility and reactivity at the surface-solution interface. STM is thus a tool for investigating favourable surface sites involved in geochemical processes of mineral alteration (Hochella et al, 1989).

Atomic resolution STM now makes possible the direct investigation of dislocations at crystal surfaces (Zheng et al, 1988).

Absorption of Molecules on Surfaces

Absorption, adsorption and desorption of gas molecules is of paramount importance for catalytic processes. The presence of molecules on a conducting or semiconducting surface results in a local variation of the work function. STM may therefore directly visualise the absorption sites and their distribution on solid surfaces.

22.3.3.2 Variants of STM

The field of surface structure analysis is widened to non-conductors by the use of AFM. Polymer surfaces as well as non-conducting crystal surfaces and even biological materials may thus be directly visualised at a nanometre or even atomic scale (e.g. Albrecht and Quate, 1987; Drake et al, 1989; Hansma et al, 1988).

The use of an STM layout for *surface lithography* opens up new perspectives for technological applications, in particular in the field of nanoelectronic circuits. Densities of some 1000 times higher with respect to the present microtechnique may be achieved.

Additional *spectrometric data* on a near atomic scale may be gathered by implementing tunnelling-induced photoemission and vibrational spectrometry.

STM and the various connected technologies are still in their infancy. Further developments and applications are to be expected soon.

REFERENCE CHAPTERS

Chapters 1, 9.

BIBLIOGRAPHY

[1] Christmann J. R. *Fundamentals of solid state physics*. John Wiley, New York, Chichester, 1988.
[2] Hansma P. K. (ed.) *Tunnelling spectroscopy. Capabilities. Applications and new techniques*. Plenum Press, New York, 1982.
[3] Kittel C. *Introduction to solid state physics*. 6th edn, John Wiley, New York, Chichester, 1986.
[4] Wolf E. L. *Principles of electron tunneling spectroscopy*. Oxford University Press, New York, 1985.

Appendix A

Physical Quantities. Units. Universal Constants. Notations

A.1 MAIN PHYSICAL QUANTITIES AND THEIR UNITS

The aim of this section is to define the main physical quantities which are involved in the techniques based on radiation–matter interactions.

The **SI units** (Système International) are used throughout the book, with some notified exceptions.

Less important physical quantities are defined in the lexicon, in section A.3.

A.1.1 Wavelength

Space period of a wave. Notation λ.

The SI unit is the *metre* (m), with its submultiples the *micrometre* or *micron* (μm) and the *nanometre* (nm).

The *ångström* (Å) is a non-conventional unit; however, it is very practical and commonly used for having a value of the order of magnitude of atomic dimensions and interatomic distances.

The scale relations are as follows:

$$1\,\text{Å} = 10^{-1}\,\text{nm} = 10^{-4}\,\mu\text{m} = 10^{-10}\,\text{m}$$

A.1.2 Frequency

Number of periods per second. Notation ν.

The unit is the *hertz* (1 Hz = 1 period/s), with its most common multiples *kilohertz* (1 kHz = 10^3 Hz) and *megahertz* (1 MHz = 10^6 Hz).

Frequency is related to wavelength through the relation $\lambda\nu = v$ (where v is the propagation velocity).

A.1.3 Electrical Potential

Potential at a point: work done in bringing a unit positive charge from infinity to the point. Notation V.

The SI unit is the *volt* (V), with its most common multiples *kilovolt* (1 kV = 10^3 V), *megavolt* (1 MV = 10^6 V) and *gigavolt* (1 GV = 10^9 V).

A.1.4 Energy

It expresses the energy conveyed by elemental particles (photons, electrons, neutrons, ions). Notation E.

The SI unit is the *joule* (J). It is far too large a unit to express energies as commonly used for material analysis in this book. The most commonly used and most useful energy unit is the *electronvolt* (eV), thanks to its straightforward physical meaning (energy acquired by an electron accelerated through a potential difference of 1 V) and to its value which is of the order of magnitude of binding energies of atomic electrons. It is related as follows to the SI unit:

$$1 \text{ eV} = 1.602 \times 10^{-19} \text{ J}$$

The following multiples are the most commonly used: **kiloelectronvolt** (1 keV = 10^3 eV), **megaelectronvolt** (1 MeV = 10^6 eV), **gigaelectronvolt** (1 GeV = 10^9 eV).

A.1.5 Wave Vector

Vector with its direction along the direction of propagation, with its length equal to the reciprocal of wavelength. Notation **k**, with

$$k = |\mathbf{k}| = 1/\lambda$$

Its length may also be defined by $k = 2\pi/\lambda$.

In the case of a particle wave, Eq. (1.2) results in the wave vector being proportional to the momentum $\mathbf{p} = m\mathbf{v}$:

$$\mathbf{k} = \frac{\mathbf{p}}{h} \tag{A.1}$$

A.1.6 Amplitude

Spatial component of a wave function. Notation G.

This is characterised by its absolute value, or magnitude, and its phase. It may therefore be expressed by a complex number or by a vector (Fresnel construction). It has the dimension of a length. It is often measured with respect to the incident amplitude, taken as unit; it then becomes a dimensionless number.

A.1.7 Intensity

The number of radiation particles emitted into the solid angle unit or passing through a unit surface per unit of time. Notation I.

The corresponding SI units are respectively the watt/steradian or the watt/m^2. In practice, intensities are commonly measured by a detector count rate.

Intensity equals the squared amplitude:

$$I = |G|^2 = GG^* \qquad (A.2)$$

A.1.8 Pressure

The pressure which intervenes in the foregoing chapters is the gas pressure inside the vacuum enclosures.

The SI unit is the *pascal* (1 Pa = 1 N/m²). The *hectopascal* is commonly used for expressing atmospheric pressure in meteorology or vacuum techniques.

A practical and often used unit in vacuum technique is the *torr* which corresponds to 1 mm Hg. Older units are the bar and the atmosphere. All these units are related as follows:

$$1 \text{ hPa} = 10^2 \text{ Pa} = 0.75 \text{ Torr} = 1 \text{ mbar} = 0.987 \times 10^{-3} \text{ atm}$$

A.2 UNIVERSAL CONSTANTS

The present list has been limited to universal constants used in the book. They are expressed in SI units. They have been printed in roman type, for them to be easily distinguished from variable quantities.

Light velocity in vacuum	$c = 2.99793 \times 10^8$ m s^{-1}
Electron charge	$e = 1.6021 \times 10^{-19}$ C
Electron rest mass	$m_0 = 9.1091 \times 10^{-31}$ kg
Neutron rest mass	$M_n = 1.6749 \times 10^{-27}$ kg
Proton rest mass	$M_p = 1.6726 \times 10^{-27}$ kg
Mass ratio proton/electron	$M_p/m_0 = 1836.1$
Planck constant	$h = 6.6257 \times 10^{-34}$ J s
Boltzmann constant	$k = 1.3806 \times 10^{-23}$ J K^{-1}
Avogadro number (molecules/mole)	$N = 6.0224$ mol^{-1}

A.3 LEXICON OF NOTATIONS AND SYMBOLS

The most used symbols are listed below, sorted according to their related headings.

Only notations with a general validity in various chapters are listed. Some specific notations are merely defined in the text.

Certain notations may appear in several chapters. The field covered by the book is very wide and the number of available symbols is limited. As a result, a similar notation may sometimes be used, if required, for different quantities in different chapters, provided that no ambiguities arise.

Vectorial quantities have been printed in **bold face**. Scalar quantities have been printed in *italics*.

A.3.1 Crystallography

Lattices

Whenever possible, notations are based on recommendations of the IUC (International Union of Crystallography), in [3].

a, b, c	direct lattice basis vectors
α, β, γ	angles of above vectors
a*, b*, c*	reciprocal lattice basis vectors
$\alpha^*, \beta^*, \gamma^*$	angles of above vectors
(hkl)	Miller indices of a set of direct lattice planes (prime to each other)
$((hkl))$	diffraction domain around a reciprocal lattice point hkl
$[hkl]^*$	indices of a reciprocal lattice line (prime to each other)
hkl	coordinates of a reciprocal lattice point (integers)
$[uvw]$	indices of a direct lattice line (prime to each other)
$(uvw)^*$	Miller indices of a set of reciprocal lattice planes (prime to each other)
uvw	coordinates of a direct lattice point (integers)
$d(hkl)$	interplanar spacing of direct lattice planes (hkl)
$d^*(uvw)$	interplanar spacing of reciprocal lattice planes $(uvw)^*$
$m = uh + vk + wl$	equation of a set of lattice planes (hkl) or $(uvw)^*$, with m an integer, the level of the planes of a set (hkl) or $(uvw)^*$: $m=0$ plane passing through origin (zero level); $m=1$ first plane from the origin (first level), etc.
$n(uvw)$	interpoint spacing of a direct lattice line $[uvw]$
$n^*(hkl)$	interpoint spacing of a reciprocal lattice line $[hkl]^*$
$\mathbf{r}_n = \mathbf{r}(uvw) = u\mathbf{a} + v\mathbf{b} + w\mathbf{c}$	direct lattice vector
$\mathbf{r}^*_g = \mathbf{r}^*(hkl) = h\mathbf{a}^* + k\mathbf{b}^* + l\mathbf{c}^*$	reciprocal lattice vector
$Q(hkl) = |\mathbf{r}^*(hkl)|^2$	quadratic form; squared reciprocal vector
$v_0 = \mathbf{abc} = \mathbf{a}(\mathbf{b} \wedge \mathbf{c})$	volume of direct lattice unit cell
$v^*_0 = \mathbf{a^*b^*c^*} = \mathbf{a^*}(\mathbf{b^*} \wedge \mathbf{c^*})$	volume of reciprocal lattice unit cell

Relations between direct and reciprocal lattice are stated in Appendix B.

Crystal

A, B, C	metrical lengths } of a parallel-faced crystal along **a**,
U, V, W	numerical lengths } **b, c**, with $A = Ua, B = Vb, C = Wc$
B	Burgers vector of a dislocation
$N(S)$	number of atoms per surface unit
$N(V)$	number of atoms per volume unit
ρ	displacement vector of a lattice defect
$v = \mathbf{ABC} = \mathbf{A}(\mathbf{B} \wedge \mathbf{C})$	parallel-faced crystal volume (or any volume of matter involved in an interaction)

A.3.2 Material Physics. General Radiation–Matter Interaction

A	atomic mass
B	brightness of radiation source
χ	absorption factor ($\chi = \mu/\sin\theta$, where θ is the emergence angle)
ΔE	energy loss during an interaction
δE	energy dispersion
E_0, E	energy of a radiation particle before and after interaction
Φ	potential barrier at the surface of a solid; $e\Phi$ is the electron work function
I_0, I	intensity of a radiation before and after interaction
l	mean free path
λ_0, λ	wavelength before and after interaction
μ	mass absorption coefficient
ν	frequency
$\mathbf{k}_0, \mathbf{k}$	wave vector before and after interaction (with $k_0 = 1/\lambda_0$; $k = 1/\lambda$)
ρ	density
σ	interaction cross-section
T	absolute temperature (in Kelvin)
$V(\mathbf{r}) = V(xyz)$	electrical potential at a point $\mathbf{r}(xyz)$
Z	atomic number

A.3.3 Diffraction and Scattering

$A_J(\mathbf{R}), A_J(hkl)$	atomic scattering amplitude of an atom J, corresponding to a scattering vector $\mathbf{R}$, for a reflection hkl
$B = 8\pi^2\sigma^2$	temperature factor (Debye–Waller factor), with σ the mean quadratic deviation of the atom
δ	optical path difference
$f(\mathbf{r})$	scattering power at any point $\mathbf{r}(xyz)$; structure function
$f_J(\mathbf{R}), f_J(hkl)$	X-ray atomic scattering factor of an atom J, corresponding to a scattering vector $\mathbf{R}$, for a reflection hkl (dimensionless)
$F(\mathbf{R}), F(hkl), F(\mathbf{g})$	structure factor (amplitude scattered by a unit cell) at a scattering vector $\mathbf{R}$, for a reflection hkl, for a reflection $\mathbf{g}$ (the last mentioned is specific for electron diffraction)
ϕ	phase
$\mathbf{g} = \mathbf{r}_g^* + \mathbf{s}\,[\mathbf{s} \in ((hkl))]$	diffraction vector for a diffraction deviation $\mathbf{s}$ (used in electron diffraction for designating an excited reflection hkl)
$G_0, G(0), G(\mathbf{R}), G(hkl), G(\mathbf{g})$	respectively incident, transmitted, scattered amplitude (scattering vector $\mathbf{R}$), diffracted amplitude (reflection hkl or $\mathbf{g}$)

$\gamma(\mathbf{r})$	size function
$\Gamma(\mathbf{R})$	size factor
hkl	reflection on a set of lattice planes; the reflection indices equal the Miller indices (first reflection order) or are multiples of Miller indices (higher orders)
$((hkl))$	diffraction domains around a reciprocal lattice point hkl
I_0, $I(O)$, $I(\mathbf{R})$, $I(hkl)$, $I(\mathbf{g})$	intensities (notations similar to G)
$\mathbf{k}_0$, $\mathbf{k}$	respectively incident and scattered wavevector (with $k_0 = 1/\lambda_0$; $k = 1/\lambda$)
$L(\mathbf{R})$	form factor
$\|L(\mathbf{R})\|^2$	interference function expressing the effect of interferences on scattered intensities
m	level of planes of a set (hkl) or $(uvw)^*$; $m = 0$ plane passing through the origin (zero level); $m = 1$ first plane from the origin (first level), etc.
n	diffraction (or reflection) orders on a set of lattice planes; multiplying the Miller indices by n results in the reflection indices
$p(\mathbf{u})$	distribution function (without central peak)
$q(\mathbf{u}) = f(\mathbf{r})*f(-\mathbf{r}) = \mathrm{TF}\{I(\mathbf{R})\}$	distribution function (with central peak), convolution square of the structure, Fourier transform (FT) of intensity
$Q(hkl) = \|\mathbf{r}^*(hkl)\|^2$	quadratic form; squared reciprocal vector
$\mathbf{R} = \mathbf{k} - \mathbf{k}_0$	scattering vector or diffraction vector
$\rho(\mathbf{r})$	electron density at a point $\mathbf{r}(xyz)$
$\mathbf{s}(s_x, s_y, s_z)$	diffraction deviation (*Resonanzfehler*, error of resonance), such as $\mathbf{R} = \mathbf{r}_g^* + \mathbf{s}$
σ	scattering cross-section (similarly mean quadratic deviation of an atom at a temperature T, see B)
θ	Bragg diffraction angle (2θ is the angle between the incident and the diffracted beam)
$\theta_E = \Delta E / 2E_0 = \Delta k / k_0$	characteristic inelastic scattering angle
$\Psi(\mathbf{r})$	wave function, solution of the Schrödinger equation, wave amplitude at a point $\mathbf{r}$

A.3.4 Spectrometry

A	detector amplifying coefficient
α	ionisation degree of a sputtered element (SIMS)
c_J	mass fraction of an element J
$c(at)$	atomic fraction of an element
c_{min}	minimum detectable mass fraction
δE	energy resolution of a spectrometer
δW	variation of binding energy W of an electron, by chemical effect

$\Delta E = E_0 - E$	radiation energy loss
E_0, E	respectively energy of the incident radiation, of the measured radiation
E	electric field
ε	detector efficiency
H	magnetic field
I_0, I	respectively intensity of the incident radiation, of the measured radiation
$\chi = \mu/\sin\theta$	absorption factor (θ is the emergence angle)
l	mean free path (analysing depth)
M	mean atomic mass of a specimen
M_{min}	minimum detectable mass
μ	mass absorption coefficient
$N = nt_c$	integrated count during count time t_c on an element line
n	count rate on an element line (n total; n_0 noise; $n_c = n - n_0$ characteristic; n_1 pure element)
N_{min}	minimum detectable number of atoms
$N(S)$	number of atoms per surface unit
$N(V)$	number of atoms per volume unit
ω	fluorescent yield
Ω	detection solid angle
p	line emission probability
R	electron backscattering factor (fraction of non-backscattered electrons, $R < 1$)
ρ	specimen density
S	electron stopping factor (derivative of energy with respect to the penetration distance)
	ion sputtering yield (SIMS)
σ	interaction cross-section (of excitation, of energy loss, etc.)
t	specimen depth of analysis or thickness
t_c	count time
t_m	detector dead time
$\tau = \varepsilon\Omega/4\pi$	transmission factor of a detector–spectrometer system
$U_X = E_0/W_X$	excitation overvoltage corresponding to the level X of an element
W_m	mean excitation threshold of an element
W_X	excitation energy of a level X (binding energy of an electron of this level)
$X = K, L1, L2, \ldots$	notation of energy levels corresponding to atomic shells K, L, $\ldots$
Z	atomic number

512   *Structural and Chemical Analysis of Materials*

A.3.5 Electron Microscopy

$A(R)$	amplitude component of the transfer function in the objective back focal plane, at a distance R from the optical axis
α	angular objective aperture
α_0	convergence angle (illumination angle)
B	Burgers vector of a dislocation
$c(\mathbf{r})$	contrast function at a point **r** of the image
C_c	chromatic aberration coefficient
C_s	spherical aberration coefficient
d	resolution
d_s	beam diameter on the object
D	defocusing
f	focal length
$\phi(\mathbf{r})$	phase; transparency phase component at an object point **r**
g	lens magnification (e.g. objective lens)
G	total magnification
$\gamma(R)$	phase component of the transfer function in the objective back focal plane, at a distance R from the optical axis
$K = L\lambda$	diffraction constant (equal to $f\lambda$ in the back focal plane of a lens of focal length f)
L	diffraction length (equal to the focal length in the back focal plane of a lens)
λ	wavelength associated with an electron
r_c, r_d, r_s, r_{tot}	respectively radius of the confusion disc of chromatic, diffraction, spherical, total aberration
ρ	displacement vector (lattice defects)
$s(\mathbf{r})$	absorption component of the transparency function at an object point **r**
t	object thickness (thin specimen)
$t(\mathbf{r}) = \text{TF}\{T(R)\}$	confusion function at a point **r** in the objective image plane
$t_g = \pi v_0 / \lambda F(\mathbf{g})$	extinction distance
$T(R) = A(R)\exp[-i\,\gamma(R)]$	transfer function in the objective back focal plane, at a distance R from the optical axis
V_0	acceleration voltage of incident electrons

BIBLIOGRAPHY

[1] Barford N. C. *Experimental measurements. Precision, error and truth*. 2nd edn. John Wiley, New York, Chichester.
[2] Bureau International des Poids et Mesures *Le système international d'unités*. Sèvres, 1970.
[3] Hahn T. (ed.) *International tables for crystallography*. Vol. A: *Space-group symmetry*. International Union of Crystallography. D. Reidel, Kluwer Acad., Dordrecht, 1987.
[4] Massey B. S. *Measures in science and engineering*. John Wiley, New York, Chichester, 1986.

Appendix B

Reciprocal Space. Reciprocal Lattice

In the most general case, a crystal lattice is determined by a set of axes $Oxyz$ such that:

1. The lengths a, b, c of the basis vectors $\mathbf{a}$, $\mathbf{b}$, $\mathbf{c}$ are different from each other and different from a unit vector;
2. The angles α, β, γ between axes are different from each other and different from a right angle.

Crystallographic calculations are facilitated by using the reciprocal lattice, with axes that are normal to the corresponding direct lattice planes.

However, the main importance of the reciprocal lattice lies in the field of diffraction. Whereas the object is located in direct space (object space), the diffraction effects are located in the corresponding reciprocal space (diffraction space or Fourier space). The Fourier transform allows passage from one to the other.

Notation and symbols: Any elements related to reciprocal space are usually noted by (*); capitals are sometimes used as well.

B.1.1 Relations Defining Reciprocal Space and Reciprocal Lattice

Reciprocal Space

A point of the ***object space***, commonly called ***direct space*** or ***real space***, is located by a vector $\mathbf{r}$ as follows:

$$\mathbf{r} = x\mathbf{a} + y\mathbf{b} + z\mathbf{c} \tag{B.1}$$

where xyz are numerical coordinates, and $\mathbf{abc}$ basis vectors.

A point of the ***reciprocal space*** is located by a vector $\mathbf{r}^*$ as follows:

$$\mathbf{r}^* = x^*\mathbf{a}^* + y^*\mathbf{b}^* + z^*\mathbf{c}^* \tag{B.2}$$

where $x^*y^*z^*$ are numerical coordinates, and $\mathbf{a}^*\mathbf{b}^*\mathbf{c}^*$ basis vectors.

Relative orientation of the sets of axes.

$$\mathbf{a}^* \perp (\mathbf{b,c}) \qquad \mathbf{b}^* \perp (\mathbf{a,c}) \qquad \mathbf{c}^* \perp (\mathbf{a,b})$$

Length of basis vectors.

$$a^* = \frac{1}{a \cos(\mathbf{a}^\wedge\mathbf{a}^*)} = \frac{\mathbf{b} \wedge \mathbf{c}}{v^*_0} \qquad b^* = \frac{1}{b \cos(\mathbf{b}^\wedge\mathbf{b}^*)} = \frac{\mathbf{c} \wedge \mathbf{a}}{v^*_0} \qquad c^* = \frac{1}{c \cos(\mathbf{c}^\wedge\mathbf{c}^*)} = \frac{\mathbf{a} \wedge \mathbf{b}}{v^*_0}$$

with $v^*_0 = \mathbf{a}^* \cdot (\mathbf{b}^* \wedge \mathbf{c}^*)$.

The above conditions may be written as scalar products. The basic defining relations are then written respectively

$$\mathbf{ab}^* = \mathbf{ac}^* = \mathbf{ba}^* = \mathbf{bc}^* = \mathbf{ca}^* = \mathbf{cb}^* = 0$$

$$\mathbf{aa}^* = \mathbf{bb}^* = \mathbf{cc}^* = 1 \tag{B.3}$$

Dimensions in reciprocal space are the inverses of dimensions in direct space.

Reciprocal Lattice

Corresponding lattices are defined for a crystalline medium.

A point N of the **crystal lattice** (or **direct lattice**, or **real lattice**) is located by a vector $\mathbf{r}_n$ such that

$$\overrightarrow{ON} = \mathbf{r}_n = u\mathbf{a} + v\mathbf{b} + w\mathbf{c} \tag{B.4}$$

where *uvw* are integers, numerical coordinates of a direct lattice point.

A lattice point N of the **reciprocal lattice** is located by a vector $\mathbf{r}^*_g$ such that

$$\overrightarrow{OG} = \mathbf{r}^*_g = h\mathbf{a}^* + k\mathbf{b}^* + l\mathbf{c}^* \tag{B.5}$$

where *hkl* are integers, numerical coordinates of a reciprocal lattice point.

B.1.2 Basic Relations between Direct and Reciprocal Lattice

Direct Lattice Vector and Reciprocal Lattice Vector

Relations (B.1–B.3) result in

$$\mathbf{r}_n \mathbf{r}^*_g = uh + vk + wl = m \quad (m \text{ integer}; \ -\infty < m < +\infty) \tag{B.6}$$

This relation expresses:

1. The equation of a set of direct lattice planes (*hkl*), i.e. the condition for a direct lattice point *uvw* to be located in a plane (*hkl*) at level *m*.
2. The equation of a set of reciprocal lattice planes (*uvw*)*, i.e. the condition for a direct lattice point *hkl* to be located in a plane (*uvw*)* at level *m*.

Orientation of Lattice Lines and Planes

Consider a set of direct lattice planes (hkl) and a reciprocal lattice line with equal indices $[hkl]^*$.

Consider a direct lattice vector $\mathbf{r}_n$ in the plane (hkl) passing through the origin $(m=0)$. Its components uvw meet the equation of the plane, i.e. $uh+vk+wl=0$. According to Eq. (B.6), this results in

$$\mathbf{r}_n \mathbf{r}_g^* = 0 \Rightarrow \mathbf{r}_g^* \perp \mathbf{r}_n$$

$\mathbf{r}_g^*$ being normal to any vector of plane (hkl), it is normal to this plane, i.e.

$$[hkl]^* \perp (hkl)$$
$$[uvw] \perp (uvw)^*$$
(B.7)

Result

1. A reciprocal lattice line $[hkl]^*$ is normal to the set of direct lattice planes with the same indices.
2. Due to reciprocity, a direct lattice line $[uvw]$ is normal to the set of reciprocal lattice planes with the same indices.

Interplanar and Interpoint Distances

Consider a direct lattice vector $\mathbf{r}_n$ having its end point in the first plane (hkl) from the origin $(m=0)$. Its components uvw then meet the plane equation $uh+vk+wl=1$. According to Eq. (B.6), this results in

$$\mathbf{r}_n \mathbf{r}_g^* = 1 = [r_n \cos(\mathbf{r}_n{}^{\wedge}\mathbf{r}_g^*)] r_g^*$$

The factor $r_n \cos(\mathbf{r}_n{}^{\wedge}\mathbf{r}_g^*)]$ represents the projection of vector $\mathbf{r}_n$ on the normal to the set of planes (hkl), i.e. on the reciprocal line $[hkl]^*$. It thus equals the interplanar distance $d(hkl)$, leading to

$$d(hkl)n^*(hkl) = 1$$
$$d^*(uvw)n(uvw) = 1$$
(B.8)

where $d(hkl)$ and $d^*(uvw)$ respectively are the interplanar distance of direct lattice planes (hkl) and reciprocal lattice planes $(uvw)^*$, and $n(uvw)$ and $n^*(hkl)$ respectively the interpoint distance of direct lattice lines $[uvw]$ and reciprocal lattice lines $[hkl]^*$.

Result. The interplanar distance of direct lattice planes (hkl) is the inverse of the interpoint distance of reciprocal lattice lines with the same indices.

Due to reciprocity, the interplanar distance of reciprocal lattice planes $(uvw)^*$ is the inverse of the interpoint distance of direct lattice lines with the same indices.

Relations (B.7) and (B.8) are very useful for crystallographic calculations. Relation (B.8) is often used in crystal diffraction analysis.

B.1.3 Reciprocal Lattices with General Constants

The above properties are preserved when replacing integer 1 in Eqs (B.3) and (B.8) by a general number K. The resulting reciprocal lattices are similar; their scale is defined by the constant K. In diffraction analysis, this constant, called the *diffraction constant*, is related to the radius L of the reflecting sphere and to the wavelength by the following relation:

$$K = L\lambda \qquad (B.9)$$

Expressing measurements in reciprocal space by lengths results in K having the dimensions of a surface.

Expressing measurements in reciprocal space by the reciprocal of lengths results in K being dimensionless (e.g. measuring direct space in Å and reciprocal space in Å^{-1}).

Appendix C

Basic Properties of Fourier Transforms

C.1.1 Definition

Consider a function (or distribution) $f(\mathbf{r})$ related to a point $\mathbf{r}(xyz)$ of *direct space xyz*. The Fourier transform (FT) of $f(\mathbf{r})$ results in a function $F(\mathbf{R})$, related to a point $\mathbf{R}(XYZ)$ in *Fourier space* or *reciprocal space* XYZ, according to the following relations:

$$F(\mathbf{R}) = \int_\infty f(\mathbf{r}) \exp(2\pi i \mathbf{r}\mathbf{R}) d\mathbf{r}^3$$
$$\mathbf{r}\mathbf{R} = xX + yY + zZ \tag{C.1}$$

$F(\mathbf{R})$ is called the *Fourier transform* of $f(\mathbf{r})$. It is expressed as $F(\mathbf{R}) = \text{FT}\{f(\mathbf{r})\}$.

The relationships between direct and reciprocal space have been surveyed in Appendix B.

C.1.2 Inverse Fourier Transform

Function $f(\mathbf{r})$ is reciprocally derived from function $F(\mathbf{R})$ by the inverse transform

$$f(\mathbf{r}) = \int_\infty F(\mathbf{R}) \exp(-2\pi i\, \mathbf{r}\mathbf{R})\, d\mathbf{R}^3$$
$$f(\mathbf{r}) = \text{FT}\{F(\mathbf{R})\} \tag{C.2}$$

C.1.3 Additivity

The FT is an additive function, i.e.

$$F = \text{FT}\{f\}; \quad G = \text{FT}\{g\} \Rightarrow F + G = \text{FT}\{f + g\}; \quad aF + bG = \text{FT}\{af + bg\} \tag{C.3}$$

where a and b are scalar numbers.

C.1.4 Translation

$$FT\{f(\mathbf{r})\} = F(\mathbf{R}) \Rightarrow FT\{f(\mathbf{r}+\mathbf{s})\} = F(\mathbf{R})\exp(2\pi i\ \mathbf{sR}) \qquad (C.4)$$

where **s** is a translation vector.

A translation **s** leaves the magnitude of the FT unchanged; however, it undergoes a phase shift $(2\pi\ \mathbf{sR})$.

C.1.5 Changing Sign. Symmetry and Antisymmetry

$$F(\mathbf{R}) = FT\{f(\mathbf{r})\} \Rightarrow F(-\mathbf{R}) = FT\{f(-\mathbf{r})\} \qquad (C.5)$$

Transform of a Real Dissymmetrical Function

Consider $f(\mathbf{r})$ a real dissymmetrical function, i.e. $f(\mathbf{r}) \neq f(-\mathbf{r})$. Its FT is a complex function as follows:

$$\begin{aligned} FT\{f(\mathbf{r})\} &= F(\mathbf{R}) &&= \alpha + i\beta \\ FT\{f(-\mathbf{r})\} &= F(-\mathbf{R}) = \alpha - i\beta = F^*(\mathbf{R}) \end{aligned} \qquad (C.6)$$

Transform of a Real Symmetrical Function

Consider $f(\mathbf{r})$ a real centrosymmetrical function, i.e. $f(\mathbf{r}) = f(-\mathbf{r})$. Its FT is a real function as follows:

$$F(\mathbf{R}) = F(-\mathbf{R}) \Rightarrow \alpha + i\beta = \alpha - i\beta \Rightarrow \beta = 0 \qquad (C.7)$$

Transform of a Real Antisymmetrical Function

Consider $f(\mathbf{r})$ a real antisymmetrical function, i.e. $f(\mathbf{r}) = -f(-\mathbf{r})$. Its FT is an imaginary function as follows:

$$F(\mathbf{R}) = -F(-\mathbf{R}) \Rightarrow \alpha + i\beta = -(\alpha - i\beta) \Rightarrow \alpha = 0 \qquad (C.8)$$

C.1.6 Transform of a Product. Convolution

General Form

Consider

(a) the functions: $f(\mathbf{r})$ and $g(\mathbf{r})$;
(b) their FTs: $F(\mathbf{R}) = FT\{f(\mathbf{r})\}$ and $G(\mathbf{R}) = FT\{g(\mathbf{r})\}$.

The FT of the product $p(\mathbf{r}) = f(\mathbf{r})g(\mathbf{r})$ is the *convolution product* (or simply *convolution*) of functions $F(\mathbf{R})$ and $G(\mathbf{R})$.

Basic Properties of Fourier Transforms

The convolution product is expressed as follows:

$$\text{FT}\{p(\mathbf{r})\} = P(\mathbf{U}) = \int_{\infty} F(\mathbf{R})G(\mathbf{U}-\mathbf{R})\mathrm{d}\mathbf{R}^3 \tag{C.9}$$

Convolution is noted by ($*$):

$$\text{FT}\{p(\mathbf{r}) = f(\mathbf{r})g(\mathbf{r})\} = P(\mathbf{U}) = F(\mathbf{R})*G(\mathbf{R}) \tag{C.10}$$

The product $P(\mathbf{U})$ is a function of an auxiliary variable $\mathbf{U}(X'Y'Z')$, superposed to $\mathbf{R}(XYZ)$ in reciprocal space XYZ.

Reciprocally, the transform of $P(\mathbf{U})$ is written

$$\text{FT}\{P(\mathbf{U}) = F(\mathbf{R})G(\mathbf{R})\} = p(\mathbf{u}) = f(\mathbf{r})*g(\mathbf{r}) \tag{C.11}$$

The product $p(\mathbf{u})$ is a function of an auxiliary variable $\mathbf{u}(x'y'z')$ superposed to $\mathbf{r}(xyz)$ in direct space.

Result. The FT of the product of two functions equals the convolution product of the FTs of those two functions.

The FT of the convolution product of two functions equals the product of the FTs of those two functions.

Alternative Form

Consider:

(a) the functions: $f(\mathbf{r})$ and $g(-\mathbf{r})$;
(b) their FTs: $F(\mathbf{R}) = \text{FT}\{f(\mathbf{r})\}$ and $G(-\mathbf{R}) = \text{FT}\{g(-\mathbf{r})\}$.

The FT of the product $p'(\mathbf{r}) = f(\mathbf{r})g(-\mathbf{r})$ is written

$$\begin{aligned} P'(\mathbf{U}) &= \text{FT}\{p'(\mathbf{r})\} = F(\mathbf{R})*G(-\mathbf{R}) \\ P'(\mathbf{U}) &= \int_{\infty} F(\mathbf{R})G(\mathbf{R}-\mathbf{U})\mathrm{d}\mathbf{R}^3 \end{aligned} \tag{C.12}$$

Reciprocally, the transform of $P(\mathbf{U})$ is written

$$\begin{aligned} p'(\mathbf{u}) &= \text{FT}\{P'(\mathbf{U}) = F(\mathbf{R})G(-\mathbf{R})\} = f(\mathbf{r})*g(-\mathbf{r}) \\ p'(\mathbf{u}) &= \int_{\infty} f(\mathbf{r})g(\mathbf{r}-\mathbf{u})\mathrm{d}\mathbf{r}^3 1 \end{aligned} \tag{C.13}$$

Result. For a given value $\mathbf{u}$ of the auxiliary variable, the convolution product $p(\mathbf{u})$ is the sum of all products of the values of function f at points $\mathbf{r}$ by the values of function g at points $(\mathbf{r}-\mathbf{u})$, when $\mathbf{r}$ traces all points in direct space.

As a result, function $p(\mathbf{u})$ goes through a maximum any time the vector $\mathbf{u}$ connects a maximum of function g to a maximum of function f.

The inverse operation to convolution is *deconvolution*. It consists of extracting functions $f(\mathbf{r})$ and $g(\mathbf{r})$, while knowing $p(\mathbf{u})$ or $p'(\mathbf{u})$.

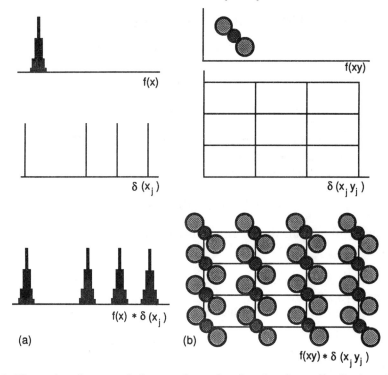

Figure C.1. Illustrating the convolution product of a function by a distribution. (a) One-dimensional distribution; (b) periodic two-dimensional distribution. Display of a crystal projection as a periodic distribution of the projections of the structure function along the lattice points.

Properties

A convolution product has similar properties to a conventional product, i.e. it is commutative, distributive, associative, i.e.

$$f*g = g*f$$
$$f*(ag + bh) = a(f*g) + b(f*h) \quad \text{(C.14)}$$
$$f*(g*h) = (f*g)*h$$

Convolution of a Continuous Function by a Distribution

Consider:

(a) a function $f(\mathbf{r})$ and its transform $F(\mathbf{R}) = \text{FT}\{f(\mathbf{r})\}$;
(b) a distribution $\delta(\mathbf{r}_j)$ and its transform $\Delta(\mathbf{R}) = \text{FT}\{\delta(\mathbf{r}_j)\}$.

$$\Delta(\mathbf{R}) = \text{FT}\{\delta(\mathbf{r}_j)\} = \sum_j \exp(2\pi i \mathbf{r}_j \mathbf{R}) \quad \text{(C.15)}$$

For a distribution, function $\delta(\mathbf{r}_j)$ is zero for any values of $\mathbf{r}$, except for discrete values $\mathbf{r}_j$ which determine the point positions and where it becomes infinite. The Fourier integral therefore becomes a three-dimensional series.

Basic Properties of Fourier Transforms

According to the results above, carrying out the convolution

$$p'(\mathbf{u}) = f(\mathbf{r}) * \delta(\mathbf{r}_j)$$

is equivalent to translating function $f(\mathbf{r})$ along vectors that define the points of distribution $\delta(\mathbf{r}_j)$ (Figure C.1).

Applications of Convolution

Crystal Function. A crystal may be represented as a *structure function* translated along a three-dimensional *lattice*. It may therefore be expressed as the convolution product of the structure function $f(\mathbf{r})$ by the lattice point distribution $\delta(\mathbf{r}_n)$, as outlined in Figure C.1.

Optical Imaging. The diffraction function at a point of the back focal plane of an objective lens (see Ch. 19) is the product of the object diffraction function and the entrance pupil diffraction function.

The image function, as observed in the Gaussian plane, is the FT of the diffraction function. It may thus be expressed as the convolution of the *stigmatic image* (FT of the object diffraction function, i.e. identical to the object) by the *confusion function* (FT of the aperture diffraction function, i.e. image of a point through the objective lens). Any stigmatic image point is thus replaced by the confusion pattern (Figure C.2). The width of this pattern determines the resolution of an optical system.

In a similar way the actually observed emission spectrum, the output of a spectrometer, represents the convolution product of the emitted line spectrum and the instrumental dispersion function of the spectrometer (e.g. Figure 10.8).

Operating a *deconvolution* theoretically enables extraction of the stigmatic image or the physical spectrum from the instrumental data, provided that the instrumental effect can be converted into a mathematical function (e.g. objective lens transfer function, see Ch. 19).

C.1.7 Convolution Square

Consider the convolution product provided when replacing G by F in Eq. (C.13):

$$q(\mathbf{u}) = \mathrm{FT}\{F(\mathbf{R})F(-\mathbf{R})\} = f(\mathbf{r}) * f(-\mathbf{r})$$

$$q(\mathbf{u}) = \int_\infty f(\mathbf{r})f(\mathbf{r}-\mathbf{u})\mathrm{d}\mathbf{r}^3$$

(C.16)

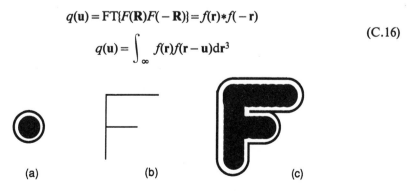

Figure C.2. Outline of the image provided by an objective lens. (a) image of a point (Airy pattern); (b) stigmatic image; (c) observed image, convolution of (a) by (b).

The function $q(\mathbf{u})$ is called a convolution square.

Result. The convolution square $q(\mathbf{u})$ goes through a maximum each time $\mathbf{u}$ equals a vector connecting two maxima of function $f(\mathbf{r})$, in either direction. Function $q(\mathbf{u})$ is thus centrosymmetric, i.e. to any vector $+\mathbf{u}$ corresponds a vector $-\mathbf{u}$.

The convolution square of a solid structure is a function with maxima determined by interatomic vectors, i.e. a *distribution function* (see Figures 3.1 and 5.11). Because of the fact that it is expressed by the FT of scattered intensity, it is accessible to measurement, whatever the nature of material, crystalline or non-crystalline.

BIBLIOGRAPHY

[1] Duffieux P. M. *The Fourier transform and its applications to optics*. 2nd edn. John Wiley, New York, Chichester, 1982.
[2] Lipson H. S., Taylor C. A. *Fourier transforms and X-ray diffraction*. Bell, London, 1958.
[3] Roddier F. *Distributions et transformée de Fourier*. Ediscience, Paris, 1971.
[4] Steward E. G. *Fourier optics. An introduction*, 2nd edn. John Wiley, New York, Chichester, 1987.
[5] Woolfson M. M. *An introduction to X-ray crystallography*. Cambridge University Press, 1970.

Appendix D

Spectrometric Tables

D.1 COMPARATIVE PERFORMANCES OF SPECTROMETRIC ANALYSING TECHNIQUES

The main performances of analysing techniques, dealt with in this book, are surveyed in Table D.1, and defined by the following quantities (see Ch. 13 and Appendix A):

c_{min} minimum detectable mass fraction
d diameter of analysing area
t depth of analysis or specimen thickness
v analysing volume
N_{min} minimum detectable number of atoms
M_{min} minimum detectable mass

Table D.1. Abbreviations are explained in Appendix E. The values displayed in this table are mere orders of magnitude, provided by current facilities, in good operating conditions. They may be improved by special layouts, e.g. by using high-intensity primary radiation (synchrotron radiation in the case of X-rays). They are liable to progress with the development of new techniques

Method	Ch.	c_{min}	d (μm)	t (μm)	v (μm^3)	N_{min}	M_{min} (g)	Z	Accuracy (%)
XRF	14	10^{-6}	10^4	10^3	10^{11}	10^{15}	10^{-7}	>11	0.1
EPMA	15	10^{-4}	1	1	1	10^7	10^{-15}	>4(WDS) >11(EDS)	1
EDS (thin specimen)	15	5×10^{-3}	2×10^{-2}	5×10^{-2}	2×10^{-5}	10^3	10^{-19}	>11	5–15
XPS	16	10^{-3}	10^3	10^3	10^3	10^{10}	10^{-12}	>3	10–30
AES-E	16	10^{-2}	1	10^{-3}	10^{-3}	10^6	10^{-16}	>3	10–30
EELS	17	10^{-4}	2×10^{-2}	5×10^{-2}	2×10^{-5}	10^2	10^{-20}	All	5–30
SIMS	18	10^{-8}	10^2	10^{-4}	1	10^{16}	10^{-8}	All	10–30

D.2 ELECTRON BINDING ENERGIES IN ATOMS

Line positions in characteristic emission or absorption spectra, the bases of qualitative analysis, are determined by electron energy levels (or binding energies) in the atoms of the specimen (see section 7.1). Knowing those values as accurately as possible is thus a preliminary condition for any elemental analysis.

Table D.2. Electron binding energies in electronvolts for elements up to $Z=30$ and some heavier elements (for the latter, only the K, L and M levels are displayed). The values correspond to Siegbahn in [1]

Z Element	1s 1/2 K	2s 1/2 L1	2p 1/2 L2	2p 3/2 L3	3s 1/2 M1	3p 1/2 M2	3p 3/2 M3	3d 3/2 M4	3d 5/2 M5
1 H	14								
2 He	25								
3 Li	55								
4 Be	111								
5 B	188			5					
6 C	284			7					
7 N	399			9					
8 O	532	24		7					
9 F	686	31		9					
10 Ne	867	45		18					
11 Na	1072	63		31	1				
12 Mg	1305	89		52	2				
13 Al	1560	118	74	73	1				
14 Si	1839	149	100	99	8	3			
15 P	2149	189	136	135	16	10			
16 S	2472	229	165	164	16	8			
17 Cl	2823	270	202	200	18	7			
18 A	3203	320	247	245	25	12			
19 K	3608	377	297	294	34	18			
20 Ca	4038	438	350	347	44	26		5	
21 Sc	4493	500	407	402	54	32		7	
22 Ti	4965	564	461	455	59	34		3	
23 V	5465	628	520	513	66	38		2	
24 Cr	5989	695	584	575	74	43		2	
25 Mn	6539	769	652	641	84	49		4	
26 Fe	7114	846	723	710	95	56		6	
27 Co	7709	926	794	779	101	60		3	
28 Ni	8333	1008	872	855	112	68		4	
29 Cu	8979	1096	951	931	120	74		2	
30 Zn	9659	1194	1044	1021	137	87		9	
41 Nb	18986	2698	2465	2371	469	379	363	208	205
42 Mo	20000	2866	2625	2520	505	410	393	230	227
45 Rh	23220	3412	3146	3004	627	521	496	312	307
74 W	69525	12099	11542	10205	2820	2575	2281	1872	1810

Spectrometric Tables

A direct and accurate measurement of energy levels has been made possible by high-resolution photoelectron spectrometry (see 16.1). Table D.2 provides an extract of the data as measured by Siegbahn [1].

These tables lead to direct determination of the positions of photoelectron emission lines (ESCA, XPS), of X-ray absorption edges (XAS, EXAFS) and of electron energy loss edges (EELS).

They allow calculation of energies or wavelengths of characteristic X-ray and Auger emission lines, for any possible transitions (see section 7.4). Complete tables of emission lines of all elements are to be found in specialised treatises (e.g. for X-rays in the ASTM tables [2]). The bibliographies of specialised chapters give further information.

BIBLIOGRAPHY

[1] Siegbahn K. *E.S.C.A.—Atomic, molecular and solid state structure studied by means of electron spectroscopy*. Almquist-Wicksells, Uppsala, 1967.

[2] *X-ray emission line wavelength and two theta tables*. ASTM Data series DS 37. Philadelphia, 1965.

Appendix E

Abbreviations and Acronyms

E.1 ABBREVIATIONS

The following abbreviations are made use of in this book. Some of them are used throughout the various chapters, others are limited to a specific field. Some have become widely used acronyms which could also figure in list E.2.

CMA	*cylindrical mirror analyser*
CRT	*cathode ray tube*
DS	*Debye–Scherrer* (powder camera)
ED	*equivalent dose* (radioprotection)
ES	*Ewald sphere* (more commonly *reflecting sphere*)
FOLZ	*first-order Laue zone*
FT	*Fourier transform*
FZ	*Fresnel zone*
GM	*Geiger–Müller* (radiation counter)
HOLZ	*high-order Laue zone*
HSA	*hemispheric analyser*
MCA	*multichannel analyser*
PM	*photomultiplier*
QF	*quality factor* (radioprotection)
SI	*Système International* (international system of units)
Si(Li)	*lithium-doped silicon semiconducting diode*
ZOLZ	*zero-order Laue zone*

E.2 ACRONYMS

The development of a great number of various techniques based on radiation–matter interactions has led to the advent of numerous acronyms to abbreviate their designation.

Only designations which have a direct or indirect link with the field covered by the book have been listed.

AEM	*Auger electron microscopy*; this designation is sometimes used for *analytical electron microscopy*, leading to ambiguities
AES	*Auger electron spectrometry*; sometimes specified:
AES-I	*ion excitation*
AES-E	*electron excitation*
AES-X	*X-ray excitation*
AFM	*atomic force microscopy*
CBD (CBED)	*convergent beam electron diffraction*
EBIC	*electron beam induced current*
ED	*electron diffraction*
EDS (EDX)	*energy dispersive X-ray spectrometry*
EELS	*electron energy loss spectrometry*
EMMA	*electron microscope microanalyser*
EPMA	*electron probe microanalysis*
ESCA	*electron spectroscopy for chemical analysis*
EXAFS	*extended X-ray absorption fine structure*
EXELFS	*extended electron energy loss fine structure*
FEEM	*field electron emission microscopy*
FIM	*field ion microscopy*
HEED	*high-energy electron diffraction*
HREM	*high-resolution electron microscopy*
HVEM	*high-voltage electron microscopy*
IMMA	*ion microprobe mass analysis*
LEED	*low-energy electron diffraction*
LEELS	*low-energy electron loss spectrometry*
PIXE	*particle-induced X-ray emission*
PSL	*photostimulable luminescence*
RHEED	*reflection high-energy electron diffraction*
SAM	*scanning Auger electron microscopy*
SEELS	*slow electron energy loss spectrometry* (or LEELS for *low energy . . .*)
SEM	*scanning electron microscopy*
SEXAFS	*surface EXAFS*
SIMS	*secondary ion mass spectrometry*
SNMS	*secondary neutral mass spectrometry*
STEM	*scanning transmission electron microscopy*
STIPE	*scanning tunnelling inverse photoemission*
STM	*scanning tunnelling microscopy*
STOM	*scanning tunnelling optical microscopy*
TEM	*transmission electron microscopy*
TOF	*time of flight (mass spectrometer)*
UPS	*ultraviolet photoelectron spectrometry*
WDS (WDX)	*wavelength dispersive X-ray spectrometry*
XANES	*X-ray absorption near edge structure*
XAS	*X-ray absorption spectrometry*
XPS	*X-ray photoelectron spectrometry*
XRD	*X-ray diffraction*
XRFA (XRF)	*X-ray fluorescence analysis*

Appendix F

Radioprotection

F.1 BIOLOGICAL ACTION OF RADIATION USED IN MATERIAL ANALYSIS

Among the various types of radiation used for material analysis, in facilities surveyed in this book, the only ones which may be directly dangerous for humans are X-rays and neutrons.

Primary radiation effects take place by atom ionisation in cells, directly with X-rays (Ch. 7), indirectly through nuclear reactions with neutrons (see section 10.4). This primary effect is followed by secondary effects through generation of free radicals reacting with molecules in the tissues.

Moderate radiation doses result in an alteration of the blood formula, notably in a decrease of the number of erythrocytes (red blood cells). A massive local irradiation leads to radiodermatitis. A general irradiation results in more or less severe lesions, depending on the doses, with a time of latence ranging from a few days to several years.

Due to their low absorption by material shields, *thermal neutrons* pose difficult protection problems. However, neutron diffraction facilities are localised in a few specialised laboratories, where any protection rules have been fulfilled.

The same statement can be made for *synchrotron X-ray sources*.

Conventional X-ray sources (*X-ray tubes*) are common in any laboratories dealing with material analysis. X-rays used for this aim are usually more dangerous than medical X-rays, due to their lower energy, i.e. their higher absorption by human tissues.

Electrons are not dangerous by themselves. They are strongly absorbed by the thinnest vacuum enclosure where they are necessarily confined. It should, however, be taken into account that any electron impact (e.g. on specimens, on apertures, on photographic film) generates X-rays with energies up to the electron energy. The X-ray sources which are to be shielded in an efficient way are therefore:

(a) any X-ray facilities for diffraction and spectrometry (XRD, XRFA, XPS, EXAFS, etc.):
(b) any high-energy electron facilities (TEM, SEM, HEED, RHEED, EELS, EXELFS, etc.), and more particularly high-voltage facilities (HVEM).

Low-energy electrons (e.g. in LEED, AES, XPS) generate readily absorbed soft X-rays.

F.1.1 Radiation Doses

Various quantities are used to express radiation doses.

Exposure Dose

The exposure dose is defined as the total electric charge (of one sign) generated by ionisation in a mass unit of matter. It is used for X-rays and γ-rays. The exposure unit is the **röntgen** (R), defined for dry air. Its value amounts to

$$1\,R = 2.58 \times 10^{-4}\,C/kg\ \text{dry air}$$

Absorbed Dose

The absorbed dose D is defined as the energy absorbed by the unit mass of the irradiated medium.

The basic unit is the **rad** (radiation absorbed dose), with the following value with respect to the SI energy unit:

$$1\,\text{rad} = 10^{-2}\,J/kg = 100\,erg/g$$

A submultiple is the **millirad** (1 mrad = 10^{-3} rad).

A more recently introduced SI unit is the **gray** (which could also be called *hectorad*: 1 gray = 100 rad).

Equivalent Dose

This is the most commonly used quantity for protection purposes. For the same energy, radiation of different types results in different effects on the human tissues. The equivalent dose (ED) is defined with respect to the dose of X-rays generating a similar effect. It is expressed as the product of the absorbed dose and a *quality factor QF* specific to the considered radiation:

$$ED = D \times QF$$

The basic ED unit is the **rem** (*röntgen equivalent for man*), defined as follows:

$$1\,\text{rem} = 1\,\text{rad} \times FQ$$

A submultiple is the **millirem** (1 mrem = 10^{-3} rem).

A more recently introduced SI unit is the **sievert** (which could also be called *hectorem*: 1 sievert = 100 rem).

$$1\,\text{sievert} = 1\,\text{gray} \times FQ$$

Values of the Quality Factor

$FQ = 1$ X-rays, γ-rays, high-energy β-rays
$FQ = 1\text{--}10$ neutrons, low-energy β-rays (depending on energy)
$FQ = >10$ α-rays

For X-rays, the biological dose in sieverts is thus equal to the absorbed dose in grays.

Limit Dose

Radiation doses are additive over about a year. Admissible limit doses have been defined on an international scale. They depend on the professional category and the exposed parts of the body.

Examples of limit doses usually allowed for laboratory workers are as follows:

(a) for the entire body: 5 rem/year; 2.5 mrem/h;
(b) for hands only: 60 rem/year; 30 mrem/h.

The limit dose generally allowed for ordinary persons is 500 mrem/year, i.e. 5×10^{-3} sieverts/year.

For comparison, it may be noted that the average natural dose of radioactivity collected by a human body is some 80–100 mrad/year. However, depending on the region inhabited, this dose may be multiplied by a factor of 5 in granitic zones and may reach up to several tens of rads (i.e. up to 1000 times the mean value) in the vicinity of uranium-rich rocks).

In addition to this natural dose, an average dose of 60–100 mrad is currently added by medical radiology.

F.2 PROTECTIVE MEASURES

Protection against radiation in laboratory facilities may be either in the form of passive protection through shielding or in the form of active protection by the operators themselves.

F.2.1 Passive Protection. Radiation Shielding

Current laboratory protection is implemented by shielding any parts of a facility liable to generate X-rays, by means of highly absorbing materials (e.g. metallic lead for X-ray tubes, lead or barium glass for electron microscope observing windows and CRT viewing screens).

The European so-called total protection standard corresponds to a residual dose smaller than 0.75 mrad/h in the direct vicinity of the X-ray emitting facility.

F.2.2 Active Protection

Informing and educating people who are working at X-ray producing facilities should complement passive protection. It seems obvious that any direct contact with a primary X-ray beam is to be prohibited (eyes are particularly sensitive). Even an exposure to scattered radiation is dangerous when prolonged. Whereas commercial facilities are subjected to severe protection standards and are therefore reasonably secure, laboratory-built or modified facilities may be dangerous. The law of variation of radiation intensity according to the inverse squared distance should likewise be taken into account for locating an X-ray facility. Inquiries have shown that the lack of information for laboratory workers about radiation effects is a prominent source of X-ray irradiation accidents.

Education is to be completed by periodic check-ups of any persons in contact with X-ray producing facilities, similar to those implemented in nuclear centres, i.e. carrying of personal dosimeters, checked every month, and analysing of the blood formula once a year.

References

In addition to scientific journals, recent developments in structural and chemical analysis of materials using the foregoing techniques may be found in the proceedings of specialised conferences, for instance in: *European Congress on Electron Microscopy, EUREM; International Congress on Electron Microscopy; International Union of Crystallography; Scanning microscopy; SIMS Conferences; X-Ray Optics and Microanalysis.*

ALBRECHT T. R., QUATE C. F. (1987) Atomic resolution imaging of a non conductor by atomic force microscopy. *J. Appl. Phys.*, **62**, 2599–602.

AMOURIC M., MERCURIOT G., BARONNET A. (1981) On computed and observed HRTEM images of perfect mica polytypes. *Bull. Minéral.*, **104**, 298.

ANDERSON C. A., HINTHORNE J. R. (1973) Thermodynamic approach to the quantitative interpretation of sputtered ion mass spectra. *Analyt. Chem.*, **49**, 2023.

BACON G. E. (1952) A neutron-diffraction study of magnesium aluminium oxide. *Acta Crystal*, **5**, 684.

BADAUT D., RISACHER F. (1979) Authigenic smectite on diatom frustules in Bolivian saline lakes. *Geoch. Cosmoch. Acta*, **47**, 363.

BALK L. J. (1988) Scanning acoustic microscopy. *Adv. in Electronic and Electron Phys.*, **71**, 1–73.

BALTZINGER C., COUSANDIER R., BURGGRAF C. (1983) High resolution photoelectron spectroscopy applied to characterization of iron compounds in surface layers. *Acta Univ. Wratislaviensis*, **455**, 43.

BETHE H. (1930) Zur Theorie des Durchgangs schneller Korpuskularstrahlen durch Materie. *Ann. Physik.*, **5**, 49.

BARONNET A., ONRUBIA Y. (1988) Correlations entre microcomposition et défauts d'intercalation dans les silicates en couches. *J. Microsc. Spectrosc. Electron.*, **13**, 385.

BETHE H. (1930) Zur Theorie des Durchgangs schneller Korpuskularstrahlen durch Materie. *Ann. Physik.*, **5**, 49.

BETHE H. (1933) Quantenmechanik der Ein- und Zweielektronenprobleme. *Hb. Phys.*, **24/1**, 273.

BINH V. T. (1988) Characterization of microtips for STM. *Surf. Science*, **202**, L 539.

BINNIG G., ROHRER H. (1983) Scanning tunneling microscope. *Surf. Science*, **126**, 236.

BINNIG G., ROHRER H. (1985) Le microscope à balayage à effet tunnel. *Pour la Science*, No. 96, 22.

BINNIG G., ROHRER H. (1987) Scanning tunneling microscopy—from birth to adolescence. *Rev. Mod. Phys.*, **59**, 615.

BINNIG G., QUATE C. F., GERBER C. (1986) Atomic force microscope. *Phys. Rev. Lett.*, **56**, 930.

BINNING G., ROHRER H., GERBER C., WEIBEL E. (1982a) Tunneling through a controlable vacuum gap. *Appl. Phys. Lett.*, **40**, 178.

BINNIG G., ROHRER H., GERBER C., WEIBEL E. (1982b) Surface studies by scanning tunneling microscopy. *Phys. Rev. Lett.*, **49**, 57.

BINNIG G., ROHRER H., GERBER C., WEIBEL E. (1983) 7×7 reconstruction on Si(111) resolved in real space. *Phys. Rev. Lett.*, **50**, 120.

BISHOP H. E. (1965) Some electron backscattering measurements for solid targets. *X-Ray Optics and Microanal.*, 4th Int. Cong., Orsay 1965, p. 153.

BISHOP H. E. (1967) Electron scattering in thick targets. *J. Appl. Phys.*, **18**, 703.

BOERSCH H. (1966) Inelastic scattering of electrons in solids. *6th Int. Cong. Electron Microsc., Kyoto 1966*, **1**, 9.

BOUVY G. (1972) Spectrométrie d'électrons ESCA. *J. de Microscopie*, **14**, 235.

BRESSE J. F. (1981) Courant induit (EBIC) et contraste de potentiel dans les dispositifs semi-conducteurs. *J. Microsc. Spectrosc. Electron.*, **6**, 17.

BROGLIE L. DE (1924) Recherches sur la théorie des quanta. Thèse de Doctorat, Univers. Paris (publ. 1925, *Ann. Phys.*, **3**, 22).

BROLL N. (1985a) Analyse quantitative par fluorescence X. Théorie et pratique de la méthode des coefficients d'influence fondamentaux. Thèse de Doctorat, Univ. Louis Pasteur, Strasbourg.

BROLL N. (1985b) Fluorescence X. Amélioration de la sensibilité de mesure et de détection des éléments légers. *C. R. Colloque. Rayons X, Grenoble*, p. 284.

BROLL N. (1986) Quantitative X-ray fluorescence analysis. Theory and practice of the fundamental coefficient method. *X-Ray Spectrom.*, **15**, 271.

BROLL N., TERTIAN R. (1983) Quantitative X-ray fluorescence analysis by use of fundamental influence coefficients. *X-Ray Spectrom.*, **12**, 30.

BURGGRAF C., GOLDSZTAUB S. (1962) Dispositif de mesure des intensités des faisceaux électroniques diffractés. *J. de Microscopie*, **1**, 411.

BUXTON B. F., EADES J. A., STEEDS J. W., RACKHAM G. M. (1976) The symmetry of electron diffraction zone axis pattern. *Phil. Trans.*, **281**, 171.

CARRIERE B., DEVILLE J. P. (1983) Tests de calibration d'analyseurs XPS. *Le Vide*, **215**, 119.

CARRIERE B., DEVILLE J. P., HUMBERT P. (1985) Les informations chimiques obtenues par spectroscopie Auger. *J. Microsc. Spectrosc. Electron.*, **10**, 29.

CASTAING R. (1951) Application des sondes électroniques à une méthode d'analyse ponctuelle chimique et cristallographique. Thèse de doctorat, Univ. Paris, publ. ONERA, No. 75.

CASTAING R. (1960) Electron probe microanalysis. *Adv. Electron. Phys.*, **13**, 317.

CASTAING R., DESCAMPS J. (1955) Sur les bases physiques de l'analyse ponctuelle par spectrographie X. *J. Phys. Radium*, **16**, 304.

CASTAING R., HENOC J. (1966) Répartition en profondeur du rayonnement caractéristique. *X-Ray Optics and Microanal., 4th Int. Cong., Orsay 1965*. Hermann, Paris, p. 120.

CASTAING R., HENRY L. (1962) Filtrage magnétique des vitesses en microscopie électronique. *C.R. Acad. Sci.*, **255**, 76.

CASTAING R., SLODZIAN G. (1962) Microanalyse par émission ionique secondaire. *J. Microsc.*, **1**, 395.

CAZAUX J. (1985) Performance of electron spectroscopies in surface microanalysis. *J. Microsc. Spectrosc. Electron.*, **10**, 583.

CAZAUX J. (1988) Microscopie Auger. Résolution latérale et apport d'autres techniques. *J. Microsc. Spectrosc. Electron.*, **13**, 315.

CAZAUX J., CHAZELAS J., CHARASSE M. N., HIRTZ J. P. (1988) Line resolution in the subten nanometer range in SAM. *Ultramicroscopy*, **25**, 31.

CHAMPNESS P. E., CLIFF G., LORIMER G. W. (1981) Quantitative analytical electron microscope. *Bull Minéral.*, **104**, 236.

CLAISSE F. (1956) Accurate X-ray fluorescence analysis without internal standard. *Quebec Dept. Mines*, P.R. 327.

CLAISSE F., QUINTIN M. (1967) Generalization of the Lachance-Traill method for the correction of the matrix effect in X-ray fluorescence analysis. *Can. Spectrosc.*, **12**, 129.

CLIFF G., LORIMER G. W. (1975) The quantitative analysis of thin specimen. *J. Microsc.*, **103**, 203.

COLLIEX C. (1985) L'instrument STEM. Une introduction au microscope électronique à balayage en transmission. *J. Microsc. Spectrosc. Electron.*, **10**, 313.

COLLIEX C., JOUFFREY B. (1972) Diffusion inélastique des électrons dans un solide par excitation de niveaux profonds. *Phil. Mag.*, **25**, 491.

COOMBS J. H., GIMZEWSKI J. K., REIHL B., SASS J. K., SCHLITTLER R. R. (1988) Photon emission experiments with the scanning tunneling microscope. *IBM Research Report*.

COSSLETT V. E. (1969) Energy loss and chromatic aberration in electron microscopy. *Z. Angew. Phys.*, **27**, 138.

COUSANDIER R. (1988) Développement d'un ensemble d'étude de composés binaires en couches minces. Application à l'étude de l'alliage or–cuivre sur silicium (111). Thèse de Doctorat, Univ. Louis Pasteur, Strasbourg.

COWLEY J. M. (1969) Image contrast in a transmission scanning electron microscope. *Appl. Phys. Lett.*, **15**, 58.

COWLEY J. M., MOODIE A. F. (1957) The scattering of electrons by atoms and crystals. A new theoretical approach. *Appl. Phys. Lett.*, **15**, 58.

CREWE A. V. (1966) Experiments with a field emission electron gun. *6th Int. Cong. Electron Microsc., Kyoto 1966*, **1**, 629.
CREWE A. V., ISAACSON M., JOHNSON R. (1969) A simple scanning electron microscope. *Rev. Scient. Instrum.*, **40**, 241.
CREWE A. V., WALL J., LANGMORE J. (1970) Single atom visibility. *7th Int. Cong. Electron Microsc., Grenoble 1970*, **1**, 485.
CROVISIER J. L., EBERHART J. P. (1985) Apports de la microscopie électronique à l'observation des couches d'altération à la surface de verres. *J. Microsc. Spectrosc. Electron.*, **10**, 171.
CROVISIER J. L., EHRET G., EBERHART J. P., JUTEAU T. (1983) Altération expérimentale de verre basaltique tholéitique par l'eau de mer entre 3 et 50 °C. *Sci Géol. Bull.*, **2-3**, 187.
DEVILLE J. P. (1968) La spectroscopie des électrons AUGER. *Rev. Phys. Appl.*, **3**, 351.
DEVILLE J. P. (1972) Etude des micas au moyen de la diffraction des électrons lents et de la spectroscopie des électrons AUGER. Thèse de Doctorat, Univ. Louis Pasteur, Strasbourg.
DEVILLE J. P., GOLDSZTAUB S. (1969) Etude du clivage de la muscovite par spectrométrie des électrons AUGER. *C.R. Acad. Sci.*, **B268**, 629.
DEVILLE J. P., EBERHART J. P., GOLDSZTAUB S. (1967) Observations sur la diffraction des électrons de faible énergie par un cristal de mica muscovite. *C.R. Acad. Sci.*, **B264**, 289.
DRAKE B., PRATER C. B., WEISENHORN A. L., GOULD S. A. C., ALBRECHT T. R., QUATE C. F., CANNELL D. S., HANSMA H. G., HANSMA P. K. (1989) Imaging crystals, polymers and processes in water with the atomic force microscope. *Science*, **243**, 1586-9.
DRUMMOND J. W. (1984) The ion optics of low-energy ion beams. *Vacuum*, **34**, 51.
DUNCUMB P., MELFORD D. (1966) Quantitative applications of ultra-soft X-ray microanalysis in metallurgy. *X-Ray Optics and Microanal., 4th Int. Cong., Orsay 1965*. Hermann, Paris, p. 240.
DUNCUMB P., SHIELDS P. K. (1963) The present state of quantitative X-ray microanalysis. Physical basis. *J. Appl. Phys.*, **14**, 617.
DUNCUMB P., SHIELDS-MASON P. K., DA CASA C. (1969) Accuracy of atomic number and absorption corrections in electron probe microanalysis. *X-Ray Optics and Microanal., 5th Int. Cong., Tübingen 1968*. Springer, Berlin, p. 146.
DUPOUY G., PERRIER F., VERDIER P., ARNAL F. (1964) Transmission d'électrons monocinétiques à traverse des feuilles métalliques minces. *C.R. Acad. Sci.*, **258**, 3655.
DUPOUY G., PERRIER F., DURRIEU L. (1970) Microscope électronique à trois millions de volts. *J. de Microscopie*, **9**, 575.
EBERHART J. P., TRIKI R. (1972) Description d'une technique permettant d'obtenir des coupes minces de minéraux argileux par ultra-microtomie. Application à l'étude de minéraux argileux interstratifiés. *J. de Microscopie*, **15**, 111.
ECKHARDT F. J. (1961) Über die Anwendung von Ultramikrotomschnitten bei der elektronoptischen untersuchung von Tonen. *Z. Krist.*, **116**, 36.
EGERTON R. F. (1976) Inelastic scattering and energy filtering in the transmission electron microscope. *Phil. Mag.*, **34**, 49.
EHRET G., CROVISIER J. L., EBERHART J. P. (1986) A new method for studying leached glasses: analytical electron microscopy on ultramicrotomic thin sections. *J. Non Cryst. Sol.*, **86**, 72.
ELLER G. VON (1955) Le photosommateur harmonique et ses possibilités. *B. Soc. Franç. Miner. Crist.*, **78**, 157.
FUKAMI A., ADACHI K., KATOH M. (1972) Microgrid techniques and their contribution to specimen preparation technique for high resolution work. *J. Microscopy*, **21**, 1.
GATINEAU L., MERING J. (1958) Précisions sur la structure de la muscovite. *Clay Miner. Bull.*, **3**, 238.
GERMER L. H., GOLDSZTAUB S., ESCARD J., DAVID G., DEVILLE J. P. (1966) Etude de l'intensité des électrons lents diffractés par le pyrographite. *C.R. Acad. Sci.*, **262**, 1059.
GOLDSZTAUB S., LANG B. (1963) Appareil métallique démontable pour la diffraction des électrons de faible énergie. *C.R. Acad. Sci.*, **257**, 1908.
GONI J., REMOND G. (1969) Localization and distribution of impurities in blende by cathodoluminescence. *Mineral. Mag.*, **37**, 153.
GOODMAN P., MOODIE A. F. (1974) Numerical evaluation on n-beam wave functions in electron scattering by the multislice method. *Acta Crystal.*, **A34**, 709.

HANSMA P. K., ELINGS K., MARTI O., BRACKER C. E. (1988) Scanning tunneling microscopy and atomic force microscopy. Applications to biology and technology. *Science*, **242**, 209–16.
HANSMA P. K., TERSOFF J. (1987) Scanning tunneling microscopy. *J. Appl. Phys.*, **61**, R1.
HEINRICH K. F. J. (1965) Electron probe microanalysis by specimen current measurement. *X-Ray Optics and Microanal., 4th Int. Cong., Orsay 1965.* Hermann, Paris, p. 159.
HENKE B. L. (1964) X-ray fluorescence analysis for sodium, fluorine, oxygen, nitrogen, carbon and boron. *Adv. X-Ray Anal.*, **7**, 460.
HENKE B. L. (1965) Some notes on ultrasoft X-ray fluorescence analysis, 10 to 100 Å region. *Adv. X-Ray Anal.*, **8**, 269.
HENKE B. L. (1966) Spectroscopy in the 15–150 Å ultrasoft X-ray region. *X-Ray Optics and Microanal., 4th Int. Cong., Orsay 1965.* Hermann, Paris, p. 440.
HENOC J., MORICEAU M., TIXIER R., TONG M. (1971) Caractéristiques comparées de quelques programmes de correction. *J. de Microscopie*, **12**, 12.
HEWITT P. J., JEFFERSON D. A., MILLWARD G. R., TSUNO K. (1989) A new high resolution analytical stage for the JEM-200 CX. *Jeol News*, **27E** (1), 2.
HILLER J., BAKER R. F. (1944) Microanalysis by means of electrons. *J. Appl. Phys.*, **15**, 663.
HOCHELLA M. F., EGGLESTON C. M., ELINGS V. B., PARKS G. A., BROWN G. E., WU C. M., KJOLLER K. (1989) Mineralogy in two dimensions: scanning tunneling microscopy of semiconducting minerals with implications for geochemical reactivity. *Amer. Miner.*, **74**, 1233–46.
HOWIE A., WHELAN M. J. (1961) Diffraction contrast of electron microscope images of crystal lattice defects. 2. The development of a dynamical theory. *Proc. Roy. Soc.*, **A263**, 217.
JONG W. F. DE, BOUMAN J. (1938) Das Photografieren von reziproken Kristallnetzen mittels Röntgenstrahlen. *Z. Krist.*, **98**, 456.
KATZ W., NEWMAN J. G. (1987) Fundamentals of secondary ion mass spectrometry. *M.R.S., Bull.*, **12**, 40.
KLEIN J., GAUTHIER S., ROUSSET S., SACKS W. (1987) Expériences récentes en microscopie tunnel. *J. Microsc. Spectrosc. Electron.*, **12**, 403.
KNOLL M., RUSKA E. (1932) Das Elektronenmikroskop. *Z. Phys.*, **78**, 318.
KOSSEL W., MOELLENSTEDT G. (1939) Elektronen-Interferenzen im konvergenten Bündel. *Ann. Physik.*, **36**, 113.
LACHANCE G. R., TRAILL R. J. (1966) A practical solution to the matrix problem in X-ray analysis. 1. Method. 2. Application to a multicomponent alloy system. *Canad. Spectrosc.*, **11**, 43, 63.
LANDAU L. (1944) On the energy loss of fast particles by ionization. *J. Phys. USSR*, **8**, 201.
LANDER J. J. (1953) Auger peaks in the energy spectra of secondary electrons from various materials. *Phys. Rev.*, **91**, 1382.
LANGERON J. P. (1988) Distribution énergétique des électrons rétrodiffusés en microanalyse Auger. *J. Microsc. Spectrosc. Electron.*, **13**, 331.
LENZ F. (1954) Zur Streuung mittelschneller Elektronen in kleinste Winkel. *Z. Naturforsch.*, **9a**, 185.
LORIMER G. W. (1987) Quantitative X-ray microanalysis of thin specimen in the transmission electron microscope; a review. *Mineral. Mag.*, **51**, 49.
LORIMER G. W., CLIFF G. (1984) Quantitative X-ray analysis of thin specimen. In *Quantitative Electron Microscopy, Proc. Scott. Summer School in Phys.*, p. 305.
MACEWAN D. M. C., RUIZ AMIL A., BROWN G. (1961) Interstratified clay minerals. In *X-ray Identification and Crystal Structures of Clay Minerals*, Miner. Soc. London, p. 393.
MAUGUIN C. (1928) Etude des micas au moyen des rayons X. *B. Soc. Franç. Miner. Crist.*, **51**, 285.
MENTER, J. W. (1958) Observations on crystal lattices and imperfections by transmission electron microscopy through thin films. *Proc. 4th Int. Congr. Electron Microsc.*, **1**, 320.
MERING J., OBERLIN A. (1967) Electron-optical study of smectites. *Clays Clay Miner., Proc. 15th Conf., Pittsburgh*, p. 3.
METHEREL A. J. F. (1973) Diffraction of electrons by perfect crystals. *Electron Microscopy in Material Science, Int. Scool Electron Microsc., Ettore Majorana.*

NORRISH K., TAYLOR R. M. (1962) Quantitative analysis by X-ray diffraction. *Clay Miner. Bull.*, **5**, 98.

OBERLIN A., TERRIERE G. (1972) Utilisation des techniques de contraste de diffraction dans l'étude d'un mélange de phases carbonées très désorganisées. *J. de Microscopie*, **14**, 1.

O'KEEFE M. A., BUSECK P. R., IIJIMA S. (1978) Computed crystal images for high-resolution electron microscopy. *Nature*, **274**, 322.

PALMBERG P. W., RHODIN T. N. (1968) Auger electron spectroscopy of fcc metal surfaces. *J. Appl. Phys.*, **39**, 2425.

PALMBERG P. W., BOHN C. K., TRACY J. C. (1969) High sensitivity AUGER electron spectrometer. *Appl. Phys. Lett.*, **15**, 254.

PETERSON S. W., LEVY H. A. (1953) A single crystal neutron diffraction study on heavy ice. *Phys. Rev.*, **92**, 1082.

PHILIBERT J. (1963) A method for calculating the absorption correction in electron-probe microanalysis. *X-Ray Optics and Microanal., 3rd Int. Cong., Stanford, 1962*. Academic Press, New York, p. 379.

PHILIBERT J., WEINRYB E. (1962) Etude de l'émission électronique secondaire dans la microsonde de Castaing. *C.R. Ac. Sc.*, **255**, 2757.

PREWETT P. D., JEFFERIES D. K., COCKHILL T. D. (1981) Liquid metal sources of gold ions. *Rev. Scient. Instrum.*, **52**, 562.

PREWETT P. D., JEFFERIES D. K., MCMILLAN D. J. (1984) A liquid metal source of caesium ions for secondary ion mass spectrometry. *Vacuum*, **34**, 107.

RIECKE W. D. (1961) Über die Genauigkeit der Übereinstimmung von ausgewähltem und beugendem Bereich bei der Feinbereichs-Elektronenbeugung im Le Pooleschen Strahlengang. *Optik*, **18**, 278.

RIECKE W. D. (1962a) Feinstrahl-Elektronenbeugung mit dreistufigem Kondensor und langbrennweitiger letzter Kondensorstufe. *Optik*, **19**, 81.

RIECKE W. D. (1962b) Über eine neue Methode zur Herstellung von Elektronenbeugungsdiagrammen extrem kleiner, ausgewählter Bereiche aus elektronenmikroskopischen Durchstrahlungspräparaten. *Optik*, **19**, 273.

RIMSKY A. (1952) Appareil permettant la photographie directe de l'espace réciproque. *B. Soc. Franç. Miner. Crist.*, **75**, 500.

ROSE H. J., ADLER I., FLANAGAN F. J. (1962) Use of La_2O_3 as a heavy absorber in the X-ray fluorescence analysis of silicate rocks. *B. Geol. Surv. USA*, **450B**, 80.

ROSE J. H., ADLER I., FLANAGAN F. J. (1963) X-ray fluorescence analysis of the light elements in rocks and minerals. *Appl. Spectrosc.*, **17**, 81.

ROSE H. J., CUTTITTA F., LARSON R. R. (1965) Use of X-ray fluorescence in determination of selected major constituents in silicates. *B. Geol. Surv., USA*, **525B**, 155.

RUCH C. (1985) Energy dispersive X-ray fluorescence spectrometry with direct excitation at picogram levels. *Analyt. Chem.*, **57**, 1691.

RUCH C. (1986) Analyse d'éléments traces par fluorescence X. Nouveaux développements. Thèse de Doctorat, Univ. Louis Pasteur, Strasbourg.

RUSKA E. (1934) Über Fortschritte im Bau und in der Leistung des magnetischen Elektronenmikroskopes. *Z. Phys.*, **87**, 580.

RUSKA E. (1987) The development of the electron microscope and electron microscopy. *Rev. Mod. Phys.*, **59**, 627.

RUSTE J. (1979) Principes généraux de la microanalyse quantitative appliquée aux éléments légers. *J. Microsc. Spectrosc. Electron.*, **4**, 123.

SAREL H. Z. (1967) Cylindrical capacitor as an analyser. *Rev. Scient. Instrum.*, **38**, 1210.

SCHEIBNER E. J., THARP L. N. (1967) Inelastic scattering of low energy electrons from surfaces. *Surf. Science*, **8**, 247.

SCHERZER O. (1949) The theoretical resolution of the microscope. *J. Appl. Phys.*, **20**, 20.

SCHNEIR J., HANSMA P. K. (1987) Scanning tunneling microscopy and lithography of solid surfaces covered with non polar liquids. *Langmuir*, **3**, 1025-7.

SHERMAN J. (1955) The theoretical derivation of fluorescent X-ray intensities from mixtures. *Spectrochem. Acta*, **7**, 283.

SLODZIAN G. (1964) Etude d'une méthode d'analyse locale chimique et isotopique utilisant l'émission ionique secondaire. *Ann. Physique*, **13**, 9.

SMITH D. P. E., KIRK M. D., QUATE C. F. (1987) Molecular images and vibrational spectroscopy of sorbic acid with the scanning tunneling microscope. *J. Chem. Phys.*, **86**, 6034–8.

SONODA M., TAKANO M., MIYAHARA J., KATO H. (1983) Computed radiography utilizing scanning laser stimulated luminescence. *Radiology*, **148**, 833–8.

SPRINGER G. (1967) Investigations into the atomic number effect in electron probe microanalysis. *N. Jahrb. Miner. Monatsh.*, **9/10**, 304.

STADELMANN P. A. (1987) EMS. A software package for electron diffraction analysis and HREM image simulation in material science. *Ultramicroscopy*, **21**, 131–46.

STADELMANN P. A., BUFFAT P. A. (1989) Simulation of high resolution electron microscope images and 2-dimensional convergent beam electron diffraction patterns. In KRAKOW W. and O'KEEFE (eds), *Computer Simulation of Electron Microscope Diffraction and Images*, The Minerals, Metals and Materials Society, pp. 159–69.

STEEDS J. W., VINCENT R. (1983) Use of high symmetry zone axis in electron diffraction in determining crystal point and space groups. *J. Appl. Cryst.*, **16**, 317.

STERN E. A. (1974) Theory of the extended X-ray-absorption fine structure. *Phys. Rev.*, **B10**, 3027.

STERN E. A. (1982) Multielectron contribution to near-edge X-ray absorption. *Phys. Rev. Lett.*, **49**, 1353.

STERN E. A., SAYERS D. E., LYTLE F. W. (1975) Extended X-ray absorption fine structure technique. Determination of physical parameters. *Phys. Rev.*, **B11**, 4836.

STORMS H. A., BROWN K. F., STEIN J. D. (1977) Evaluation of a caesium positive ion source for secondary ion mass spectrometry. *Analyt. Chem.*, **49**, 2023.

STUCK R., SIFFERT P. (1984) Secondary ion mass spectrometry (SIMS). *Progr. in Crystal Growth and Characterization*, **8**, 11.

TANAKA M., SAITO R., SEKII H. (1983a) Point group determination by convergent beam electron diffraction. *Acta Crystal.*, **A39**, 357.

TANAKA M., SEKII H., NAGASAVA T. (1983b) Space group determination by dynamic extinction in convergent beam electron diffraction. *Acta Crystal.*, **A39**, 825.

TATLOCK D. B. (1966) Rapid model analysis of some felsic rocks from calibrated X-ray diffraction patterns. *B. Geol. Surv. USA*, 1209.

TESSIER D. (1984) Etude expérimentale de l'organisation des matériaux argileux. Thèse de Doctorat, INRA, Paris.

THOMAS P. M. (1964) A method for correcting for atomic number effects in electron-probe microanalysis. *U.K. At. En. Author. Res. Group, Rept. R.*, p. 4593.

TIXIER R., PHILIBERT J. (1969) Analyse quantitative d'échantillons minces. *X-Ray Optics and Microanal., 5th Int. Cong., Tübingen 1968*. Springer, Berlin, p. 180.

UYEDA N., KOBAYASHI T., ISHIZUKA K., FUJIYOSHI Y. (1981) Improved molecular structure images of chlorinated copper phtalocyanine with atomic level resolution. *Jeol News*, **19E** (1), 2.

VERNET J. P., GAUTHIER A. (1962) La technique des coupes minces appliquée à l'étude de l'halloysite au microscope électronique. *C.R. Ac. Sc.*, **254**, 2608.

VICARIO E., PITAVAL M. (1972) Contraste de diffraction en microscopie à balayage. Principes et applications. *J. de Microscopie*, **13**, 296.

WAUGH A. R., BAYLY A. R., ANRERSON K. (1984) The application of liquid metal ion sources to SIMS. *Vacuum*, **34**, 103.

WEBER R. E., PERIA W. T. (1967) Use of LEED apparatus for the detection and identification of surface contaminants. *J. Appl. Phys.*, **38**, 4355.

WELDAY E. E., BAIRD A. K., MCINTYRE D. B., MADLEM K. W. (1964) Silicate sample preparation for light-element analysis by X-ray spectrography. *Amer. Mineral.*, **49**, 889.

WIEWIORA A. (1982) Oblique texture method in transmission X-ray diffractometry of clays and clay minerals. *9th Conf. Clay Miner. Petrol., Zvolen*, p. 43.

WILLIAMS P., LEWIS R. K., EVANS C. A., HANLEY P. R. (1977) Evaluation of a caesium primary ion source on an ion microprobe mass spectrometer. *Analyt. Chem.*, **49**, 1399.

WOOD E. A. (1964) Vocabulary of surface crystallography. *J. Appl. Phys.*, **35**, 1306.

YADA K. (1967) Study of chrysotile asbestos by a high resolution electron microscope. *Acta Crystal.*, **35**, 1306.

YASKO R. N., WHITMAYER R. D. (1971) Auger electron energies (0–2000 eV) for elements of atomic number 5–103. *J. Vac. Sci. Technol.*, **8**, 733.

ZHENG N. J., WILSON I. H., KNIPPING U., BURT T. M., KRINSLEY D. H., TSONG I. S. T. (1988) Atomically resolved scanning tunneling microscopy images of dislocations. *Phys. Rev.* **B38**, 12780–2.

Index

Main entries are printed in **bold**.

aberrations **123**, 396
absorbed current imaging 473
absorbed dose 530
absorption 10, **102**
absorption (electrons) **104**
absorption (neutrons) **107**
absorption (X-rays) **102**
   coefficient 102
   correction 291, 318
   edge **103**, 351
   factor 292, 318
   fine structure 103
   function 318, 354
accuracy of analysis 301, 321, 329
addition method (XRF) 299
adsorption layer 268
AFM (atomic force microscopy) **502**
Airy function 395
amplitude 506
analysing area 523
analysing volume 488, 523
analysis
   accuracy 301, 321, 329
   elemental 284, 304
   qualitative **276**, 287, 307, 367, 377
   quantitative **279**, 290, 314, 381, 471
   surface 339
analytical electron microscope 482, **488**
angular scanning 474
anomalous scattering 65, 212
astigmatism **124**, 399
atomic energy levels **8**, 88
atomic fraction 277
atomic number contrast 486
atomic scattering amplitude
   for electrons **44**
   for neutrons **46**
   for X-rays **42**
atomic scattering factor for X-rays **42**
atomic theory 7
Auger electron 88, **98**
Auger electron spectrometry (AES) 270, **341**, 488

Auger scanning microscopes (SAM) **344**, 471
Avogadro number 507

backscattered electrons 463, 469
backscattering
   factor **316**, 346
   yield 466, 473
beam coherence 127
bend fringes (contours) 260, 263, **411**
beta-filtering 104, **114**
Bethe relation 106
Boltzmann constant 507
bond
   lengths 362
   strengths 362
Bragg condition **51**, 227
Bragg fringes 260, 263, **411**
braking effect 19, 87, **91**
bremsstrahlung 19, 87, **91**
bright field image **405**, 486
Bunn chart 201
Burgers vector 418

carbon film 444
cathode sputtering 475
cathodoluminescence **472**, 488
CBD (convergent beam electron diffraction) 212, 213, **258**, 486
centrosymmetry 58
channeltron **151**
characteristic scattering angle 19
chemical shift 96, 100, 340, 345
chemical thinning 446
chromatic aberration **124**, 397
clay minerals 206, 209
cleavage 446
CMA (cylindrical mirror analyser) **151**, 337, 343
column approximation 409
composition contrast 466
Compton effect **14**, 89
concentration profiles
   depth 376

concentration profiles (*cont.*)
  surface 311, 344
condenser-objective lens **126**, 485
confusion function 431, **521**
contamination 265, 337, 374
contrast 431
convergent beam electron diffraction (CBD) 212, 213, **258**, 486
convolution **518**
  product 66
  square 66, **521**
coordination 352, 361, 362
core levels 13
correction algorithm (XRF) 300
counting statistics **158**
critical point sublimation 475
cross-over 120
cryoshield 402
crystal
  defects 243, **415**
  deformation 244
  lattice 48
  orientation 172
  potential projection 433
  structure **48**, 202, **211**, 250
  surface 268
  thickness 263, 415
cylindrical mirror analyser (CMA) **151**, 337, 343

dark field image **405**, 486
de Broglie wave 3
Debye–Scherrer camera 185
Debye–Waller temperature factor 47, 59
deconvolution 67, **521**
depth concentration profile 376, 385
depth of field 399, 470
depth of focus 401
detection limit **277**, 287, 309, 339, 368, 379, 488, 490, 523
detector parameters 32
deviation parameter 85
diamond-type glide plane 73
diffraction 48
  constant 230, 248, 254, **516**
  contrast 404
  deviation **63**, 78
  domains 62
  double 237
  symbol 213
dilution method (XRF) **296**, 307
Dirac peaks 60
direct
  lattice 48
  space 29
dislocations 418
disorder 207, 224, 361

displacement effect 17
displacement energy 449
displacement vector 67, 244, 415
distribution function **25**, **41**, 66, 210, 356, 522
distribution map (elements) 312, 344, 471
dosimeter 531
double diffraction 237
dynamic disorder 361
dynamic theory (of diffraction) 75, **83**

EBIC (electron beam induced current) 474
EDS (energy dispersive spectrometer) **146**, 287, 306, 488, 523
EELS (electron energy loss spectrometry) 362, 483, 488, 523
efficient thickness 293
elastic interaction **11**
elastic scattering **11**
electroacoustic imaging 474
electromagnetic waves 3
electron 5
  absorption **104**
  binding energies **87**, 524
  charge 507
  interaction **15**
  irradiation 449, 453, 475
  mass 507
  penetration depth 77
  shells **8**, 88
  transitions **88**, 93, 99
electron beam induced current (EBIC) 474
electron detector **149**, 463, 486
electron diffraction 65, 77, **226**, 408
  convergent beam **258**
  high energy **226**
  low energy **265**, 342
electron energy **5**
  dispersion 121
  levels **8**, 524
  losses 122
electron energy loss spectrometry (EELS) 362, 483, 488, 523
electron gun **118**
electron lenses **122**, 394
electron microscope microanalysis (EMMA) 483
electron microscopy
  analytical 482, **488**
  high resolution 220, **428**
  high voltage **451**
  scanning **458**
  scanning transmission **482**
  scanning tunnelling 493, **497**
  transmission **391**
electron probe 122, 459, 468, 484, 490
electron probe microanalysis (EPMA) **304**, 523

electron source **118**, 393, 459, 484
electron spectrometers **151**, 337
electrostatic prism 152
element distribution maps 490
EMMA (electron microscope microanalysis) 483
energy 506
   loss 11, 105
   transfer 11
energy and wavelength **4**
energy discrimination 137, **138**, 147, 192
energy dispersive spectrometer (EDS) **146**, 287, 306, 488, 523
energy filter 486
energy loss edges 363, **366**
enhancement rate (fluorescence) **294**, 319
EPMA (electron probe microanalysis) **304**, 523
equivalent dose (ED) 530
escape depth 99, 334, 341
escape peak 145, **147**
Ewald construction 39, **54**
Ewald sphere (reflecting sphere) 39, **54**, 227
EXAFS (extended X-ray absorption fine structure) 104, **351**
excitation 13, **88**
   cross-section 95, **280**
   ratio 95
EXELFS (extended electron energy loss fine structure) **367**, 483, 488
exposure dose 530
extended electron energy loss fine structure (EXELFS) **367**, 483, 488
extended X-ray absorption fine structure (EXAFS) 104, **351**
extinction distance **82**, 86, 410, 413
extinction rules (diffraction) 67

Fano factor 146
Fermi level 339, 493
fibre texture **204**, 254
field electron emission microscope (FEEM) 497
field emission 121, 460, 484, 494, **496**
field ion emission microscopy (FIM) 497
fine structure 352
fluorescence (X-rays) 93, **284**
   correction **295**, 319
   elemental analysis **284**
   yield **95**, 286, 489
focal series 399
focusing
   electron spectrometer **151**, 152, 337
   monochromator **116**, 190
   powder camera **190**
   X-ray spectrometer **143**, 287, 305
FOLZ (first-order Laue zone) 259

form factor **31**, 59
Fourier
   filtering 361
   inversion 29
   series 218
   transform 29, 217, 360, **517**
Fourier–Bragg projection **219**, 434
Fraunhofer diffraction 79
Frenkel defect 450
frequency 506
Fresnel construction 58
Fresnel diffraction 80
   fringes 487
   zones 81
Friedel law **65**, 213, 264
fundamental coefficients (XRF) 300
fusion technique (XRF) 302

gas discharge ion source **128**
gas ionisation detectors **135**
gas multiplication factor 136
geometrical theory (of diffraction) **50**
glide planes 71
grain (or texture) correction 295
grain size 186, **203**
gray 530
Guinier camera 191

heavy atom method 217
hemispheric analyser (HSA) **152**, 337
high-energy electron diffraction **226**
high polymers 211
high-resolution electron microscopy 220, **428**
high-voltage electron microscopy (HVEM) **451**
holed foils 445
HOLZ (high-order Laue zones) 259
hot cathode ion source **129**
Hull chart 201
HVEM (high-voltage electron microscopy) **451**

illumination coherence 393
image contrast 404, 466
inelastic
   interaction **11**
   scattering **11**
intensity 506
intensity (diffraction) 59
intensity weighted reciprocal lattice 39, **60**, 178
interaction **10**
   cross-section 12
   electrons **15**
   ions **20**
   neutrons **20**
   X-rays **13**
interference function 38, **60**, 354, 358

internal potential 268
internal standard 298
interplanar distance 515
interpoint distance 515
inverse photoelectric effect 502
ion
   distribution image 375
   emission microscopy 372
   implantation **21**, 386
   interaction **20**
   mass spectrometry **155**
   probe 372
   source **128**
   species 378
   sputtering **21**, 373
ionisation 88
   degree 377, 382
irradiation damage 490
isotopic abundance 378

Johansson focusing 117

Kikuchi pattern **239**, 260, 474
kinematic theory (of diffraction) **55**, 79, 250, 252, 408
Kossel pattern 243, 260
Koster–Kronig transitions 99

lanthanum hexaboride (LaB_6) cathode **120**, 460
lattice
   image 487
   parameters 196
   resolution 435
   types 68
Laue condition
   real lattice **50**
   reciprocal lattice 53
Laue groups **172**, 212, 213
Laue method **166**
Laue–Scherrer relation 203
Laue zones **233**, 259
LEED (low-energy electron diffraction) **265**, 342
light velocity 507
limit dose 531
limiting sphere (resolution sphere) 30, **55**, 217
Lorentz factor 216
low-energy electron diffraction (LEED) **265**, 342
lyophilisation 475

macrodiffraction 230
magnetic field contrast 473
magnetic prism 152
magnification 402, 469

mass
   fraction 277
   resolution 157
   spectrometer **155**, 270, 374
matrix
   correction 296, 471, 489
   effect 196, **280**, 290, 315, 324, 345, 382
Mauguin abacus 168
mean free path 13, 77, 105
mean grain size 203
metal ion source **130**
microbeam selection 232
microchannel plate 151
microdiffraction 231
microgrids 445
microprobe 125, 304
minimum detectable
   mass 523
   mass fraction 277, 523
   number of atoms 523
mixed layers 209
moiré pattern 422
monochromator
   neutrons 127
   X-rays **115**, 357
Monte-Carlo method 104, 321
multiple dark field 407
multiplicity factor **185**, 252
multislice method 440

nanoanalysis 488
nanodiffraction 232
nanoprobe 125, 232, 259, 261, 304, 428
nanoscopy 428
neutron **5**
   detection **157**
   diffraction **222**
   interaction **20**
   mass 507
   source 127
non-crystalline solids 210
non-radiative (or radiationless) transition **98**

objective condenser **126**, 259, 484
order–disorder transformation **207**, 224
overvoltage ratio 95

particle beams 3
Patterson series (function) 217
Pauli's exclusion principle 8
Pendellösung 415
phase contrast transfer function **431**, 453
phase object 432
phonons 17, 47
photoelectron 14, **89**
photoelectron spectrometry 90, **334**
photons 3

photosumming 219
piezoelectric translator 499
pile-up effect 147
p-i-n diode **140**
Planck constant 507
plasma thermodynamics 382
plasmons **19**, 90, 337
point group symmetry 211, 264
polarisation factor 58, 216
polycrystalline material 184, 250
ponctualized structure 58
poorly crystalised materials 209
post-ionisation 381, 383
potential 505
potential barrier 493
potential contrast 473
powder diffraction file **194**
powder diffractometer **191**
powder method **184**
precession camera 181
preferential orientation **204**, 254
pressure (vacuum) 507
primary X-ray radiation **93**
probability of radiative transition **95**
proportional counter **138**
proton mass 507
pseudo-Kikuchi pattern 242, 260

quadratic forms 197
quadrupole mass analyser **156**
qualitative analysis **276**, 287, 307, 369, 377
quality factor 530
quantitative analysis **279**, 290, 314, 381, 471
quantitative surface analysis 340, 345
quantum numbers **8**

rad 530
radiation protection **529**
　doses 530
　quality factor 530
　standard 531
radiative transition **93**
ratiocontrast 486
Rayleigh criterion 396
reciprocal lattice 29, **513**
reciprocal space **513**
reflecting planes 52
reflecting sphere (Ewald sphere) 39, **54**, 227
reflection orders 52
relaxation (atomic) 88
relaxation (diffraction conditions) **62**, 78, 227
relief map 502
rem 530
replicas 447
resolution (diffraction)
　condition **52**
　sphere (limiting sphere) 30, **55**, 217

resolution (imaging) **394**, 435, 439, 453, 459, 468, 483, 488, 498, 502
resolution test 399
retarding field (or retarding voltage) spectrometer 342
retigraph 181
röntgen 530
rotating target tubes 113
rotation method **173**
Rowland circle 117

SAM (scanning Auger microscope) **344**, 458
scanning Auger microscope (SAM) **344**, 471
scanning electron microscope (SEM) **458**, 483
scanning (electron probe) 306, 311, 459, 483
scanning transmission electron microscopy (STEM) **482**
scanning tunnelling inverse photoemission (STIPE) 502
scanning tunnelling microscopy (STM) **493**, 497
scanning tunnelling optical microscopy (STOM) 502
scattering **11**
　amplitude **28**
　cross-section 12
　intensity **34**
　power **24**, 57
　vector **28**
Scherzer defocus 396, 439
Schrödinger equation 7, 494
scintillation counter **139**, 149, 463, 464
screw axis 70
secondary electron 19, **90**, 463
secondary EXAFS 361
secondary ion mass spectrometry (SIMS) **372**, 523
secondary neutral mass spectrometer (SNMS) 381
secondary radiation 11, 14, **87**, 89, 92
secondary radiators 285
SEELS (slow electron energy loss spectrometry) 343
Seemann–Bohlin camera 190
selected area electron diffraction 231
selection errors 231
selection rules 93
selective dark field **407**, 487
selective reflection 52
SEM (scanning electron microscope) **458**
semiconductor detector **140**, 150, 464
SEXAFS (secondary EXAFS) 361
shake-off 337
shake-up 337
SI units 4, **505**
Si(Li) detector **140**
Si(Li) spectrometer **147**, 471, 488
Siegbahn notation 94
sievert 530

SIMS (secondary ion mass spectrometry) **372**, 523
single crystal diffractometer 183
size factor 27, 31
slow electron energy loss spectrometry (SEELS) 343
small angle scattering 210
Soller slits 114
space group symmetry **67**, 179, 213, 249, 264
spatial frequency spectrum 430
spatial resolution 307
specimen preparation 233, 301, 322, 344, 383, 445, 474, 490, 500
specimen support film 443
specimen thickness 263, 292, 324, 415, 523
spectrometer
   electron **151**, 337, 342
   ions **155**, 374
   parameters 133
   X-rays **143**, 286, 305, 471, 488
spectrometry 132, **275**
spherical aberration **123**, 396
spin doublet 94
spinel 224
sputtering yield 280, 373, 377
stacking faults 417
static disorder 361
STEM (scanning transmission electron microscopy) **482**
stereographic projection 172
stigmator 399
STIPE (scanning tunnelling inverse photoemission) 502
STM (scanning tunnelling microscopy) 493, **497**
STOM (scanning tunnelling optical microscopy) 502
structure
   analysis 202, **211**, 250, 362
   element 67
   factor **56**
   function 25, 57, 521
   image **432**, 439
   refinement 218
superlattice 207, 269
surface
   analysis 333, 339, 345, 362, 375
   conductivity 475
   lithography 503, 504
   structure 265, 503
   topography 466, 503
synchrotron 113, **118**, 357
systematic absences 67, 213

temperature factor (Debye–Waller factor) 47, 59
texture goniometer **207**
thermal neutrons 5

thermionic electron emission 119
thickness fringes 263, **411**
thin specimen analysis 323
Thomson factor 32
time of flight mass analyser **156**
topographical contrast 466
Townsend avalanche 138
trace analysis 300
transition probability 280
translation lattices 68
transmission
   electron diffraction **226**
   electron microscopy **391**
transmission factor 134
transparency function 429
tunnel effect 121, **494**
tunnelling distance 495
turbostratic texture 256
two-dimensional groups 264
two-wave approximation 76, **84**, 250, 252, 260, 413

ultramicrotomy 446
unitary structure factor 58
units 4, **505**
universal constants **507**

vacuum 402, 460
   gauge 128, 129
   pumps 121
   units 507

wave function 7
wavelength **4**, 505
wavelength dispersive spectrometer (WDS) **143**, 286, 305
wave vector **10**, 506
WDS (wavelength dispersive spectrometer) **143**, 286, 305
weakly scattering object **429**, 483
wehnelt 112
Weissenberg camera 181
wigglers 118
work function 89, 99, 120, 339, **493**, 497
Wulff diagram 173
Wyckoff positions 215

XPS (X-ray photoelectron spectrometry) **334**, 523
X-ray
   absorption **102**
   collimator 114
   detector **134**
   diffraction **165**

fluorescence 93, **284**
interaction **13**
monochromator **115**, 357
scattering 209
source 97, **111**
X-ray absorption spectrometry **351**
X-ray fluorescence spectrometry (XRF) **284**, 523
X-ray microanalysis 97, **304**, 471, 488
X-ray photoelectron spectrometry (XPS) **334**, 523
X-ray spectrometer **143**, 286, 305, 471, 488
X-ray tube **111**, 357
XRF (X-ray fluorescence spectrometry) **284**, 523

ZAF method **316**, 320
zeolites 224
ZOLZ (zero-order Laue zone) 259
zone axis 169